*Johann Gasteiger, Thomas Engel (Eds.)*
**Chemoinformatics**

## Related Titles from WILEY-VCH

Hans-Dieter Höltje, Wolfgang Sippl,
Didier Rognan, Gerd Folkers

**Molecular Modeling**

2003
ISBN 3-527-30589-0


Helmut Günzler, Hans-Ulrich Gremlich

**IR Spectroscopy**

**An Introduction**

2002
ISBN 3-527-28896-1


Jure Zupan, Johann Gasteiger

**Neural Networks in Chemistry
and Drug Design**

**An Introduction**

1999
ISBN 3-527-29778-2 (HC)
ISBN 3-527-29779-0 (SC)


Siegmar Braun, Hans-Otto Kalinowski,
Stefan Berger

**150 and More Basic
NMR Experiments**

**A Practical Course**

1998
ISBN 3-527-29512-7

*Johann Gasteiger, Thomas Engel (Eds.)*

# Chemoinformatics

A Textbook

WILEY-
VCH

WILEY-VCH GmbH & Co. KGaA

Editors:

**Prof. Dr. Johann Gasteiger**
Computer-Chemie-Centrum and Institute
of Organic Chemistry
University of Erlangen-Nürnberg
Nägelsbachstraße 25
91052 Erlangen
Germany

**Dr. Thomas Engel**
Computer-Chemie-Centrum and Institute
of Organic Chemistry
University of Erlangen-Nürnberg
Nägelsbachstraße 25
91052 Erlangen
Germany

**Library of Congress Card No.: applied for**
A catalogue record for this book is available
from the British Library.

**Bibliographic information published by
Die Deutsche Bibliothek**
Die Deutsche Bibliothek lists this publica-
tion in the Deutsche Nationalbibliografie;
detailed bibliographic data is available in the
Internet at http://dnb.ddb.de

ISBN 3-527-30681-1

**Composition**   Hagedorn Kommunikation
GmbH, Viernheim
**Printing**   betz-druck gmbH, Darmstadt
**Bookbinding**   Litger & Dopf Buchbinderei
GmbH, Heppenheim
Printed in the Federal Republic of Germany.

*J. Gasteiger*

To Uli, Nina, Julia and Michael


*T. Engel:*

To my family and Gabi

If you want to build a ship, don't drum up people to collect wood and don't assign them tasks and work, but rather teach them to long for the endless immensity of the sea.

*Antoine de Saint-Exupery*

# Foreword

Mathematically speaking, two chicken halves give one chicken. This is true with respect to mass. Such a chicken, however, is not a living one. Moreover, the two halves must be mirror images to form the shape of a chicken. Obviously, numbers are only half of the story, information comes in disparate guises. Chemistry, the science of materials and their transformations, exhibits a broad diversity of information, which by now encompasses an enormous body of knowledge about chemical structures, properties, and reactions. However, despite all the achievements during the last two centuries, which changed early chemical craftsmanship into a sophisticated natural science, chemistry is still devoid of the all-round theory, having, for examples the potential to predict precise structure-activity or structure-function relationships. It would explain, for instance, why palytoxin $C_{129}H_{223}N_3O_{54}$, being isolated from a Hawaiian coral in 1979, is one of the most poisonous natural substances. With its 64 chiral carbon atoms and six olefinic bonds, offering more than $10^{21}$ possible isomers, the total synthesis of palytoxin carboxylic acid may be compared with the first climbing of Mount Everest (Kishi, et al. at Harvard University, 1989).

Comparable efforts are needed to master the flood of information and accumulated knowledge in chemistry today. While until 1960 the number of natural and laboratory-produced compounds had almost linearly increased to roughly one million in about 150 years, its growth expanded exponentially from then on, reaching 18 million in 2000. This is just one aspect of the revolution in chemistry brought about by the rapid advancement of computer technology since 1965. Methods of physics, mathematics and information science entered chemistry to an unprecedented extent, which furnished laboratories with powerful new instrumental techniques. Also, a broad variety of model-based or quantum-mechanical computations became feasible, which were thought impossible a few decades ago. For example, the computer modeling of water transport through membranes mediated by aquaporins yields the time dependence of the spatial position of typically $10^5$ atoms on a picosecond scale up to 10 ns (Grubmüller et al., MPI Göttingen). Such huge data arrays can be searched for and accessed via computer networks and then evaluated in a different context („data mining"). Furthermore, new chemical techniques, such as combinatorial synthesis, have high data output. Overall it can be stated that, particularly for the chemical and pharmaceutical industries, researchers now spend more time in digesting data than in generating them, whereas the reverse was true a few years ago.

In the 1970's, chemists increasingly encountered varying aspects of the triumvirate "chemistry-information-computer" (CIC) while conducting their research. Common to all was the use of computers and information technologies for the generation of data, the mixing of data sources, the transformation of data into information and then information into knowledge for the ultimate purpose of solving chemical problems, e.g. organic synthesis planning, drug design, and structure elucidation. These activities led to a new field of chemical expertise which had distinctly different features compared with the traditional archiving approach of chemical information, which has been established about 200 years ago and comprises primary journals, secondary literature, and retrieval systems like Chemical Abstracts.

In the 1980s, computer networks evolved and opened a new era for fast data flow over almost any distance. Their importance was not generally recognized in the chemical community at the beginning. The situation may be characterized with words from the late Karl Valentin: "A computer network is something that one does not want to be in the need to have, nevertheless simply must want to have, because one always might be in need to use it. (Ein Datennetz ist etwas was man eigentlich nie brauchen müssen möchte, aber doch einfach wollen muß, weil man es immer brauchen tun könnte.)" In 1986 Johann Gasteiger, also in Munich, coinitiated the Task Force CIC of the German Chemical Society (GDCh). In the same year, he started the CIC Workshops on Software Development in Chemistry which found overwhelming acceptance. Until today these annual meetings have served as a forum for the presentation and dissemination of recent results in the various CIC fields, including chemical information systems. The Task Force CIC later merged with GDCh Division Chemical Information under the name "Chemie-Information-Computer (CIC)".

The 1990s saw the advent of the Internet, which boosted the use of computer networks in chemistry. In its sequel, at the turn of the century, the work of the forerunners at the intersection of chemistry and computer science eventually received recognition in its entirety as a new interdisciplinary science: "Chemoinformatics" had come of age. It encompasses the design, creation, organization, management, retrieval, analysis, dissemination, visualization, and use of chemical information (G. Paris). Its cousin, Bioinformatics, which was developed somewhat earlier, generally focuses on genes and proteins, while chemoinformatics centers on small molecules. Yet the distinction is fuzzy, e.g. when the binding of small molecules to proteins is addressed. From the viewpoint of the life sciences, the borderline may blur completely.

A young scientific discipline grows with its students. For students, in turn, the efforts pay dividends. The demand from industrial employers increases steadily for chemoinformaticians, and so the field is expected to become big business. The question therefore arises how and where chemoinformatics can be learned. One good place to go is Erlangen. Johann Gasteiger and his group were practicing chemoinformatics for 25 years without even knowing its name. In 1991 he received the Gmelin Beilstein medal of the German Chemical Society (GDCh) and in 1997, he received the "Herman Skolnik Award" of the Division of Chemical Information of the American Chemical Society in recognition for his many achievements in the CIC field. With the present comprehensive textbook on chemoinformatics for undergraduate and graduate students, Gasteiger and his group lay a solid foundation-stone for many more "Erlangens" in the world. My warm recommendation goes with this book, in particular to my academic colleagues. It seems to be the right time for universities to begin to teach chemoinformatics on a broad scale. Subject to local reality this may be conducted as part of a diploma course of study in chemistry, as a master's study after a BS in chemistry, or as a full course of study in chemoinformatics, like that available in bioinformatics. In all cases, this book and its supplementary material would provide the adequate basis for teaching as well as for self-paced learning.

Dieter Ziessow
Chairperson
"Chemie-Information-Computer (CIC)"
of the German Chemical Society (GDCh)

# Contents

# Preface

Computers have penetrated nearly every aspect of daily life. Clearly, scientists and engineers took the lead in this process by applying computers for solving problems. In chemistry, it was realized quite early on that the huge amount of information available can only be handled by electronic means, by storing this information in databases. Only in this way can the huge number (35 million) of chemical compounds known at present be handled. Thus, already in the 1960s, work on chemical databases was initiated. Furthermore, many relationships between the structures of compounds and their physical, chemical, or biological properties are highly complex, asking either for highly sophisticated computations or for analysis of a host of related data to make predictions on such properties. Chemical societies in many countries have recognized the importance of computers in their field and have founded divisions that focus on the use of computers in chemistry.

From the very beginning, however, it could be observed that there was a split between theoretical chemists using computers for quantum mechanical calculations and chemists using computers for information processing and data analysis. The American Chemical Society has two divisions, the Division of Computers in Chemistry and the Division of Chemical Information. In Germany there is the Theoretische Chemie group associated with the Deutsche Bunsen-Gesellschaft für Physikalische Chemie and the Division Chemie-Information of the Gesellschaft Deutscher Chemiker. In fact, in 1989 this Division changed its name to Chemie-Information-Computer (CIC) to recognize the growing importance of using computers for processing chemical information. A small group of scientists within this division was quite active in spreading the message of using computers in chemistry. Two workshops were initiated in 1987, one on *Software Development in Chemistry* and one on *Molecular Modeling*, ever since these workshops have been held on a yearly basis.

On another level, the German Federal Minister of Research and Technology (BMFT; later renamed BMBF) initiated programs in the 1980s to found so-called Fachinformationszentren (FIZ) and, in addition, to build databases. Chemists can consider themselves fortunate that the experts and politicians recognized the importance of databases in chemistry. Thus, some of the internationally most highly recognized databases were initiated: the Beilstein Database for organic compounds, the Gmelin Database for inorganic and organometallic compounds, the ChemInform RX reaction database, and the SpecInfo database for spectroscopic information.

In spite of all these activities it must nevertheless be observed that chemists have only gradually accepted the computer as a much needed tool in their daily work. But they gradually – or grudgingly? – did accept it: databases are used routinely for retrieving information, and quantum chemical or molecular mechanics programs are used – mostly a *posteriori* – to further an understanding of chemical observations. Furthermore, since the advent of combinatorial chemistry and high-throughput screening it has become increasingly clear that the flood of information produced by these techniques can only be handled by computer methods.

Thus, computers will continue to penetrate every aspect of chemistry and we have to prepare the next generation of chemists for this process. In fact, we will see that the various types of computer applications in chemistry will increasingly be used in concert to solve chemical problems. Therefore, a unified view of the entire field is needed; the various approaches to using computers in chemistry have to be ordered into a common framework, into a discipline of its own: Chemoinformatics.

With this textbook we present the first comprehensive overview of chemoinformatics, current material that can be integrated into chemistry curricula or can serve on its own as a basis for an entire course on chemoinformatics.

This textbook can build on 25 years of research and development in my group. First of all, I have to thank all my co-workers, past and present, that have ventured with me into this exciting new field. In fact, this textbook was written nearly completely by members of my research group. This allowed us to go through many text versions in order to adjust the individual chapters to give a balanced and homogeneous presentation of the entire field. Nevertheless, the individual style of presentation of each author was not completely lost in this process and we hope that this might make reading and working through this book a lively experience. Writing these contributions on top of their daily work was sometimes an arduous task. I have to thank them for embarking with me on this journey.

We also want to thank the Federal Minister of Education and Research (BMBF) for funding a project "Networked Education in Chemistry", administered by FIZ CHEMIE, Berlin. Within this project we are developing eLearning tools for chemoinformatics.

In addition, we thank Dr. Gudrun Walter of Wiley-VCH, for encouraging us to embark on this project and Dr. Romy Kirsten for the smooth collaboration in processing our manuscripts.

We just hope that this Textbook will generate interest in chemoinformatics for a wider audience, and will make them excited about this field much in the same way as we are excited about it.

Erlangen, May 2003 *Johann Gasteiger*

# Addresses of the Authors

**Burkard, Ulrike**,
ulrike.burkard@chemie.uni-erlangen.de
**Clark, Tim**,
clark@chemie.uni-erlangen.de
**Engel, Thomas**,
thomas.engel@chemie.uni-erlangen.de
**Gasteiger, Johann**,
gasteiger@chemie.uni-erlangen.de
**Herwig, Achim**,
achim.herwig@chemie.uni-erlangen.de
**Kleinöder, Thomas**,
thomas.kleinoeder@chemie.uni-
erlangen.de
**Lanig, Harald**,
lanig@chemie.uni-erlangen.de
**Lekishvili, Giorgi**,
lekishvili@python.qartu.com
**Sitzmann, Markus**,
markus.sitzmann@chemie.uni-
erlangen.de
**Spycher, Simon**,
simon.spycher@chemie.uni-erlangen.de
**Terfloth, Lothar**,
lothar.terfloth@chemie.uni-erlangen.de

All:
Computer-Chemie-Centrum
and Institute of Organic Chemistry
Friedrich-Alexander-University
of Erlangen-Nuernberg
Naegelsbachstrasse 25
91052 Erlangen
Germany

**Bangov, Ivan**,
bangov@mol-net.de
**Pförtner, Matthias**,
pfoertner@mol-net.de
**Sacher, Oliver**,
sacher@mol-net.de
**Schwab, Christof**,
schwab@mol-net.de

All:
Molecular Networks GmbH
Naegelsbachstrasse 25
91052 Erlangen
Germany


**Aires de Sousa, Joao**
Departamento de Quimica, Fac. Ciencias
e Tecnologia, Univ. Nova de Lisboa,
2829-516 Caparica
Portugal
jas@mail.fct.unl.pt

**Hemmer, Markus**
Creon Lab Control AG
Europaallee 27-29
50226 Frechen
Germany
markus.hemmer@creonlabcontrol.com

**Kochev, Nikolay,**
**Monev, Valentin**
Plovdiv University
Department of Analytical Chemistry
Tzar Assen Str. 24
Plovdic 4000
Bulgaria
nick@argon.acad.bg

**Oellien, Frank**
Intervet Innovation GmbH
Zur Propstei
55270 Schwabenheim
Germany
frank.oellien@intervet.com

**Tomczak, Jaroslaw**
Aventis Pharma Deutschland GmbH
DI&A Lead Generation
Chemoinformatics
Industriepark Hoechst, G879
65926 Frankfurt / Main
Jaroslaw.Tomczak@aventis.com

**Vogt, Jürgen**
Section of Spectra and Structure
Documentation
University of Ulm
D-89069 Ulm
Germany
Juergen.Vogt@chemie.uni-ulm.de

**Voigt, Kristina**
GSF - National Research Center
for Environment and Health, Institute
of Biomathematics and Biometry,
85764 Neuherberg
Germany,
kvoigt@gsf.de

**Yan, Ai-Xia**
Department of Chemistry
Central Chemistry Laboratory
South Parks Road
Oxford OX1 3QH
UK
aixia.yan@chem.ox.ac.uk

**From the right:** Ulrike Burkard (Chapter 9), Achim Herwig (Section 2.6), Ivan Bangov (Chapter 6), Tim Clark (Section 7.4), Harald Lanig (Section 7.2 and 7.3), Thomas Engel (Chapter 2 and 5), Johann Gasteiger (Chapter 1, 3, Sections 7.1 and 10.3.1), Lothar Terfloth (Chapter 8), Markus Sitzmann (Section 10.3.2), Simon Spycher (Section 10.1), Thomas Kleinöder (Section 10.1), Christof Schwab (Section 2.9 and 2.13).

**Not in the picture:** Giorgi Lekishvili (Chapter 4), Frank Oellien (Chapter 9.9), Oliver Sacher (Section 5.13), Jaroslaw Tomczak (Section 2.4.4 and 2.9.7), Jürgen Vogt (Section 5.7), Kristina Voigt (Section 5.17), Joao Aires de Sousa (Section 8.5 and 8.6), Matthias Pförtner (Section 10.3.2), Ai-Xia Yan (Section 10.1), Michael Wünstel (Appendix A1).

# 1

# Introduction

*Johann Gasteiger*

## 1.1
## The Domain of Chemistry

Chemoinformatics is a fairly new name for a discipline that, as we shall soon see, has been around for quite a while. Different people sometimes give rather divergent definitions of chemoinformatics. Before we discuss these different viewpoints, let us, for the time being, accept a rather broad and general definition:

**Chemoinformatics is the application of informatics methods to solve chemical problems.**

Then, we have to reflect primarily on the domain of the science of chemistry:

**Chemistry deals with compounds, their properties and their transformations.**

Thus, two objects have to be considered, *compounds* and *chemical reactions*, the static and dynamic aspects of chemistry.

The entire living and material world consists of *compounds* and mixtures of compounds. Basic chemicals, such as ethylene, are produced in many millions of tons each year and are converted into a wide variety of other chemicals. Complicated molecular structures are synthesized by Mother Nature, or by chemists having taken up the challenge posed by Nature. However, we also have materials such as glues which are composed of mixtures of rather ill-defined polymers.

In this book we shall deal largely with those compounds that can be described by a clearly defined molecular structure. The representation of polymers and of mixtures of compounds will only be mentioned in passing.

Compounds are transformed into each other by *chemical reactions* that can be run under a variety of conditions: from gas-phase reactions in refineries that produce basic chemicals on a large scale, through parallel transformations of sets of compounds on well-plates in combinatorial chemistry, all the way to the transformation of a substrate by an enzyme in a biochemical pathway. This wide range of reaction conditions underlines the complicated task of understanding and predicting chemical reaction events.

It is true that the structure, energy, and many properties of a molecule can be described by the Schrödinger equation. However, this equation quite often cannot be solved in a straightforward manner, or its solution would require large amounts of computation time that are at present beyond reach. This is even more true for chemical reactions. Only the simplest reactions can be calculated in a rigorous manner, others require a series of approximations, and most are still beyond an exact quantum mechanical treatment, particularly as concerns the influence of reaction conditions such as solvent, temperature, or catalyst.

How come then, that although the laws of chemistry are too complicated to be solved, chemists still can do their jobs and make compounds with beautiful properties that society needs, and chemists run reactions from small-scale laboratory experiments to large-scale reactors in chemical industry?

The secret to success has been to learn from data and from experiments. Chemists have done a series of experiments, have analyzed them, have looked for common features and for those that are different, have developed models that made it possible to put these observations into a systematic ordering scheme, have made inferences and checked them with new experiments, have then confirmed, rejected, or refined their models, and so on. This process is called inductive learning (Figure 1-1), a method chemists have employed from the very beginnings of chemistry.

In this manner, we have learned the laws and rules of nature, of compounds and their reactions. Thus, enough knowledge was accumulated to found an entire industry, the chemical industry, which produces a cornucopia of chemicals having a wide range of properties that allow us to maintain our present standard of living.

However, this process of inductive learning is still not over; we are still far away from understanding and predicting all chemical phenomena. This is most vividly illustrated by our poor knowledge of the undesired side effects of compounds, such as toxicity. We still have to strive to increase our knowledge of chemistry.

*This is where chemoinformatics comes in!*

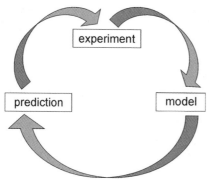

**Figure 1-1.** Inductive learning.

## 1.2
## A Chemist's Fundamental Questions

Chemists' major task is to make compounds with desired properties. Society at large is not interested in beautiful chemical structures, but rather in the properties that these structures might have. The chemical industry can only sell properties, but it does so by conveying these properties through chemical structures. Thus, the first fundamental task in chemistry is to make inferences about which structure might have the desired property (Figure 1-2).

**Figure 1-2.** Structure–Property/Activity Relationships.

This is the domain of establishing *Structure–Property* or *Structure–Activity Relationships* (SPR or SAR), or even of finding such relationships in a quantitative manner (QSPR or QSAR).

Once we have an idea which structure we should make to obtain the desired property, we have to plan how to synthesize this compound – which reaction or sequence of reactions to perform to make this structure from available starting materials (Figure 1-3).

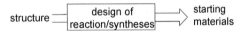

**Figure 1-3.** The design of a reaction or sequence of reactions to make a structure from available starting materials.

This is the domain of *synthesis design*, and the *planning of chemical reactions*.

Once a reaction has been performed, we have to establish whether the reaction took the desired course, and whether we obtained the desired structure. For our knowledge of chemical reactions is still too cursory: there are so many factors influencing the course of a chemical reaction that we are not always able to predict which products will be obtained, whether we also shall obtain side reactions, or whether the reaction will take a completely different course than expected. Thus we have to establish the structure of the reaction product (Figure 1-4). A similar problem arises when the degradation of a xenobiotic in the environment, or in a living organism, has to be established.

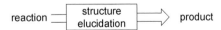

**Figure 1-4.** Structure elucidation.

This is the domain of *structure elucidation*, which, for most part, utilizes information from a battery of spectra (infrared, NMR, and mass spectra).

All three tasks are generally too complicated to be solved from first principles. They are, therefore, tackled by making use of prior information, and of information that has been condensed into knowledge. The amount of information that has to be processed is often quite large. At present, more than 41 million different compounds are known: all have a series of properties, physical, chemical, or biological; all can be made in many different ways, by a wide range of reactions; all can be characterized by a host of spectra. This immense amount of information can be processed only by electronic means, by the power of the computer.

*This is where chemoinformatics comes in!*

## 1.3
## The Scope of Chemoinformatics

It was realized quite some decades ago that the amount of information accumulated by chemists can, in the long run, be made accessible to the scientific community only in electronic form; in other words, it has to be stored in databases. This new field, which deals with the storage, the manipulation, and the processing of chemical information, was emerging without a proper name. In most cases, the scientists active in the field said they were working in "Chemical Information". However, as this term did not make a distinction between librarianship and the development of computer methods, some scientists said they were working in "Computer Chemistry" to stress the importance they attributed to the use of the computer for processing chemical information. However, the latter term could easily be confused with Computational Chemistry, which is perceived by others to be more limited to theoretical quantum mechanical calculations.

There were some clear signs that the situation was changing. In 1975 the *Journal of Chemical Documentation* changed its name to *Journal of Chemical Information and Computer Sciences*. In 1973 a seminal NATO Advanced Study Institute Summer School was held in Noordwijkerhout, The Netherlands, that for the first time brought together a broad range of scientists who came from different areas of chemistry but who were all developing computer methods to manage and make sense of chemical information. The title of the Summer School was: "*Computer Representation and Manipulation of Chemical Information*". The groups attending this conference worked on building chemical structure databases and on developing software for molecular modeling, for organic synthesis design, for analyzing spectral information, and for chemometrics. Suddenly it was realized that a new field had emerged that had implications in many areas of chemistry.

Since then, the application of methods of computer science, or informatics as it is called in many languages, to the solution of chemical problems has ventured into many more areas of chemistry. So broad are the applications of informatics, and its foundation, mathematics, in chemistry that no longer can any single conference cover the entire field.

The term "Chemoinformatics" appeared only quite recently. Here are some of the first citings:

■ *"The use of information technology and management has become a critical part of the drug discovery process. Chemoinformatics is the mixing of those information resources to transform data into information and information into knowledge for the intended purpose of making better decisions faster in the area of drug lead identification and organization."*
K. Brown, *Annual Reports in Medicinal Chemistry* **1998,** *33,* 375–384

■ *"Chemoinformatics – A new name for an old problem."*
M. Hann, R. Green, *Current Opinion in Chemical Biology* **1999,** *3,* 379–383

■ *"Chem(o)informatics is a generic term that encompasses the design, creation, organization, management, retrieval, analysis, dissemination, visualization, and use of chemical information."*
G. Paris (August 1999 Meeting of the American Chemical Society), quoted by W. Warr at http://www.warr.com/warrzone.htm

In this book, we want to build on the long history of applying informatics methods to chemical problems, and to pay tribute to the scientists who started out decades ago to develop this interdisciplinary field.

In this sense, we are rather attached to the broad definition of "Chemoinformatics":

**The application of informatics methods to solve chemical problems**

and have made this broad definition the basis for the conception of this Textbook.

The term has different spellings: Chemoinformatics and Cheminformatics. Searches in the database of the Chemical Abstracts Service have shown an approximately equal number of hits for both terms, with Cheminformatics gaining ground somewhat in recent years. Here, we use the spelling "Chemoinformatics" without trying to put forward reasons for that choice.

Having settled on a definition of chemoinformatics, it is time for us to reflect on the distinction between chemoinformatics and bioinformatics. The objects of interest of bioinformatics are mainly genes and proteins. But genes, DNA and RNA, and proteins are chemical compounds! They are objects of high interest in chemistry. Chemists have made substantial contributions to the elucidation of the structure and function of nucleic acids and proteins. The message is clear: there is no clearcut distinction between bioinformatics and chemoinformatics!

Clearly, by tradition, chemoinformatics has largely dealt with small molecules, whereas bioinformatics has started to move from genes to proteins, compounds that have been objects of interest in chemoinformatics from the very beginning. The structure and function of proteins, the binding of a ligand to its receptor protein, the conversion of a substrate to a product within its enzyme receptor, the catalysis of a biochemical reaction by an enzyme – these are all areas where chemoinformatics and bioinformatics should work together to further our insight and knowledge:

*We will make real progress in understanding the structure, the properties, and the function of proteins, DNA, and RNA only if bioinformatics and chemoinformatics work together!*

This is particularly true in drug design. Genomics methods are being developed to identify protein targets for novel drug candidates. Drugs, on the other hand, will always be fairly small molecules, and chemoinformatics methods are being developed to find new lead structures and to optimize them into drug candidates (Figure 1-5).

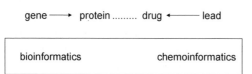

**Figure 1-5.** The cooperation of bioinformatics and chemoinformatics.

In this book, we concentrate largely on methods for the computer manipulation of small and medium-sized molecules, molecules of up to a few hundred or thousand atoms. We do this to develop an understanding of the methods available for the processing of information on chemical compounds and reactions. However, many of these methods can also be applied to macromolecules such as proteins and nucleic acids.

## 1.4
## Learning in Chemoinformatics

Having settled on a definition of chemoinformatics – the use of informatics tools to solve chemical problems – we have to address the question of how informatics can assist in answering a chemist's fundamental questions, as outlined in Section 1.2. In essence, these fundamental questions all boil down to having to predict a property, be it physical, chemical, or biological, of a chemical compound or an ensemble of compounds, such as the starting materials of a chemical reaction. In order to make predictions one has to have passed through a process of learning. There exist two different types of learning: deductive and inductive.

In *deductive learning* one must have a fundamental theory that allows one to make inferences and to calculate the property of interest.

Such a fundamental theory does exist for chemistry: quantum mechanics. The dependence of the property of a compound on its three-dimensional structure is given by the Schrödinger equation. Great progress has been made both in the development of the theory and in advances in hardware and software technology, which now allows the calculation of many interesting properties of chemical compounds of fairly reasonable size with high accuracy. However, there are large areas of interest in chemistry that are still beyond a theoretical treatment, either for lack of development of the underlying theory, or for requiring too much computation time with present-day computer technology.

Examples of this are:

- the prediction of the course of an organic reaction in a certain solvent, at a given temperature, using a specific catalyst, or
- the prediction of the biological activity of a certain compound.

There is, however, another type of learning: *inductive learning*. From a series of observations inferences are made to predict new observations. In order to be able to do this, the observations have to be put into a scheme that allows one to order them, and to recognize the features these observations have in common and the essential features that are different. On the basis of these observations a model of the principles that govern these observations must be built; such a model then allows one to make predictions by analogy.

Inductive learning has been the major process of acquiring chemical knowledge from the very beginnings of chemistry – or, to make the point, alchemy. Chemists have done experiments, have made measurements on the properties of their compounds, have treated them with other compounds to study their reactions, and have run reactions to make new compounds. Systematic variations in the structure of compounds, or in reaction conditions, provided results that were ordered by developing models. These models then allowed predictions to be made.

A point in case is provided by the bromination of various monosubstituted benzene derivatives: it was realized that substituents with atoms carrying free electron pairs bonded directly to the benzene ring (OH, $NH_2$, etc) gave o- and p-substituted benzene derivatives. Furthermore, in all cases except of the halogen atoms the reaction rates were higher than with unsubstituted benzene. On the other hand, substituents with double bonds in conjugation with the benzene ring ($NO_2$, CHO, etc.) decreased reaction rates and provided m-substituted benzene derivatives.

This led to the introduction of the concepts of inductive and resonance effects and to the establishment of the mechanism of electrophilic aromatic substitution. It should be emphasized that the concepts of inductive and resonance effect have not been derived from theory but have been introduced to "explain" the experimental observations, to put them into a systematic framework of an ordering scheme.

In the endeavor to deepen understanding of chemistry, many an experiment has been performed, and many data have been accumulated. Chapter 6, on databases, gives a vivid picture of the enormous amount of data that have been determined and made accessible. The task is then to derive knowledge from these data by inductive learning. In this context we have to define the terms, data, information, and knowledge, and we do so in a generally accepted manner.

- *Data*: any observation provides data, which could be the result of a physical mea-surement, a yes/no answer to whether a reaction occurs or not, or the determi-nation of a biological activity.
- *Information*: if data are put into context with other data, we call the result in-formation. The measurement of the biological activity of a compound gains in value if we also know the molecular structure of that compound.
- *Knowledge*: obtaining knowledge needs some level of abstraction. Many pieces of information are ordered in the framework of a model; rules are derived from a sequence of observations; predictions can be made by analogy.

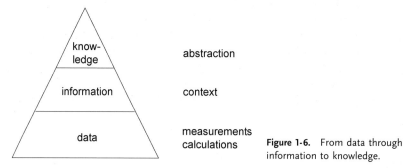

**Figure 1-6.** From data through information to knowledge.

Figure 1-6 illustrates this hierarchy in going from data through information to knowledge.

In the case of chemoinformatics this process of abstraction will be performed mostly to gain knowledge about the properties of compounds. Physical, chemical, or biological data of compounds will be associated with each other or with data on the structure of a compound. These pieces of information will then be analyzed by inductive learning methods to obtain a model that allows one to make pre-dictions.

## 1.5
## Major Tasks

The knowledge acquisition process outlined in Section 1.4 requires certain tasks to be performed.

### 1.5.1
### Representation of the Objects

First, the objects of investigation, chemical compounds or chemical reactions, have to be represented. Chemical compounds will mostly be represented by their molec-ular structure in various forms of sophistication. This task is addressed in Chapter 2. The representation of chemical reactions is dealt with in Chapter 3. The vast number of compounds known can only be managed by storing them in databases. Chapter 4 deals with storing and finding chemical compounds in databases.

## 1.5.2
### Data

The essence of chemistry is to produce compounds, or mixtures of compounds, with a wide range of physical, chemical, or biological properties. Our society could not exist any longer in its present form without the achievements of chemists: plastics and fibers, colors and dyestuffs, drugs, agrochemicals, washing powder, glues, etc.. The properties have to be measured in order to put them on a quantitative basis as a prerequisite for further optimization. Chapter 5 deals with the wide variety of data to be handled, and their processing to prepare them for inductive learning. The massive amounts of data accumulated over the years, and being accumulated with increasing speed, can only be managed by storing them in databases. Chapter 6 therefore deals with databases containing chemical information.

## 1.5.3
### Learning

The two ways of learning – deductive and inductive – have already been mentioned. Quite a few properties of chemical compounds can be calculated explicitly. Foremost of these are quantum mechanical methods. However, molecular mechanics methods and even simple empirical methods can often achieve quite high accuracy in the calculation of properties. These deductive methods are discussed in Chapter 7.

Inductive methods for establishing a correlation between chemical compounds and their properties are the theme of Chapter 9. In many cases, the structure of chemical compounds has to be pre-processed in order to make it amenable to inductive learning methods. This is usually achieved by means of structure descriptors, methods for the calculation of which are outlined in Chapter 8.

After approaches to the solution of the major tasks in chemoinformatics have thus been outlined, these methods are put to work in specific applications. Some of these applications, such as structure elucidation on the basis of spectral information, reaction prediction, computer-assisted synthesis design or drug design, are presented in Chapter 10.

## 1.6
### History of Chemoinformatics

The field of chemoinformatics was not founded, nor was it formally installed. It slowly evolved from several, often quite humble, beginnings. Scientists in various fields of chemistry struggled to develop computer methods in order to manage the enormous amount of chemical information and to find relationships between the structures and properties of a compound. In the 1960s some early developments became evident which led to a flurry of activities in the 1970s.

### 1.6.1
### Structure Databases

The storage and searching of chemical structures and associated information in databases are probably the earliest beginnings of what may now be called chemoinformatics. Work at the National Bureau of Standards, Washington DC, in 1957 showed that chemical structures can be retrieved through user-defined substructures by atom-by-atom searching. From 1960 onwards, the National Science Foundation funded the Chemical Abstracts Service to develop methods for the storage and searching of both structural and textual information in databases. Concomitantly with this, Swiss and German chemical companies such as BASF, Hoechst, and Thomae developed methods for storing their in-house chemical information. In the UK, ICI built a database of several hundred thousand structures based on Wiswesser Line Notation. Work in Sheffield, UK, and at the National Institutes of Health introduced fragment screening to enhance the speed of substructure searching.

### 1.6.2
### Quantitative Structure–Activity Relationships

The work by Hammett and Taft in the 1950s had been dedicated to the separation and quantification of steric and electronic influences on chemical reactivity. Building on this, from 1964 onwards Hansch started to quantify the steric, electrostatic, and hydrophobic effects and their influences on a variety of properties, not least on the biological activity of drugs. In 1964, the Free–Wilson analysis was introduced to relate biological activity to the presence or absence of certain substructures in a molecule.

### 1.6.3
### Molecular Modeling

In the late 1960s, Langridge and co-workers developed methods, first at Princeton, then at UC San Francisco, to visualize 3D molecular models on the screens of cathode-ray tubes. At the same time Marshall, at Washington University St. Louis, MO, USA, started visualizing protein structures on graphics screens.

### 1.6.4
### Structure Elucidation

The DENDRAL project initiated in 1964 at Stanford was the prototypical application of artificial intelligence techniques – or what was understood at that time under this name – to chemical problems. Chemical structure generators were developed and information from mass spectra was used to prune the chemical graphs in order to derive the chemical structure associated with a certain mass spectrum.

Structure elucidation systems that utilized information from several spectroscopic techniques were initiated in the late 1960s at Toyohashi, Japan, and at the University of Arizona.

### 1.6.5
### Chemical Reactions and Synthesis Design

In 1967, work was presented from a Sheffield group on indexing chemical reactions for database building. In 1969, a Harvard group presented its first steps in the development of a system for computer-assisted synthesis design. Soon afterwards, groups at Brandeis University and TU Munich, Germany, presented their work in this area.

These early roots of chemoinformatics matured into a tree that is still growing and blossoming. In fact, many of the approaches initiated in the 1960s and early 1970s have been developed into systems that are widely used and are still being refined. Some of the research groups from the early days are still actively pursuing further developments, and many a new group has joined these fields with new ideas and new systems.

### 1.7
### The Scope of this Book

This book is conceived as a textbook for application in teaching and self-learning of chemoinformatics. We aim to present a comprehensive overview of the field of chemoinformatics for students, teachers, and scientists from other areas of chemistry, from biology, informatics, and medicine. Those interested in a more in-depth presentation and analysis of the topics in this Textbook are referred to an accompanying set of four volumes,

*Handbook of Chemoinformatics – From Data to Knowledge*

where many of the issues are presented in greater detail by leading experts in the various fields.

Clearly, some of the subjects touched upon deserve entire books of their own. This is particularly true for the methods discussed in Chapters 7 and 9.

And such books do exist for

- molecular mechanics,
- quantum mechanics,
- free-energy relationships,
- chemometrics,
- neural networks,
- fuzzy logic,
- genetic algorithms, and
- expert systems.

References cited in the corresponding chapters of this book direct the interested reader to these books that provide much more detail in greater depth than is possible here. We can present only the major foundations, methods, and uses of these subjects as we have deemed necessary.

Some aspects, such as the computer representation and manipulation of proteins and nucleic acids, could not be covered. Even the modeling of the interactions of small molecules with proteins, as dealt with in docking software or software for *de novo* design could not be included in the Textbook, although chapters in the Handbook do treat these subjects.

Furthermore, here we emphasize the *chemical* concepts and aspects in chemoinformatics. We had to refrain from introducing those aspects that are more concerned with the informatics side, such as

- the theory of algorithms,
- programming languages, and
- database management systems.

This is not to say that we deem these topics not to be important. On the contrary, we think that those interested in chemoinformatics should strive to obtain a basic knowledge of these subjects. We even think that all professionals in natural sciences and engineering should in future obtain a minimum of training in these fields during their studies. However, presentation here of those aspects of informatics would go beyond the scope of this book.

The chapters in this Textbook have been written by different authors. In order to ensure somehow that the material is not too heterogeneous, we decided that these authors were largely to be members of our research group, so that intensive discussions between the authors could shaped the book; in this way we have tried to balance the presentations, with cross-references binding the chapters together.

With nearly all the authors coming from our research group there is certainly a strong bias towards our own work and research. However, clearly, this is the work we know most about and with which we could best illustrate the ideas and methods that have gone into solving the problems faced in chemoinformatics. We hope readers will therefore excuse any bias. We have tried to bring a more balanced presentation to the volumes of the Handbook by inviting contributions from a large international group of authors.

**1.8**
**Teaching Chemoinformatics**

Chemoinformatics has matured to a scientific discipline that will change – and in some cases has already changed – the way in which we perceive chemistry. The chemical and, in particular, the pharmaceutical industry are in high need of chemoinformatics specialists. Thus, this field has to be taught in academia, both in specialized courses on chemoinformatics and by integrating chemoinformatics into regular chemistry curricula.

In fact, chemoinformatic curricula have already been initiated at the University of Sheffield, UK, the University of Manchester Institute of Science and Technology, UK, Indiana University, USA, and the Université de Strasbourg, France.

In addition to this Textbook, a web page has been established which provides interesting information including list of URL's and figures (as PDF) to download, and details on chemoinformatics curricula. Moreover, it can also be used as a forum to obtain important information on the book, to find corrections of errors (if any), and to discuss aspects of chemoinformatics. The URL is *http://www2. chemie.uni-erlangen.de/publications/ci-book/index. html*

# 2

# Representation of Chemical Compounds

*T. Engel*

## Learning Objectives

- To understand different kinds of conventional nomenclature of chemical compounds
- To know how to transform a chemical structure into a language for computer representation and manipulation
- To be able to represent the constitution in an unambiguous and unique manner
- To learn more about connection tables and matrix representations of chemical structures
- To become familiar with structure exchange formats such as Molfile and SDfile
- To find out how stereochemistry can be represented
- To know how to generate 3D structures and how to represent and handle them with the computer
- To be introduced to molecular surfaces and to different models for visualization
- To recognize which programs can be used for graphical input and visualization of molecular structures

## 2.1
## Introduction

Chemistry, like any scientific discipline, relies heavily on experimental observations, and therefore on data. Until a few years ago, the usual way to publish information on recent scientific developments was to release it in books or journals. In chemistry, the enormous increase in the number of compounds and the data concerning them resulted in increasingly ineffective data-handling, on the side of the producers as well as the users. One way out of this disaster is the electronic processing, by computer methods, of this huge amount of data available in chemistry. Compared with other scientific disciplines that only use text and numbers for data transfer, chemistry has an additional, special challenge: molecules. The molecular species consist of atoms and bonds that hold them together. Moreover, compounds can be interconverted into other compounds by chemical reactions. Therefore, chemical information not only comprises text and data but also has to characterize chemical compounds with these special properties and their reactions.

At first, names were given to compounds to characterize them. In most cases, these were trivial names, which are still in use to a large extent. Very soon, symbols were used to shorten long names (see Section 2.2). The systematic classification of compounds according to their properties, the structure theory (developed around 1850), and experimental techniques induced an improved understanding of the structure of a molecule. This eventually led to the assignment to compounds of the well-known structure diagrams and the 3D arrangement of the molecules (Figure 2-1).

The 2D graphical representation of chemical structures in structure diagrams can be considered to be the universal "natural language" of chemists. These structure diagrams are models and are designed to make the molecules more conceivable. In such a model, the atoms are typified by their atomic symbols, and the bonding electrons by lines. However, the chemical structure diagram is an incomplete and highly simplified representation of a molecule. It only explains the topology (which atoms are connected by which bond type) (see Section 2.4) and not the 3D arrangement (topography) of the atoms in a molecule (see Section 2.9). Therefore, 3D representations of molecules require additional information, e.g., the position of the atoms in space or the angles and distances between the atoms in the molecule. Even more complex information has to be provided if properties (such as electrostatic potential) are to be mapped onto a surface of a 3D molecular arrangement (see Section 2.10). This kind of hierarchy in complexity of structure representation is illustrated in Figure 2-1 and is described in this chapter.

One of the major tasks in chemoinformatics is to represent chemical structures and to transfer the various types of representation into application programs. A first and basic step to "teaching" computer chemistry is to transform the molecular structure into a language amenable to computer representation and manipulation. Basically, computers can only handle bits of 0 and 1. Thus, coding is the basis for transferring data. In general, coding is considered as a form of ciphering with the help of different symbols in a certain system under observed rules. Writing is such a system of symbols, which permits the spoken natural language to be reprocessed again and again. Thus, writing is a kind of coding of the language. More about ontology and taxonomy can be found in Ref. [1].

In chemistry, chemical structures have to be represented in machine-readable form by scientific, artificial languages (see Figure 2-2). Four basic approaches are introduced in the following sections: trivial nomenclature; systematic nomenclature; chemical notation; and mathematical notation of chemical structures.

Before discussing the different ways of representing a chemical compound, some terms have to be defined concerning the reproducibility or transformation of structures and notations.

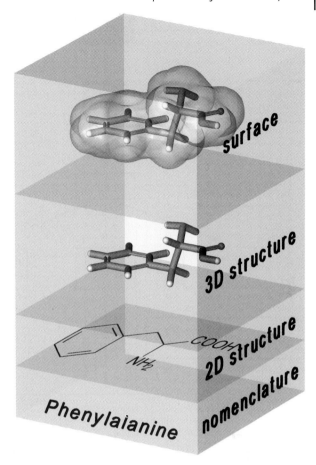

**Figure 2-1.** Hierarchical scheme for representations of a molecule with different contents of structural information.

The major significance of chemical nomenclature or notation systems is that they denote compounds in order to reproduce and transfer them from one coding to another, according to the intended application. As each coding may not include all the pieces of information in the other coding, or may have interpretable coding rules, the transformation is not always unambiguous and unique.

A nomenclature or notation is called unambiguous if it produces only one structure. However, the structure could be expressed in this nomenclature or notation by more than one representation, all producing the same structure. Moreover, "uniqueness" demands that the transformation results in only one – unique – structure or nomenclature, respectively, in both directions.

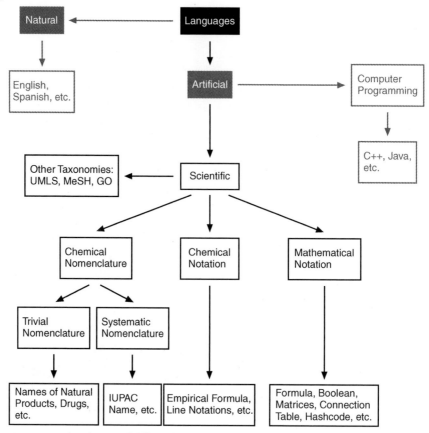

**Figure 2-2.** Classification of different languages. Examples of the languages are given in the bottom line of boxes.

## 2.2
## Chemical Nomenclature

Nomenclature is the compilation of descriptions of things and technical terms in a special field of knowledge, the vocabulary of a technical language. In the history of chemistry, a systematic nomenclature became significant only rather late. In the early times of alchemy, the properties of the substance or its appearance played a major role in giving a compound a name. Libavius was the first person who tried to fix some kind of nomenclature in *Alchemia* in 1597. In essence, he gave names to chemical equipment and processes (methods), names that are often still valid in our times.

2.2.1
## Development of Chemical Nomenclature

The first concepts of elements and atoms emerged as early as the 5th century bc. At that time, fire, water, soil, and air represented the pillars of the four-element apprenticeship. Metals were extracted as pure elements (in the present sense), even if they were not yet recognized as such. They were designated by astronomical and astrological symbols.

In 1814, J.J. Berzelius succeeded for the first time in systematically naming chemical substances by building on the results of quantitative analyses and on the definition of the term "element" by Lavoisier. In the 19th century, the number of known chemical compounds increased so rapidly that it became essential to classify them, to avoid a complete chaos of trivial names (see Section 2.2.4).

2.2.2
## Representation of Chemical Elements

Until the 17th century such fundamental chemical terms as element, compound, or mixture had no clear definitions. The development of atomic theory, and thus the concomitant improved insight into the nature of matter, gave these terms a clearer basis. In the 18th century, the characteristics of the elements were investigated in more detail. One result was the first table of equivalent weights, drawn up by Dalton in 1805. With the gradual acceptance of atomic theory, the relationships between atomic weight and the characteristics of chemical elements were recognized. Attempts were made to group elements with similar characteristics. Then in 1870, L. Meyer and D.I. Mendeleev, independently and along different lines of reasoning, succeeded in compiling the elements in a periodic table.

### 2.2.2.1  Characterization of Elements

Nowadays, chemical elements are represented in abbreviated form [2]. Each element has its own symbol, which typically consists of the initial upper-case letter of the scientific name and, in most cases, is followed by an additional characteristic lower-case letter. Together with the chemical symbol, additional information can be included such as the total number of protons and neutrons in the nucleus, the atomic number (the number of protons in the nucleus); thus isotopes can be distinguished, e.g., $^{238}_{92}U$. The charge value and, finally, the number of atoms which are present in the molecule can be given (Figure 2-3). For example, dioxygen is represented by $O_2$.

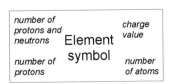

**Figure 2-3.**
Scientific notation of the chemical elements.

2.2.3
### Representation of the Empirical Formulas of (Inorganic) Compounds

At the end of the 18th century Lavoisier, Morveau, Berthollet, and Fourcroy developed a systematic method of naming compounds by a stem name and a specifying part, e.g., *sulfate de cuivre*. In addition, the quantitative composition was incorporated in the name. After Berzelius's introduction of the element symbols, the number of atoms in a molecular formula was indicated, but on the upper right-hand side (e.g., $SO^2$), not where it is put now: ever since Liebig, in 1834, suggested putting this index on the lower right-hand side of the element symbol (e.g., $SO_2$), this usage has become generally accepted.

#### 2.2.3.1  Present-Day Representation
In empirical formulas of inorganic compounds, electropositive elements are listed first [3]. The stoichiometry of the element symbols is indicated at the lower right-hand side by index numbers. If necessary, the charges of ions are placed at the top right-hand side next to the element symbol (e.g., $S^{2-}$). In ions of complexes, the central atom is specified before the ligands are listed in alphabetical order; the complex ion is set in square brackets (e.g., $Na_2[Sn(OH)_4]$).

2.2.4
### Representation of the Empirical Formulas of Organic Compounds

Organic and inorganic chemistry separated in the first half of the 19th century. Berzelius established in 1806 the term "organic chemistry" for the compounds that occur in living species. By the end of the 19th century, the need for a uniform nomenclature was felt quite urgently because of the rapidly increasing number of organic compounds. This problem was addressed in 1892 at the *International Conference of Reforming Nomenclature* in Geneva, where a standardized systematic nomenclature was recommended. This was made possible by the development of structural chemistry in the 19th century. This basic "Geneva Nomenclature" was further developed in the *Commission on Nomenclature of Organic Chemistry*, formed by the IUPAC (International Union of Pure and Applied Chemistry) in 1922. The nomenclature is still maintained and supported today [4–7].

#### 2.2.4.1  Present-Day Representation
The elements of an organic compound are listed in empirical formulas according to the Hill system [8] and the stoichiometry is indicated by index numbers. Hill positioned the carbon and the hydrogen atoms in the first and the second places, with heteroatoms following them in alphabetical order, e.g., $C_9H_{11}NO_2$. However, it was recognized that different compounds could have the same empirical formula (see Section 2.8.2, on isomerism). Therefore, fine subdivisions of the empirical

Empirical formula            Structure diagram            Condensed formula

$C_9H_{11}NO_2$                       $C_6H_5CH_2CH(NH_2)CO_2H$

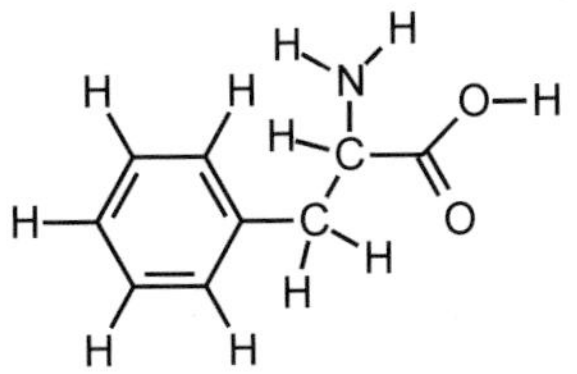

**Figure 2-4.** Different representations of phenylalanine.

formulas were developed that indicate the presence of certain structural units and functional groups.

As an example, the empirical formula of phenylalanine may be split into a more extended form that shows the presence of a phenyl ring, as well as an amino and a carboxylic acid group (the "condensed" form in Figure 2-4).

## 2.2.5
### Systematic Nomenclature of Inorganic and Organic Compounds

The systematic IUPAC nomenclature of compounds tries to characterize compounds by a unique name. The names are quite often not as compact as the trivial names, which are short and simple to memorize. In fact, the IUPAC name can be quite long and cumbersome. This is one reason why trivial names are still heavily used today. The basic aim of the IUPAC nomenclature is to describe particular parts of the structure (fragments) in a systematic manner, with special expressions from a vocabulary of terms. Therefore, the systematic nomenclature can be, and is, used in database systems such as the Chemical Abstracts Service (see Section 5.4) as index for chemical structures. However, this notation does not directly allow the extraction of additional information about the molecule, such as bond orders or molecular weight.

Clearly, this is not the place to give a comprehensive introduction to IUPAC nomenclature. Interested readers should consult the literature cited in Refs. [4–6]. Our aim here is to provide some basic understanding of the IUPAC system.

There are two basic rules for the nomenclature of organic compounds. First, the number of carbon atoms in the longest continuous aliphatic chain of carbon atoms has to be indicated. Branching of the skeleton, and the presence of rings, have to be specified by prefixes. Then, the characteristic or so-called functional groups have to be specified by a prefix and/or suffix in order to indicate the family to which the compound belongs (Figure 2-5). All the different substituents are listed in the name in alphabetical order. The more complex the structures are, the more rules are necessary to assign a unique name. There are many books on IUPAC nomenclature for each area of chemistry (inorganic, organic, polymers, etc.).

In the example of Figure 2-5, the compound has the trivial name phenylalanine but the IUPAC name is 2-amino-3-phenylpropanoic acid, which indicates a carbon

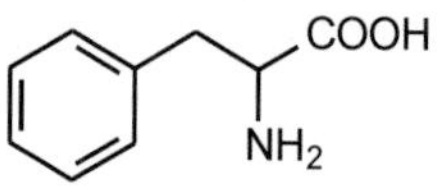

**Figure 2-5.** Phenylalanine, also known by the IUPAC name: 2-amino-3-phenylpropanoic acid.

chain of length three (propan-) of the acid (-propanoic acid) with an additional two structural units, the phenyl- and the amino- group at different positions on the carbon chain: phenyl at carbon atom number 3 and amino at carbon atom number 2, with the counting beginning with the carbon atom of the acid group (COOH).

Neither a trivial name nor the systematic nomenclature, which both represent the structure as an alphanumerical (text) string, is ideal for computer processing. The reason is that various valid compound names can describe one chemical structure (Figure 2-6). As a consequence, the name/structure correlation is unambiguous but not unique. Nowadays, programs can translate names to structures, and structures to names, to make published structures accessible in electronic journals (see also Chapter II, Section 2 in the Handbook).

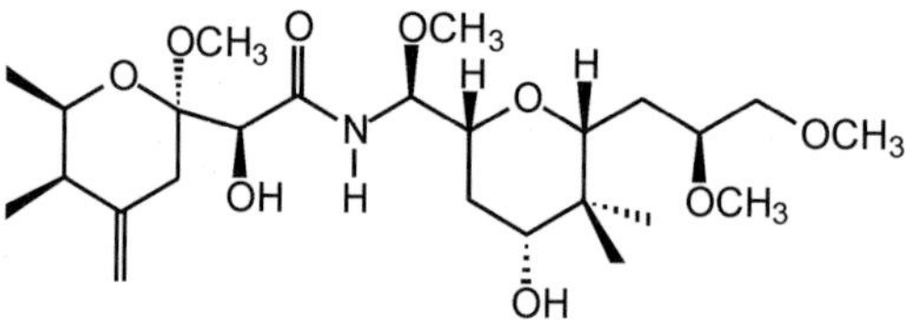

- D-*manno*-**Nonitol**, 2,6-anhydro-3,5,7-trideoxy-1-C-{[hydroxy-(tetrahydro-2-methoxy-5,6-dimethyl-4-methylene-2H-pyran-2-yl)acetyl]amino}-5,5-dimethyl-1,8,9-tri-O-methyl-,{2R-[2α,2[S*(S*)], 5β,6β]}-
- **2H-Pyran-2-acetamid**, N-[[6-(2,3-dimethoxypropyl)tetrahydro-4-hydroxy-5,5-dimethyl-2H-pyran-2-yl]methoxymethyl]tetrahydro-α-hydroxy-2-methoxy-5,6-dimethyl-4-methylene]-

**Figure 2-6.** Various logical compound names can describe one chemical structure.

Evaluation of chemical nomenclature systems for representing a chemical structure.

| Advantages | Disadvantages |
| --- | --- |
| *Trival names* | |
| • short, concise, and easy to memorize | • many available |
| • widespread | • no clear systematics |
| • unambiguous | • no evidence of stereochemistry |
| *IUPAC nomenclature* | |
| • standardized systematic classification | • extensive nomenclature rules |
| • include stereochemistry | • alternative names are allowed |
| • widespread | • complicated names |
| • unambiguous | |
| • allow reconstruction | |

## 2.3
## Line Notations

Line notations represent the structure of chemical compounds as a linear sequence of letters and numbers. The IUPAC nomenclature represents such a kind of line notation. However, the IUPAC nomenclature [6] makes it difficult to obtain additional information on the structure of a compound directly from its name (see Section 2.2).

The first line notations were conceived before the advent of computers. Soon it was realized that the compactness of such a notation was well suited to be handled by computers, because file storage space was expensive at that time. The heyday of line notations were between 1960 and 1970. A chemist, trained in this line notation, could enter the code of large molecules faster than with a structure-editing program.

In Sections 2.3.1–2.3.4, only the four most popular line notations, Wiswesser (WLN), ROSDAL, SMILES, and Sybyl (SLN), are discussed. Whereas WLN is now almost obsolete, SMILES is quite an important representation and is widely used (Figure 2-7).

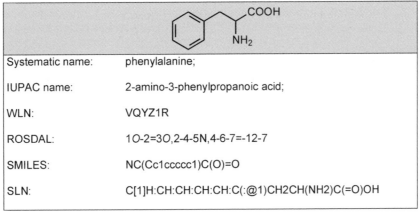

| Systematic name: | phenylalanine; |
| IUPAC name: | 2-amino-3-phenylpropanoic acid; |
| WLN: | VQYZ1R |
| ROSDAL: | 1O-2=3O,2-4-5N,4-6-7=-12-7 |
| SMILES: | NC(Cc1ccccc1)C(O)=O |
| SLN: | C[1]H:CH:CH:CH:CH:C(:@1)CH2CH(NH2)C(=O)OH |

**Figure 2-7.** Different line notations for the structure diagram of phenylalanine.

## 2.3.1
## Wiswesser Line Notation

The Wiswesser Line Notation (WLN) was introduced in 1946, in order to organize and to systematically describe the cornucopia of compounds in a more concise manner. A line notation represents a chemical structure by an alphanumeric sequence, which significantly simplifies the processing by the computer [9–11]. In many cases the WLN uses the standard symbols for the chemical elements. Additionally, functional groups, ring systems, positions of ring substituents, and posi-

**Table 2-1.** WLN coding of some important structural units.

| Class | Structural unit | WLN coding |
|---|---|---|
| Hydrogen | H | H |
| Alkanes | $C_nH_{2n+2}$ | n (e.g., $CH_3CH_2$; $CH_2CH_2$) |
| Alkenes; alkines | >C=C< ; —C≡C— | U; UU |
| Branched chains | >C< , >CH— | X, Y |
| Aromatic rings Substituted derivatives | <br>f, b, e, d, c (benzene ring) | R<br>R  B, C, D, E, F |
| (Hetero)cyclic hydrocarbons | (pentagon) ; (pyrrolidine with N) | L.n.J; T.n.J<br>L: beginning of a carbocyclic ring;<br>T: beginning of a heterocyclic ring;<br>n: number of atoms of the ring system;<br>J: termination of the ring system |
| Alkyl halides | –X (X = F, Cl, Br, I) | F, G, E, I |
| Alcohols; ethers | –OH; –O– | Q; O |
| Ketones; aldehydes | –CO–; –CO–H | V; VH |
| Carboxylic acids; esters | –COOH; –CO–O– | VQ; VO |

tions of condensed rings are assigned to individual letters or combinations of symbols (Table 2-1). This concise, linear representation of a chemical structure facilitates such tasks as the search for particular functional groups and for fragments in a molecule (see Figure 2-8). Thus, the machine retrieval of WLN characterizes the parts of a molecule.

The simple WLN uses 40 symbols from the following character sets [12]:

- capital letters: A–Z are used for elements, atom groups, branches, and ring positions;
- numbers: 0–9 indicate the length of an alkyl chain or the ring number;
- Special characters " & ", " / ", " - " and " " (blank) indicate rings and substitution position.

The great advantage of WLN codes is their compactness. Both compactness and unambiguity are achieved only by a complex set of rules, which make the notation difficult to code and error-prone. Since much information had been stored in the WLN code (functional groups, fragments, etc.), much effort was spent in the devel-

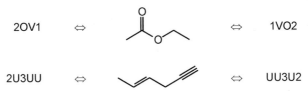

2OV1 ⇔ ⇔ 1VO2

2U3UU ⇔ ⇔ UU3U2

**Figure 2-8.** Different WLN codes could be obtained from a structure (ambiguity) but, based on the rules, only the one on the right-hand side is allowed, in order to achieve unambiguity.

opment of programs for the conversion of a WLN code into a connection table, and *vice versa*. In effect, these problems were never completely solved (see Section 2.4).

Two simple examples (Figure 2-8) should illustrate the problem of finding a unique coding (see Section 2.5.2). Although a series of different sequential arrangements of the symbols is conceivable, only one sequence, called unambiguous, is allowed as WLN code.

### 2.3.1.1  Applications

The WLN was applied to indexing the Chemical Structure Index (CSI) at the Institute for Scientific Information (ISI) [13] and the *Index Chemicus* Registry System (ICRS) as well as the Crossbow System of Imperial Chemical Industries (ICI). With the introduction of connection tables in the Chemical Abstracts Service (CAS) in 1965 and the advent of molecular editors in the 1970s, which directly produced connection tables, the WLN lost its importance.

### 2.3.2
### ROSDAL

The ROSDAL (Representation of Organic Structures Description Arranged Linearly) syntax was developed by S. Welford, J. Barnard, and M.F. Lynch in 1985 for the Beilstein Institute. This line notation was intended to transmit structural information between the user and the Beilstein DIALOG system (Beilstein-Online) during database retrieval queries and structure displays. This exchange of structure information by the ROSDAL ASCII character string is very fast.

The ROSDAL syntax is characterized by a simple coding of a chemical structure using alphanumeric symbols which can easily be learned by a chemist [14]. In the linear structure representation, each atom of the structure is arbitrarily assigned a unique number, except for the hydrogen atoms. Carbon atoms are shown in the notation only by digits. The other types of atoms carry, in addition, their atomic symbol. In order to describe the bonds between atoms, bond symbols are inserted between the atom numbers. Branches are marked and separated from the other parts of the code by commas [15, 16] (Figure 2-9). The ROSDAL linear notation is unambiguous but not unique.

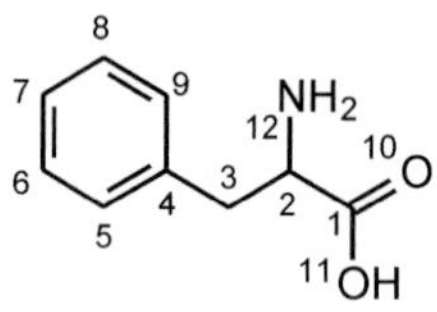

a) 1-2-3-4=5-6=7-8=9-4,1=10O,1-11O,2-12N;

b) 1-2-3-4-=9-4,1-11O,1=10O,2-12N;

**Figure 2-9.** A possible ROSDAL code for phenylalanine in a) a complete and b) a compressed notation.

The sequence for setting up a ROSDAL notation is:

1. The structure diagram is drawn and the atoms are arbitrarily numbered (each atom is assigned a unique number).
2. Atomic symbols are usually written directly behind the index of an atom.
3. Usually only the indices of the carbon atoms are written, not the symbols; hydrogen atoms can have, but do not need, an atom number
4. Bond types are described as follows:
   " – " for a single bond
   " = " for a double bond
   " # " for a triple bond
   " ? " for any connection
5. Simplifications are allowed, such as writing alternating bonds as " –= ".
6. Commas separate branches and substituents.

### 2.3.2.1 Applications

ROSDAL is used in the Beilstein-DIALOG system [17] as a data exchange format. The code can represent not only full structures and substructures but also some generic structures.

A structure drawn by a molecular editor (such as ISIS Draw) can be translated by the data conversion program AutoNom into a IUPAC name, and *vice versa*, by exchanging structure information through a ROSDAL string [18, 19].

### 2.3.3
### The SMILES Coding

In 1986, David Weininger created the SMILES (Simplified Molecular Input Line Entry System) notation at the US Environmental Research Laboratory, USEPA, Duluth, MN, for chemical data processing. The chemical structure information is highly compressed and simplified in this notation. The flexible, easy to learn language describes chemical structures as a line notation [20, 21]. The SMILES language has found widespread distribution as a universal chemical nomenclature for the representation and exchange of chemical structure information, independently of software or hardware architecture.

Compared with WLN and ROSDAL, SMILES uses only six basic rules to convert a structure into a character string (Table 2-2).

The basic SMILES rules are:

1. Atoms are represented by their atomic symbols.
2. Hydrogen atoms automatically saturate free valences and are omitted (simple hydrogen connection).
3. Neighboring atoms stand next to each other.
4. Double and triple bonds are characterized by " = " and " # ", respectively.
5. Branches are represented by parentheses.
6. Rings are described by allocating digits to the two "connecting" ring atoms.

More details about SMILES can be found in the Handbook or in Ref. [22].

SMILES has seen many extensions since 1988. The present definition, with examples, can be found at: *http://www.daylight.com/dayhtml/smiles/index.html*.

Other related coding languages are derived from enhancements of SMILES (XSMILES, SMARTS, SMIRKS, STRAPS, CHUCKLES, CHORTLES, CHARTS [22]). Each of them was designed to represent special molecular structures or to allow particular applications (polymers, mixtures, reactions, or database-handling). A special extension of SMILES is USMILES (sometimes described as Broad SMILES) [23–25]. This "Unique SMILES" of Daylight is a canonical representation of a structure. This means that the coding is independent of the internal atomic numbering and results always in the same canonical, unambiguous, and unique description of the compound, granted by an algorithm (see Section 2.5.2).

### 2.3.3.1 Applications

The compact textual coding requires no graphical input and additionally permits a fast transmission. These are important advantages of using SMILES in chemical applications via the Internet and in online services. SMILES is also used for the input of structures in the Daylight Toolkit [22].

### 2.3.4
### Sybyl Line Notation

Sybyl Line Notation (SLN) is a language used to represent molecular structures, including common organic molecules, macromolecules, polymers, and combinatorial libraries [26]. SLN was developed and is distributed by Tripos Inc. It is more or less a modification of SMILES. Its main distinction from SMILES is that all hydrogen atoms must be specified, because no assumptions are made regarding standard valences. Furthermore, structure fragments, substructure queries (Markush structures; see Section 2.7.1), and combinatorial libraries can be represented with SLN. All these features make this line notation suitable for database storage as well for data exchange between various programs.

**Table 2-2.** SMILES syntax.

| SMILES code | Chemical structure | Compound name |
|---|---|---|

*Atoms:* Atoms are represented by their atomic symbols. Ambiguous two-letter symbols (e.g., Nb is not NB) have to be written in square brackets. Otherwise, no further letters are used. Free valences are saturated with hydrogen atoms.

| | | |
|---|---|---|
| C | $CH_4$ | methane |
| [Fe+2] or [Fe++] | $Fe^{2+}$ | iron (II) cation |

*Bonds:* Single, double, triple, and aromatic (or conjugated) bonds are indicated by the symbols " - ", " = ", " # " and " : ", respectively; single and aromatic bonds should be omitted.

| | | |
|---|---|---|
| C=C | $H_2C=CH_2$ | ethene |
| O=CO | HCOOH | formic acid |

*Disconnected structures in the molecule:* Individual parts of the compound are separated by a period. The period indicates that there is no connection between atoms or parts of a molecule. The arrangement of the parts is arbitrary.

| | | |
|---|---|---|
| [Na+].[OH-] | NaOH | sodium hydroxide |

*Branches:* Branches are indicated within parentheses.

| | | |
|---|---|---|
| CC(=O)O | | acetic acid |
| CC(C)C(=O)O | | isobutyric acid |

*Cyclic structures:* Rings are described by breaking the ring between two atoms and then labeling the two atoms with the same number.

| | | |
|---|---|---|
| C1CCCCC1 | | cyclohexane |

*Aromaticity:* Aromatic structures are indicated by writing all the atoms involved in lower-case letters.

| | | |
|---|---|---|
| o1cccc1 | | furan |
| c12c(cccc1)cccc2 <br> same as <br> c1cc2ccccc2cc1 | | naphthalene |

**Table 2-3.** Basic SLN syntax without description of attributes and macro atoms.

| SLN | Chemical structure | Compound name |
|---|---|---|

*Atoms:* Atoms are represented by their atomic symbols. The first letter is upper-case, and in two-letter symbols the second letter is lower-case. Hydrogen atoms must be specified.

| | | |
|---|---|---|
| CH4 | $CH_4$ | methane |
| NH2 | $-NH_2$ | amine |

*Bonds:* Single bonds are omitted; double, triple, and aromatic bonds are indicated by the symbols " = ", " # " and " : ", respectively. In contrast to SMILES, aromaticity is not an atomic property, but a property of bonds. A period indicates the start of a new part of the structure.

| | | |
|---|---|---|
| HC(=O)OH | HCOOH | formic acid |
| Na.OH | NaOH | sodium hydroxide |

*Branches:* Branches are indicated by parentheses.

| | | |
|---|---|---|
| CH3C(=O)OH | | acetic acid |

*Cyclic structures:* Ring closures are described by a bond to a previously defined atom which is specified by a unique ID number. The ID is a positive integer placed in square brackets behind the atom. An " @ " indicates a ring closure.

| | | |
|---|---|---|
| C[15]H2CH2CH2CH2CH2CH2@15 | | cyclohexane |
| O[6]:CH:CH:CH:CH:@6 | | furan |

SLN is easy to learn and its use is intuitive. The language uses only six basic components to specify chemical structures. Four of them are listed in Table 2-3 and can be compared directly with the SMILES notation of Section 2.3.3.

Besides specifications on atoms, bonds, branches, and ring closure, SLN additionally provides information on *attributes* of atoms and bonds, such as charge or stereochemistry. These are also indicated in square [ ] or angle < > brackets behind the entity (e.g., *trans*-butane: $CH_3CH=[s=t]CHCH_3$). Furthermore, *macro atoms* allow the shorthand specification of groups of atoms such as amino acids, e.g., Ala, Protein2, etc. A detailed description of these specifications and also specifications for 2D substructure queries or combinatorial libraries can be found in the literature [26].

### 2.3.4.1 **Applications**

SLN is used in many Internet applications for fast data exchange. A number of commercial software packages, such as Alchemy 2000 [27], ChemDraw [28], and CLIFF [29], and of course Tripos products such as CONCORD [30], operate with this line notation.

Evaluation of line notations for representing a chemical structure.

| Advantages | Disadvantages |
| --- | --- |
| *Wiswesser LN* | |
| • concise linear code | • large number of complex rules |
| • unambiguous | • coding prone to errors |
| • simple substructure search | • difficult to translate into a connection table |
| • includes stereochemistry | • only those substructures contained in the |
| • unique if rules are followed | coding can be retrieved in a substructure |
| | search |
| | • no support for coding reactions |
| *ROSDAL* | |
| • simple code, easy to learn | • no support for coding reactions |
| • fast data exchange format | • not unique |
| • includes stereochemistry | |
| • unambiguous | |
| *SMILES* | |
| • simplest linear code | • not unique (except Unique SMILES) |
| • easy to learn | • some problems with aromaticity perception |
| • fast data exchange format | |
| • supports Markush, stereochemistry, | |
| and reaction coding | |
| • unambiguous | |
| *Sybyl LN* | |
| • simple code, easy to learn | • not unique |
| • Markush and macro atom definitions | • aromaticity has to be normalized |
| • includes stereochemistry | • no valence rules |
| • fast data exchange format | • no support for coding reactions |
| • unambiguous | |

## 2.4
## Coding the Constitution

Chemists have been used to drawing chemical structures for more than a hundred years. Nowadays, structures are not only drawn on paper but they are also available in electronic form on a computer for publications, for presentations, or for the input and output with computer programs. For these applications, well-known software such as ISIS/Draw (MDL [31]) or ChemWindow (Bio-Rad Sadtler [32]) are used (see Section 2.12). The structures generated with these programs are "pictures" carrying much information for chemists, but they cannot be used directly by the computer in this form. In order to process a chemical structure on the computer, the structure drawing has to be converted into another form of representation. Two types of representation were introduced in Sections 2.2 and 2.3, nomenclature and line notations.

## 2.4.1
### Graph Theory

Another approach applies graph theory. The analogy between a structure diagram and a topological graph is the basis for the development of graph theoretical algorithms to process chemical structure information [33–35].

In mathematical terms, the structure diagrams drawn by a chemist, can be considered as ordinary graphs. Graphs consist of nodes (vertices), which are the atoms, and edges, which are the bonds. In organic chemistry, these graphs are often simplified by representing the carbon atoms only as the point where connecting lines meet, the edges (bonds) (Figure 2-10). Such a graph is called topological graph because it only shows the linkages between atoms and the kind of bonds between them. This type of structure representation contains no data about the 3D structure, the topography, of a molecule.

Usually, a structure diagram is an undirected (the bonds have no direction) and labeled graph (the nodes are characterized by atom symbols) (for definitions of terms, see Table 2-4). In graph theory, a graph carries no geometric information. A weighted graph has numbers or symbols assigned to the nodes. Two nodes can have several edges between them (in chemistry, multiple bonds) (Figure 2-11).

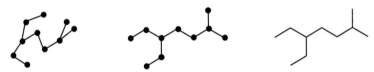

**Figure 2-10.** Different graph-theory representations of an identical diagram. In graph theory only the connections are important, not the length of the edges or the angles between them.

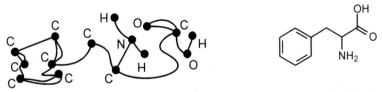

**Figure 2-11.** Phenylalanine can be represented in graph theory as a labeled, weighted graph with different atom and bond types (as on the left-hand side).

**Excursion: The Königsberg Bridge Problem**

At the time of Leonhard Euler (the 18th century), Königsberg had seven bridges across the Pregel river. Some of the townspeople wanted to know if there was a path through the town that allowed one to cross each of the seven bridges exactly once and finish at the starting point.

Euler's task was to find such a way. He first reduced the problem to its essentials by substituting a dot for each piece of land and a line for each bridge, then connecting the corresponding dots. This object was called a graph (sketched in Figure 2-12, right).

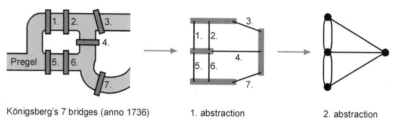

Königsberg's 7 bridges (anno 1736)    1. abstraction    2. abstraction

**Figure 2-12.** Schematic illustration of the Königsberg bridge problem, and abstraction to a graph.

Then, Euler solved the problem in a general way by proving that a graph can be traversed if it is connected and if each node has an even degree (an even number of lines emanating from it; see Table 2-4). In fact, the graph in Figure 2-12 is connected, but all four nodes have an odd number of lines coming from them.

Thus, the graph theory introduced by Euler in 1736 proved that it was not possible to walk through Königsberg by crossing each bridge exactly once and ending up at the point where the path was started [36].

### 2.4.1.1  Basics of Graph Theory

Graphs are used in mathematics to describe a variety of problems and situations [37]. The methods of graph theory analyze graphs and the problems modeled by them. The transfer of models and abstractions from other sciences (computer science, chemistry, physics, economics, sociology, etc.) to graph theory makes it possible to process them mathematically because of the easily understandable basics of graph theory.

Some fundamental definitions of graph theory are given in Table 2-4.

**Table 2-4.** Some basic definitions for graph theory.

| Graph theory term | Graph |
|---|---|
| Nodes (dots) are *adjacent* when they are connected by the same edge. | |
| If the nodes of a graph are marked (e.g., with digits), the graph is termed *labeled*. In the example, node **1** is adjacent to node **2** but not to node **3**. | |
| The *degree* (or valency) of a node is determined by the number of distinct edges that end in a given node. (e.g., nodes **1** and **3** have the degree **1**, and node **2** has the degree **2**). | |
| An edge is *incident* to the two joining nodes (e.g. *a* is incident to **1** and **2**). | |
| A graph is *connected* if at least one edge is between all the nodes. Thus, from any given node in a connected graph, all the other nodes can be reached. | |
| Conversely, a *disconnected graph* (null graph) contains isolated nodes without edges (in chemistry, these may be mixtures of compounds or collections of substructures). | |
| A *digraph* (or *directed graph*) has directed edges between two nodes (e.g., a weighted orientation). | |
| Structure diagrams are *undirected graphs*. | |
| A graph is *complete* if all the nodes are connected (adjacent) to all the other nodes. | |
| A graph is *planar* if it can be drawn on a plane without edges crossing, with intersections only at the edges (independently of how it is drawn). For example, cubane can be drawn as a planar graph. | |
| *Euler path*: A connected graph can be traversed in one path (which ends at the node where it began) if all nodes have an even degree (see the Königsberg bridge problem, Section 2.4.1). | |

**Table 2-4.** (cont.)

| Graph theory term | Graph |
|---|---|

*Euler circuit* (the house of Santa Claus): a graph can be drawn in one path if the degree of all the nodes is a multiple of 2 and two nodes have an odd degree (for starting and ending the path). The drawing has to start at one of the nodes with an odd degree.

*Isomorphism*: If a labeled graph has *n* defined nodes, it can be represented by *n*! labeled graphs. In the example with n = 3, the six graphs are isomorphic.

see Fig. 2-41 in Section 2.5.2.2

2.4.2
## Matrix Representations

A graph can also be represented as a matrix. Thus, quite early on, matrix representations of molecular structures were explored. Their major advantage is that the calculation of paths and cycles can be performed easily by well-known matrix operations.

The matrix of a structure with $n$ atoms consists of an array of $n \times n$ entries. A molecule with its different atoms and bond types can be represented in matrix form in different ways depending on what kind of entries are chosen for the atoms and bonds. Thus, a variety of matrices has been proposed: adjacency, distance, incidence, bond, and bond–electron matrices.

In the following matrices hydrogen atoms are sometimes not shown, because their numbers and positions can be calculated from organic structures on the basis of the valence rules of the other atoms.

A second notion is that each atom is described twice – in a column and in a row. Matrices in which all elements are shown twice are called redundant. A non-redundant matrix contains each element only once (e.g., only the top right or bottom left triangle of the matrix, as can be seen later).

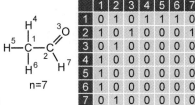

**Figure 2-13.** Adjacency (7 × 7) matrix of ethanal.

### 2.4.2.1 Adjacency Matrix

The adjacency matrix of a molecule consisting of $n$ atoms is a square ($n \times n$) matrix with the entries giving all the connectivities of the atoms. The intersection of a row and a column obtains a value of 1 if the corresponding atoms are connected. If there is no bond between the atoms being considered, the position in the matrix obtains the value 0. Thus, this matrix representation is a Boolean matrix with bits (0 or 1) (Figure 2-13).

As can be seen in Figure 2-13, the diagonal elements of the matrix are always zero and it is symmetric around the diagonal elements (undirected, unlabeled graph). Thus, it is a redundant matrix and can be reduced to half of its entries (Figure 2-14b). For clarity, all zero entries are omitted in Figures 2-14b–d.

With such a matrix representation, the storage space is dependent only on the number of nodes (atoms) and independent of the number of bonds. As Figure 2-14 demonstrates, all the essential information in an adjacency matrix can also be found in the much smaller non-redundant matrix. But the adjacency matrix is unsuitable for reconstructing the constitution of a molecule, because it does not provide any information about the bond orders.

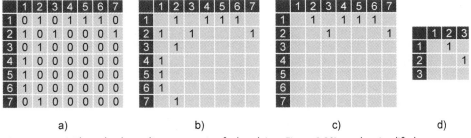

**Figure 2-14.** a) The redundant adjacency matrix of ethanal (see Figure 2-13) can be simplified step by step by b) omitting the zero values, c) reducing it to the top right triangle, and, finally, d) omitting the hydrogen atoms.

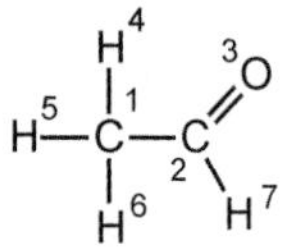

a)

| | C1 | C2 | O3 | H4 | H5 | H6 | H7 |
|---|---|---|---|---|---|---|---|
| C1 | 0 | 1.400 | 2.190 | 1.022 | 1.023 | 1.022 | 2.106 |
| C2 | 1.400 | 0 | 1.123 | 1.999 | 1.982 | 1.999 | 1.022 |
| O3 | 2.190 | 1.123 | 0 | 2.349 | 2.708 | 2.995 | 1.859 |
| H4 | 1.022 | 1.999 | 2.349 | 0 | 1.668 | 1.661 | 2.895 |
| H5 | 1.023 | 1.982 | 2.708 | 1.668 | 0 | 1.668 | 2.562 |
| H6 | 1.022 | 1.999 | 2.955 | 1.661 | 1.668 | 0 | 2.336 |
| H7 | 2.106 | 1.022 | 1.859 | 2.895 | 2.566 | 2.336 | 0 |

b)

| | C1 | C2 | O3 | H4 | H5 | H6 | H7 |
|---|---|---|---|---|---|---|---|
| C1 | 0 | 1 | 2 | 1 | 1 | 1 | 2 |
| C2 | 1 | 0 | 1 | 2 | 2 | 2 | 1 |
| O3 | 2 | 1 | 0 | 3 | 3 | 3 | 2 |
| H4 | 1 | 2 | 3 | 0 | 2 | 2 | 3 |
| H5 | 1 | 2 | 3 | 2 | 0 | 2 | 3 |
| H6 | 1 | 2 | 3 | 2 | 2 | 0 | 3 |
| H7 | 2 | 1 | 2 | 3 | 3 | 3 | 0 |

**Figure 2-15.** Distance matrices of ethanal with a) geometric distances in Å and b) topological distances. The matrix elements of b) result from counting the number of bonds along the shortest walk between the chosen atoms.

#### 2.4.2.2 Distance Matrix

The elements of a distance matrix contain values which specify the shortest distance between the atoms involved. Distances can be expressed either as geometric distances (in Å) or as topological distances (in number of bonds) (Figure 2-15a,b).

#### 2.4.2.3 Atom Connectivity Matrix

Both the adjacency and distance matrices provide information about the connections in the molecular structure, but no additional information such as atom type or bond order. One type of matrix which includes more information, the Atom Connectivity Matrix (ACM), was introduced by Spialter and is discussed in Ref. [38]. This approach was eventually abandoned but is listed here because it was quite a unique approach.

#### 2.4.2.4 Incidence Matrix

The incidence matrix is an $n \times m$ matrix where the nodes (atoms) define the columns $(n)$ and the edges (bonds) correspond to the rows $(m)$. An entry obtains the value of 1 if the corresponding edge ends in this particular node (Figure 2-16).

#### 2.4.2.5 Bond Matrix

The bond matrix is related to the adjacency matrix but gives information also on the bond order of the connected atoms. Elements of the matrix obtain the value of 2 if there is a double bond between the atoms, e.g., between atoms 2 and 3

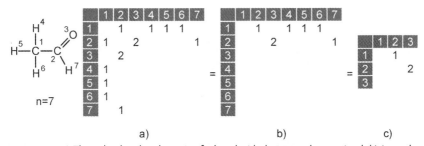

|   | C1 | C2 | O3 | H4 | H5 | H6 | H7 |
|---|----|----|----|----|----|----|----|
| a | 1  | 1  | 0  | 0  | 0  | 0  | 0  |
| b | 0  | 1  | 1  | 0  | 0  | 0  | 0  |
| c | 1  | 0  | 0  | 1  | 0  | 0  | 0  |
| d | 1  | 0  | 0  | 0  | 1  | 0  | 0  |
| e | 1  | 0  | 0  | 0  | 0  | 1  | 0  |
| f | 0  | 1  | 0  | 0  | 0  | 0  | 1  |

a)

|   | C1 | C2 | O3 | H4 | H5 | H6 | H7 |
|---|----|----|----|----|----|----|----|
| a | 1  | 1  |    |    |    |    |    |
| b |    | 1  | 1  |    |    |    |    |
| c | 1  |    |    | 1  |    |    |    |
| d | 1  |    |    |    | 1  |    |    |
| e | 1  |    |    |    |    | 1  |    |
| f |    | 1  |    |    |    |    | 1  |

b)

n=7; m=6

|   | C1 | C2 | O3 |
|---|----|----|----|
| a | 1  | 1  |    |
| b |    | 1  | 1  |

c)

**Figure 2-16.** a) The redundant incidence matrix of ethanal can be compressed by b) omitting the zero values and c) omitting the hydrogen atoms. In the non-square matrix, the atoms are listed in columns and the bonds in rows.

in the example shown. Otherwise the value can be 0, 1, or 3 for other bonding combinations. This representation is redundant, as well (Figure 2-17).

### 2.4.2.6 Bond–Electron Matrix

The bond–electron matrix (BE-matrix) was introduced in the Dugundji–Ugi model [39]. It can be considered as an extension of the bond matrix or as a modification of Spialter's atom connectivity matrix [38]. The BE-matrix gives, in addition to the entries of bond values in the off-diagonal elements, the number of free valence electrons on the corresponding atom in the diagonal elements (e.g., O3 = 4 in Figure 2-18).

In essence, a BE-matrix lists all the valence electrons of the atoms in a molecule, both the ones involved in bonds and those associated as free electrons with an atom. A BE-matrix has a series of interesting mathematical properties that directly reflect the chemical information:

**Figure 2-17.** a) The redundant bond matrix of ethanal with the zero values omitted. b) It can be compressed by reduction to the top right triangle. c) Omitting the hydrogen atoms provides the simplest non-redundant matrix representation.

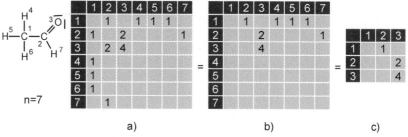

**Figure 2-18.** a) The redundant BE-matrix of ethanal with the zero values omitted. b) It can be compressed to the top right triangle. c) Omitting the hydrogen atoms provides the simplest non-redundant matrix representation.

- The sum $s_i$ over all entries of a row $b_{ji}$ or column $b_{ij}$ indicates the number of valence electrons of atom $i$ (Eq. (1)).

$$s_i = \sum_j b_{ij} = \sum_j b_{ji} \tag{1}$$

This term is also called the row/column sum (see Figure 2-19). In the example, carbon atom 2 has $1 + 2 + 1 = 4$ valence electrons.
- The sum over all entries of the BE-matrix ($S$) gives the total number of valence electrons in the molecule (Eq. (2)).

$$S = \sum_i \sum_j b_{ij} \tag{2}$$

In the example, ethanal has 36 valence electrons.
- If the number of valence electrons thus calculated does not agree with the standard number of valence electrons in an atom, this atom carries a charge. In this case, the diagonal element $b_{ii}$ has more or fewer valence electrons than the nominal value $b_{ii}^0$ of the respective atom $i$. The charge value, $\Delta b$, can be determined by subtracting the sum of the row values from the nominal value (Eq. (3)).

$$\Delta b = b_{ii}^0 - b_{ii} \tag{3}$$

The cross sum $\hat{s}_i$, which is the sum over all the entries in a row and a column of atom $i$ ($= 2s_i$ according to Eq. (1)) with the diagonal element $b_{ii}$ of atom $i$ counted only once, indicates the total number of valence electrons in the orbitals of atom $i$ (Eq. (4)).

$$\hat{s}_i = 2s_i - b_{ii} \tag{4}$$

In the example in Figure 2-19, the oxygen atom 3 has $2 + 4$ (row) $+ 2 + 4$ (column) $- 4$ (diagonal element) $= 8$ electrons. This shows that the oxygen atom obeys the octet rule.

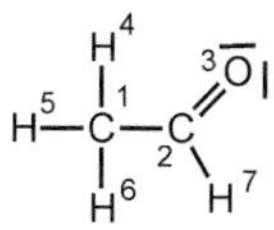

| | 1 | 2 | 3 | 4 | 5 | 6 | 7 | row sum | Element |
|---|---|---|---|---|---|---|---|---|---|
| 1 | | 1 | | 1 | 1 | 1 | | 4 | C |
| 2 | 1 | | 2 | | | | 1 | 4 | C |
| 3 | | 2 | 4 | | | | | 6 | O |
| 4 | 1 | | | | | | | 1 | H |
| 5 | 1 | | | | | | | 1 | H |
| 6 | 1 | | | | | | | 1 | H |
| 7 | | 1 | | | | | | 1 | H |
| column sum | 4 | 4 | 6 | 1 | 1 | 1 | 1 | **36** | |
| cross sum | 8 | 8 | 8 | 2 | 2 | 2 | 2 | | |

n=7

**Figure 2-19.** The BE-matrix of ethanal allows one to determine the number of valence electrons (the sum of each row) on the atoms and to validate the octet rule.

The BE-matrix also provides the basis for the matrix representation of chemical reactions, as discussed in detail in Section 3.5.

Evaluation of matrix representations of chemical structures.

| Advantages | Disadvantages |
|---|---|
| *General:* | |
| • the molecular graph is completely coded (each atom and bond is represented) | • the number of entries in the matrix grows with the square of the number of atoms ($n^2$) |
| • matrix algebra can be used | • no stereochemistry included |
| *Adjacency matrix* | |
| • describes connections of atoms | • no bond types and bond orders |
| • contains only 0 and 1 (bits) | • no number of free electrons |
| *Distance matrix* | |
| • describes geometric distances | • no bond types or bond orders |
| | • no number of free electrons |
| | • cannot be represented by bits |
| *Incidence matrix* | |
| • describes connections and bonds | • no bond types and bond orders |
| • contains only 0 and 1(bits) | • no number of electrons |
| *Bond matrix* | |
| • describes connections and bond orders of atoms | • no number of free electrons |
| | • cannot be represented by bits |
| *Bond–electron matrix* | |
| • describes connections, bond orders, and valence electrons of the atoms | • cannot be represented by bits |

### 2.4.3
### Connection Table

A major disadvantage of a matrix representation for a molecular graph is that the number of entries increases with the square of the number of atoms in the molecule. What is needed is a representation of a molecular graph where the number of entries increases only as a linear function of the number of atoms in the molecule. Such a representation can be obtained by listing, in tabular form only the atoms and the bonds of a molecular structure. In this case, the indices of the row and column of a matrix entry can be used for identifying an entry. In essence, one has to distinguish each atom and each bond in a molecule. This is achieved by a list of the atoms and a list of the bonds giving the connections between the atoms. Such a representation is called a connection table (CT).

A connection table has been the predominant form of chemical structure representation in computer systems since the early 1980s and it is an alternative way of representing a molecular graph. Graph theory methods can equally well be applied to connection table representations of a molecule.

There are many ways of presenting a connection table. One is first to label each atom of a molecule arbitrarily and to arrange them in an atom list (Figure 2-20). Then the bond information is stored in a second table with indices of the atoms that are connected by a bond. Additionally, the bond order of the corresponding connection is stored as an integer code (1 = single bond, 2 = double bond, etc.) in the third column.

Both tables, the atom and the bond lists, are linked through the atom indices. An alternative connection table in the form of a redundant CT is shown in Figure 2-21. There, the first two columns give the index of an atom and the corresponding element symbol. The bond list is integrated into a tabular form in which the atoms are defined. Thus, the bond list extends the table behind the first two columns of the atom list. An atom can be bonded to several other atoms: the atom with index 1 is connected to the atoms 2, 4, 5, and 6. These can also be written on one line. Then, a given row contains a focused atom in the atom list, followed by the indices of all the atoms to which this atom is bonded. Additionally, the bond orders are inserted directly following the atom in-

| Atom list | |
|:---:|:---:|
| 1 | C |
| 2 | C |
| 3 | O |
| 4 | H |
| 5 | H |
| 6 | H |
| 7 | H |

| Bond list | | |
|:---:|:---:|:---:|
| 1st atom | 2nd atom | bond order |
| 1 | 2 | 1 |
| 2 | 3 | 2 |
| 2 | 7 | 1 |
| 1 | 4 | 1 |
| 1 | 5 | 1 |
| 1 | 6 | 1 |

**Figure 2-20.** A connection table: the structure diagram of ethanal, with the atoms arbitrarily labeled, is defined by a list of atoms and a list of bonds.

| atom index | element | 1st index of atom | bond order | 2nd index of atom | bond order | 3rd index of atom | bond order | 4th index of atom | bond order |
|---|---|---|---|---|---|---|---|---|---|
| 1 | C | 2 | 1 | 4 | 1 | 5 | 1 | 6 | 1 |
| 2 | C | 1 | 1 | 3 | 2 | 7 | 1 | | |
| 3 | O | 2 | 2 | | | | | | |
| 4 | H | 1 | 1 | | | | | | |
| 5 | H | 1 | 1 | | | | | | |
| 6 | H | 1 | 1 | | | | | | |
| 7 | H | 2 | 1 | | | | | | |

**Figure 2-21.** Redundant connection table of ethanal.

dices of these connected atoms. The bond orders are 1 for a single bond, 2 for a double bond, etc. In our example, atom 1 (which is carbon) is connected to carbon atom 2 via a single bond and to the hydrogen atoms 4, 5, and 6 via single bonds (Figure 2-21).

Since each bond connects two atoms, each of these atoms is defined twice in such a connection table (see Figure 2-21). Similarly to the matrix representations of Section 2.4.2, the connection table contains redundant information as well, which can be eliminated. Besides duplicates, the hydrogen atoms could also be omitted in "standard" organic compounds. Programs which have to treat hydrogen atoms can deduce them from the open valences of the corresponding atoms according to the conventions of the valence bond rules (considering the bond order). In any other case, e.g., in organometallics, metal complexes, some inorganics, or whenever a hydrogen atom plays a major role, the hydrogen atoms have to be included.

By listing each bond only once and by omitting hydrogen atoms a non-redundant, compressed connection table is obtained (Figure 2-22) which is important for saving storage space. Nevertheless, the information about the chemical structure is kept intact.

| atom index | element | 1st index of atom | bond order | 2nd index of atom | bond order |
|---|---|---|---|---|---|
| 1 | C | 2 | 1 | | |
| 2 | C | | | 3 | 2 |
| 3 | O | | | | |

| atom index | element | 1st index of atom | bond order |
|---|---|---|---|
| 1 | C | 2 | 1 |
| 2 | C | 3 | 2 |
| 3 | O | | |

**Figure 2-22.** Non-redundant connection table of ethanal. Only non-hydrogen atoms are considered; bonds with the lowest indices are counted once (see Figure 2-21).

A connection table can be extended by adding other lists, such as lists of the free electrons and/or with the charges on the atoms of the molecule. Thus, in effect, all the information in a BE-matrix can also be stored in a connection table [40].

If the indexing of the atoms is changed, the CT will have a different appearance. Thus, the representation of a chemical structure in a CT is unambiguous but not unique, which can only be achieved by canonicalization (see below).

Almost all chemical information systems work with their own special type of connection table. They often use various formats distinguishing between internal and external connection tables. In most cases, the internal connection tables are redundant, thus allowing maximum flexibility and increasing the speed of data processing. The external connection tables are usually non-redundant in order to save disk space. Although a connection table can be represented in many different ways, the core remains the same: the list of atoms and the list of bonds. Thus, the conversion of one connection table format into another is usually a fairly straightforward task.

Evaluation of the representation of chemical structures by connection tables.

| Advantages | Disadvantages |
| --- | --- |
| • the graph is completely coded <br> • concise code <br> • the number of entries grows linearly with the number of atoms <br> • atom types, connections, and bond orders are described separately <br> • extensions allow addition of information on free electrons, stereochemistry, etc. <br> • widely used representation of chemical structure information | • in most compact codes, hydrogen atoms are omitted and can be derived only indirectly <br> • needs more than bits as entries |

### 2.4.4
### Input and Output of Chemical Structures

As was said in the introduction (Section 2.1), chemical structures are the universal and the most natural language of chemists, but not for computers. Computers work with bits packed into words or bytes, and they perceive neither atoms nor bonds. On the other hand, human beings do not cope with bits very well. Instead of thinking in terms of 0 and 1, chemists try to build models of the world of molecules. The models are conceptually quite simple: 2D plots of molecular structures or projections of 3D structures onto a plane. The problem is how to transfer these models to computers and how to make computers understand them. This communication must somehow be handled by widely understood input and output processes. The chemists' way of thinking about structures must be translated into computers' internal, machine representation through one or more intermediate steps or representations (see Figure 2-23). The input/output processes defined

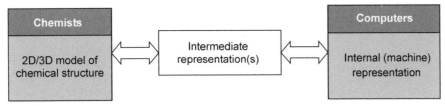

**Figure 2-23.** Transformation of representations of chemical structures between chemists and computers.

in this way have a twofold nature: they are physical actions undertaken by chemists and logical transformation(s) of the initial structural models.

In modern times, i.e., since the 1980s, when high-quality computer graphics became available, the transformation of representations has been seemingly quite straightforward. The chemist has two basic possibilities for expressing models of molecular structures (molecular structure input). One can use existing structural representations for the manual creation of a data file or other kinds of textual interaction with a computer, but also a graphical editor (or generally speaking a graphical user interface) for drawing molecular structures (see Section 2.12). The drawn molecular structure will next be converted automatically into one of the structural representations or directly into a format understood only by the computer and the software that is being used. The molecular structure expressed in the selected representation is also translated by the computer into an internal format. The reverse process is the output of molecular structures.

Three structural representations are available to the chemist:

- nomenclature: IUPAC names, trivial names, registration identifiers (see Section 2.2);
- line notations: Wiswesser line notation (WLN), ROSDAL, Sybyl line notation, SMILES, etc. (see Section 2.3);
- connection tables (see Section 2.4.3).

Interconversions of these representations are also usually possible.

At the moment, most scientists and students, both in companies and at universities, use similar tools for the encoding processes described. But this was not the case in the past, when research was much more differentiated and fascinating with regard to the chemist–computer interaction. Some of these old-fashioned results are still impressive or some are still used, even now, as the following examples illustrate.

In the "ancient times" (the 1950s), data were transferred to computers by using punched cards. But already in 1959 Ascher Opler from Dow Chemical Company reported the use of a light pen for graphical entry of chemical structures into a computer. Light pens were also used in the Chemical Abstracts Service in the 1970s.

In 1962 special formula reading machines [41] were constructed and used at BASF Ludwigshafen. They scanned formulas drawn on transparent grid sheets and recorded the pulses on punched cards (which were later replaced by punched tapes and finally by magnetic tapes). Some of these machines run for almost two decades.

A year later, a novel method of encoding chemical structures via typewriter input (punched paper tape) was described by Feldmann [42]. The constructed typewriter had a special character set and recorded on the paper tape the character struck and the position (coordinates) of the character on the page. These input data made it possible to produce tabular representations of the structure.

In 1964 Douglas Cart Engelbart (Stanford Research Institute in California) developed the mouse as an input device and Bill English built its prototype from a carved block of wood with a single red button. Shortly thereafter, the mouse was used at the Lister Hill Center of the National Institutes of Health for the input of chemical structures.

The first known application of an optical character recognition system in processing of chemical information dates back to 1973 [43]. Scientists in the Cambridge Crystallographic Data Centre were using an IBM typewriter equipped with an OCR-B golf ball. The typed sheets were processed by a service bureau using specialized OCR hardware, yielding a magnetic tape, which formed the basic input to the system.

Another example of dealing with molecular structure input/output can be found in the early 1980s in Boehringer Ingelheim. Their CBF (Chemical and Biology Facts) system [44] contained a special microprocessor-controlled semigraphic device for entering molecular structures. Moreover, their IBM-type printer chain unit had been equipped with special chemical characters and it was able to print chemical formulas.

Talking about output of chemical structures, there is a common conviction that 3D molecular graphics was born in the 1980s. But the first known system for the interactive display of molecular structures was devised in 1964 by Cyrus Levinthal and his colleagues from MIT [45–47]. The system displayed protein structures on a monochrome oscilloscope as wire frames (see Section 2.11). The illusion of the 3D view was achieved by constantly rotating the structure on the screen. Although the system was used to study short-range interaction between atoms and online manipulation of chemical structures, the importance of molecular visualization and its applications seemed to be underappreciated at that time [46]. The Early Interactive Molecular Graphics Movie Gallery [48] contains the original visualizations of several macromolecules converted to digital formats and viewable on modern computers.

Another interesting output "generator" is ORTEP (Oak Ridge Thermal-Ellipsoid Plot Program) [49], written by Carroll K. Johnson in 1965. In addition to plotting molecular structures with thermal ellipsoids, the program was and is able to generate stereoscopic images automatically. After nearly 40 years the program is still being widely used and its newest version can be obtained from the website URL in Ref. [49].

There have been plenty of other examples of similar developments in the area of molecular structure input/output, especially during the third quarter of the 20th century. Even if they are no longer applicable, studying them can be a fascinating expedition into the origins of modern GUIs (graphical user interfaces).

### 2.4.5
### Standard Structure Exchange Formats

In chemistry, numerous software programs are available to handle structure information on molecules. The scope of the programs leads from drawing structure diagrams (see Section 2.12), through collecting data from instruments (e.g., HPLC, NMR, etc.; see Chapter 4) to expert systems (Section 9.2) that process the data and produce new information. All of these systems have one task in common: to save data in a file. Many organizations and software suppliers have developed their own connection table format and quite a few have made provisions for the import or export of other file formats. The processing of data, from data to information and finally to knowledge, usually asks for the interaction and cooperation of several different software systems and databases. In this process, the exchange of chemical structure information plays a pivotal role; the internal file format of one software system has to be understood by another, i.e., converted into its internal file format. This exchange process is usually handled through an external, ASCII, file format.

As many different file formats have been developed since the early 1970s, the need for a standard chemical structure format has been increasingly felt. Various attempts have been made by different groups of the chemical community to define and push such a format, but none has achieved unanimous acceptance.

Parallel to that the MDL Molfile format (see the Tutorial in Section 2.4.6) developed at Molecular Design Limited (now MDL Information Systems, Inc.) became a *de facto* standard file format [50].

The release of this initially proprietary format to the community at large in 1982 led to its acceptance as a general exchange format for chemical datasets. Several extensions have been made to the MDL Molfile format, leading to the SDfile, RGfile, Rxnfile, or RDfile, with each one having special additional information on one or several molecules [51]. The historical development led to the situation that the Molfile is not a rigidly standardized format. Molfiles from different sources may differ in some details, depending on the software that generated them. Thus, the extension of the file name is always *.mol but the structure of the file format can be different from the MDL Molfile (e.g., Sybyl-, MSI-Molfile, etc.).

Besides the MDL Molfile format, other file formats are often used in chemistry; SMILES has already been mentioned in Section 2.3.3. Another one, the PDB file format, is primarily used for storing 3D structure information on biological macromolecules such as proteins and polynucleotides (Tutorial, Section 2.9.7) [52, 53]. CIF (Crystallographic Information File) [54, 55] is also a 3D structure information file format with more than three incompatible file versions and is used in crystallography. CIF should not be confused with the Chiron Interchange Format, which is also extended with *.cif. In spectroscopy, JCAMP is applied as a spectroscopic exchange file format [56]. Here, two modifications can be distinguished; the JCAMP–DX and the JCAMP–CS format. Whereas JCAMP–CS is an alternative to the Molfile and contains structure data, JCAMP–DX contains spectroscopic data. And last but not least, CML [57–60], which is an extension of XML (www.xml.org), tries to unify all the chemical information available, not only for Internet processing. An overview of the different file formats mentioned here is provided in Table 2-5.

The different internal and external file formats make it necessary to have programs which convert one format into another. One of the first conversion programs for chemical structure information was Babel (around 1992). It supports almost 50 data formats for input and output of chemical structure information [61]. CLIFF is another file format converter based on the CACTVS technology and which supports nearly the same number of file formats [29]. In contrast to Babel, the program is more comprehensive: it is able to convert chemical reaction information, and can calculate missing atom coordinates [29].

During the process of conversion, a program may drop some information produced by other software because the format conventions cannot handle this additional information. For example, when the JACMP format is converted to a Molfile, its content is reduced to structural data only, without spectra data. In other cases, a

**Table 2-5.** The most important file formats for exchange of chemical structure information.

| File format | Suffix | Comments | Support | Ref. |
|---|---|---|---|---|
| MDL Molfile | *.mol | Molfile; the most widely used connection table format | *www.mdli.com* | 50 |
| SDfile | *.sdf | Structure-Data file; extension of the MDL Molfile containing one or more compounds | *www.mdli.com* | 50 |
| RDfile | *.rdf | Reaction-Data file; extension of the MDL Molfile containing one or more sets of reactions | *www.mdli.com* | 50 |
| SMILES | *.smi | SMILES; the most widely used linear code and file format | *www.daylight.com* | 20, 21 |
| PDB file | *.pdb | Protein Data Bank file; format for 3D structure information on proteins and polynucleotides | *www.rcsb.org* | 53 |
| CIF | *.cif | Crystallographic Information File format; for 3D structure information on organic molecules | *www.iucr.org/iucr-top/cif/* | 55 |
| JCAMP | *.jdx, *.dx, *.cs | Joint Committee on Atomic and Molecular Physical Data; structure and spectroscopic format | *www.jcamp.org/* | 56 |
| CML | *.cml | Chemical Markup Language; extension of XML with specialization in chemistry | *www.xml-cml.org* | 57–59 |

program may append additional information calculated or derived from the information present, such as molecular weight, number of hydrogen atoms, etc.

## 2.4.6
### Tutorial: Molfiles and SDfiles

As pointed out in Section 2.4.5, there are many file formats for storing information about molecular structures. Nevertheless, only some of them have widely been accepted by the chemoinformatics community and are used as standard formats for the exchange of information on chemical structures and reactions. This is particularly true for the Molfile and SDfile formats first described by Dalby *et al.* from Molecular Design Limited (MDL) [51]. However, as the file formats are undergoing constant development, one should rather refer to the file format specification available from the MDL website [50]. Originally intended only for programs developed by MDL, Molfile and SDfile formats quickly became *de facto* industry standards for the storage and exchange of information on molecular structures and properties.

A Molfile describes a single molecular structure which can contain disjointed fragments. In turn, an SDfile (SD stands for structure–data) contains structure and data (properties) for any number of molecules, which makes it especially convenient for handling large sets of molecules – for example for data transfer between databases or from databases to data analysis tools. MDL also designed three other, but in some ways internally similar, file formats for handling queries (RGfiles), single reactions (RXNfiles), and reactions as well as molecules together with their associated data (RDfiles). All the MDL file formats are referred together as CTfiles (chemical table files) and described extensively in Refs. [50, 51].

CTfiles originated in the time of punched cards and therefore their format is quite restrictive. For example, blanks usually are significant and several consecutive spaces cannot simply be replaced by a single one. Spaces may correspond to missing entries, empty character positions within entries, spaces between entries, or zeros in the case of numerical entries. Thus, every piece of data has a precise and fixed location within a line in a data file. Moreover, the line length of CTfiles is restricted to 80 characters.

The default filename extension for Molfiles is ".mol", and for SDfiles ".sdf" or ".sd" are used.

### 2.4.6.1 Structure of a Molfile
In order to understand the Molfile format let us look at a sample file and recognize its fundamental structure. For simplicity, some less important details will be omitted in the discussion. For a complete description, users are referred to MDL's CTfile format specification [50].

Figures 2-24 and 2-25 present the structure of the ethanal molecule and a corresponding Molfile, respectively. The file was extracted from the Enhanced

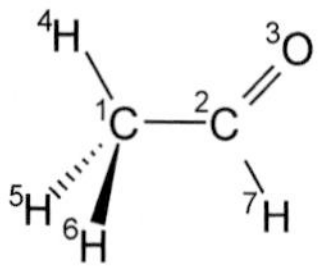

**Figure 2-24.** Structure of ethanal.

| | | |
|---|---|---|
| *1.* | `NSC7594 acetaldehyde` | |
| *2.* | `JTtclserve09180215543D 0  0.00000    0.00000NCI NS` | Header block |
| *3.* | | |
| *4.* | `7 6 0 0 0 0 0 0 0 0999 V2000` | Counts line |
| *5.* | `0.0000   0.0000   0.0000 C  0 0 0 0 0 0 0 0 0 0 0 0` | |
| *6.* | `1.5000   0.0000   0.0000 C  0 0 0 0 0 0 0 0 0 0 0 0` | |
| *7.* | `2.1200 -1.0200 -0.0200 O  0 0 0 0 0 0 0 0 0 0 0 0` | |
| *8.* | `-0.3567 -0.4872 -0.8834 H  0 0 0 0 0 0 0 0 0 0 0 0` | Atom block |
| *9.* | `-0.3567 -0.5215  0.8636 H  0 0 0 0 0 0 0 0 0 0 0 0` | |
| *10.* | `-0.3567  1.0086  0.0198 H  0 0 0 0 0 0 0 0 0 0 0 0` | |
| *11.* | `2.0245  0.9324  0.0183 H  0 0 0 0 0 0 0 0 0 0 0 0` | |
| *12.* | `1 2 1 0 0 0 0` | |
| *13.* | `2 3 2 0 0 0 0` | |
| *14.* | `1 4 1 0 0 0 0` | |
| *15.* | `1 5 1 0 0 0 0` | Bond block |
| *16.* | `1 6 1 0 0 0 0` | |
| *17.* | `2 7 1 0 0 0 0` | |
| *18.* | `M END` | Properties block |

(Connection table (Ctab))

**Figure 2-25.** Molfile representing the ethanal molecule shown in Figure 2-24.

NCI Database Browser [62, 63] and only slightly modified for the purposes of this exercise.

Each Molfile consists of two parts: the so-called *header block* specific to Molfiles (lines 1–3) and a *connection table – Ctab* (lines 4–18), which is fundamental to all MDL's CTfile formats.

The first line of the header block contains the molecule name and does not require any particular format (a rare case). If no name is available the line is blank. For ethanal two names are specified: its identifier within the NCI database (NSC 7594) and the regular name.

The second line, however, has a strict format and contains general information about the user's name, the program used to generate this file, and the date and

| Description | User's first initials | Name of the program that created this file | Date/time, when the file was created | Dimensional code | Scaling factors | | Energy | Internal registry number |
|---|---|---|---|---|---|---|---|---|
| **Column** | 12 | 1<br>34567890 | 2<br>1234567890 | 12 | 34 | 3<br>5678901234 | 4<br>567890123456 | 5<br>789012 |
| **Data** | JT | tclserve | 0918021554 | 3D | 0 | 0.00000 | 0.00000 | NCI NS |

**Figure 2-26.** Second line of the Molfile's header block from the sample file.

time when the file was created. The date and time information is formed of concatenated two-digit values representing the month (09 in the example), day (18), year (02), hour (15), and minute (54), respectively. It specifies also whether 2D or 3D atomic coordinates are given and includes other miscellaneous data and comments. Figure 2-26 shows a more detailed explanation of this line from the Molfile analyzed. Please note that "tclserve" in the second entry stands for tclserver – the core part of the CACTVS system [64, 65] (see Section 2.12) used in the Enhanced NCI Database Browser. This line may also be empty.

The third line of the header block is usually empty or contains comments.

Lines 4–18 form the connection table (*Ctab*), containing the description of the collection of atoms constituting the given compound, which can be wholly or partially connected by bonds. Such a collection can represent molecules, molecular fragments, substructures, substituent groups, and so on. In case of a Molfile, the *Ctab* block describes a single molecule.

The first line of the connection table, called the *counts line* (see Figure 2-27), specifies how many atoms constitute the molecule represented by this file, how many bonds are within the molecule, whether this molecule is chiral (1 in the chiral flag entry) or not, etc. The last-but-one entry (number of additional properties) is no longer supported and is always set to 999. The last entry specifies the version of the Ctab format used in the current file. In the case analyzed it is "V2000". There is also a newer V3000 format, called the *Extended Connection Table*, which uses a different syntax for describing atoms and bonds [50]. Because it is still not widely used, it is not covered here.

| Description | Number of atoms | Number of bonds | Number of atom lists | (obsolete) | Chiral flag | Other properties ignored for Molfiles | Number of additional properties | Current Ctab version |
|---|---|---|---|---|---|---|---|---|
| **Column** | 123 | 456 | 789 | 1<br>012 | 345 | 2<br>678901234567890 3 | 123 | 456789 |
| **Data** | 7 | 6 | 0 | 0 | 0 | 0  0  0  0  0 | 999 | V2000 |

**Figure 2-27.** Counts line of the Molfile's header block from the sample file.

| Description | Cartesian coordinates (x, y, z) | | | (space) | Atom symbol | Mass difference | Charge | 9 miscellaneous properties |
|---|---|---|---|---|---|---|---|---|
| **Column** | 1 | 2 | 3 | | | | | 4 |
| | 1234567890 | 1234567890 | 1234567890 | 1 | 234 | 56 | 789 | 012... |
| **Data** | 0.0000 | 0.0000 | 0.0000 | | C | 0 | 0 | 0... |
| | 1.5000 | 0.0000 | 0.0000 | | C | 0 | 0 | 0... |
| | 2.1200 | −1.0200 | −0.0200 | | O | 0 | 0 | 0... |

**Figure 2-28.** Structure of the atom block for the non-hydrogen atoms of ethanal.

All of the seven atoms declared in the counts line above are described next in an *atom block*. Each atom is represented by a single row, which specifies its Cartesian coordinates, atomic symbol, difference from mass in the periodic table, charge, and nine other properties, which are usually set to their default values (0s) in Molfiles. The Cartesian coordinates define a two- or 3D molecular model, as declared in the second line of the file. 2D models can be obtained, for example, from molecule editors like ISIS/Draw [50] (see Section 2.12). 3D structural data result from experiments or theoretical calculations at different theoretical levels, mostly from 3D structure generators such as CORINA [29] (see Section 2.13 in this chapter and Chapter II, Section 7.1 of the Handbook). 3D atomic coordinates can usually be recognized in the third column of the atom block, the z-coordinates. If this column only contains values of 0.0, then 2D coordinates may be stored; if it contains values different from 0.0 as in our case, the Molfile stores 3D coordinates. Figure 2-28 shows the structure of the atom block for the non-hydrogen atoms of ethanal.

Once the atoms are defined, the bonds between them are specified in a *bond block*. Each line of this block specifies which two atoms are bonded, the multiplicity of the bond (the bond type entry) and the stereo configuration of the bond (there are also three additional fields that are unused in Molfiles and usually set to 0). The indices of the atoms reflect the order of their appearance in the atom block. In the example analyzed, "1" relates to the first carbon atom (see also Figure 2-24), "2" to the second one, "3" to oxygen atom, etc. Then the two first lines of the bond block of the analyzed file (Figure 2-29) describe the single bond between the two carbon atoms $C_1$–$C_2$ and the double bond $C_2$=$O_3$, respectively.

The last part of the file presented here is a *properties block*, which can contain miscellaneous properties extensively described in Refs. [50, 51]. In most cases, however, this block is empty, except for a terminating line (line 18 in Figure 2-25).

| Description | Fist atom | Second atom | Bond type | Bond stereo | Other information |
|---|---|---|---|---|---|
| Column | 123 | 456 | 789 | 1<br>012 | 345... |
| Data | 1 | 2 | 1 | 0 | 0... |
|  | 2 | 3 | 2 | 0 | 0... |

**Figure 2-29.** Structure of the bond block for the C–C and C=O bonds in ethanal.

### 2.4.6.2 Structure of an SDfile

As mentioned in the introduction of Section 2.4.6, an SDfile contains the structural information and associated data items for one or more compounds. This makes it particularly useful, not only for exchange of data between databases, but also between computational software. Most of such programs can write the results of their calculations in this format. Within an SDfile, each molecule is represented by its Molfile with additional data items describing its non-structural properties (molecular weight, heat of formation, molecular descriptors, biological activity, etc.). The information on a molecule is terminated by a delimiter line (containing only "$$$$"). Each data item starts with a *data header* line, which reflects a molecular property name. Next, one or more rows contain the actual data; they are terminated by an empty line. Figure 2-30 shows the structure of the sulfuric diamide (sulfamide) molecule, while Figure 2-31 presents the corresponding SDfile obtained from the Enhanced NCI Database Browser [62, 63].

The file presented contains 11 data items. The header lines are property names as used by CACTVS [64, 65], and are sufficiently self-descriptive. For example, "E_NHDONORS" is the number of hydrogen bond donors, "E_SMILES" is the SMILES string representing the structure of sulfamide, and "E_LOGP" is the log$P$ value (octanol/water partition coefficient) for this substance.

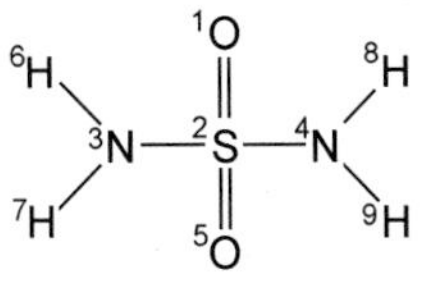

**Figure 2-30.** Structure of sulfamide (sulfuric diamide).

### 2.4.6.3 Libraries and Toolkits

There are miscellaneous libraries for molecular structure manipulation which support both reading and generating Mol- and SDfile formats. OEChem [66] from OpenEye is a commercial library for C++ programmers, which additionally contains links to Python. For the Java programming language there are two intensively developed free libraries, namely JOELib (Java-based re-implementation and exten-

```
NSC252 sulfamide
DAtclserve09180215363D 0  0.00000    0.00000NCI NS

 9  8  0  0  0  0  0  0  0  0999 V2000
   0.0000   0.0000   0.0000 O   0  0  0  0  0  0  0  0  0  0  0  0
   0.5600  -1.3400   0.0000 S   0  0  0  0  0  0  0  0  0  0  0  0
   0.0800  -2.0800   1.3600 N   0  0  0  0  0  0  0  0  0  0  0  0
   0.0800  -2.0800  -1.3600 N   0  0  0  0  0  0  0  0  0  0  0  0
   2.0200  -1.3400   0.0000 O   0  0  0  0  0  0  0  0  0  0  0  0
   0.4316  -1.5817   2.1525 H   0  0  0  0  0  0  0  0  0  0  0  0
  -0.9193  -2.0987   1.3944 H   0  0  0  0  0  0  0  0  0  0  0  0
   0.4316  -3.0161  -1.3721 H   0  0  0  0  0  0  0  0  0  0  0  0
  -0.9193  -2.0987  -1.3944 H   0  0  0  0  0  0  0  0  0  0  0  0
  1  2  2  0  0  0  0
  2  3  1  0  0  0  0
  2  4  1  0  0  0  0
  2  5  2  0  0  0  0
  3  6  1  0  0  0  0
  3  7  1  0  0  0  0
  4  8  1  0  0  0  0
  4  9  1  0  0  0  0
M  END
> <E_NSC>
252

> <E_WEIGHT>
 96.1038

> <E_NAME>
NSC252 sulfamide

> <E_NAMESET>
sulfamide (ACD/Name)
Imidosulfamic acid
Sulfamamid
Sulfamid
Sulfonyl diamid
Sulfuric diamid
Sulfuryl amid
Sulfuryl diamide

> <E_COMPLEXITY>
 72.5599

> <E_NHDONORS>
2

> <E_NHACCEPTORS>
4

> <E_NROTBONDS>
0

> <E_FORMULA>
H4N2O2S

> <E_CAS>
7803-58-9

> <E_SMILES>
NS(N)(=O)=O

> <E_LOGP>
-1.79  0

$$$$
```

Labels (right margin):
- Header block
- Connection table
- Molfile
- Data header / Data / Blank line
- Non-structural data
- Data items
- Delimiter

**Figure 2-31.** Sample SDfile for sulfamide (sulfuric diamide).

sion of the OELib library) [67] and CDK (the Chemical Development Kit) [68], and at least one commercial library – the JChem library [69] from ChemAxon Ltd.

Another interesting tool is the SDF Toolkit [70] – a set of Perl scripts for manipulating SDfiles. It provides tools for filtering SDfiles, merging them and removing duplicates, adding data from CSV (comma-separated) files to an SDfile, and so on.

## 2.5
## Processing Constitutional Information

### 2.5.1
### Ring Perception

Rings have a profound influence on many properties of a molecule: small rings introduce strain into a molecule, aromatic rings dramatically change its physical and chemical properties, rings present particular problems in syntheses, etc. Thus, a knowledge of the rings contained in a molecule is important in many applications in chemoinformatics.

In the phenylalanine example we easily recognize the existence of one ring system (Figure 2-32a).

If a system, such as a substituted adamantane (Figure 2-32b), is more complex, the process of ring perception may be quite difficult. The set of rings recognized by humans can depend on the individual and the way the structure is drawn (Figurec 2-32c and d). Therefore, a computer procedure for the recognition of rings is indispensable. Two basic approaches dominate the variety of ring perception algorithms, which are reviewed in more detail in Refs. [72–74]. One method "walks" through connection tables or matrices and produces a list of the constituent atoms and bonds of the traced rings (linear algebraic method). The other, more obvious, method uses graph theory operations on matrices, trees, and sets to identify ring systems in chemical structures.

In graph theory (see Section 2.4.1), a ring represents a connected graph, which can be traversed node by node in a single path (cycle) back to the starting point. If one edge or bond of the ring is removed, the graph is disconnected (path 1 in Figure 2-33) but not fragmented into two separate parts (as would be the case for a chain bond; see Figure 2-33 path 2 to the right).

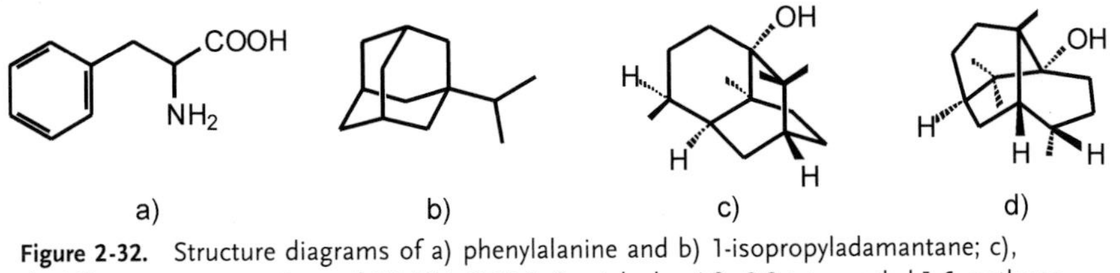

**Figure 2-32.** Structure diagrams of a) phenylalanine and b) 1-isopropyladamantane; c), d) different representations of 1R,4S,4aS,6R,8aS-octahydro-4,8a,9,9-tetramethyl-1,6-methano-naphthalen-1(2H)-ol.

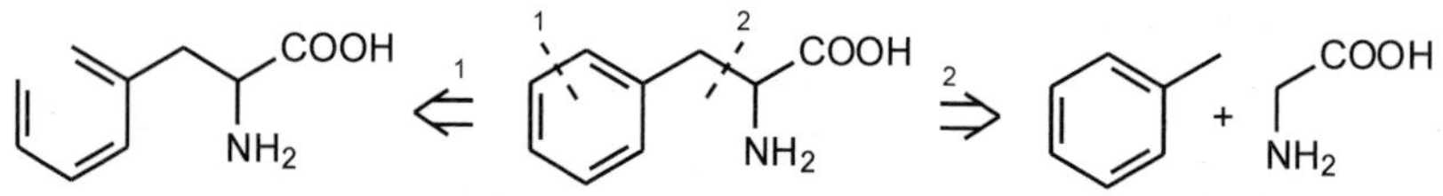

**Figure 2-33.** Any bond of a ring can be broken without fragmenting the compound (path 1 on the left-hand side), whereas breaking a bond in a chain results in two fragments (path 2 on the right-hand side).

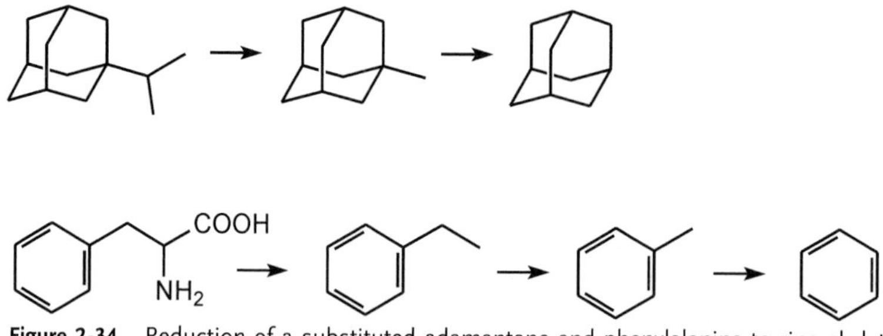

**Figure 2-34.** Reduction of a substituted adamantane and phenylalanine to ring skeletons by pruning acyclic parts of the molecules.

Thus, if we want to know the number of rings in a molecule, we have to count the number of paths in the structure that lead back to the starting point. To simplify this procedure, only nodes which are part of the ring system have to be considered. In this step of pre-processing, all nodes with degree 1 (terminal atom with one bond to a neighboring atom) are successively removed. This is done iteratively as often as possible. The end result is a ring skeleton with nodes that have a degree of 2 or more (Figure 2-34).

Another technique of pre-processing is graph reduction [72]. Here, all nodes of the ring that have a degree of 2 are merged with the corresponding neighbor, because these nodes cannot be part of a bridge to another ring system. The result is a basic graph of the structure with fewer nodes and vertices to search (Figure 2-35).

After pre-processing, an algorithm can detect all the cycles in a graph or structure. The result may be a high number of rings perceived. Adamantane, for example, has four six-membered and three eight-membered ring systems. Therefore, a definition of a suitable set of rings in a graph has to be given that is appropriate for the different requirements of each application. The art is to decide what is necessary and sufficient.

In this section only three of the various ring perception algorithms are introduced. The initial approaches are:

• Find the minimum number of rings to describe a ring system.
• Find all the possible rings.
• Find the smallest fundamental basis (SSSR).

**Figure 2-35.** Graph reduction of adamantane.

#### 2.5.1.1 Minimum Number of Cycles

The minimum number of cycles is given by the nullity or Frèrejacque number ($\mu$) according to Eq. (5). It is the difference between the number of nodes ($a$ = atoms) and the number of edges ($b$ = bonds). The value of 1 stands for the number of compounds considered (here, one compound). This minimum number corresponds to the number of chords. These are defined as nodes that turn a cyclic graph or structure into an acyclic one.

$$\mu = b - a + 1 \tag{5}$$

#### 2.5.1.2 All Cycles

The set of all cycles can be determined by the algorithm of Hanser [75]. The graph is first reduced to make the process faster. Then an odd number of edges (bonds) is removed iteratively from a node. At each iteration, the path is verified and compared to already existing ones, in order to exclude redundant cycles (Figure 2-36). In this representation adamantane has four three-membered and three four-membered ring systems, which correspond to the four six-membered and the three eight-membered ring systems mentioned above.

#### 2.5.1.3 Smallest Fundamental Basis

A fundamental set of rings is the minimum set of rings to distinguish a ring system. Let us assign a vector, $v_i$, to each cycle or ring, $r_i$, of a ring system with $k$ edges such as $v_{ij} = 1$ if $j$ is an edge of the ring $r_i$, and $v_{ij} = 0$ otherwise. These vectors generate a vector space of rings. Let $n$ be the minimum number of rings required to describe the ring system (Frèrejacque number) whose corresponding vectors $v_1...v_n$ are linearly independent. Then, these vectors $v_1...v_n$ build a basis of the vector

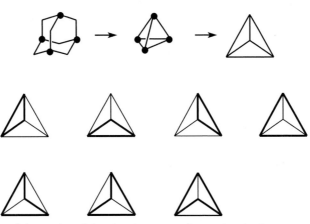

**Figure 2-36.** Identification of the number of rings in adamantane after graph reduction (the different ring systems are highlighted with bold lines). Note that a graph does not carry 3D information; thus, the two structures on the upper right-hand side are identical.

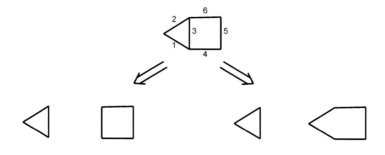

$v_1=(1,1,1,0,0,0)$  $v_2=(0,0,1,1,1,1)$     $v_1=(1,1,1,0,0,0)$   $v_3=(1,1,0,1,1,1)$

**Figure 2-37.** Two different fundamental sets of rings for bicyclo[2.1.0]pentane.

space and the set of rings $r_1...r_n$ is called a fundamental set of rings. Note that there can exist several fundamental sets of rings (see Figure 2-37).

A fundamental set is also called a smallest set. Usually such a smallest set is selected that it also contains the smallest rings: the smallest set of smallest rings (SSSR). This makes chemical sense in indane (Figure 2-38a), for example, where only the six- and the five-membered rings are of chemical significance whereas the enveloping nine-membered ring is not.

Also, the rational IUPAC nomenclature for polycyclic ring systems only selects a smallest set of rings for naming such compounds. Thus cubane obtains the rational name of pentacyclo[4.2.0.0$^{2.5}$.0$^{3.8}$.0$^{4.7}$]octane (Figure 2-38b). In effect, this nomenclature accounts for the fact that the entire ring system can be constructed from five four-membered rings. The sixth four-membered ring is linearly dependent on the other five rings (altogether, one can discern 28 rings in cubane!).

Nevertheless, there are situations where one wants to work with six four-membered rings in cubane (e.g., when considering the symmetry of the ring system). In this situation, one adds a sixth four-membered ring to obtain from the SSSR the so-called extended set of smallest rings (ESSR).

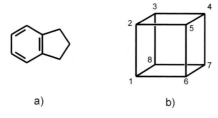

a)

b)

**Figure 2-38.**
Structure diagrams of a) indane and b) cubane.

## 2.5.2
### Unambiguous and Unique Representations

In chemical nomenclature, synthesis design, or in substructure search a general, unique representation of chemical structures is essential. An effective database handling with registry, storage, and retrieval systems requires a one-to-one correspondence of a unique and invariant notation with the respective chemical structure. The solution of this canonical coding problem and the generation of all isomers with a constitutional formula with the corresponding graph isomorphism problem will be discussed in this section.

### 2.5.2.1 Structure Isomers and Isomorphism

Organic chemistry is characterized by a cornucopia of different chemical structures. This is largely because the atoms of an organic molecule can be arranged in a variety of different bonding situations.

In many cases, quite a few different structural formulas can be produced for one and the same empirical formula (Figure 2-39).

The generation of all the structure isomers for a given empirical formula is an important task in automatic structure elucidation (see Chapter II, Section 5.3 in the Handbook). How many isomers (e.g., of $C_3H_6O$) are possible and chemically reasonable? This is a very important question in structure generation. In this process, care has to be taken that only different structures are generated, so that no isomorphic graphs are obtained. In isomorphic structures all the atoms (nodes) of two or more structures (graphs) correspond one-to-one, preserving the adjacency of the nodes. Thus, the topology of the molecules under consideration is identical. This problem is illustrated in Figure 2-40 with phenylalanine (note that stereochemistry is disregarded for the time being). The substituents –H, –COOH, and –NH$_2$ can be attached at arbitrary positions on the terminal carbon atom.

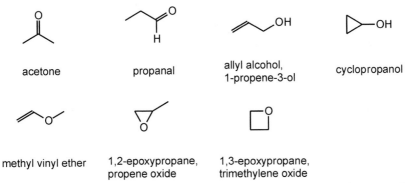

acetone      propanal      allyl alcohol, 1-propene-3-ol      cyclopropanol

methyl vinyl ether      1,2-epoxypropane, propene oxide      1,3-epoxypropane, trimethylene oxide

**Figure 2-39.** The empirical formula of $C_3H_6O$ can be expressed by seven structure diagrams and even more compound names.

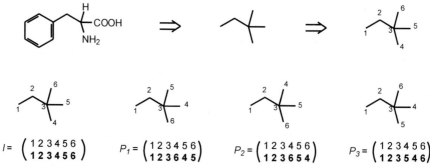

$$I = \begin{pmatrix} 1\,2\,3\,4\,5\,6 \\ 1\,2\,3\,4\,5\,6 \end{pmatrix} \qquad P_1 = \begin{pmatrix} 1\,2\,3\,4\,5\,6 \\ 1\,2\,3\,6\,4\,5 \end{pmatrix} \qquad P_2 = \begin{pmatrix} 1\,2\,3\,4\,5\,6 \\ 1\,2\,3\,6\,5\,4 \end{pmatrix} \qquad P_3 = \begin{pmatrix} 1\,2\,3\,4\,5\,6 \\ 1\,2\,3\,5\,4\,6 \end{pmatrix}$$

**Figure 2-40.** To illustrate the isomorphism problem, phenylalanine is simplified to a core without representing the substituents. Then every core atom is numbered arbitrarily (first line). On this basis, the substituents of the molecule can be permuted without changing the constitution (second line). Each permutation can be represented through a permutation group (third line). Thus the first line of the mapping characterizes the numbering of the atoms before changing the numbering, and the second line characterizes the numbering afterwards. In the initial structure (*I*) the two lines are identical. Then, for example, the substituent number 6 takes the place of substituent number 4 in the second permutation ($P_2$), when compared with the reference molecule.

Redundant, isomorphic structures have to be eliminated by the computer before it produces a result. The determination of whether structures are isomorphic or not stems from a mathematical operation called permutation: the structures are isomorphic if they can be interconverted by permutation (Eq. (6); see Section 2.8.7). The permutation $P_3$ is identical to $P_2$ if a mathematical operation ($P_x$) is applied. This procedure is described in the example using atom 4 of $P_3$ (compare Figure 2-40, third line). In permutation $P_3$ atom 4 takes the place of atom 5 of the reference structure but place 5 in $P_2$. To replace atom 4 in $P_2$ at position 5, both have to be interchanged, which is expressed by writing the number 4 at the position of 5 in $P_x$. Applying this to all the other substituents, the result is a new permutation $P_x$ which is identical to $P_1$.

$$P_x P_2 = P_3 = \begin{pmatrix} 123456 \\ 123645 \end{pmatrix}\begin{pmatrix} 123456 \\ 123654 \end{pmatrix} = \begin{pmatrix} 123456 \\ 123456 \end{pmatrix} \qquad (6)$$

Thus, the mathematical operation with all combinations of permutations shows the isomorphism of the structures.

For database handling it is necessary to compare existing database entries with new ones. Consequently, database registration and retrieval are dependent on isomorphism algorithms which compare two graphs or structure diagrams to determine whether subgraphs are identical or not.

### 2.5.2.2 Canonicalization

The representation of a chemical structure by a connection table is neither unambiguous nor unique. A molecule may be denoted with quite a variety of different connection tables describing one and the same molecule but with different numbering of the atoms (Figure 2-21). In principle, a structure with $n$ atoms can be numbered in $n!$ different manners and thus has up to $n!$ different connection tables (if the molecule has symmetry, some of these connection tables may be identical). A molecule with only three atoms, e.g., hypochlorous acid (ClOH), can be labeled and described by $3! = 6$ different atomic numberings and, consequently, by six different connection tables. These different numberings can be developed in a spanning tree (Figure 2-41).

The task is now to take one of the numberings as the standard one and to derive a unique code from it, which is called canonicalization. This can be accomplished by numbering the atoms of a molecule so that it is represented later by only one connection table or bond matrix. Such a unique and reproducible numbering or labeling of the atoms is obtained by a set of rules.

Various methods have been developed for a unique and unambiguous numbering of the atoms of a molecule and thus for deriving a canonical code for this molecule [76]. Besides eigenvalues of adjacency matrices [77], it is mainly the Morgan Algorithm that is used [79].

### 2.5.3
### The Morgan Algorithm

The Morgan Algorithm classifies all the congeneric atoms of a compound and selects invariant-labeled atoms (see Section 2.5.3.1). The classification uses the concept of considering the number of neighbors of an atom (connectivity), and does so in an iterative manner (extended connectivity, EC). On the basis of certain rules,

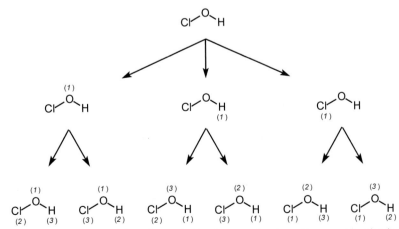

**Figure 2-41.** Six different possibilities for numbering the atoms in a hypochlorous acid molecule.

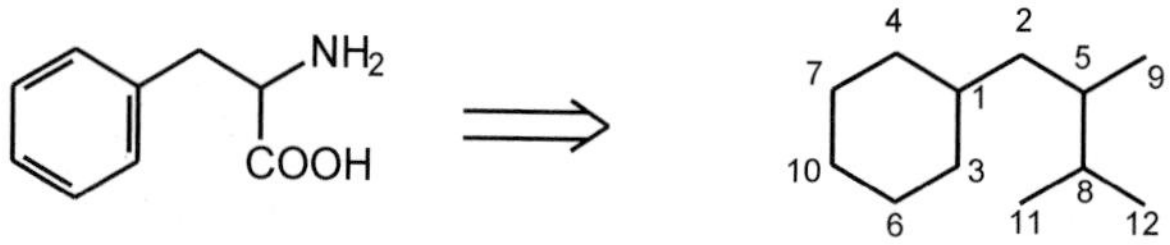

**Figure 2-42.** The Morgan Algorithm generates an unambiguous and unique numbering of phenylalanine (see Tutorial, Section 2.5.3.1).

the Morgan Algorithm produces an unambiguous and unique numbering of the atoms in a compound (Figure 2-42).

Two aspects of the Morgan Algorithm play a dominant role for structure coding:

- *The unique coding*: A molecule with $n$ atoms has $n!$ different possibilities of atom labeling, e.g., a compound with 12 atoms can be represented by about 0.5 billion connection tables. The Morgan Algorithm reduces all the different labeled connection tables to only one.
- *Consideration of stereochemistry*: The parity or "handedness" – $R/S$ or *cis/trans* – of a stereocenter can be obtained by considering the sequence of the Morgan numbers of the atoms, similarly to CIP. Then the number of pairwise interchanges is counted until the numbers are in ascending order (see Section 2.8.5).

However, it should not be concealed that in some cases the Morgan Algorithm has problems. In some structures the numbering show oscillatory behavior (for an example, see Figure 2-44). Moreover, although equivalent atoms obtain the same extended connectivity value, it cannot always be concluded that atoms with the same extended connectivity value are equivalent.

### 2.5.3.1 Tutorial: Morgan Algorithm

Many applications in chemistry, particularly registry and storage systems, require an unambiguous and unique structure representation. The canonicalization algorithm developed by Morgan in 1965 [79] is a fairly simple process that has experienced many variations and is still used in the Chemical Abstracts Service System and other databases and programs. Besides providing a unique and invariant numbering of the atoms of a structure, the Morgan Algorithm can also distinguish constitutionally equivalent atoms. The process is divided into two stages: a relaxation process that calculates the extended connectivity, and assignment of sequence numbers to the atoms.

The Morgan Algorithm is described as follows:

*Step 1: Classification of atoms by considering their neighborhood (relaxation process):* In organic structures consisting of C, N, O, H, and halogen atoms, atoms can be divided into four classes depending on the number of non-hydrogen attachments. The class number can only take the values 1 to 4 (primary to quaternary C-atom) corresponding to the degree of the node/atom. The number of hydrogen atoms is redundant and can be derived from the valence rules.

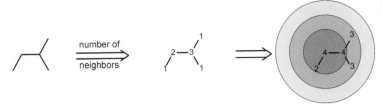

**Figure 2-43.** The EC value or the atom classification of each atom, respectively, is calculated by summing the EC values of the directly connected neighboring atoms of the former sphere (relaxation process).

In the first iteration process, the class values of the atoms of the structure show information already known (the degree of the nodes). Hence Morgan takes the neighboring atoms into account. He considers the environment of an atom by summing class values of all directly adjacent atoms. This process results in a new class value called the extended connectivity (EC) value of the atom. The new EC value expresses indirectly the neighborhood of the adjacent atoms in a second sphere (Figure 2-43).

In this simplified example of phenylalanine, in the first iteration the methyl groups are given a value of 1 in the first classification step because they contain a primary C-atom. The methylene group obtains a value of 2, and the methine carbon atom a value of 3. In the second step, the carbon atom of the methyl group on the left-hand side obtains an extended connectivity (EC) value of 2 because its neighboring atom had a value of 2 in the first classification step.

The carbon atoms of the other two methyl groups (on the right-hand side) obtain an EC value of 3 because they are adjacent to the methine carbon atom. The carbon atom of the methylene group obtains an EC value of 4 in the second relaxation process, as the sum (1 + 3) of the connectivity values of its neighboring atoms in the first iteration.

The methine carbon atom also obtains an EC value of 4 (= 1 + 1 + 2) in the second iteration. This process is repeated iteratively until the number of different EC values (c) is lower than or equal to the number of EC values in the previous iteration. Then the relaxation process is terminated. Next, the EC numbers of the previous iteration are taken for a canonical numbering and for the determination of constitutional symmetry (Figure 2-44).

As can be seen, this algorithm additionally identifies constitutionally equivalent atoms. These are atoms with the same EC value (3) in the final iteration, such as the two carbon atoms in the *ortho* positions of the phenyl ring with an EC value of 9 (Figure 2-44).

The general algorithm is:

1. The extended connectivity (EC) value of an atom of the first sphere (*1*) results from the number (*n*) of neighboring atoms (NA) according to Eq. (7):

$$EC(1) = n \cdot NA(1) \tag{7}$$

The EC value is calculated for each atom.

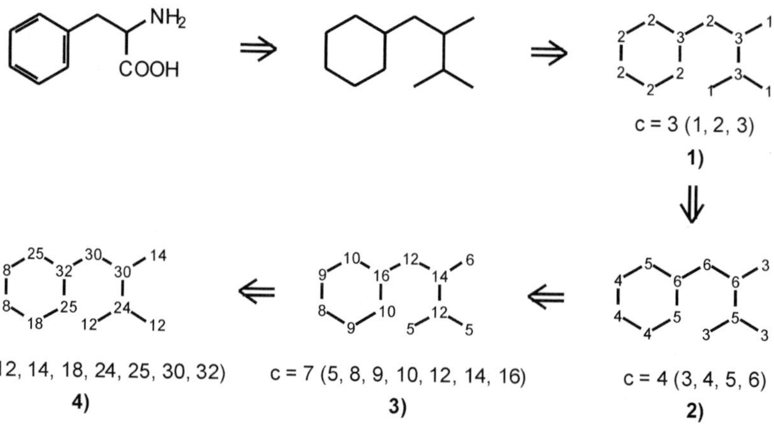

**Figure 2-44.** The EC values of the atoms of phenylalanine (without hydrogens) are calculated by considering the class values of the neighboring atoms. After each relaxation process, *c*, the number of equivalent classes (different EC values), is determined.

2. When all the EC values of the atoms have been calculated, the number of equivalent classes ($c$) for the first sphere is determined. The number of classes is equivalent to the number of different EC values.
3. In the second and higher sphere(s) the EC value for each atom is calculated by summing the EC values of the directly connected neighboring atoms of the former sphere (Eq. (8)):

$$EC(i) = n \cdot NA(i) \qquad (8)$$

4. At each sphere the number of equivalence classes ($c$) is determined.
5. This iteration is continued until the number of equivalent classes is equal to or smaller than that in the previous iteration.
6. The iteration with the highest number of equivalent classes is taken for the next step.

*Step 2: Assigning unique, invariant sequence numbers to the atoms:* The iteration where the highest number of equivalence classes first appears is taken as the starting point for the canonicalization. The atom with the highest extended connectivity value is labeled with the sequence number 1. Thus, the Morgan Algorithm focuses on the most deeply embedded atom in the structure to start the numbering. The rest of the structure is numbered from 2 to $n$, where $n$ reflects the number of atoms. From the initial atom all the first-neighbor atoms are assigned according to the magnitude of their extended connectivity value.

In the next step, the neighbor atoms of the second atom (or the current atom plus 1), which are not yet labeled, are assigned in an equivalent manner. This is done for all the atoms; arbitrary decisions are made when numbering the equivalent atoms (Figure 2-45).

**Figure 2-45.** Canonicalization starts at the atom with the highest EC value (in the example: 16), which gets the number 1. From there, all other atoms are numbered according to their EC values.

1. The atom with the highest EC value obtains sequence number 1 (and is now the current atom).
2. All the connected neighboring atoms are enumerated 2, 3, 4, etc., according to their decreasing EC values. If two or more atoms have the same EC value, the atoms are numbered serially following specific rules: atom types (C before N) or bond types (single before double), charges, etc.
3. The next highest numbered atom compared with the current atom (in this second sphere, atom 2) becomes the current one. All unnumbered atoms attached to the current atom are numbered serially according to their decreasing EC values. As in step 2, atoms with equivalent EC values are numbered serially following the specific rules.
4. This process is continued until all the atoms are canonically enumerated.

Many variations of the Morgan Algorithm were introduced, because of problems finding the terminating condition of stage 1 (oscillating number of equivalent classes [80]) or special atoms with isospectral points [81].

## 2.6
## Beyond a Connection Table

### 2.6.1
### Deficiencies in Representing Molecular Structures by a Connection Table

The concept of connection tables, as shown so far, cannot represent adequately quite a number of molecular structures. Basically, a connection table represents only a single valence bond structure. Thus, any chemical species that cannot be described adequately by a single valence bond (VB) structure with single or multiple bonds between two atoms is not handled accurately.

Benzene was probably the first compound in chemical history where the valence bond concept proved to be insufficient. Localizing the $\pi$-systems, one comes up with two equivalent but different representations. The true bonding in benzene was described as resulting from a resonance between these two representations (Figure 2-46).

**Figure 2-46.**
The representation of benzene needs two resonance structures.

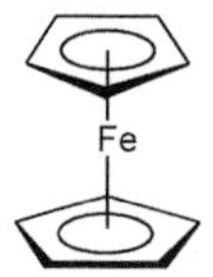

**Figure 2-47.** The bonding in organometallic complexes (e.g., ferrocene) cannot be expressed adequately by a connection table.

Ferrocene (Figure 2-47) provides a prime example of multi-haptic bonds, i.e., a situation where the electrons that coordinate the cyclopentadienyl rings with the iron atom are contained in molecular orbitals delocalized over the iron atom and the 10 carbon atoms of the cyclopentadienyl rings [82].

Representation of such a system by a connection table having bonds between the iron atom and the five carbon atoms of either one of the two cyclopentadienyl rings is totally inadequate. A few other examples of structures that can no longer be adequately described by a standard connection table are given in the Section 2.6.2.

### 2.6.2
### Representation of Molecular Structures by Electron Systems

#### 2.6.2.1 General Concepts
We describe here a new structure representation which extends the valence bond concept by new bond types that account for multi-haptic and electron-deficient bonds. This representation is called Representation Architecture for Molecular Structures by Electron Systems (RAMSES); it tries to incorporate ideas from Molecular Orbital (MO) Theory [83].

An essential feature of RAMSES is that, just as in a Hückel MO approach, the $\sigma$- and $\pi$-electron systems are separated. Each electron system is then characterized by the number of centers it extends over and the number of electrons it contains. The description of $\sigma$-skeletons thus hardly changes; a $\sigma$-bond is an electron system extending over two atoms and comprising two electrons. However, a $\sigma$-electron system may also contain fewer than two electrons, e.g., after ionization of a $\sigma$-bond in mass spectrometry. For the $\pi$-electrons, the model is no longer confined to two-center bonds. A $\pi$-electron system in RAMSES can extend over an arbitrary number of atoms and comprise up to twice as many electrons as atoms. We show below how different chemical species are represented by RAMSES and where they exceed the limits of a connection table.

#### 2.6.2.2 Simple Single and Double Bonds
A single bond (see Figure 2-48) consists of a $\sigma$-system with two atom centers and two electrons.

**Figure 2-48.** Single bonds are stored as $\sigma$-systems.

**Figure 2-49.** The $\pi$-system of a double bond.

A double bond is represented by two systems: a $\sigma$-system, as in the case of a single bond, and a $\pi$-system with two electrons on two atom centers as shown in Figure 2-49.

### 2.6.2.3 Conjugation and Aromaticity

Benzene has already been mentioned as a prime example of the inadequacy of a connection table description, as it cannot adequately be represented by a single valence bond structure. Consequently, whenever some property of an arbitrary molecule is accessed which is influenced by conjugation, the other possible resonance structures have to be at least generated and weighted. Attempts have already been made to derive adequate representations of $\pi$-electron systems [84, 85].

In the case of 1,3-butadiene, RAMSES combines the two double bonds to form a single, delocalized $\pi$-electron system containing four electrons over all four atoms (Figure 2-50a). The same concept is applied to benzene. As shown in Figure 2-50b, the three double bonds of the Kekulé representation form one electron system with six atoms and six electrons.

a)                                     b)

**Figure 2-50.** Representations of a) 1,3-butadiene and b) benzene, as examples of conjugated double bonds in RAMSES.

These examples show quite clearly the close relationship between this new structure representation embedded in RAMSES and MO Theory.

The treatment of conjugated systems in terms of electron systems that extend smoothly over all atoms allows the treatment of a variety of structural phenomena, as may be explained with a species that shows hindered rotation and with the nitro group.

An example of hindered rotation, which is neglected when only the connection table of a single VB representation is used, is the increased rotational barrier of amides (Figure 2-51a). When only the most prominent resonance structure is considered, one could assume a free rotation around the single bond between the carbon atom of the carbonyl group and the nitrogen atom. This representation neglects the important zwitterionic resonance structure, where the double bond is between the carbon and the nitrogen atom. This partial double bond character is responsible for the increased rotational barrier in amides which results in *cis*- and *trans*-isomers.

With RAMSES, the conjugation between the C=O $\pi$-system and the lone pair of the nitrogen atom in the amide group is taken into account (see Figure 2-51b).

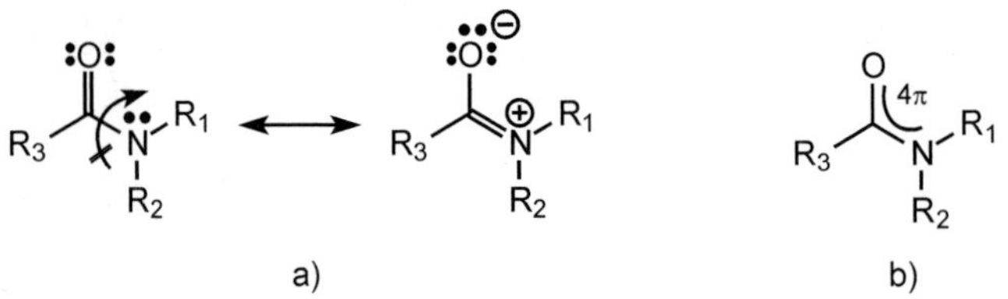

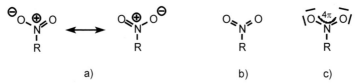

a)  b)

**Figure 2-51.** a) The rotational barrier in amides can only be explained by VB representation using two resonance structures. b) RAMSES accounts for the (albeit partial) conjugation between the carbonyl double bond and the lone pair on the nitrogen atom.

Sometimes RAMSES does not even need a charged structure when a VB representation cannot work without it. An example is the nitro group, as shown in Figure 2-52. Nitro groups are represented in a CT either by a single zwitterionic resonance structure or by a structure treating the nitrogen atom as pentavalent (see Figure 2-52b) (this structure violates the valence rules!). A correct representation of the nitro group, reflecting the symmetry in this group, needs two resonance structures (see Figure 2-52a). The RAMSES notation, however, nicely reflects the symmetry of the nitro group and does not have to put charges on any one of the three atoms (Figure 2-52c).

a)  b)  c)

**Figure 2-52.** a) Two semipolar resonance structures are needed in a correct VB representation of the nitro group. b) Representation of a nitro group by a structure having a pentavalent nitrogen atom. c) The RAMSES notation of a nitro group needs no charged resonance structures. One $\pi$-system contains four electrons on three atoms.

### 2.6.2.4 Orthogonality of $\pi$-Systems

Diphenylacteylene (tolane, shown in Figure 2-53) exists as two rotamers. In a VB representation, they cannot be distinguished without resorting to additional information, such as 3D coordinates. Using RAMSES, however, the two rotamers are distinguished by their different states of conjugation (Figure 2-54).

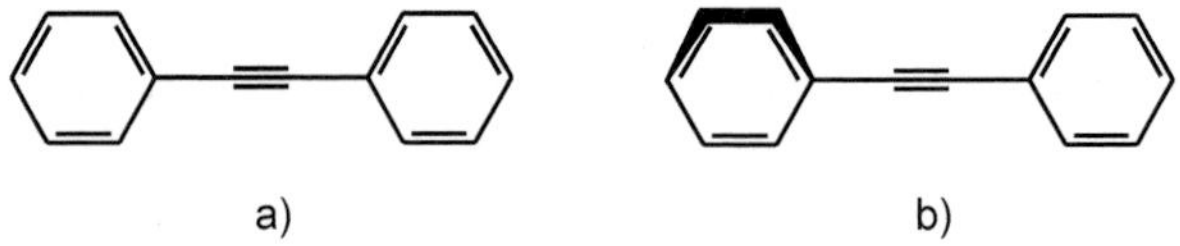

a)  b)

**Figure 2-53.** Tolane rotamers: a) one with the two aromatic rings in the same plane, and b) one with the two rings perpendicular to each other.

**Figure 2-54.** Using RAMSES, the two rotamers of tolane can be distinguished: the planar rotamer is shown on the left-hand side, and the rotamer with the ring planes perpendicular to each other is on the right.

As the usual variant of a connection table description is not affected very much by the location of free electrons in orbitals – though it is very strict about the localization of valence electrons to either bonds or atoms – discrimination between singlet and triplet carbenes (as shown in Figure 2-55) is not possible. As a consequence, the different reactivities of these two species cannot be accounted for.

With the concept of electron systems, the discrimination between singlet and triplet carbenes becomes straightforward. In both cases, the carbon atom has two $\pi$-systems. In the case of a singlet carbene, one of the $\pi$-systems is filled with two electrons and the other one is empty, whereas in the case of a triplet carbene, both $\pi$-systems house one electron (see Figure 2-55).

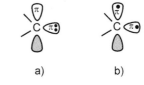

a)          b)

**Figure 2-55.** Singlet and triplet carbenes are easily distinguished in RAMSES notation.

### 2.6.2.5 Non-bonding Orbitals

The representation of non-bonding orbitals on an atom again uses the concept of $\pi$-systems, though they may have any kind of hybridization (p, sp³, etc.). In Figure 2-56 the three possibilities are shown: lone pairs, radicals, and orbitals without electrons can be accommodated by this concept.

**Figure 2-56.** Lone pairs, radicals, and orbitals without electrons are represented by a $\pi$-system with two, one, or zero electrons on the corresponding atom, respectively.

### 2.6.2.6 Charged Species and Radicals

Charged species are not a special case in RAMSES. As can be seen from Figure 2-57, $\pi$-electron systems can accommodate any number of electrons between zero and two times the number of atoms involved.

**Figure 2-57.** Charged species and radicals are represented as $\pi$-systems.

### 2.6.2.7 Ionized States

Enol ethers (Figure 2-58a) have two electron pairs on the oxygen atom in two different orbitals, one delocalized across the two carbon atoms, the other strictly localized on the oxygen atom (Figure 2-58b). Ionization from either of these two orbitals is associated with two quite different ionization potentials, a situation that cannot be handled by the present connection tables.

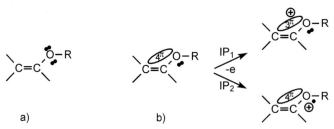

a)  b)

**Figure 2-58.** a) Enol ethers have two different ionization potentials, depending on b) the orbitals concerned.

The species produced through ionization of an electron from a σ-orbital (such as from a C–H or a C–C bond of an alkane in mass spectrometry) cannot be represented at all by a connection table, yet the RAMSES notation can account for it as shown in Figure 2-59.

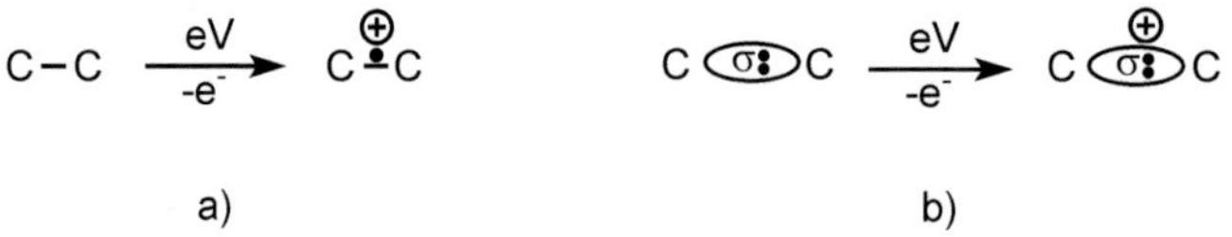

a)  b)

**Figure 2-59.** Singly occupied σ-systems are highly reactive intermediates that occur in MS experiments. They cannot be handled adequately by a) a connection table description, but are easily accommodated by b) RAMSES.

### 2.6.2.8 Electron-Deficient Compounds

Boranes are typical species with electron-deficient bonds, where a chemical bond has more centers than electrons. The smallest molecule showing this property is diborane. Each of the two B–H–B bonds (shown in Figure 2-60a) contains only two electrons, while the molecular orbital extends over three atoms. A correct representation has to represent the delocalization of the two electrons over three atom centers as shown in Figure 2-60b. Figure 2-60c shows another type of electron-deficient bond. In boron cage compounds, boron–boron bonds share their electron pair with the unoccupied atom orbital of a third boron atom [86]. These types of bonds cannot be accommodated in a single VB model of two-electron/two-centered bonds.

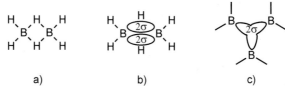

a)                    b)                    c)

**Figure 2-60.** Some examples of electron-deficient bonds: a) diborane featuring B–H–B bonds; b) diborane in a tentative RAMSES representation; c) the orbital in a B–B–B bond (which occurs in boron cage compounds).

### 2.6.2.9 Organometallic Compounds

Many organometallic structures have chemical bonds that involve more than two atom centers. This constitutes a major problem for all chemoinformatic systems that do not confine themselves to 3D coordinates.

The ideas presented above on the representation of bonding in molecular structures by electron systems can be extended to the different types of bonding in organometallic complexes. Such a system has not yet been fully elaborated but the scheme is illustrated with one example, the case of multi-haptic bonds.

Ferrocene (Figure 2-61a) has already been mentioned as a prime example of multi-haptic bonds, i.e., the electrons that coordinate the cyclopentadienyl rings with the iron atom are contained in a molecular orbital delocalized over all 11 atom centers [81], for which representation by a connection table having bonds between the iron atom and the five carbon atoms of either cyclopentadienyl ring is totally inadequate.

As a case in point, a search for the smallest set of smallest rings (SSSR) will find 10 three-membered rings in ferrocene thus represented, and will miss the two five-membered rings. A new representation has to account for all the electrons on the iron (Figure 2-61b).

**Figure 2-61.** a) The bonding in organometallic complexes (e.g., ferrocene) cannot be expressed adequately by a connection table. b) A new representation has to account for all the valence electrons of iron.

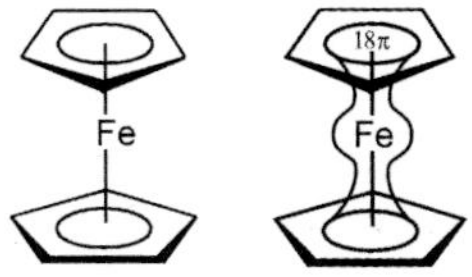

### 2.6.3
### Generation of RAMSES from a VB Representation

RAMSES is usually generated from molecular structures in a VB representation. The details of the connection table (localized charges, lone pairs, and bond orders) are kept within the model and are accessible for further processes. Bond orders are stored with the $\sigma$-systems, while the number of free electrons is stored with the atoms. Upon modification of a molecule (e.g., in systems dealing with reactions), the VB representation has to be generated in an adapted form from the RAMSES notation.

Conversion in both directions needs heuristic information about conjugation. It would therefore be more sensible to input molecules directly into the RAMSES notation. Ultimately, we hope that the chemist's perception of bonding will abandon the connection table representation of a single VB structure and switch to one accounting for the problems addressed in this section in a manner such as that laid down in the RAMSES model.

## 2.7
## Special Notations of Chemical Structures

Over and beyond the representations of chemical structures presented so far, there are others for specific applications. Some of the representations discussed in this section, e.g., fragment coding or *hash coding*, can also be seen as structure descriptors, but this is a more philosophical question. Structure descriptors are introduced in Chapter 8.

### 2.7.1
### Markush Structures

Markush structures are mainly used in patents, for protecting compounds related to an invention. The first generic claim, submitted by Markush, was granted in 1924 by the US Patent Office [87–90].

The Markush structure diagram is a specific type of representation of a series of chemical compounds. This diagram can describe not only a specific molecule but also various compound families, which is why it is also called a generic structure diagram. Markush structures have a fixed core, the body, and variable parts that, again, can contain variable parts such as frequency variation. The substituents at different attachment positions can be alternative structure classes (e.g., alkyl or aryl) or functional groups. In contrast to other representations, the substituents or the variable parts are displayed as text separately from the diagram. Thus, one structure diagram can cover immensely large classes of simple or highly complex structures (Figure 2-62).

Nowadays, Markush structures are utilized mainly in patent databases, where they describe a number of different chemical compounds. (Searching in patent databases is very important for companies to ascertain whether a new compound is

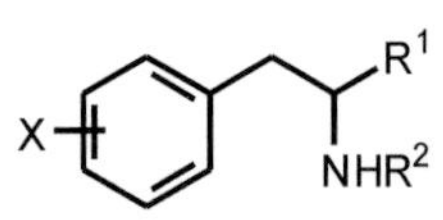

$R^1$ = H or small alkyl, halogen, OH, COOH

$R^2$ = H, $CH_3$

X = H, $(CH_2)_n CH_3$

**Figure 2-62.** The substituted phenyl derivative is an example of a typical Markush structure. Herein, a number of compounds are described in one structure diagram by fill-ins. Phenylalanine is one of these structures when $R^1$ is COOH, $R^2$ is H, and X is H.

already patented or not.) Markush structures are used in the Derwent World Patent Index, the INPI Merged Markush Service, and the Chemical Abstracts Service MARPAT.

## 2.7.2
### Fragment Coding

Fragment codes have always played an important role in chemical information systems [91, 92]. Basically, they are indexing expressions of chemical structures. Much as a document can be indexed by specified keywords, it is also possible to index a chemical structure by specific chemical characteristics. Usually, these are small assemblies of atoms, functional groups, ring systems, etc., which can be specified beforehand (Figure 2-63). Depending on the system, it is reasonable to define different fragment codes. In the first information systems the fragments were identified manually and represented on punched cards.

Fragments:
-OH
>C=O
-COOH
-NH₂
-Ph

**Figure 2-63.** An example of the possible fragments in phenylalanine.

   In contrast to canonical linear notations and connection tables (see Sections 2.3 and 2.4), fragment codes are ambiguous. Several different structures could all possess an identical fragment code, because the code does not describe how the fragments are interconnected. Moreover, it is not always evident to the user whether all possible fragments of the structures are at all accessible. Thus, the fragments more or less characterize a class of molecules; this is also important in generic structures that arise in chemical patents (see Section 2.7.1)

### 2.7.2.1   Applications
Today, fragment coding is still quite important in patent databases (see Chapter 5, Section 5.11, e.g., Derwent) where Markush structures are also stored. There, the fragments can be applied to substructure or other types of searches where the fragments are defined, e.g., on the basis of chemical properties.

## 2.7.3
### Fingerprints

A fingerprint of a chemical structure tries to identify a molecule with some special characteristics, much in the same way as a human fingerprint identifies a person. The characteristic property can, for example, be described by the structure or structural keys. These keys indicate whether or not a specific substructure or fragment exists in the molecule. The fragments of chemical structures can be coded in binary keys. Here, structural fragments are represented as sequences of 0 and 1 (bit strings), where 0 stands for a fragment which is not present in the structure; otherwise the bit is 1 when the fragment is present at least once (Figure 2-64). The characteristic sequence of a structure, called its fingerprint, has a typical length of 150–2500 bits. If a fragment library is defined with the same number of fragments (i.e., the same value of the bit length, e.g., 150), the bits could be correlated 1:1 to the fragments [93]. However, if a structure contains only a few defined fragments, then also only a few bits are set.

This ambiguous representation of chemical structures as a string allows a very efficient similarity search.

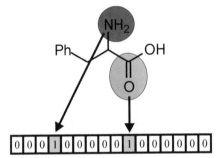

**Figure 2-64.** How an excerpt from a binary code could appear, if only –NH$_2$ and C=O are available in the fragment library.

### 2.7.3.1 "Hashed Fingerprints"

In this procedure, all bonds in the molecule are traversed, starting at an atom and proceeding through several (e.g., seven) bond lengths [93]. Thereby, one receives information about the substructures of the molecule and also about its internal relationships. Each fragment received by this procedure (hashing) is displayed in the bit string as 1. But due to the algorithm, the fingerprint received in that way may include collision entries (an asterisk in Figure 2-65). The advantage of the "normal" fragment-based fingerprint is that no fragments have to be pre-defined. Thus, a better description of the structure will be received. However, no direct correlation of the bit-entry to a substructure is possible because the fragment library is excluded.

### 2.7.4
### Hash Codes

Hash coding is an established method in computer science, e.g., in registration procedures [94, 95]. In chemoinformatics the structure input occurs as a sequence of characters (names) or numbers (which may also be obtained, e.g., from a connection table (see Section 2.4) by conversion of a structure drawing). Both names and numbers may be quite large and may not be usable as an address

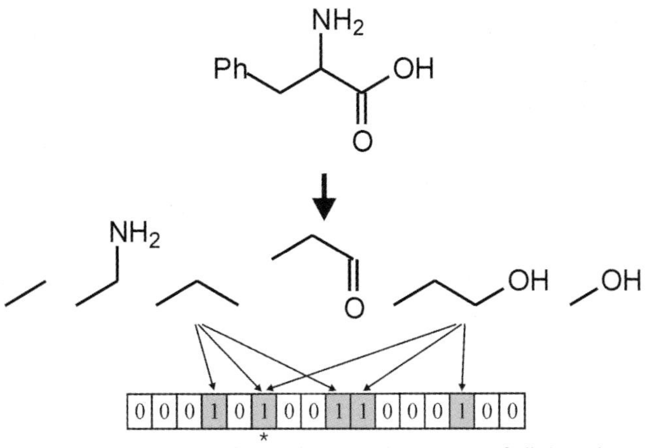

**Figure 2-65.** This fingerprint was received by hashing, whereas only one part of all the substructures is specified in the illustration. The asterisk indicates the address of a collision in the bit string, generated by the algorithm.

for the storage of chemical structures. The "hashing" chops or cuts the input into small pieces to obtain a short, acceptable number within the file length. This contains the information that will be stored (Figure 2-66). This procedure makes it possible to make the connection between the query and the information, e.g., in a database, with an extremely high transfer speed.

In more detail, the hash procedure (key transformation) computes a number of storage addresses from alphabetic, numeric, or alphanumeric keys. The data or the structures are separated into fragments, which are assigned to an identification number (ID). However, this ID of the fragments is not directly accessible in the computer. First, the ID has to be transformed into a pre-set, fixed number of characters (a hash code) by a hash algorithm. This algorithm produces a highly compressed code dependent only on the input information, e.g., molecular weight or empirical formula. However, this code includes no data information and is only used as a key to the storage address of the data entry. Thus, a hash code is a unique number which describes and identifies molecular data structures in chemistry such as atoms and bonds, or is characteristic of an individual chemical structure [96].

The ciphered code is indicated with a defined length, i.e., a fixed bit/byte length. A hash code of 32 bits could have $2^{32}$ (or 4 294 976 296) possible values, whereas one of 64 bits could have $2^{64}$ values. However, due to the fixed length, several diverse data entries could assign the same hash code ("address collision"). The probability of collision rises if the number of input data is increased in relation to the range of values (bit length). In fact, the limits of hash coding are reached with about 10 000 compounds with 32 bits and over 100 million with 64 bits, to avoid collisions in databases [97].

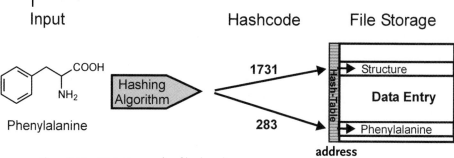

**Figure 2-66.** Operation mode of hash coding.

Thus the hash code is not used as a direct way to access data; rather it serves as an index or key to the filed data entry (Figure 2-66). Since hash coding receives unique codes by reducing multidimensional data to only one dimension, information gets lost. This loss prevents a reconstruction of the complete data from the hash code.

### 2.7.4.1 Applications

Intensive data reduction is an efficient method of managing large datasets. Generally, hash codes are used within chemical information processes such as molecule identification and recognition of identical atoms [98].

Evaluation of special notations of chemical structures.

| Advantages | Disadvantages |
|---|---|
| *Markush structures* | |
| • represents compound families | • high number of compounds |
| • widespread in patents | • less compact code |
| • manual in/output | • ambiguous |
| • convertible into other representations | • difficult to extract individual compounds |
| *Fragment coding* | |
| • characterizes parts of a molecule | • ambiguous |
| • easy classification of compounds | • not convertible to other representations |
| • code is highly compact | |
| *Fingerprints* | |
| • structural keys identify a molecule | • ambiguous |
| • code is highly compact | • not convertible to other representations |
| • represented in bits | • dependent on the fragment library |
| *Hash codes* | |
| • identification number a molecule | • structure information cannot be restored |
| • code is very highly compact | • address collision |
| • represented in bits | • not convertible to other representations |
| • unambiguous and unique | |

The hash code has specific characteristics and provides numerous possibilities for use:

- Hash codes of molecules which are already pre-computed are suitable for use in full structure searches in database applications. The compression of the code of a chemical structure into only one number also makes it possible to compute in advance the transformation results for a whole catalog. The files can be stored and kept complete in the core memory during execution of the program, so that a search can be accomplished within seconds.
- Atomic and bond hash codes are helpful in structure manipulation programs, e.g., in reaction prediction or in synthesis design [99].

## 2.8
## Representation of Stereochemistry

### 2.8.1
### General Concepts

The earlier sections have only considered the way atoms are bonded to each other in a molecule (topology) and how this is translated into a computer-readable form. Chemists define this arrangement of the bonds as the constitution of a molecule. The example in Figure 2-39, Section 2.5.2.1, shows that molecules with a given empirical formula, e.g., $C_3H_6O$, can have several different structures, which are called isomers [100]. Isomeric structures can be divided into constitutional isomers and stereoisomers (see Figure 2-67).

Constitutional isomers are molecules with different connectivities between the atoms. Either the structures have different functional groups (these are called structural isomers), or the same functional group is placed in different positions (positional isomers).

If compounds have the same topology (constitution) but different topography (geometry), they are called stereoisomers. The configuration expresses the different positions of atoms around stereocenters, stereoaxes, and stereoplanes in 3D space, e.g., chiral structures (enantiomers, diastereomers, atropisomers, helicenes, etc.), or *cis/trans* (*Z/E*) configuration. If it is possible to interconvert stereoisomers by a rotation around a C–C single bond, they are called conformers.

The next sections describe briefly the nomenclature of configurational isomers, and how this stereochemistry can be handled by computer.

### 2.8.2
### Representation of Configuration Isomers and Molecular Chirality

The stereochemistry is usually expressed in structure diagrams by wedged and hashed bonds. A wedge indicates that the substituent is in front of a reference plane and a hashed bond indicates that the substituent is pointing away from the viewer (behind the reference plane). This projection is applied both to tetrahe-

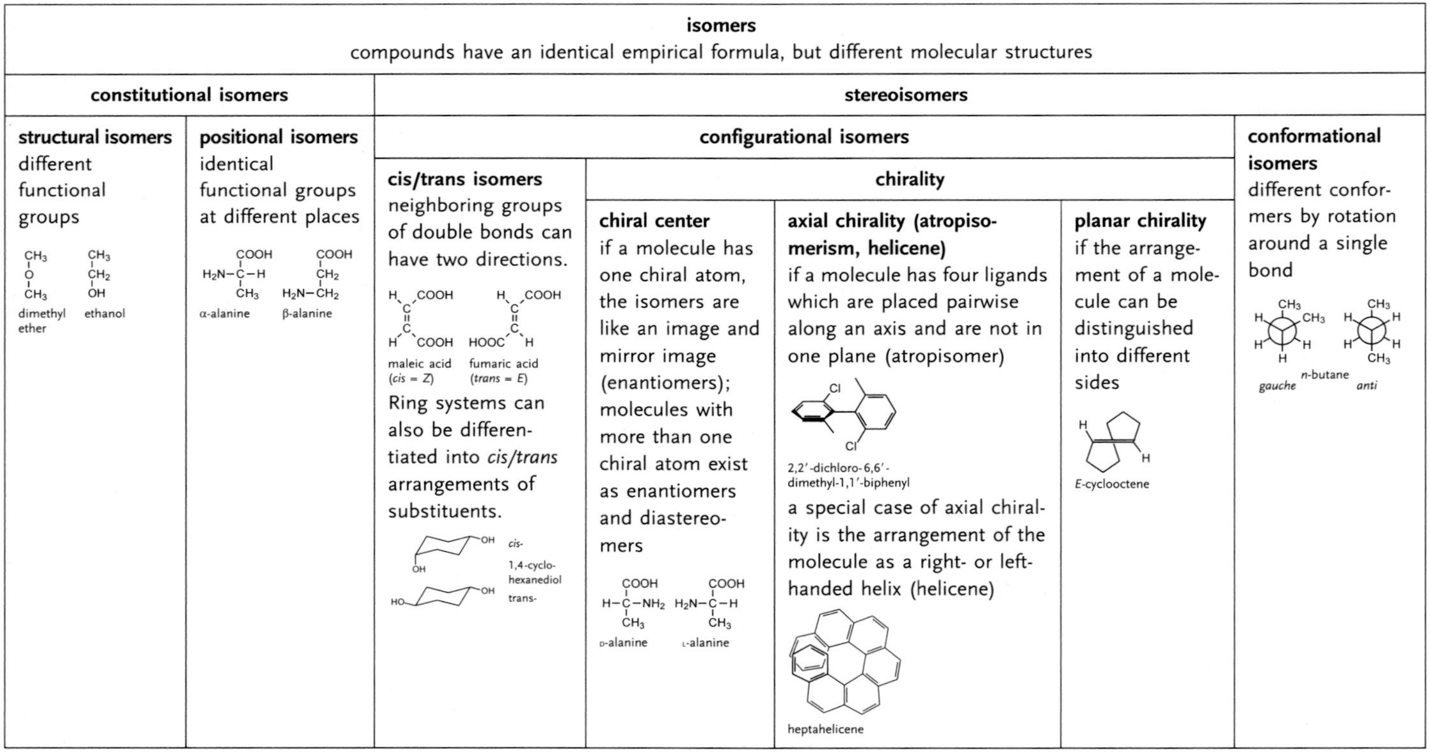

**isomers**

compounds have an identical empirical formula, but different molecular structures

| constitutional isomers | | stereoisomers | | | |
|---|---|---|---|---|---|
| | | configurational isomers | | | conformational isomers |
| structural isomers | positional isomers | cis/trans isomers | chiral center | chirality | |

| structural isomers | positional isomers | cis/trans isomers | chiral center | axial chirality (atropisomerism, helicene) | planar chirality | conformational isomers |
|---|---|---|---|---|---|---|
| different functional groups | identical functional groups at different places | neighboring groups of double bonds can have two directions. Ring systems can also be differentiated into *cis/trans* arrangements of substituents. | if a molecule has one chiral atom, the isomers are like an image and mirror image (enantiomers); molecules with more than one chiral atom exist as enantiomers and diastereomers | if a molecule has four ligands which are placed pairwise along an axis and are not in one plane (atropisomer) | if the arrangement of a molecule can be distinguished into different sides | different conformers by rotation around a single bond |

Figures and chemical structures (dimethyl ether / ethanol; α-alanine / β-alanine; maleic acid (*cis* = Z) / fumaric acid (*trans* = E); *cis*-/*trans*-1,4-cyclohexanediol; D-alanine / L-alanine; 2,2'-dichloro-6,6'-dimethyl-1,1'-biphenyl; heptahelicene; E-cyclooctene; *gauche*/*anti* n-butane) appear within the respective cells.

a special case of axial chirality is the arrangement of the molecule as a right- or left-handed helix (helicene)

**Figure 2-67.** Classification of isomeric structures of organic compounds.

**Figure 2-68.** Different possibilities for the display of stereochemistry; the one in the middle is not unambiguous and therefore is not allowed.

dral stereocenters and to stereocenters with a higher coordination number. In addition, wedged bonds are also used to indicate the spatial arrangement of substituents at ring systems.

However, care has to be taken to keep this graphical 3D information in 2D structure diagrams unambiguous. Since the reference plane can be chosen arbitrarily, there exist different possible ways of displaying the stereochemistry (Figure 2-68).

If the tetravalent carbon atom has three different substituents, the molecule is chiral and it is not possible to superimpose it onto its mirror image. Our feet are also chiral objects: the right foot is a mirror image of the left one and does not fit into the left shoe.

An example of a chiral compound is lactic acid. Two different forms of lactic acid that are mirror images of each other can be defined (Figure 2-69). These two different molecules are called enantiomers. They can be separated, isolated, and characterized experimentally. They are different chemical entities, and some of their properties are different (e.g., their optical rotation).

Molecular chirality is of extraordinary importance in many different fields because it often has dramatic consequences in observable molecular properties. For example, two molecules that are mirror images of each other (enantiomers) often have different pharmacological activity, odor, environmental impact, chemical reactivity (when another chiral object is present), or physical properties (e.g., CD spectra). Chirality is a major point in drug safety evaluation, and an increasing percentage of drugs are marketed as single enantiomers. This requires synthesis methodologies for preparing one enantiomer selectively (enantioselective synthesis), as well as analytical methods to separate, quantify, and identify opposite enantiomers (e.g., chiral chromatography).

The stereochemical analysis of chiral structures starts with the identification of stereogenic units [101]. These units consist of an atom or a skeleton with distinct ligands. By permutation of the ligands, stereoisomeric structures are obtained. The three basic stereogenic units are a center of chirality (e.g., a chiral tetravalent

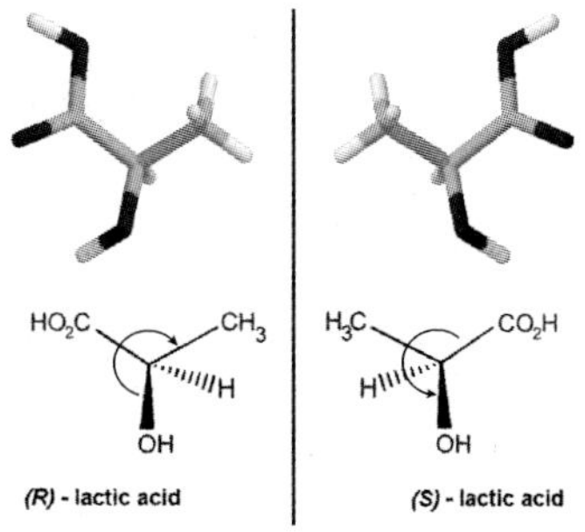

**Figure 2-69.** The two enantiomers of lactic acid: assignment of $R$ and $S$ configurations to the enantiomers of lactic acid after ranking the four ligands attached to the chiral center according to the CIP rules (OH > CO$_2$H > Me > H).

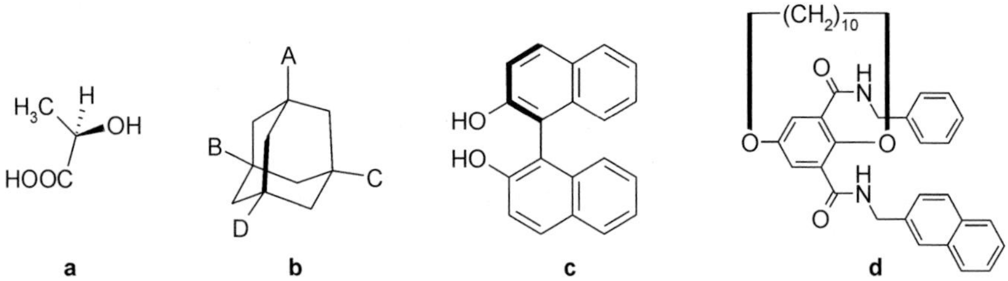

**Figure 2-70.** Examples of chiral molecules with different types of stereogenic units.

carbon atom as in Figure 2-70a), a plane of chirality (e.g., a substituted paracyclophane as in Figure 2-70d), and an axis of chirality (e.g., a chiral binaphthyl compound as shown as in Figure 2-70c).

Chiral carbon atoms are common, but they are not the only possible centers of chirality. Other possible chiral tetravalent atoms are Si, Ge, Sn, N, S, and P, while potential trivalent chiral atoms, in which non-bonding electrons occupy the position of the fourth ligand, are N, P, As, Sb, S, Se, and Te. Furthermore, a center of chirality does not even have to be an atom, as shown in the structure represented in Figure 2-70b, where the center of chirality is at the center of the achiral skeleton of adamantane.

In chemoinformatics, chirality is taken into account by many structural representation schemes, in order that a specific enantiomer can be unambiguously specified. A challenging task is the automatic detection of chirality in a molecular structure, which was solved for the case of chiral atoms, but not for chirality arising from other stereogenic units. Beyond labeling, quantitative descriptors of molecular chirality are required for the prediction of chiral properties (such as biological activity or enantioselectivity in chemical reactions) from the molecular structure. These descriptors, and how chemoinformatics can be used to automatically detect, specify, and represent molecular chirality, are described in more detail in Chapter 8.

### 2.8.2.1 Detection and Specification of Chirality

To assign the stereochemistry of an atom (molecule) unambiguously, we need a system that defines the chirality. Cahn, Ingold, and Prelog proposed such a system in the 1950s [102] and the rules (CIP rules) stand as the official way to specify the chirality of molecular structures [101–104]. The CIP rules are also in themselves a strategy for detecting chirality or, more precisely, for detecting whether two ligands attached to a stereogenic unit are different from each other. It seems a simple task, and for most cases it is. A simplified version of the rules can be put as follows:

1. Ligands are ranked in order of decreasing atomic number of the atom directly connected to the stereogenic unit.
2. If the relative order of two ligands cannot thus be decided, it is determined by a similar comparison of atomic numbers of the next atoms in the ligands, or, if

this fails, of the next: "so one works outward, always first towards atoms of higher atomic number, where there is any choice, until a decision is obtained [101]."

3. If two atoms have the same atomic number but different mass number, the atom with higher mass number comes first.
4. Double and triple bonds are counted as if they were split into two or three single bonds, respectively.

Once the four ligands around a center of chirality have been ranked, the *R* or *S* configuration can be determined. First, you rotate the molecule so that the group of lowest priority is pointing directly away from you. For the other three groups, you determine the direction of high to low priority. If the direction is clockwise, the configuration is *R* (for *rectus* = right). If the direction is anti-clockwise, the configuration is *S* (for *sinister* = left). By using this approach, one can easily assign the absolute configuration *R* to the structure in Figure 2-69 (left-hand side). However, the establishment of rules for the distinction of *any* two different ligands was a highly challenging task, and sophisticated rules were required.

Early implementations of the CIP rules for computer detection and specification of chirality were described for the LHASA [105], CHIRON [106], and CACTVS [107] software packages. Recently, several commercial molecular editors and visualizers (e.g., CambridgeSoft's ChemOffice, ACD's I-Lab, Accelrys' WebLab, and MDL's AutoNom) have also implemented the CIP rules.

Other methods have been proposed for detecting chiral carbon atoms which do not rely on the CIP system, and which have been more convenient for some specific applications [108].

### 2.8.3
### Ordered Lists

An ordered list is one method of determining the configuration at a tetrahedral carbon atom by computational methods. The four ligands (A–D) can be arranged at the stereocenter according to their priority (1–4) in 4! (i.e., in 24) ways. The resulting 24 permutations of different priorities of the ligands are listed in Figure 2-71. For example, the ligands at the chiral carbon atom have the priority D > A > C > B in line 5 on the left-hand side of Figure 2-71, which corresponds to 2 4 3 1.

These 24 arrangements can be grouped into two classes because of the symmetry of a tetrahedron. Interchanging ligands, starting with the arbitrary initial priority 1 2 3 4, leads to the other class when only one permutation of two ligands is made, whereas two permutations give an isomer of the same class. For example, transposing 1 and 2 in 1 2 3 4 leads to 2 1 3 4, an isomer from the list on the right-hand side, whereas a double transposition of the ligands in 1 2 3 4 (e.g., 1, 2 and 3, 4) leads to 2 1 4 3, an isomer of the original ordered list on the left-hand side. According to the classification into clockwise and anti-clockwise, the

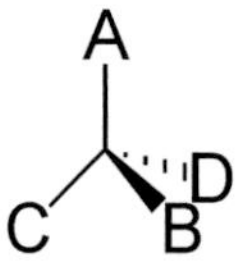

A B C D

| 1 2 3 4 |
| 1 4 2 3 |
| 1 3 4 2 |
| 2 3 1 4 |
| 2 4 3 1 |
| 2 1 4 3 |
| 3 1 2 4 |
| 3 4 1 2 |
| 3 2 4 1 |
| 4 1 3 2 |
| 4 2 1 3 |
| 4 3 2 1 |

A B C D

| 1 3 2 4 |
| 1 4 3 2 |
| 1 2 4 3 |
| 2 1 3 4 |
| 2 4 1 3 |
| 2 3 4 1 |
| 3 2 1 4 |
| 3 4 2 1 |
| 3 1 4 2 |
| 4 1 2 3 |
| 4 3 1 2 |
| 4 2 3 1 |

**Figure 2-71.** The ordered list of 24 priority sequences of the ligands A–D around a tetrahedral stereocenter. The permutations can be separated into two classes, according to the CIP rules: the *R* stereoisomer is on the right-hand side, and the *S* stereoisomer on the left.

two classes correspond to the *R* (list on the right-hand side) and *S* (list on the left-hand side) notations of a stereocenter.

But how does the computer know the ranking of the priorities of the ligands? One method considers the atomic number of the neighboring atoms used in the CIP rules. Another, introduced by Petrarca, Lynch, and Rush (PLR), allocates the priorities to the ligands A–D following the unique numbering by the Morgan Algorithm (see Section 2.5.3) [109]. Since there is not a direct correspondence between the CIP and PLR priorities, the different stereoisomers were denoted by PLR as *Y* and *X*.

A similar approach can be applied for treating the stereochemistry at double bonds.

## 2.8.4
**Rotational Lists**

Description by rotational lists was introduced by Cook and Rohde [110] in the specification of the Standard Molecular Data (SMD) format [111]. In this stereochemical approach, the basic geometrical arrangements around a stereocenter are defined in a list (e.g., square, tetrahedron, etc.). The atoms in these stereoelements are also labeled with numbers in a pre-defined way (Figure 2-72).

Thus, each stereochemical structure can be described and recognized with this rotational list if the structure is designated, e.g., in the "STEREO" block of the SMD format. The compact and extensible representation of the rotational list can include additional information, such as the name specification of the geometry or whether the configuration is absolute, relative, or racemic (Figure 2-73).

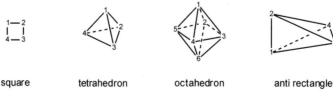

square        tetrahedron        octahedron        anti rectangle

**Figure 2-72.** A selection of stereochemical arrangements in a rotational list.

**Figure 2-73.** The "stereo part" (right) of a tetrahedral C-atom included in the connection table of an SMD file.

```
>STEREO    ABSOLUTE
] TETRAHEDRON
  3 4 1 2
] END
>END
```

## 2.8.5
### Permutation Descriptors

It has already been mentioned that most stereodescriptors occur in pairs, e.g., *R* and *S*, *cis* and *trans*, etc. This is true as long as a molecule contains only atoms with a maximum coordination number of four. A situation with pairwise descriptors can be handled in the computer by a bit, as 0 or 1. In a mathematical treatment, the sign of the permutation group, consisting of + 1 / –1, can be used instead of the 0 and 1 in computer science. If the coordination number of an atom is higher than four, the sign of the permutation group is no longer sufficient.

In order to handle the stereochemistry by permutation groups, the molecule is separated at each stereocenter into a skeleton and its ligands (Figure 2-74).

Both the skeleton and the ligands are then numbered independently. The numbering of the skeleton can be arbitrary, but once it has been decided it has to be retained, whatever geometric operations such as rotation or reflection are performed with the molecule. The numbering of the ligands has to be based on rules such as the CIP rules. In the following, we indicate the indices of the skeleton by italic numbers and the numbering of the ligands by bold numbers (see Figure 2-74). A reference stereoisomer is then defined when the ligand with index **1** is on skeleton site *1*, the ligand with index **2** on skeleton site *2*, etc. This reference isomer is shown in Figure 2.74 on the right-hand side; it obtains the descriptor (+ 1). By comparing a stereoisomer with the reference isomer, the permutation descriptor can be calculated. This is shown in Figure 2-75 for a reflection operation at the reference isomer, an inversion of the stereocenter. For the resulting stereoisomer, a simple transposition of two ligand indices has to be made in order to bring it into correspondence with the reference isomer. Thus, the stereoisomer resulting from a reflection operation obtains the descriptor $(-1)^1 = (-1)$. More details on the treat-

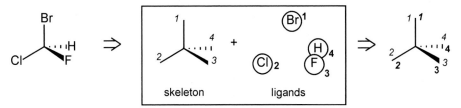

**Figure 2-74.** Basic stages for describing a stereoisomer by a permutation descriptor. At the stereocenter, the molecule is separated into the skeleton and its ligands. Both are then numbered independently, with the indices of the skeleton in italics, the indices of the ligands in bold.

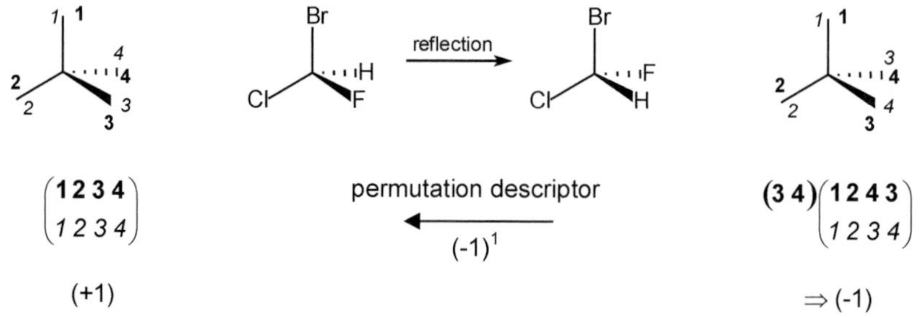

$$\begin{pmatrix} 1\,2\,3\,4 \\ 1\,2\,3\,4 \end{pmatrix}$$

(+1)

permutation descriptor

$\longleftarrow$
$(-1)^1$

$(3\,4)\begin{pmatrix} 1\,2\,4\,3 \\ 1\,2\,3\,4 \end{pmatrix}$

$\Rightarrow (-1)$

**Figure 2-75.** Determination of a permutation descriptor of a stereoisomer after reflection at the stereocenter

ment of the stereochemistry of molecules by permutation group theory can be found in Section 2.8.7.

### 2.8.6
### Stereochemistry in Molfile and SMILES

The basic idea of specifying the priority of the atoms around a stereocenter in order to obtain a stereodescriptor is also incorporated into the most widespread structure representations, the Molfile and SMILES (see Sections 2.3.3, and 2.4.6).

In both encodings, stereoisomerism is discussed with an example of chirality and *cis/trans* isomerism.

#### 2.8.6.1   Stereochemistry in the Molfile
A molecule editor can draw a chemical structure and save it, for example as a Molfile. Although it is possible to include stereochemical properties in the drawing as wedges and hashed bonds, or even to assign a stereocenter/stereogroup with its identifiers (*R/S* or *E/Z*), the connection table of the Molfile only represents the constitution (topology) of the molecule.

The additional stereoinformation has to be derived from the graphical representation and encoded into stereodescriptors, as described above. The stereodescriptors are then stored in corresponding fields of the connection table (Figure 2-76) [50, 51].

The stereochemical descriptor that is entered into a Molfile is based on the ordered list of PLR. Wipke and Dyott further developed this concept in the SEMA (Stereochemically Extended Morgan Algorithm) name project in the early 1970s [112]. The name already indicates that the atoms around the stereocenter are numbered according to the Morgan Algorithm. Then, with knowledge of the priority of the ligands and their ordering in the PLR list, the stereodescriptor can be determined. Since the stereochemistry cannot always be determined or is sometimes unknown, the descriptor has to hold more than the two values for *R* and *S*. This problem was solved with the "parity value". This value can be calculated by comparing the parity value and the ordered list. Here, indices of the

```
TH"tkserve10230216552D 0    0.00000     0.00000???

 18 17  0  0  1  0  0    0  0  0  1 V2000
    99.0000 -129.0000    0.0000 C   0  0  0  0  0  0  0  0  0  0  0  0
   144.0000 -129.0000    0.0000 C   0  0  0  0  0  0  0  0  0  0  0  0
   166.5000 -167.9711    0.0000 C   0  0  0  0  0  0  0  0  0  0  0  0
   144.0000 -206.9423    0.0000 C   0  0  0  0  0  0  0  0  0  0  0  0
   166.5000 -245.9134    0.0000 C   0  0  0  0  0  0  0  0  0  0  0  0
   144.0000 -284.8846    0.0000 C   0  0  1  0  0  0  0  0  0  0  0  0
   166.5000 -323.8557    0.0000 O   0  0  0  0  0  0  0  0  0  0  0  0
    99.0000 -284.8846    0.0000 C   0  0  0  0  0  0  0  0  0  0  0  0
    76.5000 -323.8557    0.0000 N   0  0  0  0  0  0  0  0  0  0  0  0
  1  2  1  0  0  0  0
  2  3  2  0  0  0  0
  3  4  1  0  0  0  0
  4  5  2  0  0  0  0
  5  6  1  0  0  0  0
  6  7  1  1  0  0  0
  6  8  1  0  0  0  0
  8  9  3  0  0  0  0
M  END
$$$$
```

**Figure 2-76.** A typical 2D Molfile of (2*R*,3*E*,5*E*)-2-hydroxy-3,5-heptadiene nitrile with stereochemical flags (parity values, etc.) in the gray columns. For further explanation, see the text.

atoms obtained from of the Morgan Algorithm are interchanged until they are in ascending order. For assigning the parity value, the number of permutations is essential. If the number is odd, the parity value is 1; it is 2 if the number is even. Additionally, the parity value can be 0 if no stereochemistry at the stereocenter is defined, and 3 if the configuration is not known.

In our example in Figure 2-76, the stereocenter at atom 6 (row 6, column 7 in the atom block) has a parity value of 1. This results from an odd number of permutations to bring the Morgan numbers into an ascending order. Thus, first of all, the arbitrary numbering of the atoms has to be canonicalized by the Morgan Algorithm (Figure 2-77a). Then, the Morgan numbers of the atoms next to the stereocenter are listed, starting with the lowest number (Figure 2-77b, first line). The other three Morgan numbers follow in a clockwise manner when viewed looking down the bond from the first. Now, the permutation succeeds as long as the numbers are in an ascending order. In our example, the numbers have to be exchanged only once (3 to 4), thus, the parity value corresponds to the *R*-isomer (Figure 2-77b).

Similarly the "stereobonds" can be defined and added to the bond list in the fourth column of the CT. A single bond acquires the value of 0 if it is not a "stereobond", 1 for up (a wedged bond), 4 for either up or down, and 6 for down (a hashed bond). The cis/trans or E/Z configuration of a double bond is determined by the x,y,z coordinates of the atom block if the value is 0. If it is 3, the double bond is either cis or trans. In the bond block of our example (Figure 2-76), the stereocenter is set to 1 (up) at atom 6 (row 6, column 4 in the bond block), whereas the configurations of the double bonds are determined by the x,y coordinates of the atom block.

a)

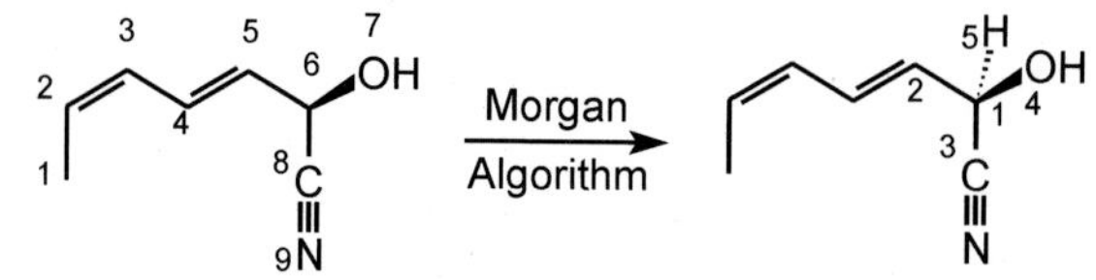

b)  start with Morgan No.:       2 4 3 5

    permutation (3? 4):          2 3 4 5

    result:                      odd (i.e. parity = 1)

**Figure 2-77.** Determination of parity value. a) First the structure is canonicalized. Only the Morgan numbers at the stereocenter are displayed here. b) The listing starts with the Morgan numbers of the atoms next to the stereocenter (**1**), according to certain rules. Then the parity value is determined by counting the number of permutations (odd = 1, even = 2).

### 2.8.6.2 Stereochemistry in SMILES

Stereochemistry can also be expressed in the SMILES notation [113]. Depending on the clockwise or anti-clockwise ordering of the atoms, the stereocenter is specified in the SMILES code with @ or @@, respectively (Figure 2-78). The atoms around this stereocenter are then assigned by the sequence of the atom symbols following the identifier @ or @@. This means that, reading the SMILES code from the left, the three atoms behind the identifiers (@ or @@) describe the stereochemistry of the stereocenter. The sequence of these three atoms is dependent only on the order of writing, and independent of the priorities of the atoms.

Stereoisomerism at double bonds is indicated in SMILES by "/" and "\". The characters specify the relative direction of the connected atoms at a double bond and act as a frame. The characters frame the atoms of a double bond in a parallel or an opposite direction. It is therefore only reasonable to use them on both sides (Figure 2-78). There are other valid representations of *cis/trans* isomers, because the characters can be written in different ways. Further details are listed in Section 2.3.3, in the Handbook or in Ref. [22].

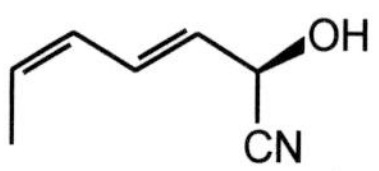

SMILES: C/C=C\C=\[C@@H](O)C#N

**Figure 2-78.** The stereochemistry of (2*R*,3*E*,5*E*)-2-hydroxy-3,5-heptadiene nitrile can be expressed in the SMILES notation with @ or (back)slashes.

## 2.8.7
## Tutorial: Handling of Stereochemistry by Permutation Groups

The preceding section gave a brief introduction to the handling of the stereochemistry of molecules by permutation group descriptors. Here we discuss this topic in more detail. The treatment is largely based on ideas introduced in Ref. [100].

The essential steps for the definition of a permutation descriptor are:

1. First, both the skeleton and the ligands at a stereocenter have to be numbered independently of each other. The sites of the skeleton can be numbered arbitrarily but then this numbering has to remain fixed all the time in any further operations. The atoms directly bonded to the stereocenter have to be numbered according to rules such as the CIP rules or the Morgan Algorithm (Figure 2-79).

*During all the operations the numbering of the skeleton must not be changed!*

2. After the skeleton and the ligands have been numbered, a permutation matrix – a mapping of the ligands onto the skeleton sites – is set up. This simple mathematical representation allows the comparison of the sites of the skeleton and the sequence of the ligands. Each ligand is positioned in the first line according to its place on the skeleton in the second line (Figure 2-80). If the ligand with index 1 sits on skeleton site 1, the ligand with index 2 is on skeleton site 2, etc., we have the reference isomer. The permutation matrix of this reference molecule obtains the descriptor (+ 1).

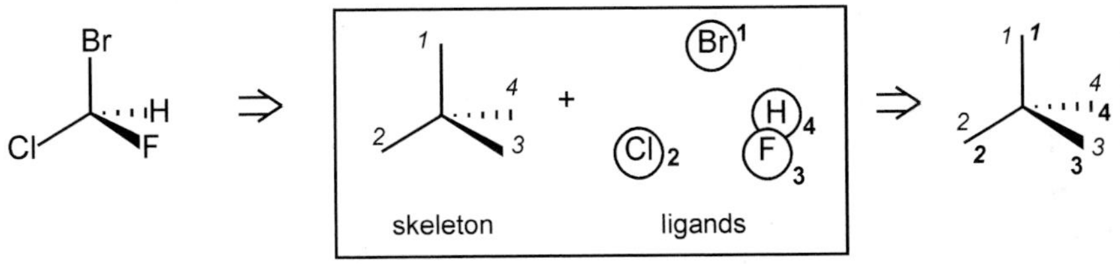

**Figure 2-79.** Basic steps for describing a stereoisomer by a permutation descriptor: at the stereocenter, the molecule is separated into the skeleton and its ligands. Both are then numbered independently, with the indices of the skeleton in italics, the indices of the ligands in bold.

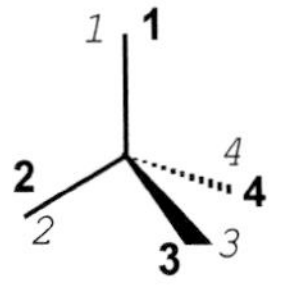

$$\begin{pmatrix} \mathbf{1\,2\,3\,4} \\ \mathit{1\,2\,3\,4} \end{pmatrix} \equiv \begin{pmatrix} \text{indices of the ligands} \\ \text{indices of the skeleton sites} \end{pmatrix}$$

**Figure 2-80.** The permutation matrix of the reference isomer: the second line gives the indices of the sites of the skeleton and the first line the indices of the ligands (e.g., the ligand with index *3* is on skeleton site **3**).

### 2.8.7.1 Stereochemistry at Tetrahedral Carbon Atoms

#### 2.8.7.1.1 Determination of the Permutation Descriptor after Rotation of the Molecule

In order to determine the permutation stereodescriptor of a stereoisomer, the permutation matrix has to be set up and brought into correspondence with the reference isomer by permutation of the ligands. Figure 2-81 shows this for the stereoisomer that is obtained from the reference isomer through rotation by 120°.

In the next step, the number of transpositions – the number of interchanges of two ligands – is determined. This is achieved by changing the indices of two ligands until the reference sequence is obtained. In this process, only the interchange of two neighboring ligand indices is allowed. Each interchange is written as a transposition operation on the top left-hand side of the permutation matrix (the matrix in the center of Figure 2-81).

In the example, we proceed from the right (rotated molecule) to the left (reference molecule). The first transposition is done with ligands **4** and **2**, in order to obtain the first part of the reference sequence "**1 2**". Then, only a permutation of ligands **3** and **4** has to be done to obtain the reference matrix on the left-hand side. Thus, in total, we have executed two transpositions (**4 2**) and (**3 4**).

The two structures in our example are identical and are rotated by only 120°. Clearly, rotation of a molecule does not change its stereochemistry. Thus, the permutation descriptor of both representations should be (+1). On this basis, we can define an equation where the number of transpositions is correlated with the permutation descriptors in an exponential term (Eq. (9)).

$$(-1)^{\text{ number of transpositions}} \cdot p_{\text{ref}} = p_{\text{isomer}} \tag{9}$$

Once again, in the case of rotation we have had two transpositions, resulting in a permutation descriptor of (+1), a result as desired: $(-1)^2 \cdot (+1) = (+1)$.

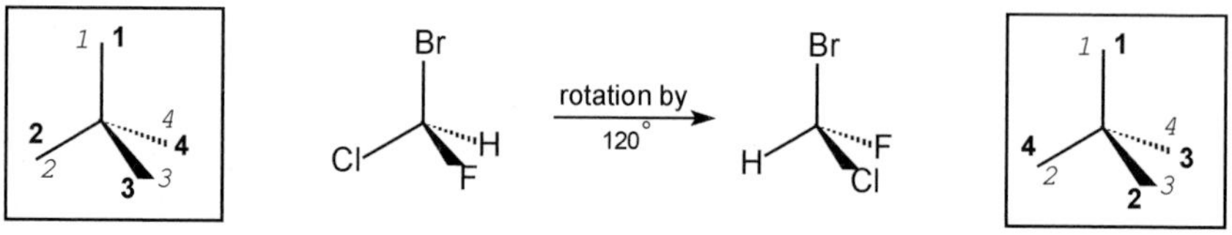

$$\begin{pmatrix} 1\,2\,3\,4 \\ 1\,2\,3\,4 \end{pmatrix} \quad \leftarrow \quad (3\,4)(4\,2)\begin{pmatrix} 1\,2\,4\,3 \\ 1\,2\,3\,4 \end{pmatrix} \quad \leftarrow \quad (4\,2)\begin{pmatrix} 1\,4\,2\,3 \\ 1\,2\,3\,4 \end{pmatrix}$$

**Figure 2-81.** The permutation matrices of two structures that differ through rotation by 120°. The permutation matrix of the rotated isomer can be brought into correspondence with the permutation matrix of the reference isomer by two interchanges of two ligands (transpositions).

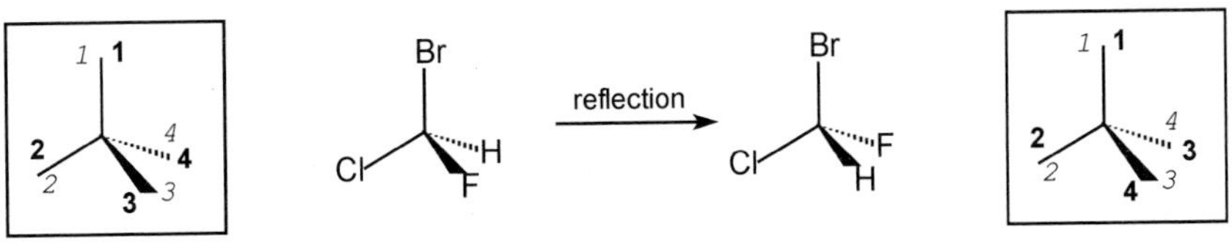

$$\begin{pmatrix} 1\,2\,3\,4 \\ 1\,2\,3\,4 \end{pmatrix}$$

$\leftarrow$

$$(4\,3)\begin{pmatrix} 1\,2\,4\,3 \\ 1\,2\,3\,4 \end{pmatrix}$$

**Figure 2-82.** The permutation matrices of two structures that differ by reflection through the plane of the drawing. One transposition is necessary to bring these matrices into correspondence.

### 2.8.7.1.2 Determination of the Reflection Permutation Descriptor

The same reference molecule is now reflected at a plane spanned by the Cl, Br, and C atoms, to give the other enantiomer. In the same manner as previously, we write down the permutation matrices of the two structures, and then determine the transpositions (Figure 2-82).

In this case, only one transposition has to be performed, resulting in a permutation descriptor of $(-1)$. According to Eq. (9): $(-1)^1 \cdot (+1) = (-1)$.

### 2.8.7.1.3 Example

Two steroids with the same constitution should be checked to see if they are stereoisomers (Eq. (10).

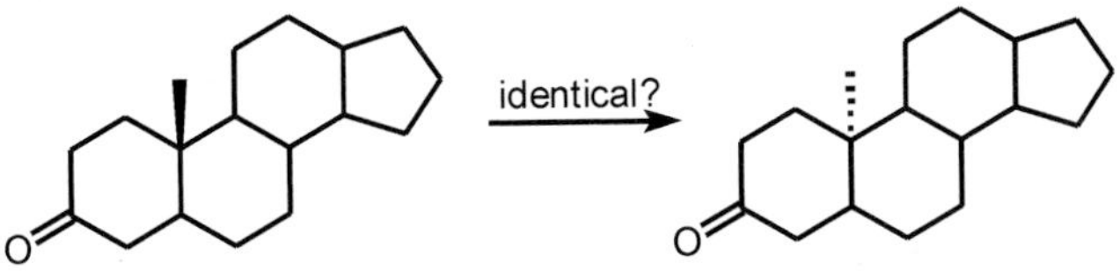

(10)

The determination of their permutation descriptors follows the procedure described above. To assign numbers to the "ligands" of the stereocenter, the Morgan Algorithm can be used. For simplification, the Morgan numbering of the "ligands" is transferred to the numbers 1 to 4 according to the numbering of the skeleton (Figure 2-83, first two lines). Then the number of transpositions is determined, using one of the steroids (the left-hand one) as the reference isomer (Figure 2-83, bottom line).

The permutation descriptor is $(-1)$ according to Eq. (9) and therefore the two molecules are enantiomers.

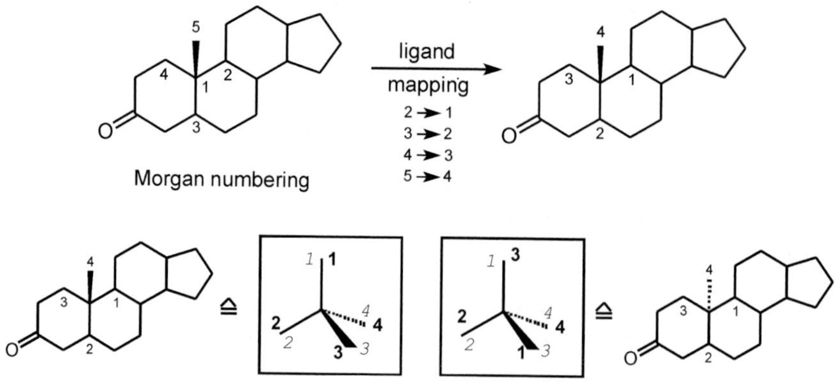

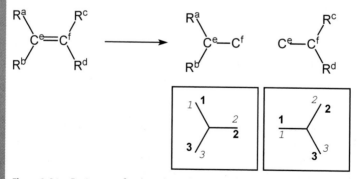

$$\begin{pmatrix} \mathbf{1\,2\,3\,4} \\ 1\,2\,3\,4 \end{pmatrix} \xleftarrow{(2\,3)(1\,3)(1\,2)} \begin{pmatrix} \mathbf{1\,3\,2\,4} \\ 1\,2\,3\,4 \end{pmatrix} \xleftarrow{(1\,3)(1\,2)} \begin{pmatrix} \mathbf{3\,1\,2\,4} \\ 1\,2\,3\,4 \end{pmatrix} \xleftarrow{(1\,2)} \begin{pmatrix} \mathbf{3\,2\,1\,4} \\ 1\,2\,3\,4 \end{pmatrix}$$

**Figure 2-83.** Example of the process to decide whether two structures are enantiomers by determining the permutation descriptor.

### 2.8.7.2 Stereochemistry at Double Bonds

The basic method for determining stereodescriptors at double bonds is quite similar to that used for tetrahedral C-atoms. Again, the permutation matrix has to be set up and brought into correspondence with a reference isomer. To this end, the double bond is first separated into two monocentric units before the skeletons and the ligands of both parts are numbered (Figure 2-84).

**Figure 2-84.** Basic steps for describing the stereochemistry at a double bond by a permutation descriptor: The double bond is split into two parts. These parts are separated into the skeleton and its ligands. Both are then numbered independently, with the indices of the skeletons in italics, the indices of the ligands in bold.

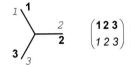

**Figure 2-85.** The permutation matrix of the reference isomer for double bonds.

One of the segments represents the reference isomer, and the permutation matrix of this one gets the descriptor (+ 1) (Figure 2-85).

#### 2.8.7.2.1 Determination of the Permutation Descriptor of a *cis* Stereoisomer

The sequence of steps is as described above: numbering of the skeletons and the ligands, establishing the mapping of the ligands onto the skeleton sites, and determining the number of transpositions. Each part of the fragment obtains a descriptor of its own according to Eq. (9) (Figure 2-86).

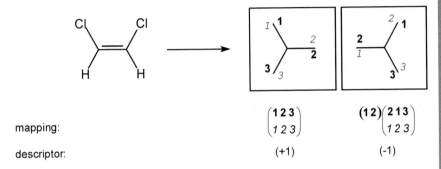

mapping:

descriptor:

$$\begin{pmatrix}1 2 3\\1 2 3\end{pmatrix}$$

$$(+1)$$

$$(1 2)\begin{pmatrix}2 1 3\\1 2 3\end{pmatrix}$$

$$(-1)$$

**Figure 2-86.** The permutation matrices of the fragments of a *cis* isomer. One transposition is necessary to bring these matrices into correspondence. The overall descriptor is obtained from those of the two separate units by multiplication: $(+1)(-1) = (-1)$.

However, the descriptors cannot be considered independently as there is no free rotation around the double bond. In order to take account of this rigidity, the descriptors of the two units have to be multiplied to fix a descriptor of the complete stereoisomer.

#### 2.8.7.2.2 Determination of the Permutation Descriptor of a *trans* Stereoisomer

The same reference unit of Figure 2-86 is also used here for the determination of the permutation descriptor of the *trans* isomer. In the same manner as above, we write down the permutation matrices of the two structures, and then determine the transpositions (Figure 2-87).

In this case, two transpositions have to be performed, resulting in a permutation descriptor of (+ 1) for the right-hand subunit.

The complete descriptor of this isomer is again obtained by multiplication of the descriptors of the two subunits. In this case, we obtain a value of (+ 1), the opposite of the value for the *cis* isomer, as desired.

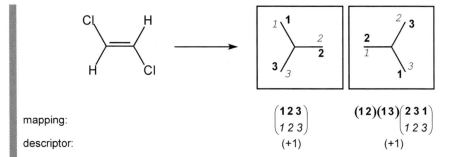

mapping:

$$\begin{pmatrix} 1\,2\,3 \\ 1\,2\,3 \end{pmatrix}$$

$$(12)(13)\begin{pmatrix} 2\,3\,1 \\ 1\,2\,3 \end{pmatrix}$$

descriptor: (+1) (+1)

**Figure 2-87.** The permutation matrices of the fragments of a *trans* isomer. Two transpositions are necessary to bring these matrices into correspondence. The overall descriptor is obtained from those of the two separate units by multiplication: $(+1)(+1) = (+1)$.

Now let us see how the descriptors behave under rotation of the entire molecule. If the *trans* isomer used before is rotated by 180° and the descriptors are calculated again, a value of (–1) is obtained two times (right-hand side, bottom line of Figure 2-88).

However, as the descriptor of the entire stereoisomer is obtained by multiplication of the individual descriptors, again a value of (+ 1) is obtained. Thus, as desired, the stereocenter of a double bond does not change through rotation of a molecule.

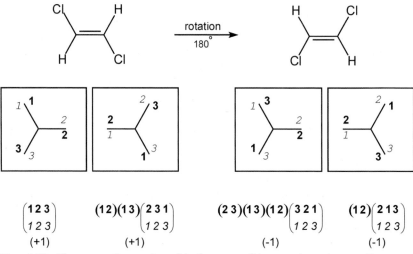

$$\begin{pmatrix} 1\,2\,3 \\ 1\,2\,3 \end{pmatrix}$$ $$(12)(13)\begin{pmatrix} 2\,3\,1 \\ 1\,2\,3 \end{pmatrix}$$ $$(23)(13)(12)\begin{pmatrix} 3\,2\,1 \\ 1\,2\,3 \end{pmatrix}$$ $$(12)\begin{pmatrix} 2\,1\,3 \\ 1\,2\,3 \end{pmatrix}$$

(+1)  (+1)  (-1)  (-1)

**Figure 2-88.** The permutation matrices of the fragments of the rotated *trans* isomers. The rotated structure (right-hand side) has two descriptors of (–1) whereas the initial structure (left-hand side) had two values of (+1). The overall descriptor of both sides is obtained by multiplication: $(+1)(+1) = (+1)$ and $(-1)(-1) = (+1)$.

**2.9**
**Representation of 3D Structures**

2.9.1
**Walking through the Hierarchy of Chemical Structure Representation**

The previous sections have dealt mainly with the representation of chemical structures as flat, two-dimensional, or topological objects resulting in a structure diagram. The next step is the introduction of stereochemistry (see Section 2.8), leading to the term "configuration" of a molecule. The configuration of a molecule defines the positions, among all those that are possible, in which the atoms in the molecule are arranged relative to each other, unless the various arrangements lead to distinguishable and isolable *stereoisomeric compounds* of one and the same molecule. A major characteristic of stereoisomeric compounds is that they have the same constitution, but are only interconvertible by breaking and forming new bonds.

To code the configuration of a molecule various methods are described in Section 2.8. In particular, the use of wedge symbols clearly demonstrates the value added if stereodescriptors are included in the chemical structure information. The inclusion of stereochemical information gives a more realistic view of the actual spatial arrangement of the atoms of the molecule under consideration, and can therefore be regarded as "between" the 2D (topological) and the 3D representation of a chemical structure.

Clearly, the next step is the handling of a molecule as a real object with a spatial extension in 3D space. Quite often this is also a mandatory step, because in most cases the 3D structure of a molecule is closely related to a large variety of physical, chemical, and biological properties. In addition, the fundamental importance of an unambiguous definition of stereochemistry becomes obvious, if the 3D structure of a molecule needs to be derived from its chemical graph. The molecules of stereoisomeric compounds differ in their spatial features and often exhibit quite different properties. Therefore, stereochemical information should always be taken into account if chiral atom centers are present in a chemical structure.

The actual 3D geometry of a molecule is called its conformation. In contrast to configurational isomers, conformational isomers can be interconverted simply by rotation around rotatable bonds. Furthermore, most molecules can adopt more than one conformation of nearly equal energy content. Each of these geometries corresponds to one of the various minima of the potential energy function of the molecule, a high-dimensional mathematical expression which correlates the geometric parameters of a chemical structure with its energy content. Which of these conformations is the preferred one is heavily influenced by the interactions of the molecule with its current environment. Significantly different conformations can be observed for one and the same molecule if it is, e.g., isolated in the gas phase, influenced by solvent effects in solution, about to take part in a chemical reaction, or part of a crystal lattice in the solid state.

Before we go into further detail on the handling of chemical structures in 3D space from the chemoinformatics point of view, it should be noted that there

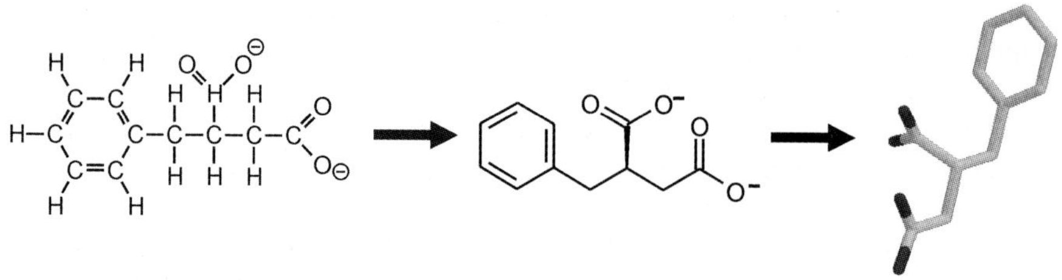

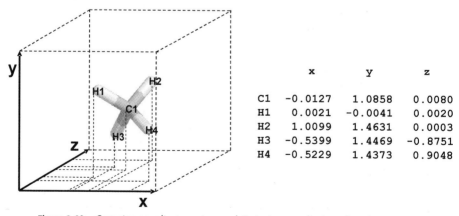

**Figure 2-89.** From the *constitution* to the *configuration* and then to the *conformation* (3D structure) of a molecule with the example of 2R-benzylsuccinate.

is a clear hierarchy in the representation of chemical compounds. Starting from the *constitution* of a molecule, the provision of stereochemical information leads to the *configuration*, a more precise description of a molecule, which can then be translated into the corresponding *3D structure*, i.e., a single conformation of the molecule under consideration, or a set of them (see Figure 2-89).

The representation of molecular surfaces, including the display of molecular surface properties, can be regarded as the next level of this hierarchy, but will be addressed in Sections 2.10 and 2.11 in this volume.

### 2.9.2
### Representation of 3D Structures

Basically, two different methods are commonly used for representing a chemical structure in 3D space. Both methods utilize different coordinate systems to describe the spatial arrangement of the atoms of a molecule under consideration. The most common way is to choose a Cartesian coordinate system, i.e., to code the $x$-, $y$-, and $z$-coordinates of each atom, usually as floating point numbers. For each atom the Cartesian coordinates can be listed in a single row, giving consecutively the $x$-, $y$-, and $z$-values. Figure 2-90 illustrates this method for methane.

|     | x        | y        | z        |
|-----|----------|----------|----------|
| C1  | -0.0127  | 1.0858   | 0.0080   |
| H1  | 0.0021   | -0.0041  | 0.0020   |
| H2  | 1.0099   | 1.4631   | 0.0003   |
| H3  | -0.5399  | 1.4469   | -0.8751  |
| H4  | -0.5229  | 1.4373   | 0.9048   |

**Figure 2-90.** Cartesian coordinate system and Cartesian coordinates of methane.

The connectivity information can be given either implicitly by approximating bonding distances between the atoms, or explicitly by a connection table (bond list) as shown in Figure 2-20 and 2-25.

It is always advisable to specify the complete connection table in addition to the 3D information. Otherwise, post-processing software has to calculate all the interatomic distances and to estimate which atoms are at a distance which lies within the range of a certain bond type, considering the atom types, charges, hybridization states, etc. of the connected atoms. In the case of non-bonding interactions (e.g., intramolecular hydrogen bonding), this might lead to difficulties or even misinterpretation and therefore to distorted 3D molecular models. Most of the standard file formats for 3D chemical structure information contain both the 3D atom coordinates and the connection table.

The second method for representing a molecule in 3D space is to use internal coordinates such as bond lengths, bond angles, and torsion angles. Internal coordinates describe the spatial arrangement of the atoms relative to each other. Figure 2-91 illustrates this for 1,2-dichloroethane.

The most common way to describe a molecule by its internal coordinates is the so-called *Z-matrix*. Figure 2-92 shows the Z-matrix of 1,2-dichloroethane.

A set of rules determines how to set up a Z-matrix properly. Each line in the Z-matrix represents one atom of the molecule. In the first line, atom 1 is defined as C1, which is a carbon atom and lies at the origin of the coordinate system. The second atom, C2, is at a distance of 1.5 Å (second column) from atom 1 (third column) and should always be placed on one of the main axes (the x-axis in Figure 2-92). The third atom, the chlorine atom Cl3, has to lie in the xy-plane; it is at a distance of 1.7 Å from atom 1, and the angle $\alpha$ between the atoms 3–1–2 is 109° (fourth and fifth columns). The third type of internal coordinate, the torsion angle or dihedral $\tau$, is introduced in the fourth line of the Z-matrix in the sixth and seventh column. It is the angle between the planes which are

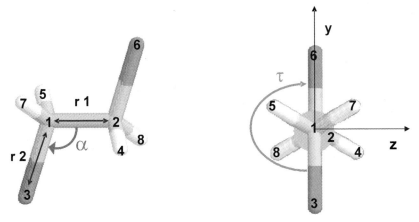

**Figure 2-91.** Internal coordinates of 1,2-dichloroethane: bond lengths *r1* and *r2*, bond angle $\alpha$, and torsion angle $\tau$.

```
C1
C2    1.5   1
C13   1.7   1   109   2
H4    1.1   2   109   1   -60    3
H5    1.1   1   109   2   180    4
C16   1.7   2   109   1    60    5
H7    1.1   1   109   2   -60    6
H8    1.1   2   109   1   180    7
```

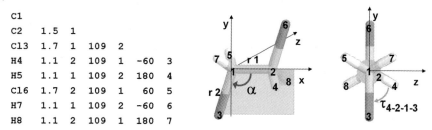

**Figure 2-92.** Z-matrix of 1,2-dichloroethane.

spanned by the atoms 4, 2, and 1, and 2, 1, and 3 (the *xy*-plane). Except of the first three atoms, each atom is described by a set of three internal coordinates: a distance from a previously defined atom, the bond angle formed by the atom with two previous atoms, and the torsion angle of the atom with three previous atoms. A total of $3N - 6$ internal coordinates, where $N$ is the number of atoms in the molecule, is required to represent a chemical structure properly in 3D space. The number ($3N - 6$) of internal coordinates also corresponds to the number of degrees of freedom of the molecule.

Z-matrices are commonly used as input to quantum mechanical (*ab initio* and semi-empirical) calculations as they properly describe the spatial arrangement of the atoms of a molecule. Note that there is no explicit information on the connectivity present in the Z-matrix, as there is, e.g., in a connection table, but quantum mechanics derives the bonding and non-bonding intramolecular interactions from the molecular electronic wavefunction, starting from atomic wavefunctions and a crude 3D structure. In contrast to that, most of the molecular mechanics packages require the initial molecular geometry as 3D Cartesian coordinates plus the connection table, as they have to assign appropriate force constants and potentials to each atom and each bond in order to relax and optimize the molecular structure. Furthermore, Cartesian coordinates are preferable to internal coordinates if the spatial situations of ensembles of different molecules have to be compared. Of course, both representations are interconvertible.

Table 2-6 gives an overview on the most common file formats for chemical structure information and their respective possibilities of representing or coding the constitution, the configuration, i.e., the stereochemistry, and the 3D structure or conformation (see also Sections 2.3 and 2.4). Except for the Z-matrix, all the other file formats in Table 2-6 which are able to code 3D structure information are using Cartesian coordinates to represent a compound in 3D space.

### 2.9.3
### Obtaining 3D Structures and Why They are Needed

The 3D structure of a molecule can be derived either from experiment or by computational methods. Regardless of the origin of the 3D model of the molecule under consideration, the user should always be aware of how the data were obtain-

**Table 2-6.** Common file formats for representing chemical structures.

| File format | Representation (coding) of | | |
| --- | --- | --- | --- |
| | constitution | configuration | 3D structure/conformation |
| MDL SDfile | yes | yes | yes |
| SMILES | yes | yes | no |
| SYBYL MOL2 | yes | not explicitly | yes |
| PDB | yes | not explicitly | yes |
| XYZ | no | no | yes |
| Z-matrix | no | not explicitly | yes |

ed. As already mentioned, molecules often adopt significantly different geometries, i.e., conformations, under various conditions, and thus can exhibit different properties (e.g., the dipole moment). Therefore, the choice of a certain 3D structure may heavily influence all further investigations based on this geometry.

Experimental methods such as X-ray crystallography, electron diffraction, 2D NMR, IR, or microwave spectroscopy observe a molecule under certain physical and chemical conditions, e.g., in the solid state or in solution. Theoretical approaches rely on numerical calculations and empirical approaches of various levels of sophistication depending on the method applied in order to predict or to generate a 3D molecular model. Therefore, the deployment of machines and computers has a long tradition in this field of chemoinformatics, automatic 3D structure generation. The term "automatic 3D structure generators" describes computer programs which are capable of automatically predicting, without any intervention by the user, a 3D molecular model starting from the constitution and the stereochemical information of a molecule under consideration (see Figure 2-93). Clearly there is a need for these methods, as demonstrated by the misbalance between the approximately 270 000 compounds with an experimentally determined 3D structure and the number (about 25 million) of known compounds.

In principle, two different classes of structure generators can be distinguished, namely empirical approaches, such as fragment-based or rule- and data-based

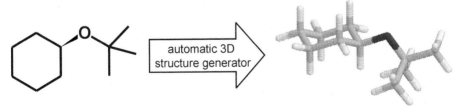

topological information
2D representation

three-dimensional
structure information

**Figure 2-93.** Automatic 3D structure generation.

methods, and theoretical methods, such as quantum mechanical (QM) and molecular mechanical (MM) calculations.

Empirical approaches are based on the implicit and explicit knowledge of chemists on the rules and principles of the geometry and energy of molecules, which have been derived from experimental data and theoretical investigations. The underlying concepts of theoretical (numerical) methods for the construction of a 3D structure are the identification of equilibria either based on the electronic structure of the molecule (QM), or force-field driven (MM). Thus, theoretical methods solve the problem of 3D structure generation *ab initio* (to a certain extent depending on the applied theory), but need at least a reasonable starting geometry. Therefore, they cannot be regarded as genuine automatic 3D structure generators (3D model builders).

But why are 3D structures needed? Part of the answer to this question has already been given. As mentioned previously, a large variety of physical, chemical, and biological properties of a molecule are strongly dependent on its 3D structure. Therefore, studies which try to correlate chemical structures with a certain property under consideration – so-called *QSAR/QSPR* studies (Quantitative/Qualitative Structure–Activity/Property Relationship) – may gain more insight into the problem under investigation if 3D structural information is used. Modeling and prediction of biological activity, virtual screening and docking experiments (prediction of receptor/ligand interactions and complexes in biological systems), or investigations to model the chemical reactivity of a compound clearly require information on the 3D structure of the molecules under consideration. In addition, the results of structure elucidation techniques which are based on experimental data, such as those obtained from X-ray crystallography, NMR or IR spectra, depend heavily on the quality of the initial geometries of the molecules during the structure refinement procedure. Furthermore, quantum mechanical or molecular mechanical calculations need at least a crude 3D molecular model as starting geometry.

### 2.9.4
### Automatic 3D Structure Generation

This section describes briefly some of the basic concepts and methods of automatic 3D model builders. However, interested readers are referred to Chapter II, Section 7.1 in the Handbook, where a more detailed description of the approaches to automatic 3D structure generation and the developed program systems is given.

Automatic 3D structure generators can be regarded as automatic model building kits. They are comparable with mechanical molecular model building kits where the 3D structure is built manually using standard units for atom types, hybridization states, bond lengths, and bond angles. However, 3D structure generators build the spatial molecular geometries fully automatically, i.e., without any interaction by the user, and not in real space but as a computer model. In general, there are three different categories of approaches to building 3D molecular models (see Figure 2-94).

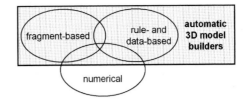

**Figure 2-94.** Classification of automatic 3D structure generators.

*Fragment-based methods* (also called template-based methods) use an incremental approach. First, they fragment the input structure implementing certain rules. Secondly, the entire 3D structure is assembled by linking appropriate predefined 3D structural fragments (3D templates) taken from a library. Although, fragment-based methods make extensive use of 3D structure information, they need at least a few explicit rules on the fragmentation of input structures, on finding the best matching analogs in the 3D template library, and on properly combining the individual templates with the entire 3D structure.

As the name already suggests, *rule-* and *data-based methods* use chemical knowledge that was obtained from theoretical investigations and experimental data on 3D structures. The results of these investigations and analyses were transformed into either explicit rules (e.g., preferred ring geometries) or implicit data (e.g., standard bond lengths). The rules and the data are implemented in the program system as a knowledge base which can be applied directly to the 3D structure generation process.

As already mentioned, *numerical methods* are not classified in the category of genuine 3D model builders. Numerical methods such as quantum mechanics (QM) and molecular mechanics (MM) apply computationally demanding mathematical optimization procedures to derive the 3D structure, and need at least a crude starting geometry (e.g., as a Z-matrix). Although, the distance geometry (DG) approach suggested and developed for 3D structure generation purposes by G. M. Crippen and T. F. Havel [114] has to be regarded as a numerical method, as it represents a special modeling technique (see also Chapter II, Section 7.2 in the Handbook). Distance geometry can generate starting coordinates for further optimization and requires less computation time than quantum or molecular mechanics (QM >> MM > DG). The most well-known "model builders" based on a distance geometry approach are DGEOM [115, 116] and the program system MOLGEO [117].

Figure 2-94 already implies that there are no sharp borders between the methods discussed above. Fragment-based methods need sets of rules on how to split the input structure into smaller subunits which are available in the 3D library and rule- and data-driven approaches use small, predefined templates (bond lengths or allowed ring geometries, at least) to build the spatial molecular model. In addition, both methods quite often apply simplified numerical optimization methods in order to refine the geometry of the models generated. It should also be mentioned that most of the programs developed in this area utilize the principles of conformational analysis (see Section 2.9.5) to identify a low-energy conformation of the molecule under consideration.

Regardless of which approach is realized, during the development of an automatic 3D structure generator several general problems have to be addressed. One major problem is the difference in conformational behavior of the cyclic and the acyclic portions of a molecule. Therefore, most 3D model builders treat rings and chains separately. Because of the ring closure condition, the number of degrees of freedom is rather restricted for ring systems compared with the open-chain portions. This geometrical constraint has to be taken into account in the 3D structure generation process. A method frequently applied to tackle this problem is to define allowed ring geometries (ring templates). These templates ensure a precise ring closure and can be chosen so that they represent a low-energy conformation for each ring size (e.g., the chair form of cyclohexane). A quite different situation is encountered for chain structures and substructures. The number of degrees of freedom, and thus the number of possible conformations, dramatically increase with the number of rotatable bonds. But which of all these conformations is the preferred one? One approach is to stretch the main chains as much as possible by setting the torsion angles to *trans* configurations, unless a *cis* double bond is specified (principle of the longest pathways, i.e., stretching the main chains as long as possible; see Figure 2-95). This method also effectively minimizes nonbonding interactions. Finally, the complete 3D model, i.e., after the cyclic and acyclic portions have been reassembled, has to be checked for steric crowding or atom overlap, and a mechanism should be implemented to eliminate such situations.

One of the major applications of automatic 3D structure generators is the 2D-to-3D conversion of large databases of molecules and compound collections in the chemical and pharmaceutical industry and in academia. Quite often, these databases, either from in-house or from chemical suppliers, contain millions of structures and their conversion requires large computational resources. Therefore, during the design and the development of a 3D model builder several criteria, such as robustness, conversion times and rates, processing of large files, and handling of a broad variety of chemical and structural types, should be taken into account and carefully addressed.

The program system COBRA [118, 119] can be regarded as a rule- and data-based approach, but also applies the principles of fragment-based (or template-based) methods extensively (for a detailed description see Chapter II, Sections 7.1 and 7.2 in the Handbook). COBRA uses a library of predefined, optimized 3D molecular fragments which have been derived from crystal structures and force-field calculations. Each fragment contains some additional information on

**Figure 2-95.** The principle of longest pathways for acyclic fragments and molecules.

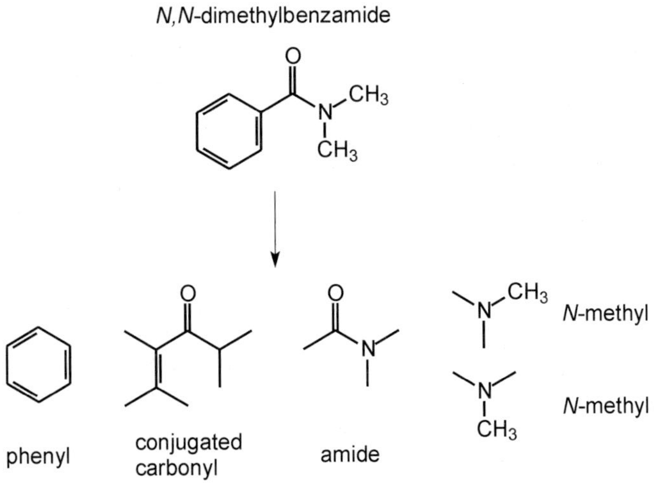

N,N-dimethylbenzamide

**Figure 2-96.** Fragmentation of N,N-dimethylbenzamide in COBRA.

the energy content and flexibility. Furthermore, a set of rules determines the fragmentation of a 2D input structure (chemical graph), the retrieval of the best matching fragments from the 3D template library, and the combination of the fragments to the entire molecular geometry in 3D space. In the first step, COBRA fragments the molecule into smaller subunits (conformational units). Figure 2-96 shows the fragmention of N,N-dimethylbenzamide. Neighboring fragments always have to have overlapping atoms.

In the second step, a 3D template taken from the library is assigned to each conformational unit. If no 3D template can be found for a specific fragment, the program is able to generalize and to search for similar fragments. If more than one template matches with the same subunit (e.g., different conformations), an internal symbolic representation of the molecule under consideration and a directed search technique, the so-called A* algorithm [120], are able to suggest a set of combinations of the templates which may lead to low-energy conformations of the entire molecule. Finally, predefined rules on combinations of templates which are known as unfavorable or impossible (in terms of geometric fit or steric crowding) are used to rank the remaining suggestions. In addition, in this step COBRA is able to derive and to store its own rules, i.e., the system is able "to learn" and to include the self-learned knowledge in the 3D structure generation process. This outstanding feature of COBRA is achieved by using symbolic logic and by adapting techniques of artificial intelligence. These rules (predefined and self-learned) are then used to evaluate which of the different 3D templates found so far may be combined to build geometrically and chemically reasonable entire 3D models.

In the next step, the suggested models are translated into 3D space by subsequently combining the templates. Again, each model is assessed and ranked according to various structural criteria, such as the geometric fit of the 3D templates and non-bonding interactions (steric clashes). If none of the solutions fulfills the criteria according to a preset level, the less criticized 3D model is used for further refinement, unless an acceptable 3D structure is found.

A major advantage of COBRA is that it uses pre-optimized 3D fragments, and therefore generates high-quality molecular models. In addition, because of the internal symbolic representation of molecules, the 3D structure generation process performs quite fast compared with numerical methods. On the other hand, the results strongly depend on the 3D templates which are present in the library and whether appropriate geometries can be assigned to the fragments. Unfortunately, the development of COBRA was discontinued and the program is not available at the moment.

A quite well-known 3D structure generator is CORINA [121–124] (for a detailed description see Chapter II, Sections 7.1 and 7.2 in the Handbook). It was originally developed to model the influence of the spatial arrangement of the atoms of a molecule on its reactivity within the reaction prediction program EROS [125, 126] and matured through a series of versions. Since then, it has found a wide range of applications, from infrared spectra simulation [127] to drug design [128]. CORINA is a rule- and data-based program system. The generation of 3D molecular models in CORINA is based on a set of rules derived from experimental data and theoretical investigations, such as X-ray crystallography, force-field calculations, and geometric considerations. In addition, a condensed set of data on bond lengths, bond angles, and ring geometries is included. The rules and data involved in the build-up process have a broad range of validity to ensure that the program is applicable to a large variety of chemical structures; in fact it covers the entire range of organic chemistry and a large variety of organometallic compounds. Figure 2-97 shows the general principles of CORINA.

In the first step, bond lengths and bond angles are assigned to standard values depending on atom types, the atomic hybridization states, and the bond order of the atom pair under consideration. Since bond lengths and angles possess only one rigid minimum, these values are taken from a table parameterized for the entire periodic table. Atoms with up to six neighbors can be handled according to the Valence Shell Electron Pair Repulsion (VSEPR) model. In addition, stereochemical information given in the input file is considered. When stereo descriptors are lacking, reasonable assumptions are made.

In the next step, the molecule is fragmented into ring systems and acyclic parts. Ring systems are then separated into small and medium-sized rings with up to nine atoms and into large and flexible systems.

For small and medium-sized ring systems consisting of fewer than ten atoms, the number of reasonable conformations is rather limited. Thus, these systems are processed by using a table of allowed single-ring conformations (ring templates). The ring templates are stored as lists of torsional angles for each ring size and number of unsaturations in the ring, ordered by their conformational energy.

The number of possible conformations rises dramatically with increasing ring size. Therefore, large ring systems cannot be handled by the methods applied to small rings. However, in rigid polymacrocyclic structures for example, an overall

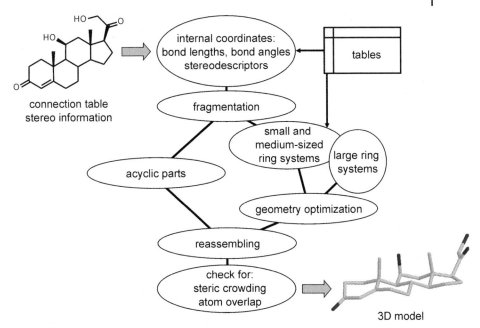

Figure 2-97. General principles of CORINA.

general outline can often be found, a so-called *superstructure*. The unsymmetric superphane molecule in Figure 2-98 shows a prism-like superstructure. This superstructure retains approximately the shape and symmetry of the complete system by reducing the original structure to the number of skeleton macrocycles and bridgehead atoms (anchor atoms).

The so-called *principle of superstructure* is implemented in CORINA for generating a 3D structure for rigid macrocyclic and polymacrocyclic systems. First, the ring system is reduced to its superstructure, preserving the essential topological features. As this superstructure contains only small rings, albeit with very long bonds, the algorithms described for small ring systems can be applied to generate a 3D model for the superstructure by using long (super-) bonds. Finally, the removed atoms are restored and a complete 3D model of the entire ring system is obtained.

After generating the entire ring system of the molecule, CORINA uses a reduced force field to optimize the ring geometries. Two simplifications lead to this so-

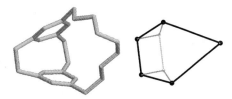

Figure 2-98. An unsymmetric superphane and its superstructure.

called *pseudo-force field*. First, ring systems are considered as being quite rigid. Thus, influences of torsional energies and steric influences of exocyclic substituents can be neglected. Only the ring skeleton needs to be optimized. Secondly, the major aim is to optimize geometries and not to calculate energies, i.e., no actual energy values need to be calculated. These assumptions result in a large reduction of the energy terms to be calculated, guaranteeing a fast convergence after a few iterations and short computation times during the optimization process.

For acyclic fragments and molecules, the principle of longest pathways has been implemented in CORINA (see Figure 2-95); i.e., since no *cis* configuration is specified, all torsions are set to *anti* in order to minimize steric interactions.

After the combination of the 3D fragments of the ring systems and of the acyclic parts, the complete 3D model is checked for overlap of atoms and for close contacts. If such situations are detected, CORINA performs a reduced conformational analysis to avoid these interactions (see Figure 2-99): first, a strategic rotatable bond within the pathway connecting the two interacting atoms is determined, depending on the topological features and double-bond character. Secondly, some rules of conformational preferences of torsional angles in open-chain portions are implemented to change the torsional angle of this bond, until the non-bonded interactions are eliminated.

By default, CORINA generates a single low-energy conformation. It outputs the 3D Cartesian coordinates of the atoms of the converted molecule and supports several standard file formats for structure information, such as MDL SDfile (and RDfile), SYBYL MOL and MOL2, or PDB file format (see also Table 2-6). It performs very fast, robustly, and with high conversion rates. For example, the conversion of the open part of the database of the National Cancer Institute (NCI, Developmental Therapeutics Program) [129] has been converted into 3D molecular models. This dataset contains approximately 250 000 structures and was processed by CORINA (version 2.6) within 2.5 h (0.04 s/molecule) on a Pentium III 1.6 GHz Linux workstation with a conversion rate of 99.5 % without a program crash or any intervention. It generates high-quality molecular models and has no limitations concerning the size of the molecule or the size of ring systems. Even large molecules, such as the fullerene dendrimer shown in Figure 2-100 with 762 non-hydrogen atoms, can be converted simply by starting from the connectivity information (CPU time: 1.5 min) [130].

A widely used 3D structure generator is CONCORD [131, 132] (for a more detailed description see Chapter II, Section 7.1 in the Handbook). CONCORD is also a rule- and data-based program system and uses a simplified force field for geometry optimization. CONCORD converts structures from 2D to 3D fairly fast

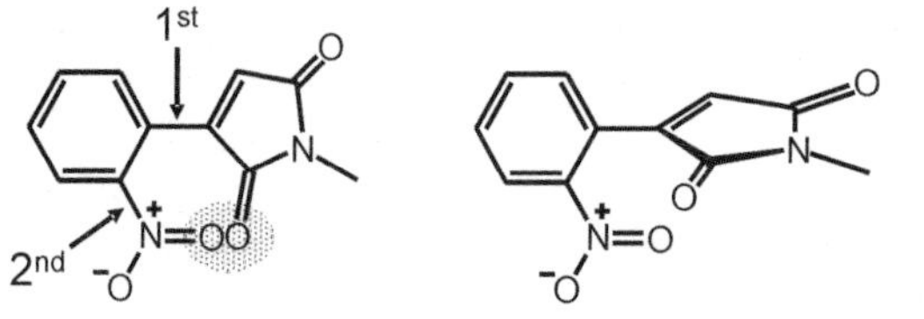

**Figure 2-99.** Elimination of non-bonded interactions (close contacts).

**Figure 2-100.** CORINA-generated
3D molecular model of a fullerene dendrimer
with 1278 atoms (762 non-hydrogen atoms).

(approximately 0.02 s/molecule on a common UNIX workstation for small and medium-sized molecules), but is restricted to a limited number of elements (H, C, N,
O, F, Si, P, S, Cl, Br, and I) and is able to process only molecules with up to 200
non-hydrogen atoms. In addition, the maximum connectivity (coordination number) of an atom is four.

2.9.5
**Obtaining an Ensemble of Conformations: What is Conformational Analysis?**

As mentioned above, most molecules can adopt more than one conformation, or
molecular geometry, simply by rotation around rotatable bonds. Thus, the different
conformations of a molecule can be regarded as different spatial arrangements of
the atoms, but with an identical constitution and configuration. They are interconvertible and mostly they cannot be isolated separately. Figure 2-101 shows a superimposition of a set of conformations of 2R-benzylsuccinate (cf. Figure 2-89).

**Figure 2-101.** Superimposition of a set of conformations of
2R-benzylsuccinate with the benzene ring fixed.

Each conformation corresponds to a single point on the potential energy surface of the molecule, which describes the dependence of the energy content on the conformational parameters of a certain geometry. Note that the energy content always depends on all $3N - 6$ internal coordinates of a molecule. Therefore, for most molecules the potential energy surface is a rather high-dimensional function, and it is sometimes referred to as a hypersurface. For example, the water molecule already contains three internal coordinates, so the energy is a function of three parameters and results in a four-dimensional potential surface.

The potential energy surface usually shows one global minimum, which corresponds to the lowest-energy conformation. In addition, various local minima, i.e., low-energy conformations can be observed. These minimum-energy structures are separated by so-called *rotational barriers*. Figure 2-102 illustrates this for the simple molecule *n*-butane. To visualize how the potential energy depends on the different geometries only one conformational parameter, the torsion angle $\tau$ between the carbon atoms C2 and C3, is taken into account. Thus, the potential energy $E$ is regarded as a function of only one variable ($E = f(\tau)$) and can be displayed as a 2D curve (in reality the system is of dimension 37).

For $\tau = 0°$, the terminal methyl groups exhibit the shortest spatial distance, which maximizes the non-bonded, repulsive interactions (torsion or Pitzer strain) between the groups. Thus, the conformation with the highest energy is found for $\tau = 0°$ (and 360°). By clockwise rotation, the first local minimum is reached at $\tau = 60°$ (*gauche* conformation). A rotational barrier of 10.9 kJ separates this geometry from the global minimum structure of butane at $\tau = 180°$, which is also known as the *trans* or *anti* conformation. Due to the symmetry of butane, a further local minimum is found at $\tau = 300°$ (*gauche*).

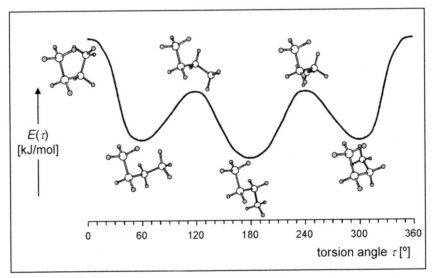

**Figure 2-102.** Dependence of the potential energy curve of *n*-butane on the torsion angle $\tau$ between carbon atoms C2 and C3.

Conformational analysis tries to model the relationship between the changes in the relative atomic positions by rotation around bonds and the influence on the properties of the molecule under consideration (such as energy or chemical reactivity). Thus, the generation and the study of the various geometries and their potential energy are of major interest in this field and led to the development of automatic conformer generators. These computer programs automatically generate sets or ensembles of conformations, starting from a given 3D molecular model. The total number of possible conformations of a molecule is called the conformational space. Because of the above-mentioned complexity and dimensionality of the potential energy surface, exploration of the entire conformational space is not feasible in most cases. Therefore, most computational methods which try to gain insight into the conformational behavior of molecules rely on conformational sampling, i.e., on a broad search in conformational space in order to find as many local minima (low-energy conformations) as possible and to identify the global minimum structure. Thus, an ensemble of conformations is obtained, ranked, e.g., by their internal energy.

## 2.9.6
### Automatic Generation of Ensembles of Conformations

Many approaches and methods for the generation of multiple conformations have been developed and published since the early 1980s. Below we describe briefly some of the basic concepts and methods of automatic conformer generation. However, interested readers are referred to Chapter II, Section 7.2 in the Handbook, where the approaches to automatic generation of ensembles of conformations and the program systems that have been developed are described in detail.

A general work flow scheme – an iterative process – for the generation of conformations is shown in Figure 2-103.

After an initial starting geometry has been generated and optimized (e.g., in a force field), the new conformation is compared with all the previously generated conformations, which are usually stored as a list of unique conformations. If a substantially different geometry is detected it is added to the list; otherwise, it is rejected. Then a new initial structure is generated for the next iteration. Finally, a preset stop criterion, e.g., that a given number of loops has been performed or that no new conformations can be found, terminates the procedure.

The major problems in the generation of conformations concern the coverage of conformational space, and the conformational diversity of the generated structures. The conformations generated should include all the relevant geometries, e.g., all the important low-energy conformations, including the global minimum; but a diverse and representative subset of all possible conformations resulting from a broad conformational sampling, should be obtained. In other words, the output structures should not be too geometrically similar, but should cover a broad region of the entire conformational space. In addition, the problems discussed in Section 2.9.4 on 3D structure generation also have to be addressed, and must be solved properly when developing an automatic conformer generator; examples

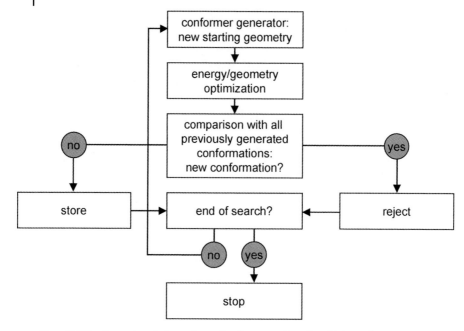

**Figure 2-103.** General work flow scheme for the generation of multiple conformations.

are the different conformational behavior of the cyclic and the acyclic portions of a molecule, or the check for and the elimination of close contacts or atom overlap. During the development of an automatic conformer generator most of the program intelligence is dedicated to the module which actually generates the different starting geometries (the topmost box in Figure 2-103). To generate a new conformation of the molecule under consideration, the spatial arrangement of the atoms relative to each other has to be changed while retaining the constitution and configuration: that is, the internal coordinates vary from one conformation to the next.

One of the oldest techniques applied to explore the conformational space of a molecule is *systematic generation* (sometimes also called systematic searches) [133]. As the name already implies, systematic methods change the coordinates (internal or Cartesian) in a predefined, regular, and stepwise manner. One of the simplest methods is the so-called *grid search*, which systematically changes the torsion angle of each rotatable bond of an input structure through 360° using a constant increment $n$ ($n$ = 2, 3, 4, ...). Figure 2-104 illustrates this procedure for $n$-butane by rotating the central bond between the carbon atoms 2 and 3 using an increment of $n$ = 6, i.e., by applying a grid of 60°. Starting from the lowest-energy conformation ($\tau$ = 180°; see Figure 2-102) the torsion angle is changed in steps of 60°.

The grid search technique can easily be applied to acyclic systems. Ring systems can be treated as "pseudo-acyclic" by cutting one ring bond. The major drawback of this technique is that, because of the restricted number of degrees of freedom in

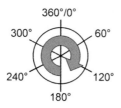

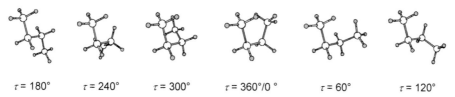

$\tau = 180°$      $\tau = 240°$      $\tau = 300°$      $\tau = 360°/0\ °$      $\tau = 60°$      $\tau = 120°$

**Figure 2-104.** Grid search using an increment of $n = 6(60°)$ to generate a set of conformations for *n*-butane.

cyclic systems, most of the geometries thus generated will not fulfill the ring closure condition. Therefore, each newly generated conformation has to be checked to see whether the two atoms which are connected by the cut bond are at a distance which is within the range at which a bond can be formed again, before it is submitted to the next step (e.g., geometry optimization). Another problem inherent in grid search techniques is that the total number of possible conformations, N, increases exponentially with the number of rotatable bonds, $k$ ($N = (360°/n)^k$). However, small and medium-sized molecules typically often have five or more rotatable bonds (not including rotatable bonds in ring systems). A simple system with five rotatable bond processed in a grid of 30° ($n = 12$) will already lead to a total number, N, of 248 832 conformations. All of these starting geometries have to be optimized and compared in the following, computationally demanding steps, which is an almost unfeasible task. However, if a coarser grid (e.g., 60°, $n = 6$, or 90°, $n = 4$) is applied, important geometries may be lost.

As already mentioned and shown in Figure 2-103, after a conformation is generated and geometry-optimized (e.g., in a force field) it has to be compared with all previously generated conformations. If the current conformation represents a totally new geometry, it is added to the list of unique conformations; otherwise it is discarded. A commonly used metric method to compare the geometry of conformations is to determine their minimum *RMS* (root mean square) deviation either in Cartesian space, $RMS_{XYZ}$ in [Å] (see Eq. (11)), or for internal coordinates, for example, in torsion angle space, $RMS_{TA}$ in [°]. In Eq. (11), N is the number of non-hydrogen atoms over which the $RMS_{XYZ}$ is calculated and $d_i$ is the distance between the Cartesian coordinates of the *i*th corresponding atom pair in the two conformations, when they are superimposed (see also Fig. 2-105).

$$RMS_{XYZ} = \sqrt{\frac{\sum_{i}^{N}(d_i)^2}{N}} \qquad (11)$$

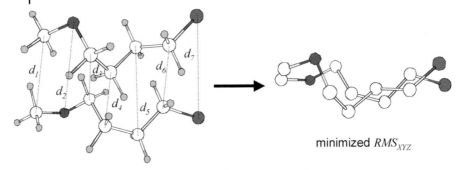

minimized $RMS_{XYZ}$

**Figure 2-105.** Derivation of the $RMS_{XYZ}$ deviation of two conformations.

Figure 2-105 illustrates the derivation of the $RMS_{XYZ}$ deviation of two conformations. All corresponding non-hydrogen atoms are superimposed and finally the minimized $RMS_{XYZ}$ deviation is obtained by translating and rotating both geometries against each other. Note that during the minimization step the internal coordinates of both conformations are kept rigid.

Usually, two conformations are regarded as geometrically different if their minimized $RMS$ deviation is equal to or larger than 0.3 Å in Cartesian space ($RMS_{XYZ}$), or 30° in torsion angle space ($RMS_{TA}$), respectively.

A technique used quite often to explore the conformational space of molecules is *random or stochastic generation* (sometimes also called random searches) [134, 135]. In contrast to systematic approaches, random methods generate conformational diversity, not in a predictable fashion, but randomly. To obtain a new starting geometry either the Cartesian coordinates are changed (e.g., by adding random numbers within a certain range to the $x$-, $y$-, and $z$-coordinates of the atoms) or the internal coordinates are varied (e.g., by assigning random values to the torsion angle of the rotors) in a random manner. Again, as discussed for the systematic techniques, ring portions can be treated as "pseudo-acyclic", including whether the ring closure condition is fulfilled. After the new conformation has been optimized and compared with all the previously generated conformations, it can be used as the starting point for the next iteration. In addition, a frequently used criterion for selecting a new starting geometry in random techniques is the so-called *Metropolis Monte Carlo* scheme [136]. Thereby, a newly generated (and optimized) conformation is used as a starting geometry for the next iteration only if it is lower in energy than the previous one or if it has a higher statistical probability (calculated by the Boltzmann factor of their energy difference). Otherwise, the previous structure is taken as the starting point. Thus, the selection of starting conformations is biased towards lower-energy structures, but also allows "jumps" into high-energy regions of the molecular hypersurface. Because of the random changes, stochastic methods are able to access a completely different region of the conformational space from one iteration step to the next. On the one hand, this ensures a broad sampling of the conformational space, but on the other, an artificial stop criterion has to be defined as random methods do not have a "natural" end point. Usually, random generations are stopped if the same conformation has been generated several times or if a user-defined number of iterations has been performed.

Other methods which are applied to conformational analysis and to generating multiple conformations and which can be regarded as random or stochastic techniques, since they explore the conformational space in a non-deterministic fashion, are genetic algorithms (GA) [137, 138] and simulation methods, such as molecular dynamics (MD) and Monte Carlo (MC) simulations [139], as well as simulated annealing [140]. All of these approaches and their application to generate ensembles of conformations are discussed in Chapter II, Section 7.2 in the Handbook.

*Rule- and data-based methods* can also be applied to generate ensembles of conformations. The program system COBRA has already been introduced in Section 2.9.4 of this chapter. In order to generate an ensemble of a molecule's "most different" conformations, the developers of COBRA combined the $A^*$ algorithm with a clustering technique [120]. The $A^*$ algorithm is able to identify reasonable combinations of the 3D templates which have been assigned to the conformational units of a molecule after the fragmentation. The clustering method is based on a pairwise comparison of the conformations in torsion angle space. The algorithm starts with the lowest-energy conformation obtained by the procedure as already described. Then, the geometry which is most different, i.e., has the greatest distance in torsion space from the first conformation, is stored as the second conformation. In the next step, the third conformation is determined by searching for the geometry furthest from the first two conformations. This procedure is repeated unless a user-defined number of structures has been generated or no 3D templates are left to combine.

The basic principles of the 3D structure generator CORINA have been presented in Section 2.9.4 of this chapter. It has also been mentioned that for small and medium-sized ring systems consisting of up to nine ring atoms, a table of allowed single-ring conformations, so-called *ring templates*, which have been derived from statistical and empirical data, is used to generate precise ring geometries. These ring templates are stored as lists of torsional angles depending on the number of unsaturated bonds in the ring. They are characterized and ordered by a strain energy value representing the conformational energy. Figure 2-106 gives an example of the conformations of a saturated six-membered ring (e.g., cyclohexane) and a six-membered ring with one double bond (e.g, cyclohexene) as they are stored in

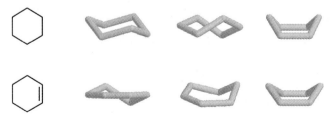

**Figure 2-106.** Ring templates for a saturated six-membered ring and a six-membered ring with one double bond as implemented in the ring conformation table of the 3D structure generator CORINA.

the ring conformation table. In the case of fused or bridged ring systems, all possible, but structurally reasonable combinations of the different ring templates are generated by checking the compatibility of the torsion angles of the bonds common to the two neighboring rings. The different ring geometries can be used either for a conformational analysis in order to find the lowest-energy conformation or to output a set of ring conformations.

Rule- and data-based approaches can also be applied to explore the conformational space of the acyclic parts of a molecule. The conformer generator ROTATE [141,142], which has been implemented in the group of the CORINA developers, as well as the conformational analysis package MIMUMBA [143], are using a set of rules and data which results from a statistical analysis of the conformational preferences of the open-chain portions in small molecule crystal structures (see also Chapter II, Section 7.2 in the Handbook). This knowledge is stored in the so-called *Torsion Angle Library* (TA Library) and has been derived from the Cambridge Structural Database (CSD) [144]. The TA Library contains over 900 entries of torsion angle fragments (torsion patterns) consisting of four atoms and their adjacent neighbors and a single bond in the center. For each pattern, the distributions of torsion angles $\tau$ as they have been observed in the crystal structures are stored as histograms. Figure 2-107 illustrates the derivation of the TA Library from the CSD.

The histograms contain the implicit information on the conformational behavior of molecules in a structured molecular environment with varying intermolecular directional forces and dielectric conditions in the different crystal packings. They are used to derive a set of preferred torsion angles for the rotatable bonds of the molecule under consideration in the following way. After all the rotatable bonds in an input structure have been identified, an appropriate torsion angle histogram from the TA Library is assigned to each rotor. The histograms are then transformed into empirical potential energy functions, simply by the assumption that torsion angle values which are sparsely or even not populated in the histogram refer to a high energy content of the fragment under consideration, and conversely, torsion angle values which are well populated correspond to a low energy content.

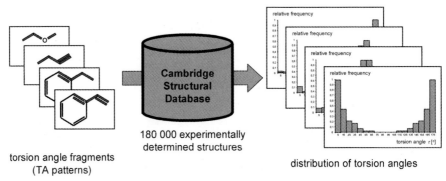

**Figure 2-107.** Derivation of the torsion angle library (TA Library).

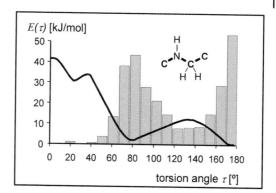

**Figure 2-108.** Derivation of a symbolic potential energy function from the torsion angle distribution of a torsion fragment.

Figure 2-108 shows the correspondence between a histogram and the derived empirical energy function for the torsion angle fragment C–N(H)–C(H)(H)–C.

These symbolic energy functions are used to select a set of preferred torsion angles for the rotatable bond under consideration (initial torsion angles) and all possible combinations of the initial torsion angles of all the rotors are generated. After a new conformation has been generated, each rotatable bond is geometry-optimized by applying the empirical energy function and is only accepted if no steric problems can be detected. In addition, duplicate conformations are rejected.

As the number of conformations increases exponentially with the number of rotatable bonds, for most molecules it is not feasible to take all possible conformations into account. However, a balanced sampling of the conformational space should be ensured if only subsets are being considered. In order to restrict the number of geometries output, while retaining a maximum of conformational diversity, ROTATE offers the possibility of classifying the remaining conformations, i.e., similar conformations can be combined into classes. The classification is based on the *RMS* deviation between the conformations, either in Cartesian ($RMS_{XYZ}$ in [Å]) or torsion space ($RMS_{TA}$ in [°]). The *RMS* threshold, which decides whether two conformations belong to the same class, is adjustable by the user and each class is finally represented by one conformation.

The distance geometry (DG) approach and its application to 3D structure generation have already been mentioned in Section 2.9.4 of this chapter. Distance geometry can also be applied to conformational analysis, since the algorithm is able to generate sets of different molecular geometries starting from a single, but very crude, spatial representation of a molecule (the so-called *distance matrix*; see also Chapter II, Section 7.2 in the Handbook). Furthermore, structural information which has been derived from experiments (e.g., interatomic distances obtained from 2D NMR spectroscopy) can be included in the coordinate generation process of the distance geometry approach. The resulting conformations will then fulfill the experimentally determined geometrical restrictions [145].

2.9.7
## Tutorial: 3D Structure Codes (PDB, STAR, CIF, mmCIF)

### 2.9.7.1 Introduction

In 1971 the Protein Data Bank – PDB [146] (see Section 5.8 for a complete story and description) – was established at Brookhaven National Laboratories – BNL – as an archive for biological macromolecular crystal structures. This database moved in 1998 to the Research Collaboratory for Structural Bioinformatics – RCSB. A key component in the creation of such a public archive of information was the development of a method for efficient and uniform capture and curation of the data [147]. The result of the effort was the PDB file format [53], which evolved over time through several different and non-uniform versions. Nevertheless, the PDB file format has become the standard representation for exchanging macromolecular information derived from X-ray diffraction and NMR studies, primarily for proteins and nucleic acids. In 1998 the database was moved to the Research Collaboratory for Structural Bioinformatics – RCSB.

However, as the years have gone by it has been recognized that the PDB format was unable to express adequately many different aspects of experimental data for macromolecules. Moreover, the PDB representation has a fixed format with many non-coherent rules and the order of data in the file is strongly defined. This makes PDB files complicated for parsing and manual handling (the latter is nearly impossible). And finally, changes in the PDB file format specification enforcet the reformatting of existing files to make them conform with the new format specification.

An alternative and much more flexible approach is represented by the STAR file format [148, 149], which can be used for building self-describing data files. Additionally, special dictionaries can be constructed, which specify more precisely the contents of the corresponding data files. The two most widely used such dictionaries (and file formats) are the CIF (Crystallographic Information File) file format [150] – the International Union of Crystallography's standard for representation of small molecules – and mmCIF [151], which is intended as a replacement for the PDB format for the representation of macromolecular structures.

This section provides only a very brief overview of the file formats introduced above. Taking into account that the mmCIF file format specifications alone already comprise about 1700 different entries [152], a detailed description of these file formats would occupy an entire book.

### 2.9.7.2 PDB File Format

#### 2.9.7.2.1 General Remarks
PDB files were designed for storage of crystal structures and related experimental information on biological macromolecules, primarily proteins, nucleic acids, and their complexes. Over the years the PDB file format was extended to handle results from other experimental (NMR, cryoelectron microscopy) and theoretical methods

of 3D structure determination also. PDB files contain first of all the Cartesian co-ordinates of the atoms forming the molecule, primary and secondary structures (in the case of proteins), bibliographic citations, and crystallographic structure factors as well as X-ray diffraction or NMR experimental data. PDB files can store data for a single molecule, one molecule with associated water molecules, multiple mole-cules, or an ensemble of alternative models for a single molecule (from NMR ex-periments). The current PDB file format version is 2.2 and its specification can be found in Ref. [53].

Each line of a PDB file must contain exactly 80 printable ASCII characters (selected from the lower and upper case Latin letters, the digits 0–9, and the symbols ' - = [ ] \ ; ', . / ~ ! @ # $ % ∧ & * ( ) _ + { } | : " < > ? as well as the space and end-of-line indicator. The end-of-line indicator is a system-specific line-feed character in Unix, a sequence of carriage return and line-feed characters in DOS and Windows, and a carriage return in MacOS. If the actual data occupy less than 80 characters, or columns using the punched card analogy, then spaces are appended up to the 80th column (inclusive). Each line is self-identifying – its first six characters contain a left-justified and blank-filled record name (key-word), which must be one of the predefined names (described in later sections). Columns 7–70 contain data (built from fields), and columns 71–80 are usually empty in older PDB files, but can contain other information in files conforming to newer PDB specifications. The format and contents of the fields within the re-cord are dependent on the keyword. Furthermore, all records in a PDB file must appear in a predefined order. This ordering is presented with an example later this tutorial.

This record/field terminology allows the treatment of a PDB file as an ordered collection of record types.

### 2.9.7.2.2 Types of Records

All the records defined for PDB files can be grouped into six categories on the basis of how many times a given record can appear in a PDB file and how many lines it may occupy.

### Single Record

Each record in this category can appear only once in a PDB file and it occupies ex-actly one line. Examples of such records are HEADER – the starting record of each PDB file discussed in detail below, END – the last (terminating) record, and CRYST1 – describing the crystallographic cell.

### Single Continued Record

This record can appear only once in a file and it may occupy more than one line. The second and subsequent lines contain a continuation field, which is a right-jus-tified integer followed by a blank character. This number is incremented by one for each additional line of the record. Several records belong to this category, e.g., AUTHOR (contains names of the people responsible for the contents of the file), KEYWDS (contains a list of keywords describing the macromolecule),

COMPND (describes the macromolecular contents of the entry), and EXPDTA (identifies the technique used to determine the 3D structure of the compound experimentally).

**Multiple Record**

Records in this category make multiple appearances as single lines without continuation and they are used to form lists (of atoms, hydrogen bonds, etc.) The two most frequently used records belonging to this category, namely ATOM and CONECT, are presented and discussed in more detail later.

**Multiple Continued Record**

Multiple continued records exist in a multiple manner in an entry and can occupy more than one line. The continuation field is similar to that of the single continued record category. To this category belong the HETATM record, storing atomic coordinates for atoms within "non-standard" groups (e.g., water molecules present in protein crystals, which are highly hydrated), as well as FORMUL and HETNAM, representing chemical formulas and names of the "non-standard" groups, respectively.

**Grouping Record**

Three record types are used to group other records: TER indicates the end of a chain, while MODEL/ENDMDL surround groups of ATOM, HETATM, TER and similar records.

**Other Records**

To this category belong only two record types: JRNL and REMARK, which in turn have their own detailed inner structures. Examples of the two records types are presented later.

### 2.9.7.2.3 Order of Records

All the records constituting a PDB file must appear in a strictly defined order, collected into sections. The order of the sections, together with a short description and sample record names, is presented in Table 2-7 (source: Ref. [53]).

### 2.9.7.2.4 Analysis of a Sample PDB File

Having looked at the general structure of PDB files, let us now examine a sample PDB file. The file represents the structure of α-conotoxin PNI1 polypeptide (PDB ID: 1pen) and was retrieved from the Protein Data Bank [53]. Figure 2-109 shows the 3D structure of the molecule.

The molecule is built up of 16 amino acids, but the file also contains the positions of the oxygen atoms of 12 water molecules contained within the unit cell. To keep the example simple, only the most important parts of the file are presented and discussed here. Each part of the file is annotated with corresponding row and column numbers. The complete file can be obtained from the PDB [53] or from this book's website [153].

**Table 2-7.** Ordered list of sections in PDB files (souce: Ref. [53]).

| No. | Section | Description | Sample records |
|-----|---------|-------------|----------------|
| 1 | Title | summary descriptive remarks | HEADER, TITLE, COMPND, SOURCE, KEYWDS, AUTHOR, JRNL |
| 2 | Remark | bibliography, refinement annotations | REMARKs 1, 2, 3 and others |
| 3 | Primary structure | peptide and/or nucleotide sequence and the relationship between the PDB sequence and that found in the sequence database(s) | SEQRES |
| 4 | Heterogen | description of non-standard groups | HET, HETNAM, FORMUL |
| 5 | Secondary structure | description of secondary structure | HELIX, SHEET, TURN |
| 6 | Connectivity | chemical connectivity annotations | SSBOND, LINK, HYDBND |
| 7 | Miscellaneous features | features within the molecule | SITE |
| 8 | Crystallographic | description of the crystallographic cell | CRYST1 |
| 9 | Coordinate transformation | coordinate transformation operators | ORIGXn, SCALEn, MTRIXn, TVECT |
| 10 | Coordinate | atomic coordinate data | MODEL, ATOM, SIGATOM |
| 11 | Connectivity | chemical connectivity | CONECT |
| 12 | Bookkeeping | summary information, end-of-file marker | MASTER, END |

The first line of the file (see Figure 2-110) – the HEADER record – holds the molecule's classification string (columns 11–50), the deposition date (the date when the data were received by the PDB) in columns 51–59, and the PDB IDcode for the molecule, which is unique within the Protein Data Bank, in columns 63–66. The second line – the TITLE record – contains the title of the experiment or the analysis that is represented in the entry. The subsequent records contain a more detailed description of the macromolecular content of the entry (COMPND), the biological and/or chemical source of each biological molecule in the entry (SOURCE), a set of keywords relevant to the entry (KEYWDS), information about the experiment (EXPDTA), a list of people responsible for the contents of this entry (AUTHOR), a history of modifications made to this entry since its release (REVDAT), and finally the primary literature citation that describes the experiment which resulted in the deposited dataset (JRNL).

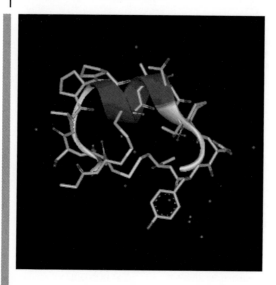

**Figure 2-109.** 3D molecular structure of α-conotoxin PNI1 polypeptide (PDB ID: 1pen).

```
            1         2         3         4         5         6         7         8
   12345678901234567890123456789012345678901234567890123456789012345678901234567890
 1 HEADER    NEUROTOXIN                                    29-JAN-96   1PEN
 2 TITLE     ALPHA-CONOTOXIN PNI1
 3 COMPND    MOL_ID: 1;
 4 COMPND    2 MOLECULE: ALPHA-CONOTOXIN PNIA;
 5 COMPND    3 CHAIN: NULL;
 6 COMPND    4 ENGINEERED: YES
 7 SOURCE    MOL_ID: 1;
 8 SOURCE    2 SYNTHETIC: YES;
 9 SOURCE    3 ORGANISM SCIENTIFIC: CONUS PENNACEUS
10 KEYWDS    NEUROTOXIN, ACETYLCHOLINE RECEPTOR, POSTSYNAPTIC,
11 KEYWDS    ANTAGONIST, ACETYLCHOLINE RECEPTOR INHIBITOR
12 EXPDTA    X-RAY DIFFRACTION
13 AUTHOR    S.-H.HU,J.GEHRMANN,L.W.GUDDAT,P.F.ALEWOOD,D.J.CRAIK,
14 AUTHOR    2 J.L.MARTIN
15 REVDAT    1   21-APR-97 1PEN      0
16 JRNL         AUTH   S.H.HU,J.GEHRMANN,L.W.GUDDAT,P.F.ALEWOOD,
17 JRNL         AUTH 2 D.J.CRAIK,J.L.MARTIN
18 JRNL         TITL   THE 1.1 A CRYSTAL STRUCTURE OF THE NEURONAL
19 JRNL         TITL 2 ACETYLCHOLINE RECEPTOR ANTAGONIST, ALPHA-CONOTOXIN
20 JRNL         TITL 3 PNIA FROM CONUS PENNACEUS
21 JRNL         REF    STRUCTURE (LONDON)          V.   4   417 1996
22 JRNL         REFN   ASTM STRUE6 UK ISSN 0969-2126              2005
```

**Figure 2-110.** Title section of the analyzed PDB file.

Records forming the remark section (Figure 2-111) have their own complicated syntax. For their detailed description readers are referred to Ref. [53]. Note that numbers appearing after the REMARK keyword are not continuation numbers, but integral parts of the keyword. "REMARK 1" lists important publications related to the described structure, "REMARK 2" specifies the highest resolution, in Ångström, that was used for building the model. The content of the "REMARK 3" record depends on the experiment type. In case of X-ray diffraction it has a well-defined format and contains information on the refinement program(s) (solving the electron density map) and the related statistics. For non-diffractional studies, the record usually contains a free text description of any refinement procedure, if any.

```
        1         2         3         4         5         6         7         8
     12345678901234567890123456789012345678901234567890123456789012345678901234567890
23   REMARK   1
24   REMARK   1 REFERENCE 1
25   REMARK   1  AUTH   R.MILLER,S.M.GALLO,H.G.KHALAK,C.M.WEEKS
26   REMARK   1  TITL   SNB: CRYSTAL STRUCTURE DETERMINATION VIA
27   REMARK   1  TITL 2 SHAKE-AND-BAKE
28   REMARK   1  REF    J.APPL.CRYSTALLOGR.              V.  27   613 1994
29   REMARK   1  REFN   ASTM JACGAR  DK ISSN 0021-8898                  0228
...
38   REMARK   2 RESOLUTION. 1.1  ANGSTROMS.
39   REMARK   3
40   REMARK   3 REFINEMENT.
41   REMARK   3   PROGRAM      : X-PLOR 3.1
42   REMARK   3   AUTHORS      : BRUNGER
43   REMARK   3
44   REMARK   3  DATA USED IN REFINEMENT.
45   REMARK   3   RESOLUTION RANGE HIGH (ANGSTROMS) : 1.1
46   REMARK   3   RESOLUTION RANGE LOW  (ANGSTROMS) : 6.1
47   REMARK   4
48   REMARK   4 1PEN COMPLIES WITH FORMAT V. 2.2, 16-DEC-1996
...
213  REMARK 999 REFERENCE: FAINZILBER, M., HASSON, A., OREN, R.,
214  REMARK 999 BURLINGAME, A.L., GORDON, D., SPIRA, W.E., ZLOTKIN, E.
215  REMARK 999 (1994). NEW MOLLUSC-SPECIFIC A-CONOTOXIN BLOCK APLYSIA
216  REMARK 999 NEURONAL ACETYLCHOLINE RECEPTORS. BIOCHEMISTRY, 30,
217  REMARK 999 9370-9 377.
...
```

**Figure 2-111.** Remark section of the analyzed PDB file.

The REMARK 4–999 records contain other optional, but predefined, remarks. As shown above, the "REMARK 4" record specifies that the analyzed file conform fully with the current (December 2002) version 2.2 of the PDB file format specification.

The SEQRES records (Figure 2-112) contain the amino or nucleic acid sequence of residues in each chain of the macromolecules being studied. The number directly following the record name is the serial number of the SEQRES record; next, the number of residues in the chain is specified (17 in this case) and the corresponding residues are listed. Residues occur in order, starting from the N-terminal residue for proteins and the 5'-terminus for nucleic acids. The last residue is a non-standard group ($NH_2$) and is described precisely in subsequent lines. Note that water molecules, whose coordinates will be specified later, are also mentioned in one of the FORMUL records.

The next important and mandatory part of every PDB file is a set of seven records forming the crystallographic and coordinate transformation sections (Figure 2-113): CRYST1, ORIGX$n$ and SCALE$n$ ($n$ = 1,2,3). CRYST1 specifies the unit cell parameters, space group, and $Z$ value. ORIGX$n$ and SCALE$n$ define

```
        1         2         3         4         5         6         7         8
     12345678901234567890123456789012345678901234567890123456789012345678901234567890
219  SEQRES   1     17  GLY CYS CYS SER LEU PRO PRO CYS ALA ALA ASN ASN PRO
220  SEQRES   2     17  ASP TYR CYS NH2
221  HET    NH2     17       1
222  HETNAM     NH2 AMINO GROUP
223  FORMUL   1  NH2      H2 N1
224  FORMUL   2  HOH   *12(H2 O1)
...
```

**Figure 2-112.** Primary structure and heterogen sections of the analyzed PDB file.

|     |        | 1           | 2           | 3           | 4           | 5           | 6           | 7           | 8          |
|-----|--------|-------------|-------------|-------------|-------------|-------------|-------------|-------------|------------|
|     |        | 12345678901234567890123456789012345678901234567890123456789012345678901234567890 |
| 229 | CRYST1 | 15.000    19.800    16.500   90.00 113.40   90.00 P 1 21 1          2 |
| 230 | ORIGX1 | 1.000000  0.000000  0.000000          0.00000 |
| 231 | ORIGX2 | 0.000000  1.000000  0.000000          0.00000 |
| 232 | ORIGX3 | 0.000000  0.000000  1.000000          0.00000 |
| 233 | SCALE1 | 0.066667 0.000000 0.028849     0.00000 |
| 234 | SCALE2 | 0.000000 0.050505 0.000000     0.00000 |
| 235 | SCALE3 | 0.000000 0.000000 0.066037     0.00000 |

**Figure 2-113.** Crystallographic and coordinate transformation sections of the analyzed PDB file.

coordinate transformation operators from orthogonal coordinates to the ones submitted and to fractional crystallographic coordinates, respectively.

The description of the crystallographic unit cell is followed by probably the most important section of the file – the Cartesian coordinates of the atoms.

The atomic coordinate data section shown in Figure 2-114 is constructed mostly from the ATOM records, whose format is described in Table 2-8.

The careful reader can recognize that there are no hydrogen atoms specified in the file presented. This has an obvious consequence: most (more than 80 %) of the macromolecular structure entries in the Protein Data Bank are determined by X-ray crystal diffraction studies, which usually cannot resolve the positions of hydrogen atoms. Another problem arising in X-ray diffraction experiments is that of reliably distinguishing nitrogen from oxygen from carbon atoms [53]. However, some newer X-ray crystal diffraction PDB files contain hydrogen positions, which were computed by theoretical modeling.

The temperature factor (together with the Cartesian coordinates) is the result of the refinement procedure as specified by the REMARK 3 record. High values of the temperature factor suggest either disorder (the corresponding atom occupied different positions in different molecules in the crystal) or thermal motion (vibration). Many visualization programs (e.g., RasMol [154] and Chime [155]) have a special color scheme designated to show this property.

As mentioned before, the TER record terminates the chain.

The HETATM records (Figure 2-115) represent atomic coordinates for atoms within "non-standard" groups (water molecules and atoms presented in HET

|     |        | 1           | 2           | 3           | 4           | 5           | 6           | 7           | 8          |
|-----|--------|-------------|-------------|-------------|-------------|-------------|-------------|-------------|------------|
|     |        | 12345678901234567890123456789012345678901234567890123456789012345678901234567890 |
| 236 | ATOM   | 1   N   GLY   1       -4.788  -8.935   3.453  1.00 11.53       N |
| 237 | ATOM   | 2   CA  GLY   1       -4.218 -10.294   3.312  1.00  9.54       C |
| 238 | ATOM   | 3   C   GLY   1       -3.815 -10.534   1.870  1.00  8.53       C |
| 239 | ATOM   | 4   O   GLY   1       -4.276  -9.836   0.965  1.00  7.01       O |
| ... |        |             |             |             |             |             |             |             |            |
| 339 | ATOM   | 104 N   CYS  16      -4.268 -18.747  -3.228  1.00  4.90       N |
| 340 | ATOM   | 105 CA  CYS  16      -5.412 -17.886  -3.502  1.00  6.31       C |
| 341 | ATOM   | 106 C   CYS  16      -6.399 -18.412  -4.535  1.00  6.50       C |
| 342 | ATOM   | 107 O   CYS  16      -7.570 -17.975  -4.499  1.00  8.67       O |
| 343 | ATOM   | 108 CB  CYS  16      -4.925 -16.490  -3.891  1.00  5.14       C |
| 344 | ATOM   | 109 SG  CYS  16      -4.058 -15.631  -2.538  1.00  6.52       S |
| 345 | HETATM | 110 N   NH2  17      -5.978 -19.313  -5.402  1.00  5.72       N |
| 346 | TER    | 111     NH2  17 |

**Figure 2-114.** Atomic coordinate data section of the analyzed PDB file.

**Table 2-8.** Format of the ATOM record.

| Columns | Definition |
| --- | --- |
| 1–6 | Record name ("ATOM"). |
| 7–11 | Atom serial number. The atoms are ordered in a standard manner, starting from the backbone (N–C–C–O for proteins) and proceeding in increasing distance (remoteness) from the $a$-carbon, along the side chain. |
| 13–16 | Atom name: the first two characters specify the chemical symbol (except for hydrogen atoms, for which a different naming convention is applied), the next character (if specified) defines a remoteness code described normally by Greek letters (A stands for alpha, B for beta, G for gamma etc.). The fourth character can contain a numeric branch designator and is usually empty. |
| 18–20 | Residue name – one of the standard amino acids (three-letter abbreviations are used), nucleic acids (one- or two-letter abbreviations), or the non-standard group designation as defined in the HET dictionary. |
| 23–26 | Residue sequence number. |
| 31–38 | Orthogonal coordinates for $X$ in Ångström. |
| 39–46 | Orthogonal coordinates for $Y$ in Ångström. |
| 47–54 | Orthogonal coordinates for $Z$ in Ångström. |
| 55–60 | Occupancy. |
| 61–66 | Temperature factor. |
| 77–78 | Element symbol, right-justified. |
| 79–80 | Charge on the atom if non-zero, otherwise blank. |

```
             1         2         3         4         5         6         7         8
    12345678901234567890123456789012345678901234567890123456789012345678901234567890
347 HETATM 112  O    HOH    17      -5.717 -22.380  -0.096  1.00 14.35           O
348 HETATM 113  O    HOH    18       3.723 -23.965  -3.804  1.00 11.74           O
...
363 CONECT 109   16  108
364 CONECT 110  106
365 MASTER       195    0    1    1    0    0    0    6  122    1    6    2
366 END
...
```

**Figure 2-115.** Last lines of the analyzed PDB file.

groups) and have a format similar to the ATOM records. The CONECT records, which come next, specify connections between atoms that are not covered by the standard residue connectivity in chains. Line 363 specifies a disulfide bridge between two cysteines, while the record in line 364 specifies a connection between the last amino acid in the chain (cysteine) and the terminating "non-standard" amino group. The MASTER record is a special record for controlling the integrity of the file. It specifies the counts of the most significant records such as REMARK: 195; atom coordinate records (ATOM and HETATM): 122, etc. The last (366th) line, containing the mandatory END record, terminates the PDB file.

As this short example shows, PDB files use different syntax for different records and both writing and reading such files require much effort. Another problem is the extensibility of this format to handle new kinds of information, which further complicates the file structure. The Protein Data Bank has been faced with the consequences – the existing legacy data comply with several different PDB formats, so they are not uniform and they are more difficult to handle [145, 155, 157]. As mentioned in Section 2.9.7.1, there is a much more flexible and general way of representing molecular structure codes and associated information – the STAR file format and the file formats based on it.

### 2.9.7.3  STAR File Format and Dictionaries

The Self-defining Text Archive and Retrieval (STAR) file format addresses primarily the problem of the inflexibility of the PDB file format, its fixed sets of allowable fields, and their strong dependence on order. To overcome the problems described, both the data structure and the actual data items within a STAR file are self-defined, which means that they are preceded by corresponding names (labels) which identify and describe the data. The data may be of any type and there is no predefined order of the data. STAR files, in contrast to PDB files, are easy to read and write manually. The whole syntax of STAR files is very simple and is defined by only a few rules:

1. A STAR file contains lines of standard visible ASCII characters.
2. Each file is a sequence of *data blocks* containing individual data items. Data blocks represent logical grouping of data that are related in some way.
3. Each *data item* is preceded by a corresponding *data name* – the label that identifies the data. A data name is a string that starts with the underscore character ("_").
4. There is a special keyword – "loop_" – which enables repetition of data. This keyword is used mainly when there are several data items with the same type of content (e.g., description of the atoms within a molecule).
5. Data formats based on the STAR file with their own dictionaries can further restrict the syntax of files to conform to their definition.
6. The set of possible data names can be restricted by a *dictionary*, which specifies which names can be used within a file to conform to the specified dictionary. Dictionaries can also be used to specify which data items will be processed by some software. In this case entries not defined by an appropriate dictionary will simply be ignored. Dictionaries are defined in so-called *dictionary definition language* (DDL). Several commonly used dictionaries are shown in Table 2-9. Two of them (CIF and mmCIF) are discussed below in greater detail.

There are also a few additional, but less important, syntax rules proposed in Ref. [148]. Users interested in a detailed specification of the STAR file format should make themselves familiar with the definitive STAR file written specification [149].

**Table 2-9.** Sample STAR-based dictionaries.

| Name | Description |
| --- | --- |
| Core CIF | Used for archiving and exchanging raw and processed data and derived structural results for single-crystal small-molecule and inorganic crystal studies. |
| | *http://www.iucr.org/iucr-top/cif/cif_core/* |
| ImgCIF/CBF | Efficient storage of 2D area detector data and other large datasets. |
| | *http://ndbserver.rutgers.edu/mmcif/cbf/* |
| mmCIF | Macromolecular crystallographic data. |
| | *http://ndbserver.rutgers.edu/mmcif/* |
| Powder CIF | Extends the core CIF dictionary by adding details of the powder diffraction experiments on single-crystal structures, using conventional X-ray diffractometers as well as synchrotron, CW neutron, TOF neutron, and energy-dispersive X-ray instruments. |
| | *http://www.iucr.org/iucr-top/cif/pd/* |
| Modulated structures CIF | Supplement to the core CIF dictionary designed to permit the description of incommensurately modulated crystal structures [157]. |
| | *http://www.iucr.org/iucr-top/cif/ms/* |
| NMR-STAR | Results of macromolecular NMR measurements. This file format is used to deposit data in BMRB (BioMagResBank – a repository for data from NMR spectroscopy of proteins, peptides, and nucleic acids. |
| | *http://www.bmrb.wisc.edu/elec_dep/Forms/complete_form_v21.txt* |
| MDB | Dictionary defining data format and structure for 3D models of biological molecules stored in the MDB database [158]. |
| | *http://www.gwer.ch/proteinstructure/mdb/* |

Readers familiar with XML (see also Chapter IV, Section 4 in the Handbook) will find a direct analogy with this extensible meta-language. Simply speaking, the STAR language, like XML, provides a generic syntax for data files, and DDL may restrict content of a STAR file in the same way as the XML Schema restricts contents of XML-based file formats. One could probably expect an analogy between DDL and the DTD (Document Type Definition), but the DTD syntax is completely different from the XML syntax, while Dictionary Definition Language uses the STAR file syntax, in the same way as the XML Schema is defined in XML.

### 2.9.7.4 CIF File Format (CCDC)

The Crystallographic Information File (CIF) is the standard archive file for crystallography of small molecules, and is recommended and supported by the International Union of Crystallography [55, 150]. Although in the late 1970s the IUCr Commissions on Crystallographic Data and Crystallographic Computing had already developed and promoted another "standard" file format – the Standard Crystallographic File Structure (SCFS) [160, 161] – this file format had drawbacks and constraints similar to those of the PDB file format, e.g., it required a predefined order of data and its data records had a fixed format which had to be defined for every data type.

The main requirement for the CIF file format was then its desired portability defined at three levels [55]: between machines; between similar applications; and across diverse applications. Portability across machine architectures usually excludes usage of binary data file formats in favor of plain text files. Although it is possible to define and implement binary data formats that can be read and written on computers with different architectures, typical binary data are stored in a compact form which is dependent on the computer architecture (mostly the word size and byte ordering). However, a plain text file is not in a universal format either. Such files can differ by character widths (e.g., 7-, 8-, 16-bit), coding schemes (e.g., ASCII and EBCDIC), end-of-line delimiters, and end-of-file markers. Nevertheless, most operating systems support reliable interconversions between text files from different sources, making plain text files highly portable between systems with different architectures.

Portability between similar applications requires that a CIF file supplies the same information to different crystallographic programs. This is achieved by storing in a CIF file not only the actual data but also keys (STAR data names), which describe uniquely the meaning of each datum. Having defined the CIF dictionary, crystallographic programs can easily interpret data in read files. For example, the Editorial Office of the *Acta Crystallographica* journal stores submitted papers with structural data in the CIF format and passes these files to miscellaneous crystallographic programs for checking the accuracy and consistency of the data. Moreover, there are libraries (enumerated in the latter part of this tutorial) for parsing CIF or other STAR-based file formats, which simplify the adoption of the CIF file format for existing or new software. And finally, portability between different applications allows further manipulation of the information stored in a CIF file. The data contained in a CIF file can be used for other calculations or simply stored in a database. Mapping between fields specified by the CIF dictionary and database tables is straightforward and has already been implemented in the CIFER [162] program, which has been used for years by the Cambridge Crystallographic Data Centre. Another interesting example of further processing of CIF files is the *CIFTEX* program [163], which translates a CIF into a TeX file [164], which is then used for-high quality typesetting. Thus CIF data files containing data for publication can be translated automatically into papers, without manual intervention.

The CIF format not only defines its dictionary of available data names, but restricts the STAR file syntax as well. For example, lines may not exceed 80 characters and data names can contain a maximum of 32 characters. Furthermore, the CIF dictionary specifies default units for numeric fields that represent numeric values with units. If the data item is not stored in the default units, then the units code is appended to the data name. Thus _cell_length_a and _cell_length_a_pm represent the dimensions of the unit cell in Ångström (the default units conforming to the SI Standard) and picometers, respectively.

The CIF file format was quickly and widely adopted by the scientific community for at least two reasons [165]: it was, and still is, endorsed by the IUCr; and submission of data to the journal *Acta Crystallographica, Section C* in a form conforming to CIF assures faster processing and hence faster publication of accepted papers. The current CIF file dictionary defines about 1200 data names, but it is still unable to represent all the details of the crystallographic measurements of macromolecules. Thus, yet another STAR-based data format is needed.

### 2.9.7.5 mmCIF File Format

However, instead of using yet another file format, it was decided that the Crystallographic Information File (CIF) dictionary would be extended to include data items relevant to macromolecular crystallographic experiments. The whole project, although initially appearing to be a small task, took seven years' work by a dedicated group [166]. The result of this effort under the auspices of the International Union of Crystallography (IUCr) was the macromolecular Crystallographic Information File (mmCIF) [151, 152]. Version 1.0 of the dictionary comprised 1700 terms and was ratified by an IUCr committee (COMCIFS). Furthermore, it was also recognized that the Dictionary Definition Language (DLL) used was too informal and it led to many ambiguities. Hence, the new version (2.0) of DDL was developed [167]. The mmCIF file format was quickly adopted by the RCSB for internal representation of structural data. However, for reasons of backward compatibility and because users are most familiar with PDB, the RCSB still distributes data using the PDB v. 2.2 format. The PDB files are produced from their corresponding mmCIF representations. It should be also noted that the informality of the PDB format restricts the possibility of conversion in the opposite direction, from PDB to mmCIF. The mmCIF website [152] is a definitive resource for a full specification of the Dictionary Definition Language, the mmCIF dictionary, and related information.

### 2.9.7.6 Software

The PDB, CIF, and mmCIF file formats are strongly supported by the existence of a large number of miscellaneous software libraries and programs that simplify their conversions, parsing, generation, verification, manipulation, and visualization. There are numerous libraries for the different programming languages: C [152, 168], C++ [168], Fortran [55, 169], Java [170], Perl [171], and even Objective C [172], which build up a common framework for most of the programs using PDB/mmCIF file formats.

In order to simplify transition from PDB to mmCIF some special tools have been developed. The CIFTr [168] and CIF2PDB [173] programs, for example, transform mmCIF files to PDB, while PDB2CIF [173] performs the reverse operation. PDB/ mmCIF files can also be converted to other data formats suitable for presentation of data. Except for the CIFTEX program [163] mentioned earlier, there are also various other interesting converters. PDB2VRML [174] builds VRML scenes on the basis of given PDB files. Similar functionality (but generating VRML files conforming to the newer versions of the file format) is offered by one of the services available at the Computer Chemistry Center website [175]. Another web service is PDB2MGIF [176], which converts submitted PDB files into animated GIF images containing molecular structures. The images generated can be next used within web pages. mmCIF data files can also be loaded directly into relational databases or translated into XML files by using the mmCIF loader [168].

Molecular structure visualization programs constitute the largest group of applications supporting the PDB/mmCIF format. The most popular are probably RasMol [154], Swiss PDB Viewer [177], PyMOL [178], ViewerLite/Pro from Accelrys [179], and many others. For web-based visualization the standard tool is Chime [155] from MDL. Another interesting example of a web viewer is canDo Shockware 3D PDB Viewer [179], which uses the Macromedia Shockwave Player plug-in [181]. There are also several applets available on the Internet. The MIME media types for PDB and CIF files are chemical/x-pdb and chemical/x-cif, respectively. On the RCSB [53] software web page one can find a list of most of the existing applications supporting the file forms discussed here.

## 2.10
## Molecular Surfaces

Two-dimensional structure diagrams and 3D molecular structures can form the basis for describing many chemical and physical properties of compounds. But all the models used so far represent only the 3D skeleton of a molecule and not the actual space requirements. In analogy to the human body, which has a skeleton and a surrounding body with a limiting surface (the skin), molecules can be seen as objects with a molecular surface. This surface separates the 3D space in an inner part of the volume filled by the molecule, and an outside part (the rest of the universe). But this picture of an exact separation through a discrete surface is only an approximation. Since molecules cannot be treated with the laws of classical mechanics, the concept of surfaces is only an analogy to macroscopic objects. In a quantum mechanical sense molecules have neither a body, nor a fixed surface. Their ingredients are atoms with nuclei made up of protons and neutrons, which are surrounded by electrons. The space that these electrons occupy is not bounded by a surface, but can be characterized by an "electron cloud" distribution. The electron density is continuous and approaches zero value at large distances from the nuclei (see Section 7.2). In particular, the distribution of electrons is significant for molecular interactions and determines the properties of the molecule. The molecular surface can express these different properties, such as electrostatic potential, atomic charges, or hydrophobicity, using colored mapping (see Section 2.11). Therefore, the spatial figure, or envelope, of the molecule has to be considered; this can be determined by various methods. Molecular surfaces can be obtained *de novo* by mathematical calculation methods, but also from experiments. Three-dimensional structure analyses such as 2D NMR or X-ray crystallography give an impression of the spatial requirements of the molecule. If only

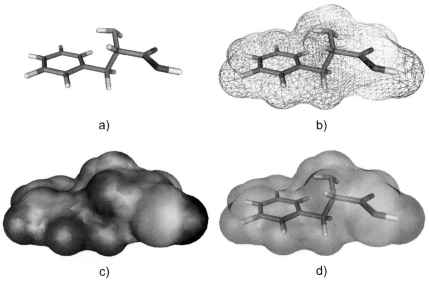

a)          b)

c)          d)

**Figure 2-116.** Graphical representations of molecular surfaces of phenylalanine: a) dots; b) mesh or chicken-wire; c) solid; d) semi-transparent.

2D information is available, e.g., in databases without experimental data, the different types of surfaces (see below) can be calculated only after a 3D structure has been determined by a 3D structure generator, which might be followed by computational refinement, e.g., with a force-field calculation.

Depending on the application, models of molecular surfaces are used to express molecular orbitals, electronic densities, van der Waals radii, or other forms of display. An important definition of a molecular surface was laid down by Richards [182] with the solvent-accessible envelope. Normally the representation is a cloud of points, reticules (meshes or chicken-wire), or solid envelopes. The transparency of solid surfaces may also be indicated (Figure 2-116).

The following models describe those definitions of molecular surfaces that are most widely used. The van der Waals surface, the solvent-accessible surface, and the Connolly surface (see below) based on Richards' definitions play a major role [182].

The interpretation of molecular surfaces is particularly important wherever molecular interactions, reactions, and properties play a dominant role, such as in drug design or in docking experiments.

2.10.1
**van der Waals Surface**

The van der Waals surface (or the hard sphere model, also known as the scale model or the corresponding space-filling model) is the simplest representation of a molecular surface. It can be determined from the van der Waals radii of all

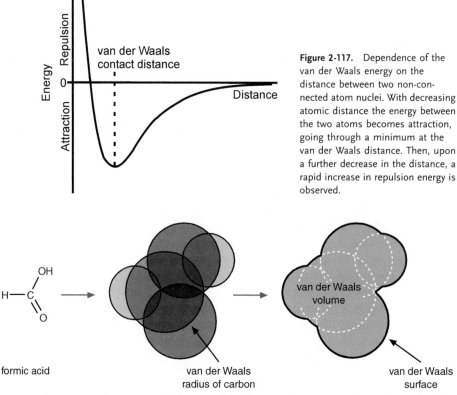

**Figure 2-117.** Dependence of the van der Waals energy on the distance between two non-connected atom nuclei. With decreasing atomic distance the energy between the two atoms becomes attraction, going through a minimum at the van der Waals distance. Then, upon a further decrease in the distance, a rapid increase in repulsion energy is observed.

**Figure 2-118.** Cross-section of the 3D model of formic acid (HCOOH). The van der Waals radius of each atom of the molecule is taken and by fusing the spheres the van der Waals surface is obtained.

the atoms. In this procedure, the van der Waals radius of each atom is first obtained from the energetically favorable distance of two non-connected atoms that are approaching each other (Figure 2-117). In a molecule, the spheres of the van der Waals radii of bonded atoms penetrate each other. In the next step, all the spheres of all the atoms are fused (Figure 2-118). The space inside this fusion area is called the van der Waals volume and the envelope of the spheres defines the van der Waals surface of the molecule. It can be calculated quite easily and rapidly.

### 2.10.2
### Connolly Surface

In contrast to the van der Waals surface, the Connolly surface [183, 184] has a smoother surface structure. The spiky and hard transition between the spheres of neighboring atoms is avoided. The Connolly surface can be obtained by rolling

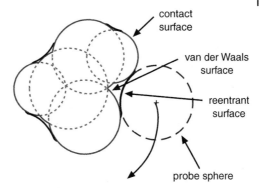

**Figure 2-119.** The Connolly surface is determined by moving a probe sphere (usually a water molecule) over the van der Waals surface. The surface thus obtained is also called the molecular or solvent-excluded surface (see Section 2.10.4 and Figure 2-120).

a spherical probe (that represents the solvent schematically) over the van der Waals surface. The radius of this sphere, which can be chosen in this procedure, is usually taken as the effective radius of a water molecule (1.4 Å). The resulting molecular surface consists of two regions (Figure 2-119):

- the convex *contact surface* segment of the van der Waals surface, with direct contact to the solvent, and
- the concave *re-entrant surface* segment where the solvent sphere has contact with two or more atom spheres of the structure.

Connolly surfaces are standard in Molecular Modeling tools, and permit the quantitative and qualitative comparison of different molecules.

### 2.10.3
### Solvent-Accessible Surface

In general, the solvent-accessible surface (SAS) represents a specific class of surfaces, including the Connolly surface. Specifically, the SAS stands for a quite discrete model of a surface, which is based on the work of Lee and Richards [182]. They were interested in the interactions between protein and solvent molecules that determine the hydrophobicity and the folding of the proteins. In order to obtain the surface of the molecule, which the solvent can access, a probe sphere rolls over the van der Waals surface (equivalent to the Connolly surface). The trace of the center of the probe sphere determines the *solvent-accessible surface,* often called the *accessible surface* or the *Lee and Richards surface* (Figure 2-120). Simultaneously, the trajectory generated between the probe and the van der Waals surface is defined as the *molecular* or *Connolly surface.*

Whereas the contact region is the basis in the Connolly method, the center of the solvent-sphere determines the shape of the molecular surface in the SAS method. In this case, the resulting surface is larger and the transition between the different atoms is more significant.

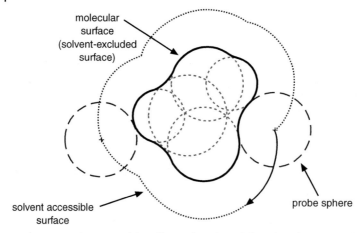

molecular
surface
(solvent-excluded
surface)

solvent accessible
surface

probe sphere

**Figure 2-120.** The center of the rolling probe sphere defines the solvent-accessible surface during movement of the probe over the van der Waals surface. Thus, the molecular surface is expanded by the radius of the solvent molecule.

### 2.10.4
### Solvent-Excluded Surface (SES)

Solvent-excluded surfaces correlate with the molecular or Connolly surfaces (there is some confusion in the literature). The definition simply proceeds from another point of view. In this case, one assumes to be inside a molecule and examines how the molecule "sees" the surrounding solvent molecules. The surface where the probe sphere does not intersect the molecular volume is determined. Thus, the SES embodies the solvent-excluded volume, which is the sum of the van der Waals volume and the interstitial (re-entrant) volume (Figures 2-119, 2-120).

The surfaces of large molecules such as proteins cannot be represented effectively with the methods described above (e.g., SAS). However, in order to represent these surfaces, less calculation-intensive, harmonic approximation methods with SES approaches can be used [185].

### 2.10.5
### Enzyme Cavity Surface (Union Surface)

The molecular surface of receptor site regions cannot be derived from the structure information of the molecule, but represents the form of the active site of a protein surrounded by a ligand. This surface representation is employed in drug design in order to illustrate the volume of the pocket region or the molecular interaction layers [186].

2.10.6
**Isovalue-Based Electron Density Surface**

Besides the expressions for a surface derived from the van der Waals surface (see also the CPK model in Section 2.11.2.4), another model has been established to generate molecular surfaces. It is based on the molecular distribution of electronic density. The definition of a limiting value of the electronic density, the so-called isovalue, results in a boundary layer (isoplane) [187]. Each point on this surface has an identical electronic density value. A typical standard value is about 0.002 au (atomic unit) to represent electronic density surfaces.

Isovalue-based surfaces are also often used for the representation of molecular orbitals.

2.10.7
**Experimentally Determined Surfaces**

A completely new method of determining surfaces arises from the enormous developments in electron microscopy. In contrast to the above-mentioned methods where the surfaces were calculated, molecular surfaces can be determined experimentally through new technologies such as electron cryomicroscopy [188]. Here, the molecular surface is limited by the resolution of the experimental instruments. Current methods can reach resolutions down to about 10 Å, which allows the visualization of protein structures and secondary structure elements [189]. The advantage of this method is that it can be applied to derive molecular structures of macromolecules in the native state.

**2.11**
**Visualization of Molecular Models**

Since the early 20th century, chemists have represented molecular information by molecular models. The human brain comprehends these representations of graphical models with 3D relationships more effectively than numerical data of distances and angles in tabular form. Thus, visualization makes complex information accessible to human understanding easily and directly through the use of images.

Diverse methods can be used for the display of a molecular model. The most widely used representations of molecules are 2D structure diagrams, which arrange the atoms in 2D space in such a way that bond lengths and bond angles are shown as undisturbed as possible and avoiding the overlap of atoms. In addition, stereochemical information may be given through wedged and hashed bonds. However, no 2D structure representation can explain a 3D molecule in its entirety. The 3D structure information is especially important for an understanding of the chemical and biological properties of a compound and the relationships between structural features and molecular functions. Often, the physical and chemical properties of a molecule are more evident when one changes between different display styles. The first molecular display type to be introduced here is structure-based. Other types of display styles are surface and field representations which can apply various color schemes.

For all the different methods of chemical visualization, a large number of special techniques are available, depending on the purpose of visualization. These software programs can be installed on a local computer or can be operated via the Internet. An overview of these programs is given in Section 2.12.3.

### 2.11.1
### Historical Review

Before the advent of computers, molecular models were made manually of wood, paper, wire, rubber, plastic, or other materials, a laborious and time-consuming process. Kendrew et al., in 1958, built the first brass-wire model of a macromolecule, myoglobin (Figure 2-121), whose structure was determined by X-ray crystallography [190]. A series of models followed over the following years. Byron Rubin and his Byron's Bender machine produced the most widespread models [191]. These wire models represented the backbone of protein structures. In the 1970s, the scientific advantages of these models were demonstrated. At that time, two proteins represented by Byrons Bender models were compared and the first indication of the existence of a superfamily of structures was recognized [192].

Despite the many advantages of these physical models, they have several essential deficiencies. As the size of the structure increases, the assembly of a model becomes progressively more unmanageable and more complex. Furthermore, data such as atom distances and atom angles can be determined only with difficulty, or not at all.

New ways to represent structure data became available through molecular modeling by computer-based methods. The birth of interactive computer representation of molecular graphics was in the 1960s. The first dynamic molecular pictures of small molecules were generated in 1964 by Levinthal in the Mathematics and Computation (MAC) project at the Electronic Systems Laboratory of the Massachusetts

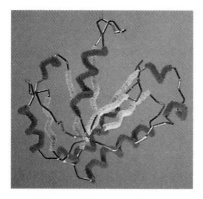

**Figure 2-121.** The first brass-wire model of a macromolecule built by Kendrew et al. in 1958 [193].

**Figure 2-122.** The first dynamic molecular display of small molecules by Levinthal was driven by the "Crystal Ball" [193].

Institute of Technology (MIT) [193]. Molecules were represented as line drawings on a homemade display (an oscilloscope) (Figure 2-122). In addition, the system had diverse peripherals with many switches and buttons which allowed the modification of the scene. The heart of the system was the so-called "Crystal Ball" which could rotate the molecule about all three orthogonal axes. This prototype cost approximately two million US dollars.

The first pure molecular graphics system followed after a few years. It was built in 1970 by Langridge at Princeton University [195]. The system was based on the Picture System 2 of Evans and Sutherland and could also display bonds and colored atoms.

A further milestone was achieved in 1977 by Richardson et al. They could for the first time visualize a complete protein structure from X-ray crystallography data [196]. A large number of structures were generated in the following years.

As the graphical capabilities of the computer systems became more powerful simultaneously the number of visualized structures increased. With the introduction of raster graphics (1974) and colored raster graphics displays (1979), other forms of molecular representations were possible [197]. CPK models could be represented and colored bonds or molecular surfaces could be visualized.

The era of the Evans and Sutherland computer systems vanished in the first half of the 1980s, when powerful and more economical workstations were introduced. In spite of advances in computer graphics and in CPU power, these workstations dominate the everyday life of molecular modeling even today.

In recent years, the rapid development of low-budget 3D-capable graphics cards makes it possible to visualize molecular models with standard PC systems. Some molecular modeling software, which was once available only for workstations, is now also offered for PCs [198].

2.11.2
**Structure Models**

The main objective of a structure model is to produce an image of a molecule that invokes 3D information although it is physically two-dimensional. Additional lighting effects (such as shadows on the objects of the structure) may enhance the impression of depth in space. In the following sections, some of the more common structure models used for small molecules or for crystal structures are discussed.

### 2.11.2.1 Wire Frame Model

The most well-known and at the same time the earliest computer model for a molecular structure representation is a wire frame model (Figure 2-123a). This model is also known under other names such as "line model" or "Dreiding model" [199]. It shows the individual bonds and the angles formed between these bonds. The bonds of a molecule are represented by colored vector lines and the color is derived from the atom type definition. This simple method does not display atoms, but atom positions can be derived from the end and branching points of the wire frame model. In addition, the bond orders between two atoms can be expressed by the number of lines.

### 2.11.2.2 Capped Sticks Model

The capped sticks model can be seen as a variation of the wire frame model, where the structure is represented by thicker cylindrical bonds (Figure 2-123b). The atoms are shrunk to the diameter of the cylinder and are used only for smoothing or closing the ends of the tubes. With its thicker bonds, the capped sticks model conveys an improved 3D impression of a molecule when compared with the wire frame model.

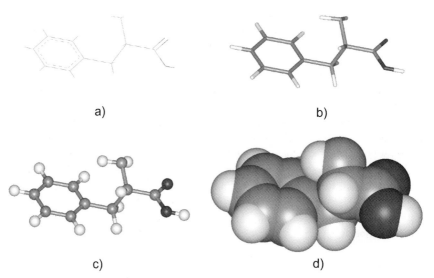

a)

b)

c)

d)

**Figure 2-123.** The most common molecular graphics representations of phenylalanine a) wire frame; b) capped sticks; c) balls and sticks; d) space-filling.

### 2.11.2.3    Balls and Sticks Model

A more convenient representation for the eyes is the balls and sticks model. The atoms are represented as small balls centered at the nuclei of the atoms, and the bonds between the balls as cylinders or "sticks" (Figure 2-123c). The size of the balls is standardized but variable and is only a fraction of the actual atomic radius in comparison with the scale of the bonds as represented by the cylinders. The color of the balls is used for displaying atomic properties such as the atom type. Similarly, the color and the number of cylinders express bond types or atomic properties, respectively. The decisive advantage of this model is based on an essentially improved 3D display of the positions of the atoms and bonds in space. Those parts of the molecule placed in the background are partly masked by the atoms and bonds that are closer to the viewer. The more voluminous balls and sticks convey this spatial impression more strongly. It is also enhanced by light and shade effects such as Gouraud shading [200].

### 2.11.2.4    Space-Filling Model

The space filling model developed by Corey, Pauling, and Koltun is also known as the CPK model, or scale model [197]. It shows the relative volume (size) of different elements or of different parts of a molecule (Figure 2-123d). The model is based on spheres that represent the "electron cloud". These atomic spheres can be determined from the van der Waals radii (see Section 2.10.1), which indicate the most stable distance between two atoms (non-bonded nuclei). Since the spheres are all drawn to the same scale, the relative size of the overlapping electron clouds of the atoms becomes evident. The connectivities between atoms, the bonds, are not visualized because they are located beneath the atom spheres and are not visible in a non-transparent display (see Section 2.10). In contrast to other models, the CPK model makes it possible to visualize a first impression of the extent of a molecule.

### 2.11.3
### Models of Biological Macromolecules

The visualization of hundreds or thousands of connected atoms, which are found in biological macromolecules, is no longer reasonable with the molecular models described above because too much detail would be shown. First of all the models become vague if there are more than a few hundred atoms. This problem can be solved with some simplified models, which serve primarily to represent the secondary structure of the protein or nucleic acid backbone [201]. (Compare the balls and sticks model (Figure 2-124a) and the backbone representation (Figure 2-124b) of lysozyme.)

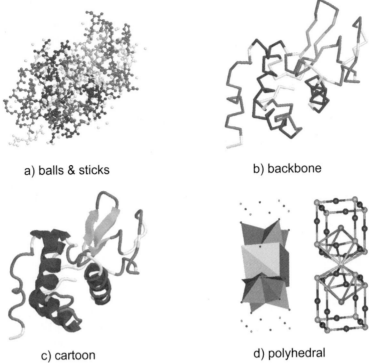

**Figure 2-124.** The most common molecular graphic representations of biological molecules (lysozyme): a) balls and sticks; b) backbone; c) cartoon (including the cylinder, ribbon, and tube model); and of inorganic molecules ($YBa_2Cu_3O_{7-x}$): d) polyhedral (left) and the same molecule with balls and sticks (right).

### 2.11.3.1 Cylinder Model

The cylinder model is used to characterize the helices in the secondary structure of proteins (see the helices in Figure 2-124c).

### 2.11.3.2 Ribbon Model

Where helical secondary structures are represented by the cylinder model, the $\beta$-strand structures are visualized by the ribbon model (see the ribbons in Figure 2-124c). The broader side of these ribbons is oriented parallel to the peptide bond. Other representations replace the flat ribbons with flat arrows to visualize the sequence of the primary structure.

### 2.11.3.3 Tube Model

The tube structure consists of small tubular formations to represent so-called coils and turns (see the tubes in Figure 2-124c).

## 2.11.4
## Crystallographic Models

It is often difficult to represent inorganic compounds with the usual structure models because these structures are based on complex crystals (space groups), aggregates, or metal lattices. Therefore, these compounds are represented by individual polyhedral coordination of the ligands such as the octahedron or tetrahedron (Figure 2-124d).

## 2.11.5
## Visualization of Molecular Properties

Knowledge of the spatial dimensions of a molecule is insufficient to understand the details of complex molecular interactions. In fact, molecular properties such as electrostatic potential, hydrophilic/lipophilic properties, and hydrogen bonding ability should be taken into account. These properties can be classified as: scalar (isosurfaces), vector field, and volumetric properties.

### 2.11.5.1  Properties Based on Isosurfaces

Molecular orbitals were one of the first molecular features that could be visualized with simple graphical hardware. The reason for this early representation is found in the complex theory of quantum chemistry. Basically, a structure is more attractive and easier to understand when orbitals are displayed, rather than numerical orbital coefficients. The molecular orbitals, calculated by semi-empirical or *ab initio* quantum mechanical methods, are represented by isosurfaces, corresponding to the electron density surfaces (Figure 2-125a).

Knowledge of molecular orbitals, particularly of the HOMO (Highest Occupied Molecular Orbital) and the LUMO (Lowest Unoccupied Molecular Orbital), imparts a better understanding of reactions (Figure 2-125b). Different colors (e.g., red and blue) are used to distinguish between the parts of the orbital that have opposite signs of the wavefunction.

Besides molecular orbitals, other molecular properties, such as electrostatic potentials or spin density, can be represented by isovalue surfaces. Normally, these scalar properties are mapped onto different surfaces (see above). This type of high-dimensional visualization permits fast and easy identification of the relevant molecular regions.

To display properties on molecular surfaces, two different approaches are applied. One method assigns color codes to each grid point of the surface. The grid points are connected to lines (chicken-wire) or to surfaces (solid sphere) and then the color values are interpolated onto a color gradient [200]. The second method projects colored textures onto the surface [202, 203] and is mostly used to display such properties as electrostatic potentials, polarizability, hydrophobicity, and spin density.

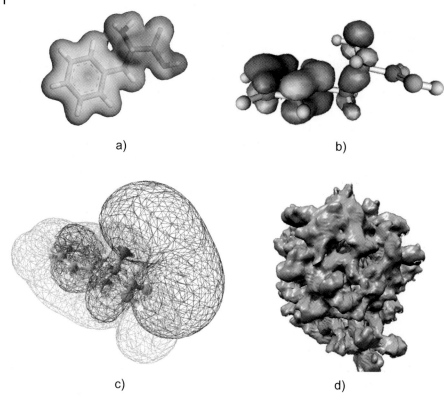

**Figure 2-125.** Different isovalue-based surfaces of phenylalanine a) isoelectronic density; b) molecular orbitals (HOMO–LUMO); c) isopotential surface; and d) isosurface of the electron cryo-microscopic volume of the ribosome of *Escherichia coli*.

### Electrostatic Potentials

Bonaccorsi et al. [204] defined for the first time the molecular electrostatic potential (MEP), which is clearly the most important and most used property (Figure 2-125c). The electrostatic potential helps to identify molecular regions that are significant for the reactivity of compounds. Furthermore, the MEP is decisive for the formation of protein–ligand complexes. Detailed information is given in Ref. [205].

### Polarizability and Hydrophobicity

These properties are also relevant if molecular interactions are considered. In contrast to electrostatic potentials, they only take effect at small distances between interacting molecular regions.

### Spin Density

Above all, spin density is most significant for radicals. Their unpaired electrons can be localized rapidly, by visualizing this property on the molecule.

**Vector Fields**

The representation of molecular properties on molecular surfaces is only possible with values based on scalar fields. If vector fields, such as the electric fields of molecules, or potential directions of hydrogen bridge bonding, need to be visualized, other methods of representation must be applied. Generally, directed properties are displayed by spatially oriented cones or by field lines.

**Volumetric Properties**

The visualization of volumetric properties is more important in other scientific disciplines (e.g., computer tomography in medicine, or convection streams in geology). However, there are also some applications in chemistry (Figure 2-125d), among which only the distribution of water density in molecular dynamics simulations will be mentioned here. Computer visualization of this property is usually realized with two or three dimensional textures [203].

A comprehensive range of interesting visualization tools is provided on the Internet under *http://www2.chemie.uni-erlangen.de/projects/ChemVis/index2.html*. There, complex factual relationships are displayed in interactive high-end graphics.

## 2.12
### Tools: Chemical Structure Drawing Software – Molecule Editors and Viewers

### 2.12.1
### Introduction

The development of molecule editors and molecule viewers is a highly active field with new software constantly entering the arena. This makes it difficult for a beginner to choose the "right" software. This section introduces the basics of molecule editors and viewers and gives a current overview of the most widely used software packages. Most drawing and visualization software currently runs on PCs with MS Windows as the operating system, with only a few programs on Macintoshes or workstations.

A chemical structure drawing package is more than a simple, conventional drawing package, such as CorelDraw or MS Paint. In fact, molecule editors are often used to create professional-looking structure diagrams or even chemical reaction equations for publication. Furthermore, a structure drawing program should incorporate the possibility of deducing additional information on the compound from the drawing in order to make the molecule amenable to processing in computational chemistry. Therefore, the electronic structure input and the further use of the structure or structure data play a dominant role in the field of chemoinformatics, particularly in database queries or molecular modeling. Herein, the usage of editors and viewers is different. Whereas an editor is used solely to create a structure code such as SMILES or a connection table from a molecule, a viewer application mainly applies these codes to visualize the structures in a more sophisticated manner (Figure 2-126).

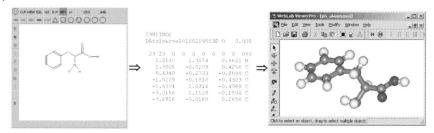

**Figure 2-126.** Molecule editor applications (left-hand side) create structure codes (e.g., a CT in the middle), which are then visualized by a viewer program (right-hand side) in a more sophisticated manner.

Software applications can mainly be divided into stand-alone programs that have to be installed on a local computer, and web-based applications. Furthermore, there are two different techniques used in web-based applications for providing the programs: applets and plug-ins. Whereas plug-ins (helper applications) have to be installed to be displayed correctly in a browser, Java applets are loaded by the browser on demand. The basic requirement for executing Java applets is that the Java Virtual Machine (JVM) must be pre-installed on the local computer, which is often the case.

### 2.12.2
### Molecule Editors

Generally, molecule editors display the structure diagrams as 2D images. Stereochemical information such as *R/S* identity can be visualized in the software by using wedged or hashed bonds, which are well known to chemists. Nevertheless, the structure still is only two-dimensional. If the 3D geometry of the molecule (the 3D coordinates of the atoms) is required, it is necessary to send the 2D structure from the editor to a 3D structure generator (see Section 2.13). Some editors already provide this 3D feature within the software. Of course simple, appropriate drawing tools, such as different bond types (simple, double, etc.) and element symbols are available in chemistry and text settings in the editors discussed here.

The most important feature of editing software is the option to save the structure in standard file formats which contain information about the structure (e.g., Mol-file, PDB-file). Most of these file formats are ASCII text files (which can be viewed in simple text editors) and cover international standardized and normalized specifications of the molecule, such as atom and bond types or connectivities (CT) (see Section 2.4). Thus, with these files, the structure can be exchanged between different programs. Furthermore, they can serve as input files to other chemical software, e.g., to calculate 3D structures or molecular properties.

The introduction of the Java programming language [206] considerably increased the interactivity of web applications and allowed chemical information to be integrated on the web in a more efficient way. Small applets could be integrated directly into a web page to add practically any desired functionality. Probably the first molecule editor in Java was an applet written by D. Bayada from the University of Leeds. The source code of this program was released and one can still find several derivatives of this editor on the Internet. Independently, the Java Molecular Editor (JME) applet was developed by P. Ertl at Ciba-Geigy (now Novartis) as part of the company's web-based chemoinformatics system [207], and is described in more detail below. Several other structure drawing programs written in Java have also been developed, and now more than 20 such programs of different complexity and quality exist.

Not all available editors can be discussed in detail in this section, because their number is rather large and the field is quite dynamic. A broad summary, with more editors, is given in Table 2-10.

### 2.12.2.1 Stand-Alone Applications

#### 2.12.2.1.1 CACTVS (Version 3.176; Win V 2.87)

The CACTVS molecule editor is a graphical input tool for molecular structures and is free of charge for non-profit use. It can be used as a stand-alone or as a dependent remote program of the CACTVS computation workbench. The software is available for all platforms (excluding Macintosh systems).

One advantage of the CACTVS editor is its ability to create chemically exact structures. Hereby, the properties of atoms (charges, chirality, etc.) and bonds (*cis/trans*) can be defined via the menu bar or a mouse click. Pre-defined structure elements of a fragment library can be dragged into the drawing area and attached to others. Molecular formulas and molecular weights are automatically calculated and visualized. In addition, elements of a selection panel or of the periodic table of the chemical elements can be chosen easily. But the ultimate purpose of the molecule editor is to read and write a structure file as one of 23 data exchange formats, including the Molfile.

*http://www2.chemie.uni-erlangen.de/software/cactvs/index.html*

#### 2.12.2.1.3 ChemDraw 7.0

ChemDraw Ultra 7.0 is among the most popular commercial chemical drawing software. It is available as a separate program or integrated into the commercial software suite ChemOffice from CambridgeSoft Inc. with Chem3D (3D molecule viewer, modeling software), ChemFinder (database manager), ChemInfo (chemical databases and catalogs), and E-Lab Notebook (organizer).

The drawing software comprises a comprehensive collection of standard tools to sketch 2D chemical structures. To specify all its facilities and tools would go far beyond the scope of this overview, but there are some nice features that are very useful for chemists so they are mentioned here briefly. One of these enables the prediction of $^1H$ and $^{13}C$ NMR shifts from structures and the correlation of atoms with NMR peaks (Figure 2-127). IUPAC standard names can be generated

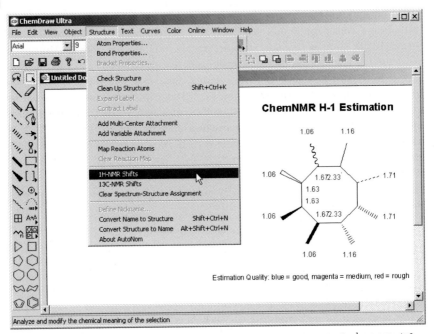

**Figure 2-127.** ChemDraw Ultra 7.0 can display eight different bond types. The ${}^1$H-NMR shift estimations for octamethylcyclooctane are displayed in this example artificially; the calculated spectrum is not shown here.

from chemical structures with Beilstein's AutoNom 2.1 program. On the other hand, for most substances a structure can be created by typing in a systematic chemical name. Stereocenters can be identified using Cahn–Ingold–Prelog rules. Additional physical properties such as log$P$, boiling point, melting point, etc., of chemical compounds can also be computed.

*http://www.cambridgesoft.com/*

### 2.12.2.1.3  Chemistry 4D-Draw (Version 7.0)

The commercial 2D structure editor, Chemistry 4D-Draw, from ChemInnovative Software Inc., includes two additional special modules besides conventional chemical drawing tools. NamExpert provides the interpretation of a compound name according to the IUPAC nomenclature to create the corresponding chemical structure. The latter can be represented in three different styles: the shorthand, Kekulé, or semi-structural formula. In contrast to NamExpert, the Nomenclature module assigns IUPAC names to drawn structures.

Drawing-, text-, and structure-input tools are provided that enable easy generation of flow charts, textual annotations or labels, structures, or reaction schemes. It is also possible to select different representation styles for bond types, ring sizes, molecular orbitals, and reaction arrows. The structure diagrams can be verified according to free valences or atom labels. Properties such as molecular weight, empirical formula, number of atoms, or percentage composition can be calculated by the Mass Calculator. Chemistry 4D-Draw supports only Molfile as a data exchange format.

*http://www.cheminnovation.com/chem4dd.html*

### 2.12.2.1.4 ChemSketch (Version 5.0)

ChemSketch is a professional software package that is available free of charge from Advanced Chemistry Development Inc. (ACD). Besides the editor, it has several modules (ACD/Dictionary, ACD/Tautomers), extensions, and add-ins concerning the calculation of physicochemical properties, input of spectra and chromatograms, naming of molecules, and a viewer.

The editor provides two modes for drawing: the structure and the draw mode. This distinction is important for differentiating chemical information from conventional drawings. Both modes are switchable and provide an extensive set of features in the menu bar to create chemical structures and reactions, or just drawings. The number of options can be quite confusing for beginners; however, one becomes accustomed to them after a short period of vocational adjustment.

The free valences of self-made structures or (manipulated) structures from many templates are checked and are automatically saturated with hydrogen atoms. Molecular properties such as empirical formula, molecular weight, percentage composition, molar refractivity, refraction index, density, and the parachor can be calculated or estimated (Figure 2-128). A wide range of bond types (e.g., electron-donating, stereo-nonspecific, and also organometallic such as the ferrocene type) are accessible. Enhanced graphics such as electron orbitals (s, p, d, f) and other drawing aids (lines, polygons, Bezier curves, arrows) can be integrated into the representation. Many operations are accessible via shortcuts (by simultaneous pressing of combinations of Control, Alt, Shift, and other keys). Online access with the ACD/I-Lab add-on permits the prediction of properties, e.g., NMR spectra, $\log P$, and $pK_a$, or IUPAC name generation of the input structures. Twelve input/output file formats are provided which cover the most common file formats used in chemistry (mol, rxn, cdx, etc.) and graphics (gif, wmf, etc.). ChemSketch has detailed Help options and numerous demonstration pages of diverse tools on the home page.

The additional integrated module, the ACD/3D Viewer, can visualize 2D structures as 3D models, after geometry optimization (see Section 2.12.3.1 and Figure 2-132, below).

*http://www.acdlabs.com/download/chemsk.html*

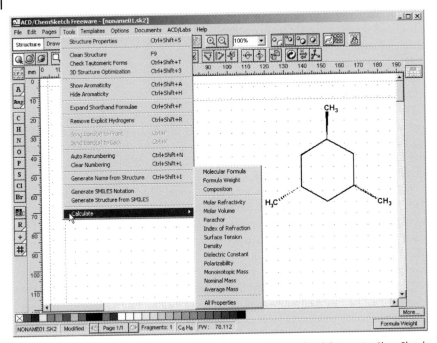

**Figure 2-128.** Screenshot showing the three possible bond types of cyclohexane in ChemSketch V 5.0, and various tools for calculating physicochemical properties.

### 2.12.2.1.5 ChemWindow (Version 6.0)

ChemWindow is a commercial editor available in the Sadtler Suite (Bio-Rad Sadtler, formerly SoftShell), which contains the additional modules SymApps (3D rendering), Database Building, and KnowItAll Informatics System (the last two are primarily tools for analyzing spectra).

Chemical structures can be created without any problems with this structure editor. The user can operate with numerous tools. Besides a library of 4500 structures, 130 glassware drawings and 250 technical symbols are accessible. All structures and drawings can be manipulated in the layout without restriction (of orientation, color, or view). In addition, electron orbitals and other drawings (lines, reaction arrows, Newman projections, text fields and tables, but no curves or polygons) can be integrated into the representation. Molecular weights, percentage composition, or concentrations of the molecule can be calculated, but no physicochemical properties.

ChemWindow supports ten (mostly internal) file formats for input and output, but only Molfile as a genuine exchange format.

Several tutorials and Help options are available on the home page.

SymApps is an additional 3D molecule-rendering module from the ChemWindow suite, designed for desktop visualization (see Section 2.12.3.1).

*http://www.chemwindow.com/cw6.html*

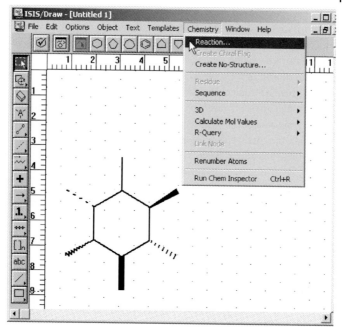

**Figure 2-129.** Isis/Draw V 2.4 provides six different bond types and various important features for chemical tasks.

### 2.12.2.1.6 **ISIS/Draw (Version 2.4)**

ISIS/Draw from MDL Information Systems, Inc., is free of charge for non-profit use. The powerful structure editor has drawing tools for 2D molecular structures, reaction equations, and publishing graphics (journal settings). Structures can easily be rotated with the "3D rotate tool" The graphical user interface provides the ability to create chemical structures intuitively (Figure 2-129). Four data formats, Molfile, Rxnfile, TGfile, and sequence-file are supported. Besides many templates (pre-defined structures), the program incorporates additional features that automatically display hydrogen atoms at free valences or check properties (bond order, etc.) of the chemical structure with the "Chem Inspector". Some additional molecular parameters can be calculated, e.g., molecular weight. Structure and reaction databases can be generated easily with the help of this structure editor.

*http://www.mdli.com/products/isisdraw.html*

### 2.12.2.2 **Web-Based Applications**

JChemPaint and the ACDStructure Drawing Applet (ACD/SDA 1.30) are introduced only briefly, whereas one of the most well-known Java editors (JME) is presented in more detail below.

JChemPaint is a chemical structure drawing applet. The noteworthy characteristic of this 2D molecule editor is that it is an open source program [208]. This means that the software and the source code of the program are freely available. Every programmer or interested person can participate and enter individual special requests for further development of the application.

Besides providing basic regular drawing features (different bond types, templates, coloring, etc.), JChemPaint can predict an approximate $^{13}$C NMR chemical shift range based on one-sphere HOSE codes (see Section 10.2.2.2). MDL Molfiles, SDfiles, SMILES, and CML (Chemical Markup Language) are supported as exchange formats. Images can be saved as bmp, gif, jpg, png, or svg files.

A larger number of features are provided by the ACDStructure Drawing Applet (ACDLabs). Both structures and reactions can be drawn, imported, and also exported. This applet supports Molfiles and has a large, integrated collection of pre-defined templates, which are extensible by the user. Additionally, gif files can be exported. It is not possible to draw or to import/export chemical reactions.

### 2.12.2.2.1 **JME Molecule Editor**

The JME Editor is a Java program which allows one to draw, edit, and display molecules and reactions directly within a web page and may also be used as an application in a stand-alone mode. The editor was originally developed for use in an in-house web-based chemoinformatics system but because of many requests it was released to the public. The JME currently is probably the most popular molecule entry system written in Java. Internet sites that use the JME applet include several structure databases, property prediction services, various chemoinformatics tools (such as for generation of 3D structures or molecular orbital visualization), and interactive sites focused on chemistry education [209].

The basic function of the JME Editor is to allow the creation and modification of molecules and reactions directly within a web page. The editor has all the standard chemical drawing and editing capabilities, including a rich set of keyboard shortcuts for adding common structural fragments easily (Figure 2-130).

A SMILES code [22], MDL Molfile [50], or JME's own compact format (one-line representation of a molecule or reaction including the 2D coordinates) of created molecules may be generated. The created SMILES is independent of the way the molecule was drawn (unique SMILES: see Section 2.3.3). Extensions to JME developed in cooperation with H. Rzepa and P. Murray-Rust also allow output of molecules in the CML format [60].

The JME can also serve as a query input tool for structure databases by allowing creation of complex substructure queries (Figure 2-130), which are automatically translated into SMARTS [22]. With the help of simple HTML-format elements the creation of 3D structure queries is also possible, as were used in the 3D pharmacophore searches in the NCI database system [129]. Creation of reac-

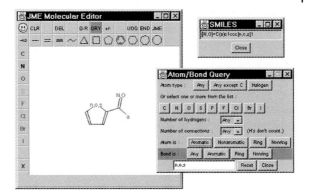

**Figure 2-130.** Creation of substructure queries by the JME editor.

tions is also supported (Figure 2-131), including generation of reaction SMILES and SMIRKS [22].

The applet may also be used in a depiction mode as a molecule viewer (without editing buttons), to visualize molecules. Internal JME format or MDL Molfiles may be viewed in this way.

When incorporated into an HTML page, the JME applet may communicate with other page elements (other applets or JavaScript objects). These functions allow the JME appearance to be altered by manipulating displayed molecules or displaying new structures. For example, JME can easily be connected with a figure in a web page and display corresponding molecules when the mouse touches the respective points on the figure. For details of JME public functions, see the JME documentation [209].

The JME applet is written in Java 1.0, which is available in all types and versions of web browsers. The size of the program classes is minimal (about 40 kB), which assures fast loading. In addition, the editor may be used in a stand-alone mode as a Java application within web pages as an applet. Thanks to the independence of the Java platform, JME can run on Windows PCs, Mac/OS machines and practically all UNIX clones, including, of course, LINUX.

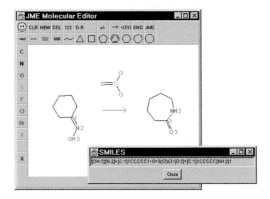

**Figure 2-131.** Input of reactions by the JME editor.

For non-commercial purposes and intranet applications the JME Editor may be obtained directly from the author [209].

### 2.12.3
### Molecule Viewers

Molecule editors represent only two-dimensional chemical structures (thus also could be considered as 2D viewers), the third dimension is visualized by 3D viewers, mainly user-interactive.

In order to represent 3D molecular models it is necessary to supply structure files with 3D information (e.g., pdb, xyz, cif, mol, etc.). If structures from a structure editor are used directly, the files do not normally include 3D data. Inclusion of such data can be achieved only via 3D structure generators, force-field calculations, etc. 3D structures can then be represented in various display modes, e.g., wire frame, balls and sticks, space-filling (see Section 2.11). Proteins are visualized by various representations of helices, $\beta$-strains, or tertiary structures. An additional feature is the ability to color the atoms according to subunits, temperature, or chain types. During all such operations the molecule can be interactively moved, rotated, or zoomed by the user.

Molecular biology database providers offer different viewers with conventional visualization options, such as Swiss-PdbViewer (SPDBV 3.7, *http://www.expasy. org/swissmod*). These are specialized interactive molecule graphic programs for viewing and analyzing protein and nucleic acid structures. A more extensive summary, listing more viewers, is given in Table 2-11.

#### 2.12.3.1   Stand-Alone Applications

Some of the stand-alone programs mentioned above have an integrated modular 3D visualization application (e.g., ChemWindow → SymApps, ChemSketch → ACD/3D Viewer, ChemDraw → Chem3D). These relatively simple viewers mostly generate the 3D geometries by force-field calculations. The basic visualization and manipulation features are also provided. Therefore, the molecular models can be visualized in various display styles, colors, shades, etc. and are scalable, movable and rotatable on the screen.

##### 2.12.3.1.1   ISIS/Draw 2.4

ISIS/Draw has no genuine molecular visualization tool. The rotate tool changes only the 2D rotate tool into a 3D rotate tool which rotates 2D structures in three dimensions. In order to visualize chemical structures in different styles and perspectives, it is necessary to paste the drawing, e.g., to the ACD/3D Viewer.

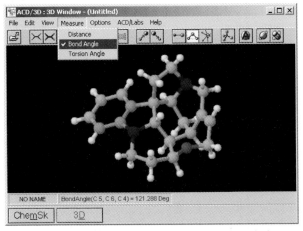

**Figure 2-132.** User-interface of the ACD/3D Viewer; the bond angle between three marked atoms is displayed on the taskbar.

### 2.12.3.1.2  ACD/3D Viewer V. 5.07

This viewer module is fully integrated in the ACD/Chemsketch drawing software which allows 3D representations to be obtained quickly via a mouse click. The 3D molecules can then be manipulated (moved, rotated, etc., but not zoomed) and displayed in different styles (sticks, balls and sticks, spheres, etc.), fonts, and colors. It is also possible to optimize the structure using a force-field calculation (CHARMM type) and to measure bond lengths, bond and torsion angles (see Figure 2-132). The 3D structure representation created can only be saved in an ACD-specific file format, or pasted into other graphics applications.

### 2.12.3.1.3  Chem3D Ultra V. 6.0

Chem3D is much more than a molecule viewer. This autonomous software module from the ChemOffice package provides simple molecule editor tools to create structures, but is mainly used as a molecular modeling tool.

### 2.12.3.1.4  SymApps V. 5.0

SymApps converts 2D structures from the ChemWindow drawing program into 3D representations with the help of a modified MM2 force field (see Section 7.2). Besides basic visualization tools such as display styles, perspective views, and light source adjustments, the module additionally provides calculations of bond lengths, angles, etc. Moreover, point groups and character tables can be determined. Animations of spinning movements and symmetry operations can also be created and saved as movie files (*.avi).

**Table 2-10.** Overview of molecule editors and their features.

| Name/version | Alchemy 2000 V2.0 | ACD/Chemsketch 5.0 | ChemDraw 7.0.1 | Chemsite 3.01 | Chemistry 4D Draw 6.0 | Chem Window 6.0 | Isis/Draw 2.4 | CACTVS 3.176 | ACD/SDA 1.30 | JChemPaint7 V1.2pre | JME | Marvin 2.10.5 |
|---|---|---|---|---|---|---|---|---|---|---|---|---|
| Description | 2D/3D editor, molecular modeling | 2D editor, chem. drawing software, 3D viewer | 2D editor | 2D editor, molecular modeling | 2D editor | 2D editor | 2D editor, database setup | 2D editor | 2D editor | 2D editor | 2D editor | 2D editor |
| **General** operating system | WIN | WIN | WIN, Mac | WIN | WIN, Mac, Java | WIN | WIN, SGI (V2.3 in Mac) | WIN, Unix | Java applet | Java software | Java applet + program | Java applet + program |
| costs | from $500 | freeware | $400 | $200 | $300 | commercial | freeware | freeware, commercial | freeware | open source | freeware | freeware |
| language | Eng. | Eng. | Eng. | 3.01 Eng. 2.4 Ger. | Eng. | Eng., Ger. | Eng. | Eng. | Eng. | Eng. | Eng. | Eng. |
| WWW plug-in | – | X | – | | | X | – | – | – | – | – | – |
| **Supported file formats** standard import/export: | 24/28, mol, mol2, pdb, SMILES, hin, | mol, rxn, sdf, chx, cml, SMILES | mol, rxn, sdf, SMILES, SLN, smd, cts | mol, pdb, z-matrix, ct, cta, ent | mol | chm, edb, mol, scf | mol, tgf, rxn | alc, cex, cbin, gau, mol, idp, msi, pdb, rdf, res, rxn, sdf, sln, sharc, smi, tgf, xyz | mol | mol, sdf, cml | mol, SMILES, cml | mol, sdf, SMILES, SMARTS, CML, pdb, xyz |
| specials | .al2 | skc, sk2, chm, vrml | cdx, DARC, jdx, rxn, | lib, syc | – | cw2, cwg, cwt | skc, Seq (sequences) | ctx, jdx, cml | – | – | JME | – |
| image output | bmp, hplg, tif, wmf, ps | bmp, gif, pcx, pdf, tif, wmf | bmp, eps, gif, ps, pict, png, tif, wmf | bmp, tga | bmp, gif, pict, wmf | gif | – | – | gif | bmp, gif, jpg, png, tif, svg | – | – |

**Table 2-10.** (cont.)

| Name/version | | Alchemy 2000 V2.0 | ACD/Chems-ketch 5.0 | Chem-Draw 7.0.1 | Chemsite 3.01 | Chemistry 4D Draw 6.0 | Chem Window 6.0 | Isis/Draw 2.4 | CACTVS 3.176 | ACD/SDA 1.30 | JChem-Paint7 V1.2pre | JME | Marvin 2.10.5 |
|---|---|---|---|---|---|---|---|---|---|---|---|---|---|
| Functionality (general) | flying toolbars | X | X | X | – | – | – | – | – | X | – | – | – |
| | shortcuts | X | X | – | – | X | – | X | X | – | – | X | X |
| | intuitive user interface | X | X | X | X | X | X | X | X | X | X | X | X |
| | drag-and-drop | X | X | X | – | X | X | X | – | – | – | – | – |
| | undo | X | X | X | X | X | X | X | X | X | X | X | X |
| | zoom | X | X | X | X | X | X | X | – | X | – | – | – |
| | OLE function | ? | X | X | – | X | X | X | – | – | – | – | – |
| Database elements | simple text settings | – | X | X | – | X | X | X | – | – | – | – | – |
| | templates | X | X | X | – | X | X | X | X | X | X | – | – |
| | IUPAC nomenclature | – | X | X | – | X | – | – | X | – | – | – | – |
| | access to on-line libraries | X | X | X | – | – | X | X | – | – | – | – | – |
| Additional visualization | conformer search | X | X | – | – | – | – | – | – | – | – | – | – |
| | Newman projection | – | – | – | – | X | – | – | – | – | – | – | – |
| | glassware | – | X | – | – | – | – | – | – | – | – | – | – |
| | 2D/3D optimization (clean-up) | X | X | X | X | X | X | X | X | X | X | – | – |

**Table 2-10.** (cont.)

| Name/version | | Alchemy 2000 V2.0 | ACD/Chemsketch 5.0 | ChemDraw 7.0.1 | Chemsite 3.01 | Chemistry 4D Draw 6.0 | Chem Window 6.0 | Isis/Draw 2.4 | CACTVS 3.176 | ACD/SDA 1.30 | JChemPaint7 V1.2pre | JME | Marvin 2.10.5 |
|---|---|---|---|---|---|---|---|---|---|---|---|---|---|
| Calculations | 3D molecule representation | X | ACD/3D | Chem3D | X | X | SymApps | – | – | X | JMol | – | – |
| | RMS fit | X | – | – | – | – | – | – | – | – | – | – | – |
| | log$P$ | X | X | X | – | – | – | – | – | – | – | – | – |
| | physicochemical properties | X | X | X | X | – | – | – | – | – | – | – | – |
| | spectrum prediction | – | X | X | – | X | X | – | – | – | X | – | – |
| | modeling/force-field (MM3) | X | X | – | X | – | – | – | – | – | – | – | – |
| | reaction processing | – | X | X | – | – | – | X | X | – | – | x | – |
| Further features | scripting capabilities | – | – | X | – | – | – | – | X | – | – | – | – |
| Validation | miscellaneous | MOPAC presentation tool | spectrum import (add-on), many additional modules | NMR spectrum estimation | surfaces; in Pro-V: point groups, membranes | 3 editions with extensions | NMR shift prediction | journal settings, | recognition of stereochemistry, checking | web page integration | open source, web page integration | web page integration | web page integration |
| URL | | + | +++ | ++ | + | + | + | ++ | ++ | ++ | ++ | +++ | ++ |
| | | http://www.scivision.com/Alchemy.html | http://www.acdlabs.com/download/ | http://www.cambridgesoft.com/ | http://www.norgwyn.com/chemsite.htm | http://www.norgwyn.com/chemsite.htm | http://www.chemwindow.com/ | http://www.mdli.com | http://www2.chemie.uni-erlangen.de/software/cactvs/index.html | http://www.acdlabs.com/download/ | http://jchempaint.sourceforge.net/index.html | www.molinspiration.com/jme | http://www.chemaxon.com/marvin/ |

Table 2-11. Overview of molecule viewers and their features.

| Name/version | | SymApps | Chem3D Ultra 7.0 | ACD/3D View | UltraMol 2003 | Rasmol | Molmol | JMol | chime 2.63 | Ortep III | ViewerLite |
|---|---|---|---|---|---|---|---|---|---|---|---|
| Description | | 3D viewer | molecular modeling, analysis, viewer | 3D viewer | 2D editor, 3D viewer | 3D viewer | 3D viewer | 3D viewer | 3D viewer | crystallographic viewer | 3D viewer |
| General | operating system | WIN | WIN | WIN | WIN | stand-alone for all OS, applet | Unix | Java-software | plug-in | WIN | WIN |
| | costs | Module of ChemWindow | $900 | add-in of ACD/ChemSketch | $300 | freeware | freeware | open source | freeware | freeware | freeware, shareware with struct. editing: $250 |
| | language | Eng., Ger. | Eng. | Eng. | Eng., Ger. | Eng. | Eng. | Eng. | Eng. | Eng. | Eng. |
| Supported file formats | standard import/export: | hin, mol, mop, mpc, pdb, xyz | 39/35 pdb, ct, alc, mol, sml, sm2, mop, dat | mol | mdm, hin, mol, pdb, ent | alc, charm, cif, mol, mol2, mopac, pdb, xyz | diana, mol2, pdb, seq | mol, sdf, cml | pdb, mol | shelx, gx, cif, spf, crystals, cssr-xr, csd-fdat, gsas, sybyl mol, xyz, pdb, rietica-(lhpm) | msv, pdb, ent, mol, sdf, mdl, car, csd, fdat, dat, msf, mol2, msi, cpd, cif, xyz, grd, smi |
| | specials | sma, smv | avi, mov | s3d | str | – | vrml | – | – | – | skc, wcv, log, vrml |

**Table 2-11.** (cont.)

| Name/version | | SymApps | Chem3D Ultra 7.0 | ACD/3D View | UltraMol 2003 | Rasmol | Molmol | JMol | chime 2.63 | Ortep III | ViewerLite |
|---|---|---|---|---|---|---|---|---|---|---|---|
| Supported file format | image output | gif | bmp, wmf, gif, png | – | bmp, jpg, vrml | bmp, epst, gif,ppm, rast | tiff, jpeg, png, bmp, ps | bmp, gif, jpg, png, tif, svg | – | hpgl, ps, bmp | bmp, gif, lpg |
| Functionality (general) | flying toolbars | X | X | – | – | – | – | – | – | – | X |
| | shortcuts | – | – | – | – | – | – | – | – | – | X |
| | intuitive user interface | X | X | X | X | X | X | X | X | X | X |
| | drag-and-drop | – | X | – | X | – | X | – | – | – | – |
| | undo | – | X | – | X | – | X | X | – | – | X |
| | zoom | X | X | – | X | X | X | X | X | X | X |
| | OLE function | X | X | – | – | – | – | – | – | – | X |
| Additional visualization | simple labeling | – | X | – | X | X | X | – | X | X | X |
| | templates | – | X | – | X | – | – | – | – | – | X |
| | structure models | X | – | X | X | X | X | X | X | – | X |
| Calculations | surface | X | X | X | X | X | X | – | X | X | X |
| | 2D/3D optimization (Clean-up) | – | – | X | X | – | X | – | – | – | X |

**Table 2-11.** (cont.)

| Name/version | | SymApps | Chem3D Ultra 7.0 | ACD/3D View | UltraMol 2003 | Rasmol | Molmol | JMol | chime 2.63 | Ortep III | ViewerLite |
|---|---|---|---|---|---|---|---|---|---|---|---|
| Calculations | distances, angles | – | X | X | – | X | X | – | X | X | X |
| | physicochemical properties | – | X | – | X | – | – | – | – | X | – |
| Further features | modeling/force-field | MM | MM2, Gamess, Gaussian, MOPAC, Mechanics | X | X | – | X | – | – | – | X |
| Validation | symmetry elements | X | | – | X | – | X | – | – | | X |
| URL | miscellaneous | animations | editor | – | editor module, DRAWmol | scriptable | Ramachandran plots | open source, web page integration | web page integration | specialized in crystallographic | high-quality representations of proteins, cascades |
| | | + | ++ | ++ | ++ | +++ | + | ++ | +++ | ++ | +++ |
| | | http://www.chemwindow.com/ | http://www.cambridgesoft.com/ | http://www.acdlabs.com/download/ | http://www.compuchem.com/ultra1.htm | http://www.umass.edu/microbio/rasmol/ | http://www.uni-koeln.de/themen/Chemie/software/Molmol/molmol.html | http://jchempaint.sourceforge.net/index.html | http://www.mdli.com/products/chimepro.html | http://www.chem.gla.ac.uk/~louis/software/ | http://www.accelrys.com/dstudio/ds_viewer/viewerlite/ |

### 2.12.3.1.5 **ViewerLite 5.0**

ViewerLite 5.0 from Accelrys (formerly Molecular Simulations, Inc.) is available free of charge, and runs on MS-Windows and Macintosh. It is a high-end molecule visualization tool for molecular models but cannot create chemical structures (Figure 2-133). ViewerLite supports a variety of popular molecule file formats: 15, such as pdb, mol, csd, mol2, msi, xyz, etc., for input; and 12 formats (pdb, mol, vrml, etc.) for export. Additionally, the viewer automatically generates the 3D geometry after pasting a 2D structure, for example from ISIS/Draw.

The models can be rotated, scaled, edited, labeled, and analyzed to provide a better understanding of 3D structures. Besides conventional molecular model visualization types such as wire frame, balls and sticks, etc., the program can represent advanced protein visualization, add missing hydrogen atoms, and rotate residues around a bond. Atoms or amino acid sequences can be colored according to element conventions, charges, or subunits. Molecule surfaces can be also generated and colored.

A number of additional features are available in the commercial version Viewer Pro, especially more import and export formats, and display and calculation options.

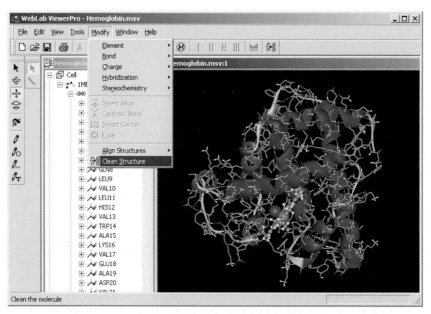

**Figure 2-133.** The ViewerLite shows an elaborate depiction of hemoglobin on the right-hand side, with the amino acids in a cascade window on the left.

## 2.12.3.2 **Web-Based Applications**

### 2.12.3.2.1 **JMol**

JMol, is a Java molecule viewer developed in tandem with the molecule editor JChemPaint in an open source project. Both programs are free of charge, and are supplied to chemists as visualization and measurement tools for 3D chemical structures. Various exchange file formats (mol, pdb, xyz, etc.) and output formats from quantum chemistry programs (MOPAC, Gaussian, etc.) can be loaded. The images can be saved as standard graphic formats (gif, jpg, ppb, bmp, png).

Additional features determine properties such as interatomic distances, bond angles, and dihedral angles from atomic coordinates. Animations of computed vibrational modes from quantum chemistry packages are also included.

*http://sourceforge.net/projects/jmol/*

### 2.12.3.2.2 **RasMol 2.7**

The 3D visualization tool from Roger Sayle (GlaxoWellcome) is free of charge and easy to use, and serves for representing molecular structures (from small molecules up to macromolecules or proteins). RasMol can be used both as a standalone program and as an external viewer for a web page. Both possibilities provide identical features in a pulldown menu which is activated by a click on the righthand mouse button (Figure 2-134).

RasMol is open source and runs on every platform (Windows, Macintosh, UNIX).

One significant characteristic of the program is the Command Line. Hereby, the visualization of the molecules is directed in the viewer with commands from a script language. The scripts, such as loop or batch files, can be saved.

*http://openrasmol.org*

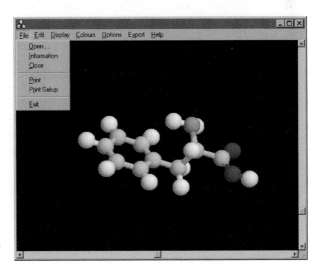

**Figure 2-134.** RasMol as a stand-alone program.

### 2.12.3.2.3 **Chime 2.6 SP4**

In contrast to RasMol, Chime is not a stand-alone program, but a plug-in for web browsers such as MS-Internet Explorer or Netscape. The user has to download and install the plug-in before it is active. Chime makes it possible to integrate and visualize 3D structures on HTML pages. The protected source code of the free-of-charge plug-in was developed by MDL Information Systems, Inc. The program is based on the open source of RasMol, which was converted to C++ and modified. Thus, these two viewers have various similarities. An advantage of Chime is that the plug-in can be included several times on web pages (RasMol: only once in a page). Furthermore, the visualization of the molecules can be controlled by using different buttons on the web page (Figure 2-135). The scripts behind these controls are integrated in the hypertext. RasMol is only controllable via direct scripting, which means that the kind of presentation has to be defined before including it on a web page. Then, the Chime presentation is fixed without direct interactivity of the plug-in. The latter is still accessible, similarly to RasMol, by pressing the right-hand mouse button, which opens a small pull-down menu where different representation styles can be chosen. New script commands provide a time-controlled movement or a specified display of a molecular orientation. Text fields (legends, annotations, etc.) can also be integrated.

*www.mdli.com/chime/index.html*

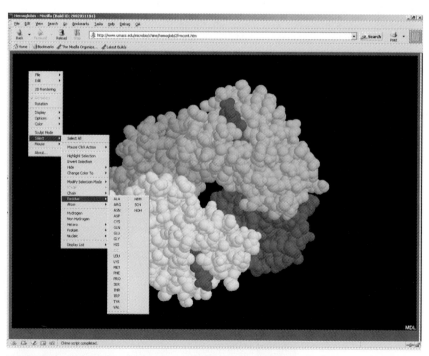

**Figure 2-135.** Hemoglobin displayed on a web-page with Chime. The display can be manipulated interactively with the mouse button.

## 2.13
## Tools: 3D Structure Generation on the Web

As already mentioned in Section 2.9, automatic 3D structure generation has a long tradition in the field of chemoinformatics. Various algorithms and approaches to addressing the problem of automatically generating 3D molecular models have been developed and published in the literature since the early 1980s. Some of the basic concepts and methods are discussed in Section 2.9 and a more detailed description is given in Chapter II, Section 7.1 in the Handbook.

The basic principles of the 3D Structure Generator CORINA were also given briefly in Section 2.9. In addition, CORINA has been made accessible to the scientific community via the Internet by the original developers of the program system. At *http://www2.chemie.uni-erlangen.de/services/3d.html* everybody is invited to test CORINA interactively. Figure 2-136 shows a screenshot of the web interface to CORINA.

The user can simply draw a chemical structure with the JME molecule editor (see Tools Section 2.12) provided on the web page and transfer it as a SMILES string to a form field (or even input a SMILES string of a chemical compound into this form field). When the submit button "Generate 3D Structure" is pressed, the structure information is sent to the web server in Erlangen, Germany, it is converted to 3D by CORINA, and the 3D coordinates are sent back to the user. Finally, the structure can be displayed either by a 3D molecule viewer (ChemSympony, Java

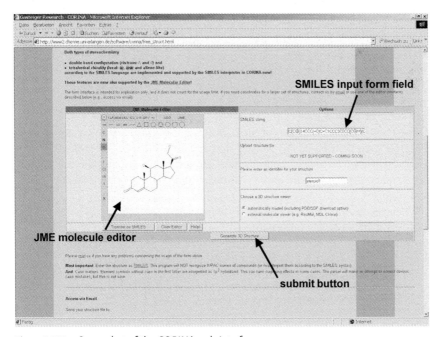

**Figure 2-136.** Screenshot of the CORINA web interface.

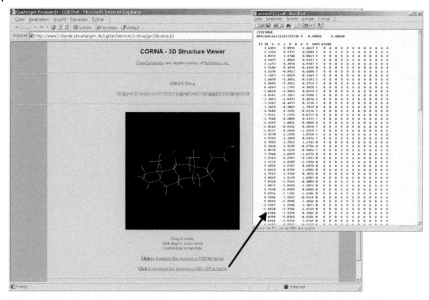

**Figure 2-137.** Screenshot of the 3D structure viewer and download options for the 3D structure generated on the CORINA web interface.

applet by NetGenics, Inc., *http://www.netgenics.com*) provided by the web server in Erlangen, or by an external viewer, e.g., RasMol or MDL Chime plug-in (see Tools Section 2.12). In addition, the structure file including the 3D coordinates can be downloaded in MDL SDfile or PDB file format (see Figure 2-137).

The company Molsoft L.L.C., La Jolla, CA, USA, also offers a 2D-to-3D converter for test purposes on the Internet. Use of the web interface of Molsoft's 2D-to-3D converter is quite similar to that of CORINA on the web. At *http://www.molsoft. com/services/2d_3d_converter.htm* the user can enter a SMILES string through a web form or draw a molecule with the JME molecule editor (see Figure 2-138). The 2D-to-3D converter is part of the ICM (Internal Coordinates Mechanics) modeling package distributed by Molsoft and is based on using MMFF atom type assignment and force-field optimization. Unfortunately, no further information or reference is given on the entire 3D model building algorithm available on the web page.

When the button "submit smiles" is pressed, the SMILES string is sent to the web server of Molsoft, converted to 3D, and the 3D structure is displayed in a Java molecule viewer on an automatically created web page (see Figure 2-139). Unfortunately, the Molsoft server does not support downloading of the 3D structures in a standard file format.

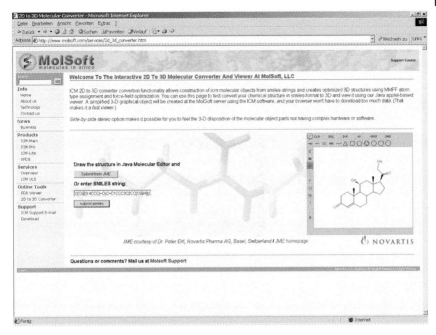

**Figure 2-138.** Screenshot of the web interface of the Molsoft ICM 2D-to-3D converter.

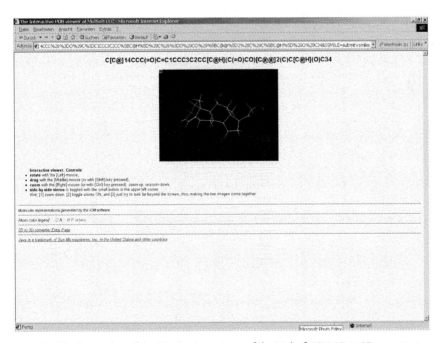

**Figure 2-139.** Screenshot of the 3D structure viewer of the Molsoft ICM 2D-to-3D converter.

**Essentials**

- Chemical structures can be transformed into a language for computer representation via line notations such as ROSDAL, SMILES, Sybyl.
- Chemical structures can also be represented and handled in matrices or connection tables.
- The constitution can be represented in an unambiguous and unique manner by canonicalization (Morgan Algorithm).
- Ring perception and equivalence of atoms and bonds are also very important tasks in for processing chemical compounds.
- Molfile, SDfile, and PDB-file are the most popular data exchange formats.
- Stereochemistry can be represented graphically in 2D structures, but also by (permutations) descriptors. It is included in all line notations and exchange formats.
- 3D structures can be generated with fragment-based, data-based, and numerical methods.
- Molecular surfaces can express various chemical and physical properties, such as electrostatic potential, atomic charges or hydrophobicity, using colored mapping.
- Chemical structures can be visualized in various different models.
- Many programs exist which can be used to generate and visualize molecular structures.

**Selected Reading**

- R.S. Cahn, O.C. Dermer, *Introduction to Chemical Nomenclature*, 5th ed., Butterworths, London, **1979**.
- For a review of different coding systems see, for example: J.E. Ash, W.A. Warr, P. Willett, *Chemical Structure Systems*, Ellis Horwood, Chichester, UK, **1991**.
- J.E. Ash, P.A. Chubb, S.E. Ward, S.M. Welford, P. Willett, *Communication Storage, and Retrieval of Chemical Information*, Ellis Horwood, Chichester, UK, **1985**.
- Graph Theory in chemistry: A.T. Balaban, *J. Chem. Inf. Comput. Sci.* **1985**, *25*, 334–343.
- *Chemical Applications of Graph Theory*, A.T. Balaban (Ed.), Academic Press, London, **1976**.
- Isomorphism: M. Randic, *J. Chem. Int. Comput. Sci.*, **1977**, *17*, 171–180.
- R.C. Read, *J. Graph Theory*, **1977**, *1*, 339–363.
- J. Zupan, *Algorithms for Chemists*, John Wiley & Sons, Chichester, UK, **1989**.
- D. Bawden, E.M. Mitchell, *Chemical Information Systems – Beyond the Structure Diagram*, Ellis Horwood, Chichester, UK, **1990**.
- P. Willett, *Three-Dimensional Chemical Structure Handling*, John Wiley & Sons, New York, **1991**.
- P.C. Jurs, *Computer Software Applications in Chemistry*, John Wiley & Sons, New York, **1996**.
- Ctfile Formats, MDL Information Systems, CA, USA, **1998**; *http://www.mdli.com/downloads/ctfile/ctfile_subs.html*

- PDB Format Description: *http://www.rcsb.org/pdb/docs/format/pdbguide2.2/ guide2.2_frame.html*
- P.G. Mezey, Molecular surfaces, in: *Reviews in Computational Chemistry*, K. Lipko-witz, D. Boyd (Eds.), VCH, Weinheim, **1990**, pp. 265–294.
- More details on all the topics in this chapter can found in the Handbook.

## Interesting Websites

- SMILES: http://www.daylight.com
- Molfile: http://www.mdli.com
- PDB Format Description: http://www.rcsb.org
- Chemical Markup Language: http://www.xml-cml.org/
- See Tables 2-10, 2-11.

## Available Software

- See Tables 2.10, 2.11.
- *http://www.cambridgesoft.com*
- *http://www.mol-net.de*
- *http://www.tripos.com*
- *http://www.mdli.com*

# References

[1] M.M. Slaughter, Universal Languages and Scientific Taxonomy in the Seventeenth Century, Cambridge University Press, Cambridge, UK, **1982**.

[2] J.W. van Spronsen, The Periodic System of Chemical Elements: A History of the First Hundred Years, Elsevier, Amsterdam, **1969**.

[3] IUPAC Nomenclature of Inorganic Chemistry: Recommendations, Part 1, G.J. Leigh (Ed.), Blackwell Scientific, London, **1990**.

[4] Nomenclature of Organic Compounds: Principles and Practice, 2nd Ed., R.B. Fox, W.H. Powell (Eds.), American Chemical Society, Washington, DC/ Oxford University Press, Oxford, New York, **2001**.

[5] G.J. Leigh, H.A. Favre, W.V. Metanomski, Principles of Chemical Nomenclature: A Guide to IUPAC Recommendations, Blackwell Science, Malden, MA, **1998**.

[6] International Union of Pure and Applied Chemistry, Commission on the Nomenclature of Organic Chemistry, The Nomenclature of Organic Chemistry. Sections A, B, C, D, E, F and H, Pergamon, Oxford, **1979**.

[7] http://www.iupac.org

[8] E.A. Hill, J. Am. Chem. Soc. **1900**, 22, 478–494.

[9] W.J. Wiswesser, J. Chem. Inf. Comput. Sci. **1985**, 25, 258–263.

[10] W.J. Wiswesser, J. Chem. Inf. Comput. Sci. **1982**, 22, 88–93.

[11] W.J. Wiswesser, A Line-Formula Chemical Notation, Y. Thomas Crowell, New York, **1954**.

[12] E.G. Smith, *The Wiswesser Line-Formula Chemical Notation*, McGraw-Hill, New York, **1968**

[13] C.E. Granito, M.D. Rosenberg, J. Chem. Doc., **1971**, 11, 251–256.

[14] J.M. Barnard, C.J. Jochum, S.M. Welford, ROSDAL: A universal structure/ substructure representation for PC–host communication, in Chemical Structure Information Systems: Interfaces Communication; and Standards, W.A. Warr (Ed.), ACS Symposium Series No. 400, American Chemical Society, Washington, DC, **1989**, pp. 76– 81.

[15] H.-G. Rohbeck, Representation of structure description arranged linearly, in Software Development in Chemistry 5, J. Gmehling (Ed.), Springer-Verlag, Berlin, **1991**, pp. 49–58.

[16] The Beilstein Online Database – Implementation, Content and Retrieval, S.R. Heller (Ed.), American Chemical Society, Washington, DC, **1990**.

[17] S. Welford, C. Jochum, Chemical structure registration for Beilstein Online, in Chem. Struct. 2, Proc. 2nd Int. Conf., **1993**, pp. 161–170.

[18] L. Goebels, A.J. Lawson, J.L. Wisniewski, J. Chem. Inf. Comput. Sci. **1991**, 31(2), 216–225.

[19] J.L. Wisniewski, J. Chem. Inf. Comput. Sci. **1990**, 30, 324–332.

[20] D. Weininger, J. Chem. Inf. Comput. Sci. **1990**, 30(3), 237–243.

[21] D. Weininger, J. Chem. Inf. Comput. Sci. **1988**, 28(1), 31–36.

[22] http://www.daylight.com

[23] J. Hinze, U. Welz, Broad SMILES, in Software Development in Chemistry 10, J. Gasteiger (Ed.), Springer-Verlag, Berlin, **1991**, pp. 59–65.

[24] R.G.A. Bone, M.A. Firth, R.A. Sykes, J. Chem. Inf. Comput. Sci., **1999**, 39(5), 846–860.

[25] D. Weininger, A. Weininger, J.L. Weininger, J. Chem. Inf. Comput. Sci. **1989**, 29(2), 97–101.

[26] S. Ash, M.A. Cline, R.W. Homer, T. Hurst, G.B. Smith, J. Chem. Inf. Comput. Sci. **1997**, 37, 71–79.

[27] http://www.scivision.com/Alchemy.html

[28] http://www.cambridgesoft.com

[29] http://www.mol-net.de

[30] http://www.tripos.com/sciTech/inSilicoDisc/chemInfo/concord.html

[31] http://www.mdli.com

[32] http://www.biorad.com

[33] A.T. Balaban, J. Chem. Inf. Comput. Sci. **1995**, 35(3), 339–350.

[34] H.P. Schultz, J. Chem. Inf. Comput. Sci. **1989**, 29(3), 227–228.

[35] O. Ivanciuc, Coding the constitution – Graph Theory in chemistry, in Handbook of Chemoinformatics, J. Gasteiger (Ed.), Wiley-VCH, Weinheim, **2003**, Chapter II, Section 4.

[36] L. Euler, Solutio Problematis ad Geometriam Situs Pertinentis, Commentarii Academiae Scientiarum Imperialis Petropolitanae 8, **1736**, pp. 128–140.

[37] A. Beck, M. Bleicher, D. Crowe, Excursion into Mathematics, Worth, **1969**.

[38] L. Spialter, J. Am. Chem. Soc. **1963**, 85(13), 2012–2013.

[39] J. Dugundji, I. Ugi, Topics Curr. Chem. **1973**, 39, 19–64.

[40] J. Gasteiger, M.G. Hutchings, B. Christoph, L. Gann, C. Hiller, P. Löw, M. Marsili, H. Saller, K. Yuki, Topics Curr. Chem. **1987**, 137, 19–73.

[41] E. Meyer, J. Chem. Inf. Comput. Sci. **1991**, 31, 68–75.

[42] A. Feldman, D.B. Holland, D.P. Jacobus, J. Chem. Doc. **1963**, 3, 187–189.

[43] F.H. Allen, O. Kennard, W.D.S. Motherwell, W.G. Town, D.G. Watson, J. Chem. Doc. **1973**, 13, 119–123.

[44] J. Becker, D. Jung, W. Kalbfleisch, G. Ohnacker, J. Chem. Inf. Comput. Sci. **1981**, 21, 111–117.

[45] http://www.umass.edu/molvis/francoeur/levinthal/lev-index.html

[46] C. Levinthal, Scientific American **1966**, 214, 42–52.

[47] C. Levinthal, C.D. Barrv et al., Computer Graphics in Macromolecular Chemistry. Emerging Concepts in Computer Graphics, D. Secrest, J. Nievergelt (Eds.), W.A. Benjamin, New York, pp. 213–253.

[48] http://www.umass.edu/molvis/francoeur/movgallery/moviegallery.html

[49] C.K. Johnson, ONRL Report No. 3794, Oak Ridge National Laboratory, Oak Ridge, TN, **1965**; abstract available at http://www.umass.edu/molvis/francoeur/ortep/ortepabstract.html

[50] Ctfile Formats, MDL Information Systems, CA, **1998**; http://www.mdli.com/downloads

[51] A. Dalby, J.G. Nourse, W.D. Hounshell, A.K.I. Gushurst, D.L. Grier, B.A. Leland, J. Laufer, J. Chem. Inf. Comput. Sci. **1992**, 32, 244–255.

[52] R.C. Bernstein, T.F. Koetzle, G.J. B. Williams, E.F. Meyer, Jr., M.D. Brice, J.R. Rodgers, O. Kennard, T. Shimanouchi, M. Tasumi, J. Mol. Biol. **1977**, 112, 535–542.

[53] PDB format description: http://www.rcsb.org

[54] S.R. Hall, F.H. Allen, I.D. Brown, Acta Crystallogr. **1991**, A47, 655–685.

[55] Macromolecular Crystallographic Information File: http://www.iucr.org/iucr-top/cif/mmcif

[56] J. Gasteiger, B.M.P. Hendriks, P. Hoever, C. Jochum, H. Somberg

[57] P. Murray-Rust, H.S. Rzepa, J. Chem. Inf. Comp. Sci., **1999**, 39, 928.

[58] P. Murray-Rust, H.S. Rzepa, J. Chem. Inf. Comp. Sci., **2001**, 41, 1113–1123.

[59] G. Gkoutos, P. Murray-Rust, H.S. Rzepa, M. Wright, J. Chem. Inf. Comp. Sci., **2001**, 41, 1124–1130.

[60] Chemical Markup Language: http://www.xml-cml.org

[61] http://openbabel.sourceforge.net

[62] W.-D. Ihlenfeldt, J.H. Voigt, B. Bienfait, F. Oellien, M.C. Nicklaus, J. Chem. Inf. Comput. Sci. **2002**, 42, 46–57.

[63] http://www2.chemie.uni-erlangen.de/services/ncidb2/index.html

[64] W.D. Ihlenfeldt, Y. Takahashi, H. Abe, S. Sasaki, J. Chem. Inf. Comput. Sci. **1993**, 34, 109–116.

[65] CACTVS Home Page: http://www2.chemie.uni-erlangen.de/software

[66] http://www.eyesopen.com/products/oechem.html

[67] http://joelib.sourceforge.net/

[68] http://cdk.sourceforge.net/

[69] http://www.jchem.com/

[70] http://cactus.cit.nih.gov/SDF_toolkit/

[71] http://www.ornl.gov/ortep/ortep.html

[72] G.M. Downs, V.J. Gillet, J.D. Holliday, M.F. Lynch, J. Chem. Inf. Comput. Sci. **1989**, 29, 172–187.

[73] G.M. Downs, Ring perception, in The Encyclopedia of Computational Chemistry, Vol. 4, P.v.R. Schleyer, N.L. Allinger, T. Clark, J. Gasteiger, P.A. Kollman, H.F. Schaeffer III, P.R. Schreiner (Eds.), John Wiley & Sons, Chichester, UK, **1998**.

[74] A.T. Balaban, P. Filip, T.S. Balaban, J. Comput. Chem. **1985**, 6, 316–329.

[75] Th. Hanser, Ph. Jauffret, G. Kaufmann, J. Chem. Inf. Comput. Sci. **1996**, 36(6), 1146–1152.

[76] J. Gasteiger, C. Jochum, J. Chem. Inf. Comput. Sci. **1979**, 19(1), 43–48.

[77] O. Ivanciuc, Canonical numbering and constitutional symmetry, in Handbook of Chemoinformatics, J. Gasteiger (Ed.); Wiley-VCH, Weinheim, **2003**, Chapter II, Section 5.1.

[78] M. Randi, J. Chem. Inf. Comput. Sci. **1975**, 15, 105.

[79] H.L. Morgan, J. Chem. Doc. **1965**, 5, 107–113.

[80] W.C. Herndon, Canonical labelling and linear notation for chemical graphs, in Chemical Applications of Topology and Graph Theory, R.B. King (Ed.), Elsevier, Amsterdam, **1983**, pp. 231–242.

[81] X. Liu, K. Balasubramanian, M.E. Munk, J. Chem. Inf. Compu. Sci. **1990**, 30, 263–269

[82] J.E. Huheey, E.A. Keiter, R.L. Keiter, Inorganic Chemistry, 4th ed., Harper-Collins, New York, **1993**, p. 672.

[83] S. Bauerschmidt, J. Gasteiger, J. Chem. Inf. Comput. Sci. **1997**, 37, 705–714.

[84] J. Gasteiger, Z. Naturforsch. **1979**, 34b, 67–75.

[85] J. Gasteiger, J. Chem. Inf. Comput. Sci. **1979**, 19, 111–115.

[86] J.E. Huheey, E.A. Keiter, R.L. Keiter, Inorganic Chemistry, 4th ed., Harper-Collins, New York, **1993**, pp. 789–806.

[87] J.M. Barnard, J. Chem. Inf. Comput. Sci. **1991**, 31, 64–68.

[88] E.S. Simmons, J. Chem. Inf. Comput. Sci. **1991**, 31, 45–53.

[89] For a general discussion of Markush structure searching, see: A.H. Berks, J.M. Barnard, M.P. O'Hara, Markush structure searching in patents, in The Encyclopedia of Computational Chemistry, Vol. 3, P.v.R. Schleyer, N. L. Allinger, T. Clark, J. Gasteiger, P.A. Kollman, H.F. Schaefer III, P.R. Schreiner (Eds.); John Wiley & Sons, Chichester, UK, **1998**, pp. 1552–1559.

[90] www.fiz-karlsruhe.de, or download: http://free.madster.com/data/free.madster.com/983/1piug165.pdf

[91] D.M. Bayada, H. Hamersma, V.J. van Geerestein, J. Chem. Inf. Comput. Sci. **1999**, 39, 1–10.

[92] D. Wild, C.J. Blankley, J. Chem. Inf. Comput. Sci. **2000**, 40, 155–162.

[93] R.A. Lewis, S.D. Pickett, D.E. Clark, Computer-aided molecular diversity analysis and combinatorial library design, in Reviews in Computational Chemistry, Vol. 16, K.B. Lipkowitz, D.B. Boyd (Eds.), Wiley-VCH, New York, **2000**, pp. 8–51.

[94] P. Willett, Similarity and Clustering in Chemical Information Systems, Research Studies Press, Letchworth, UK, **1986**.

[95] W.T. Wipke, S.K. Krishnan, G.I. Ouchi, J. Chem. Inf. Comput. Sci. **1978**, 18, 32–37.

[96] J. Zupan, Algorithms for Chemists, John Wiley & Sons, Chichester, UK, **1989**.

[97] R.G. Freeland, S.A. Funk, L.J. O'Korn, G.A. Wilson, J. Chem. Inf. Comput. Sci. (**1979**) 19, 94–97.

[98] W.-D. Ihlenfeldt, J. Gasteiger, J. Comput. Chem. **1994**, 15(8), 793–813.

[99] H.E. Helson, Structure diagram generation, in Reviews in Compuational Chemistry, Vol. 13, K.B. Lipkowitz,

D.B. Boyd (Eds.), Wiley-VCH, New York, **1999**, pp. 313–399.

[100] J. Blair, J. Gasteiger, C. Gillespie, P.D. Gillespie, I. Ugi, Tetrahedron **1974**, 30(13), 1845–1859.

[101] L.M. Masinter, N.S. Sridharan, J. Lederberg, D.H. Smith, J. Am. Chem. Soc. **1976**, 96(25), 7703–7723.

[102] V. Prelog, G. Helmchen, Angew. Chem. Int. Ed. Engl. **1982**, 21, 567–654.

[103] R.S. Cahn, C. Ingold, V. Prelog, Angew. Chem. Int. Ed. Engl. **1966**, 5, 385–419.

[104] P. Mata, A.M. Lobo, C. Marshall, A.P. Johnson, Tetrahedron: Asymmetry **1993**, 4(4), 657–668.

[105] P. Mata, A.M. Lobo, C. Marshall , A.P. Johnson, J. Chem. Inf. Comput. Sci. **1994**, 34, 491–504.

[106] S. Hanessian, J. Franco, G. Gagnon, D. Laramée, B. Larouche, J. Chem. Inf. Comput. Sci. **1990**, 30, 413–425.

[107] W.-D. Ihlenfeldt, J. Gasteiger, J. Chem. Inf. Comput. Sci. **1995**, 35, 663–674.

[108] J. Aires-de-Sousa, Representation of molecular chirality, in Handbook of Chemoinformatics, J. Gasteiger (Ed.); Wiley-VCH, Weinheim, Chapter VIII, Section 4.

[109] A.E. Petrarca, M.F. Lynch, J.E. Rush, J. Chem. Doc., **1967**, 7, 154–164.

[110] J.M. Barnard, A.P.F. Cook, B. Rohde, Storage and searching of stereochemistry in substructure search systems, in Chemical Information Systems Beyond the Structure Diagram, D. Bawden, E.M. Mitchell (Eds.), Ellis Horwood, Chichester, UK, **1990**, pp. 29–41.

[111] J.M. Barnard, J. Chem. Inf. Comput. Sci. **1990**, 30, 81–97.

[112] W.T. Wipke, T.M. Dyott, J. Am. Chem. Soc. **1974**, 96, 4834–4842.

[113] D. Weininger, SMILES – a language for molecules and reactions, in Handbook of Chemoinformatics, J. Gasteiger (Ed.); Wiley-VCH, Weinheim, **2003**, Chapter II, Section 3.

[114] G.M. Crippen, T.F. Havel, Distance geometry and molecular conformations, in Chemometrics Research Studies Series 15; D. Bawden (Ed.), Research Studies Press (Wiley), New York, **1988**.

[115] DGEOM, QCPE Program QCPE-No. 590; Quantum Chemistry Program Exchange, Indiana University, Bloomington, IN, **1995**.

[116] J.M. Blaney, J.S. Dixon, Distance geometry in molecular modeling, in Reviews in Computational Chemistry, Vol. 5, K.B. Lipkowitz, D.B. Boyd (Eds.), VCH, New York, **1994**, pp. 299–335.

[117] E.V. Goordeva, A.R. Katritzky, V.V. Shcherbukhin, N.S. Zefirov, J. Chem. Inf. Comput. Sci. **1993**, 33, 102–111.

[118] A.R. Leach, K. Prout, J. Comput. Chem. **1990**, 11, 1193–1205.

[119] A.R. Leach, D.P. Dolata, K. Prout, J. Chem. Inf. Comput. Sci. **1990**, 30, 316–324.

[120] A.R. Leach, J. Chem. Inf. Comput. Sci. **1994**, 34, 661–670.

[121] J. Sadowski, J. Gasteiger, Chem. Rev. **1993**, 7, 2567–2581.

[122] J. Sadowski, J. Gasteiger, G. Klebe, J. Chem. Inf. Comput. Sci. **1994**, 34, 1000–1008.

[123] J. Sadowski, Three-dimensional structure generation: automation, in Encyclopedia of Computational Chemistry, P.v.R. Schleyer, N.L. Allinger, T. Clark, J. Gasteiger, P.A. Kollman, H.F. Schaefer III, P.R. Schreiner (Eds.), John Wiley & Sons, Chichester, UK, **1998**, pp. 2976–2988.

[124] CORINA Version 3.0 is available from Molecular Networks GmbH, Erlangen, Germany (http://www.mol-net.de).

[125] J. Gasteiger, M.G. Hutchings, B. Christoph, L. Gann, C. Hiller, P. Löw, M. Marsili, H. Saller, K. Yuki, Topics Curr. Chem. **1987**, 137, 19–73.

[126] J. Gasteiger, W.D. Ihlenfeldt, P. Röse, Recl. Trav. Chim. Pays-Bas **1992**, 111, 270–290.

[127] J.H. Schuur, P. Selzer, J. Gasteiger, J. Chem. Inf. Comput. Sci. **1996**, 36, 334–344.

[128] a) J. Sadowski, M. Wagener, J. Gasteiger, Angew. Chem. Int. Ed. Engl. **1995**, 34, 2674–2677; b) M. Wagener, J. Sadowski, J. Gasteiger, J. Am. Chem. Soc. **1995**, 117, 7769–7775.

[129] National Cancer Institute, National Institute of Health (Developmental Therapeutics Program, DTP/2D and

3D Structural Information, http://dtp.nci.nih.gov/docs/3d_data

[130] H. Schönberger, C.H Schwab, A. Hirsch, J. Gasteiger, J. Mol. Model. **200**, 6, 379– 395.

[131] R.S. Pearlman, Chem. Des. Auto. News **1987**, 2, 1/5–6.

[132] CONCORD is available from Tripos, Inc., St.Louis, MO, USA (http://www.tripos.com).

[133] See, for example: a) A. Smellie, S.D. Kahn, S.L. Teig, J. Chem. Inf. Comput. Sci. **1995**, 35, 285–294; b) A. Smellie, S.D. Kahn, S.L. Teig, J. Chem. Inf. Comput. Sci. **1995**, 35, 295–304.

[134] a) M. Saunders, J. Am. Chem. Soc. **1987**, 109, 3150–3152; b) M. Saunders, J. Comput. Chem. **1989**, 10, 203–208.

[135] G. Chang, W.C. Guida, W.C. Still, J. Am. Chem. Soc. 1989, 111, 4379–4386.

[136] N. Metropolis, A.W. Rosenbluth, M.N. Rosenbluth, A.H. Teller, E. Teller, J. Chem. Phys. **1953**, 21, 1087–1092.

[137] D.E. Goldberg, Genetic Algorithms in Search, Optimization and Machine Learning, Addison-Wesley, New York, **1989**.

[138] See, for example: N. Naiv, J.M. Goodman, J. Chem. Inf. Comput. Sci. **1998**, 38, 317–320.

[139] T.P. Lybrand, Computer simulations of biomolecular systems using molecular dynamics and free energy perturbation methods, in Reviews in Computational Chemistry, Vol. 1, K.B. Lipkowitz, D.B. Boyd (Eds.), VCH, New York, **1990**, pp. 295–320.

[140] S. Kirkpatrick, C.D. Gelatt, M.P. Vecchi, Science **1983**, 220, 671–680.

[141] C.H. Schwab, Konformative Flexibilität von Liganden im Wirkstoffdesign. Ph.D. Thesis; University of Erlangen-Nuremberg, Erlangen, **2001**.

[142] ROTATE Version 1.1 is available from Molecular Networks GmbH, Erlangen, Germany (http://www.mol-net.de).

[143] a) G. Klebe, T. Mietzner, J. Comput.-Aided Mol. Des. **1994**, 8, 583–606

[144] a) F.H. Allen, J.E. Davies, J.J. Galloy, O. Johnson, O. Kennard, C.F. Macrae, E.M. Mitchell, G.F. Mitchell, J.M. Smith, D.G. Watson, J. Chem. Inf. Comput. Sci. **1991**, 31, 187–204; b) F.H. Allen, V.J. Hoy, Cambridge Structural

Database, in Encyclopedia of Computational Chemistry, P.v.R. Schleyer, N.L. Allinger, T. Clark, J. Gasteiger, P.A. Kollman, H.F. Schaefer III, P.R. Schreiner (Eds.), John Wiley & Sons, Chichester, UK, **1998**, pp. 155–167.

[145] See, for example: a) S.G. Grdadolnik, D.F. Mierke, J. Chem. Inf. Comput. Sci. **1997**, 37, 1044–1047; b) M. Pellegrini, S. Mammi, E. Peggion, D.F. Mierke, J. Med. Chem. **1997**, 40, 92–98.

[146] F.C. Bernstein, T.F. Koetzle, G.J. Williams, E.E. Meyer, M.D. Brice, J.R. Rogers, O. Kennard, T. Shimanouchi, M. Tasumi, J. Mol. Biol. **1977**, 112, 535–542.

[147] H.M. Berman, J. Westbrook, Z. Feng, G. Gilliland, T.N. Bhat, H. Weissig, I.N. Shindyalov, P.E. Bourne, Nucleic Acids Res. **2000**, 28, 235–242.

[148] S.R. Hall, J. Chem. Inf. Comput. Sci. **1991**, 31, 326–333.

[149] S.R. Hall, N. Spadaccini, J. Chem. Inf. Comput. Sci. **1994**, 34, 505–508.

[150] S.R. Hall, F.H. Allen, I.D. Brown, Acta Crystallogr. **1991**, A47, 655–685.

[151] P.E. Bourne, H.M. Berman, B. McMahon, K.D. Watenpaugh, J.W. Westbrook, P.M.D. Fitzgerald, Methods Enzymol., **1997**, 277, 571–590. This paper is also available on the WWW: http://www.sdsc.edu/pb/cif/papers/methenz.html

[152] http://ndbserver.rutgers.edu/mmcif

[153] http/www2.chemie.uni-erlangen.de/publications/CI-book/index.html

[154] http://www.openrasmol.org

[155] http://www.mdlchime.com/chime/

[156] T.N. Bhat, P. Bourne, Z. Feng, G. Gilliland, S. Jain, V. Ravichandran, B. Schneider, K. Schneider, N. Thanki, H. Weissig, J. Westbrook, H.M. Berman, Nucleic Acids Res. **2001**, 29, 214–218.

[157] J. Westbrook, Z. Feng, S. Jain, T.N. Bhat, N. Thanki, V. Ravichandran, G.L. Gilliland, W.F. Bluhm, H. Weissig, D.S. Greer, P.E. Bourne and H.M. Berman, Nucleic Acids Res. **2002**, 30, 245–248.

[158] G. Chapuis, J.M. Farkas., J.M. Pérez-Mato, M. Senechal, W. Steurer, C. Janot, D. Pandey, A. Yamamoto, Acta Crystallogr. **1997**, A53, 95–100.

[159] E. Migliavacca, A.A. Adzhubei, M.C. Peitsch, Bioinformatics **2001**, 17(11), 1047–52.

[160] I.D. Brown, Acta Crystallogr. **1981**, A39, 216–224.

[161] I.D. Brown, Acta Crystallogr. **1988**, A44, 232.

[162] F.H. Allen, P.R. Edgington, CIFER: A Program for CIF Generation, Cambridge Crystallographic Data Centre, Cambridge, UK, **1992**.

[163] B. McMahon, CIFTEX. A Filter for Translating a Crystallographic Information File to a TeX File, International Union of Crystallography, 5 Abbey Square, Chester, UK,. **1992**.

[164] D. Knuth The TeXBook, Addison Wesley Longman, **1988**.

[165] J.D. Westbrook, P.E. Bourne, Bioinformatics **2000**, 16, 159–168.

[166] P.E. Bourne, H.M. Berman, B. McMahon, K.D. Watenpaugh, J.D. Westbrook, P.M.D. Fitzgerald, Macromolecular crystallographic information file, in: Methods in Enzymology **1997**, 277, 571–590.

[167] J.D. Westbrook, S.R. Hall, A dictionary description language for macromolecular structure, Report NDB-110, Rutgers University, New Brunswick, NJ, **1995**.

[168] http://pdb.rutgers.edu/mmcif

[169] S.R. Hall, H.J. Bernstein, J. Appl. Crystallogr. **1996**, 29, 598–603.

[170] http://openmms.sdsc.edu

[171] http://pdb.sdsc.edu/index.html#mmcif

[172] http://www.sdsc.edu/pb/cif/OOSTAR.html

[173] http://www.bernstein-plus-sons.com/software/cif2pdb

[174] http://www.pc.chemie.tu-darmstadt.de/research/vrml/pdb2vrml.html

[175] http://www2.chemie.uni-erlangen.de/services

[176] http://www.dkfz-heidelberg.de/spec/pdb2mgif

[177] http://www.expasy.ch/spdbv

[178] http://www.pymol.org

[179] http://www.accelrys.com

[180] http://www.candomultimedia.com/medical

[181] http://www.macromedia.com

[182] B. Lee, F.M. Richards, J. Mol. Biol. **1971**, 55(3), 379–400.

[183] M.L. Connolly, J. Appl. Crystallogr., **1983**, 16, 548–558.

[184] M.L. Connolly, Science **1983**, 221, 709–713.

[185] B.S. Duncan, A.J. Olson, J. Mol. Graphics **1995**, 13, 250–257.

[186] A.N. Jain, T.G. Dietterich, R.H. Lathrop, D. Chapman, J. Comput.-Aided Mol. Design, **1994**, 8, 635–652.

[187] P.G. Mezey, Molecular surfaces, in Reviews in Computational Chemistry, K. Lipkowitz, D. Boyd (Eds.), VCH, Weinheim, **1990**, pp. 265–294.

[188] W. Kühlbrandt, A. Williams, Curr. Opin. Chem. Biol. **1999**, 3, 537–543.

[189] H. Stark, P. Dube, R. Luhmann, B. Kastner, Nature **2001**, 409(6819), 539–542.

[190] J.C. Kendrew, G. Bodo, H.M. Dintzis, R.G. Parrish, H. Wyckoff, D.C. Phillips, Nature **1958**, 181, 662–666.

[191] B. Rubin, J.S. Richardson, Biopolymers **1972**, 11(11), 2381–2385.

[192] J.S. Richardson, D.C. Richardson, K.A. Thomas, E.W. Silverton, D.R. Davies, J. Mol. Biol., **1976**, 102, 221–235.

[193] C. Levinthal, Scientific American **1966**, 214, 42–52.

[194] http://www.umass.edu/molvis/francoeur/index.html

[195] A.M. Lesk, Comput. Biol. Med. **1977**, 7, 113–129.

[196] K.M. Beem, D.C. Richardson, K.V. Rajagopalan, Biochemistry **1977**, 16, 1930–1936.

[197] T.K. Porter, Comput. Graphics (SIGGRAPH) **1979**, 13, 234–236.

[198] R. Koradi, M. Billeter, K. Wüthrich, J. Mol. Graphics **1996**, 14, 51–55.

[199] A.S. Dreiding, Helv. Chim. Acta **1959**, 42, 1339–1344.

[200] H. Gouraud, IEEE Trans. Comput. **1971**, 20(6), 623–628.

[201] J.S. Richardson, Adv. Protein. Chem. **1981**, 34, 167–339.

[202] J.F. Blinn, Comput. Graphics **1978**, 12, 286–292.

[203] M. Teschner, C. Henn, H. Vollhardt, S. Reiling, J. Brickmann, J. Mol. Graphics **1994**, 12, 98–105.

[204] R. Bonaccorsi, E. Scrocco, J. Tomasi, J. Chem. Phys. **1970**, 54(10), 5270.

[205] J.S. Murray, P. Politzer, Electrostatic potential, in Encyclopedia of Computational Chemistry, Vol.2, P. von Rague-Schleyer, N.L. Allinger, T. Clark, J. Gasteiger, P.A. Kollman, H.F. Schaefer III, P.R. Schreiner (Eds.); John Wiley & Sons, Chichester, UK, **1998**, pp. 912–920.

[206] www.java.sun.com

[207] P. Ertl, O. Jacob, J. Mol. Struct. (THEOCHEM) **1997**, 419, 113–120.

[208] http://jchempaint.sourceforge.net

[209] Representative sites which use the JME applet are listed at www.molinspiration.com/jme and the JME editor may be obtained from peter.ertl@pharma.novartis.com

# 3
# Representation of Chemical Reactions

*Johann Gasteiger*

## Learning Objectives

- To understand how to extract knowledge from reaction information
- To recognize reaction classification as an important step in learning from reaction instances
- To appreciate the reaction center and its importance in reaction searching
- To become familiar with basic models of chemical reactivity
- To know simple approaches to quantify chemical reactivity
- To be able to follow some algorithmic approaches to reaction classification
- To understand the formal treatment of the stereochemistry of reactions

## 3.1
## Introduction

Reactions represent the dynamic aspect of chemistry, the interconversion of chemical compounds. Chemical reactions produce the compounds that are sold by industry and that play a big role in maintaining the standard of living of our society: they transform the food that we take up in our body into energy and into other compounds; and they provide the energy for surviving in a hostile environment and the energy for a large part of our transportation systems.

Reactions are run under a variety of conditions, ranging from making large collections of small amounts of compounds through parallel syntheses on well-plates, through laboratory syntheses in standard three-necked flasks, all the way to the large-scale industrial processes in huge reactors. In a mass spectrometer, single molecules break up in the gas phase after ionization through electron impact; in a cell, genes control the synthesis of complex proteins.

Clearly then, the understanding of chemical reactions under such a variety of conditions is still in its infancy and the prediction of the course and products of a chemical reaction poses large problems. The *ab initio* quantum mechanical calculation of the pathway and outcome of a single chemical reaction can only be done after careful and time-consuming systematic studies that ask for large computational resources. And even then, the effect of solvents can be estimated only roughly, to say nothing of the prediction of the influence of catalysts, temperature, or pressure.

Nevertheless, chemists have been planning their reactions for more than a century now, and each day they run hundreds of thousands of reactions with high degrees of selectivity and yield. The secret to success lies in the fact that chemists can build on a vast body of experience accumulated over more than a hundred years of performing millions of chemical reactions under carefully controlled conditions. Series of experiments were analyzed for the essential features determining the course of a reaction, and models were built to order the observations into a conceptual framework that could be used to make predictions by analogy. Furthermore, careful experiments were planned to analyze the individual steps of a reaction so as to elucidate its mechanism.

Let us illustrate this with the example of the bromination of monosubstituted benzene derivatives. Observations on the product distributions and relative reaction rates compared with unsubstituted benzene led chemists to conceive the notion of inductive and resonance effects that made it possible to "explain" the experimental observations. On an even more quantitative basis, linear free energy relationships of the form of the Hammett equation allowed the estimation of relative rates. It has to be emphasized that inductive and resonance effects were conceived, not from theoretical calculations, but as constructs to order observations. The "explanation" is built on analogy, not on any theoretical method.

The objective of chemoinformatics is to assist the chemist in giving access to reaction information, in deriving knowledge on chemical reactions, in predicting the course and outcome of chemical reactions, and in designing syntheses. Specifically, the problems of accomplishing the following tasks have to be solved:

- storing information on chemical reactions,
- retrieving information on chemical reactions,
- comparing and analyzing sets of reactions,
- defining the scope and limitations of a reaction type,
- developing models of chemical reactivity,
- predicting the course of chemical reactions,
- analyzing reaction networks,
- developing methods for the design of syntheses.

Chemists usually represent reactions by a reaction equation that gives the structures of the starting materials and of the products of a reaction, and, optionally, information on reagents, catalysts, solvents, temperature, etc., as well as data on the yield of the reaction (Figure 3-1).

When is a compound to be considered as a starting material, and when as a reagent? There is certainly some arbitrariness involved in such a distinction, because both a starting material and a reagent might contribute atoms to the reaction products. Some reaction databases consider a compound to be a starting material when

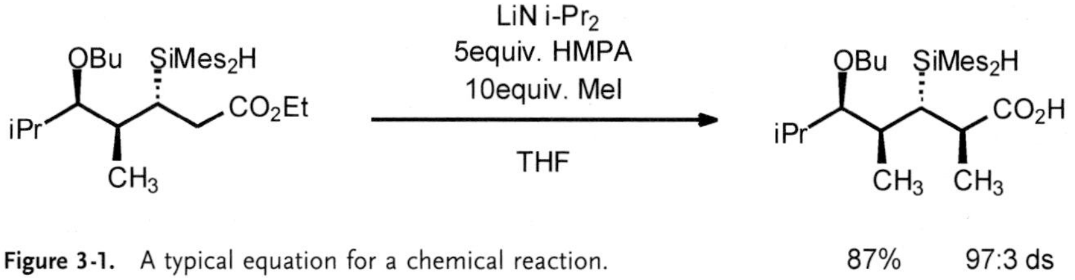

**Figure 3-1.** A typical equation for a chemical reaction.          87%          97:3 ds

it contributes at least one carbon atom to a reaction product, and take it as a re-agent otherwise.

Unfortunately, in most cases not all the available information on a reaction is given in the reaction equation in a publication, and even less so in reaction data-bases. To obtain a fuller picture of the reaction that was performed, the text describ-ing the experimental procedure in the publication (or a lab journal) would have to be consulted. Reaction products that are considered as trivial, such as water, alco-hol, ammonia, nitrogen, etc., are generally not included in the reaction equation or mentioned in the text describing the experimental work. This poses serious prob-lems for the automatic identification of the reaction center. It is highly desirable to have the full stoichiometry of a reaction specified in the equation.

When are two reactions identical? Clearly, they should have the same starting materials, and the same reaction conditions. But does a slight change in the ratio of starting materials or in reaction temperature warrant storage of those two reactions separately in a reaction database? In principle, yes, but it is never done in practice! If we really want to learn more about a given reaction and how its course can be influenced, we should know how changes in the reaction condi-tions change the course, or the yield, of a chemical reaction.

A further indication of the poor standard of information in a reaction equation is that the plus (+) symbol is used in a reaction equation for two entirely different purposes (Figure 3-2).

In the first reaction equation (Figure 3.2a), the + symbol on the right-hand side indicates that a molecule of ethanol is simultaneously, and necessarily, produced

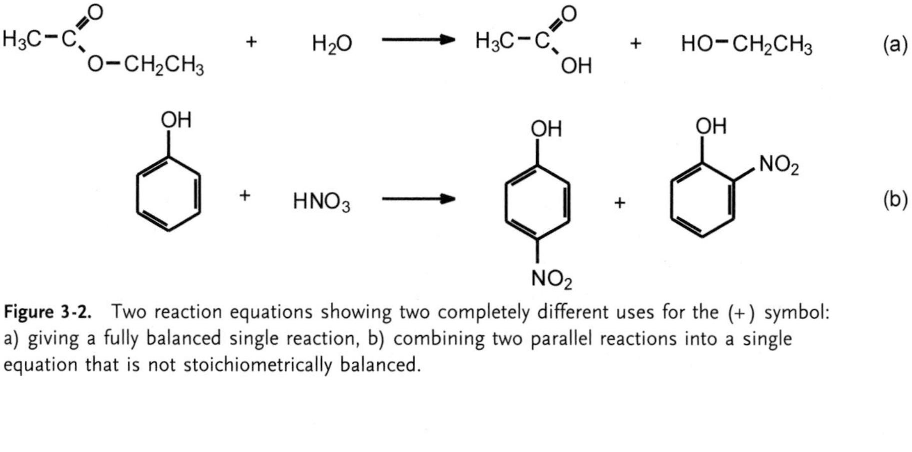

**Figure 3-2.** Two reaction equations showing two completely different uses for the (+) symbol: a) giving a fully balanced single reaction, b) combining two parallel reactions into a single equation that is not stoichiometrically balanced.

together with acetic acid: If acetic acid is produced, then ethanol also must be formed. In the second reaction equation (Figure 3.2b), the + symbol on the right-hand side actually reports that there are two parallel reactions occurring, one leading to the first product, *para*-nitrophenol, and a second one giving the second product, *ortho*-nitrophenol. It can easily be imagined that such a dual use of one and the same symbol for two different purposes poses problems in the automatic processing of chemical reaction information. In fact, in the second case (Figure 3.2b) the two parallel reactions are usually stored separately in reaction databases. However, links should be provided to indicate that these two reactions occur in parallel.

Now, chemists have acquired much of their knowledge on chemical reactions by inductive learning from a large set of individual reaction instances. How has this been done? And how can we build on these methods and knowledge and perform it in a more systematic manner by algorithmic techniques?

An important step in learning from individual reactions is the grouping of reaction instances into reaction types, the classification of reactions. In this chapter we first show different approaches that have been made in chemistry to classify reactions into reaction types (Section 3.2). We then emphasize the importance of the reaction center, and the bonds broken and made in a chemical reaction (Section 3.3). Next, we present some of the reasoning that has been put forward by chemists to rationalize the breaking and making of certain bonds in a molecule. In addition, we will discuss some simple methods for a more quantitative treatment of chemical reactivity (Section 3.4). Building on these discussions, in Section 3.5 we present a few approaches to reaction classification that have also been implemented as algorithms. Section 3.6 deals with the stereochemistry of reactions and its handling by permutation groups.

## 3.2
## Reaction Types

The first step in an inductive learning process is always to order the observations to group those objects together that have essential features in common and to separate objects that are distinctly different. Thus, in learning from individual reactions we have to classify reactions – we have to define reaction types that encompass a series of reactions with essential common characteristics. Clearly, the definition of what are essential common features is subjective and thus a variety of different classification schemes have been proposed.

Chemists usually learn about reactions according to functional groups; for example, "How can I make an aldehyde and what reactions are known for aldehydes?" This is clearly not a very good starting point for classifying reactions. The poor state of affairs in the definition of reaction types is further quite vividly illustrated by the fact that many chemical reactions are identified by being named after their inventor: Diels–Alder reaction, Michael addition, Lobry–de Bruyn–van Ekenstein rear-

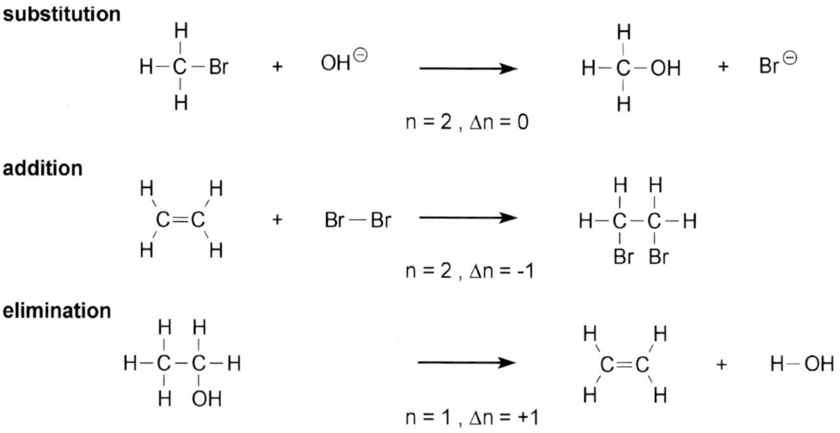

**substitution**

$$H-\underset{\underset{H}{|}}{\overset{\overset{H}{|}}{C}}-Br \quad + \quad OH^{\ominus} \quad \longrightarrow \quad H-\underset{\underset{H}{|}}{\overset{\overset{H}{|}}{C}}-OH \quad + \quad Br^{\ominus}$$

$$n = 2 , \Delta n = 0$$

**addition**

$$\underset{H}{\overset{H}{\diagdown}}C=C\underset{H}{\overset{H}{\diagup}} \quad + \quad Br-Br \quad \longrightarrow \quad H-\underset{\underset{Br}{|}}{\overset{\overset{H}{|}}{C}}-\underset{\underset{Br}{|}}{\overset{\overset{H}{|}}{C}}-H$$

$$n = 2 , \Delta n = -1$$

**elimination**

$$H-\underset{\underset{H}{|}}{\overset{\overset{H}{|}}{C}}-\underset{\underset{OH}{|}}{\overset{\overset{H}{|}}{C}}-H \quad \longrightarrow \quad \underset{H}{\overset{H}{\diagdown}}C=C\underset{H}{\overset{H}{\diagup}} \quad + \quad H-OH$$

$$n = 1 , \Delta n = +1$$

**Figure 3-3.** Representative, simple examples of a substitution, an addition, and an elimination reaction showing the number, *n*, of reaction partners, and the change in *n*, Δ*n*, during the reaction.

rangement, etc. Secretive as this grouping of reactions into name reactions might seem, chemists nevertheless associate with a certain name reaction much knowledge on its essential features, its mechanism, and its scope and limitations.

On a more rational basis, reactions can be classified according to the overall change in molecularity, the change Δ*n* in the number of molecules (*n*) participating in a reaction (Figure 3-3).

A more detailed classification of chemical reactions will give specifications on the mechanism of a reaction: electrophilic aromatic substitution, nucleophilic aliphatic substitution, etc. Details on this mechanism can be included to various degrees; thus, nucleophilic aliphatic substitutions can further be classified into $S_N1$ and $S_N2$ reactions. However, as reaction conditions such as a change in solvent can shift a mechanism from one type to another, such details are of interest in the discussion of reaction mechanism but less so in reaction classification.

**3.3
Reaction Center**

Essential features of any chemical reaction are the bonds broken and made in its course and, where applicable, the free electrons shifted. The atoms and bonds directly involved in the bond and electron rearrangement process constitute the reaction center. With the reaction center specified, studies can be done to elucidate a reaction mechanism, to determine the sequence of events in breaking and making bonds. Figure 3-4a shows the overall bond change during the elimination of hydrogen bromide from an alkyl bromide to give an alkene. The three possible mechanisms for achieving this reaction are:

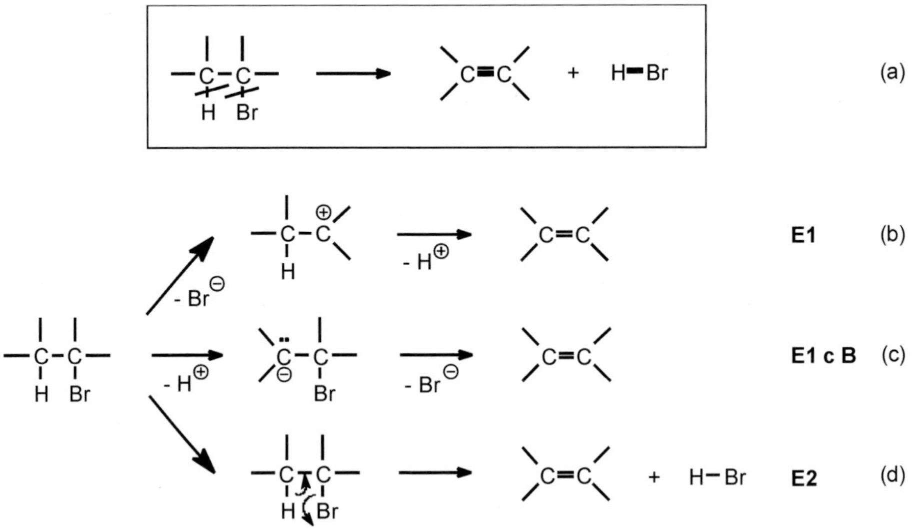

**Figure 3-4.** a) The reaction site of an elimination reaction. The bonds to be broken are crossed through, and the bonds to be made are drawn with heavy lines. b) to d): The three mechanisms to achieve this reaction.

- loss of bromide ion first, followed by deprotonation (Figure 3-4b): E1 mechanism);
- deprotonation first, followed by loss of bromide ion (Figure 3-4c): E1cB mechanism);
- simultaneous loss of a proton and a bromide ion (Figure 3-4d): E2 mechanism).

Each of these three reaction mechanisms can be observed, depending on the other groups bonded to the reaction center and the reaction conditions chosen. The elucidation of the mechanism of reactions is the primary vehicle for deepening our understanding of the driving forces of chemical reactions. Chemists have amassed a huge arsenal of methods to shed light on the sequence of events that occur and the intermediates that are involved on the pathway from the starting materials to the products of a reaction. Chemoinformatics tools can help in analyzing reaction information and thus in increasing our knowledge of the factors influencing the course and products of chemical reactions. Clearly, the deeper the analysis can be driven, the more detailed the available information. Unfortunately, basically no information on reaction mechanisms or kinetic data is stored at present in reaction databases.

Consideration of the reaction center or reaction site is of central importance in reaction searching. It does not suffice to specify the functional groups in the starting materials and in the products of a reaction when one is interested in a certain transformation. On top of that, one also has to specify that these functional groups should participate directly in the reaction – that they should be part of the reaction center.

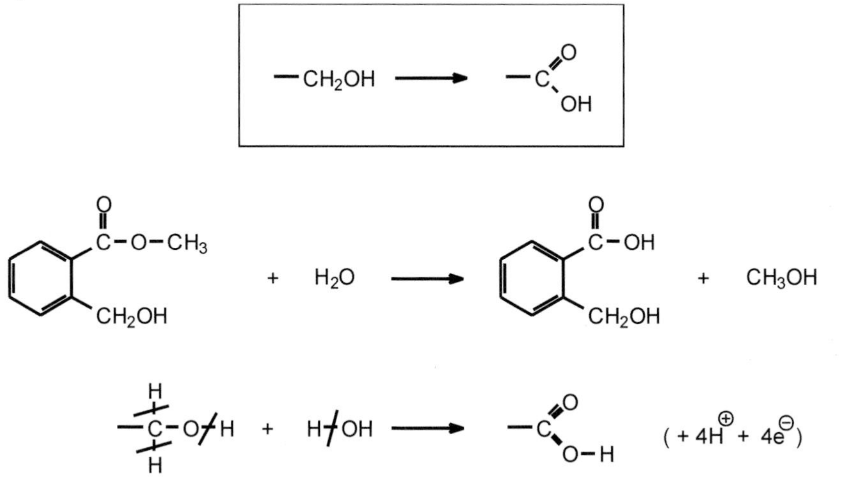

(a)

(b)

(c)

**Figure 3-5.** The search for a) oxidations of primary alcohols to carboxylic acids will obtain reaction b) as a hit, although this reaction is in reality a hydrolysis of an ester. c) The correct specification of the query to obtain reactions involving the oxidation of alcohols to carboxylic acids.

Thus, when one is interested in oxidations of primary alcohols to carboxylic acids, it does not suffice to specify that the starting material should contain a primary alcohol and the product should contain a carboxyl group (Figure 3-5a). For then the reaction in Figure 3-5b will also be retrieved as a hit, although, in reality, it constitutes the hydrolysis of an ester to an acid. Only when it is specified that the primary alcohol and the acid should be part of the reaction center by indicating which bonds should be broken and made in the course of the reaction (Figure 3-5c), will reaction Figure 3-5b not be a hit; only then will reactions that show the desired oxidation of primary alcohols to acids be obtained as hits.

The reaction center has either to be specified when inputting a reaction into a database, or it has to be determined automatically. Specification on input is time-consuming but it can benefit from the insight of the human expert, particularly so if the reaction input is done by the primary investigator as is the case in an electronic notebook. Automatic determination of reaction centers is difficult, particularly so when incomplete reaction equations are given where the stoichiometry of a reaction is not balanced (see Section 3.1). One approach is to try first to complete the stoichiometry of a reaction equation by filling in the missing molecules such as water, $N_2$, etc. and then to start with reaction center determination. A few systems for automatic reaction center specification are available. However, little has been published on this matter and therefore it is not discussed in any detail here.

**3.4**
**Chemical Reactivity**

3.4.1
**Physicochemical Effects**

Quantum chemical methods have matured to a high level of sophistication allow-ing the calculation of the energies of transition states and the geometry of reaction coordinates to a high degree of accuracy. Such calculations, however, are rather time-consuming, only allowing the investigation of single reactions one at a time.

Thus, when a large set of chemical reactions has to be investigated, an inductive learning process, deriving knowledge on chemical reactions and reactivity from a series of reactions, still has many merits. Such chemical knowledge can be put into models that then allow one to predict the course of new reactions.

Some of the concepts that chemists have introduced for the discussion of chem-ical reactivity are summarized below. Much of this will be common knowledge to readers that have studied chemistry; they can easily skip this section. However, for readers from other scientific disciplines or whose chemical knowledge has become rusty, some fundamental concepts are presented here.

Chemists have formulated a variety of concepts of a physicochemical or theore-tical nature in their endeavors to order their observations on chemical reactions and to develop insight into the effects that control the initiation and course of chemical reactions. The main effects (but not the only ones, by far) influencing chemical reactivity are described below.

3.4.1.1 **Charge Distribution**
Carbon and hydrogen atoms have similar electron-attracting power, and thus com-pounds consisting of these two kinds of atoms only, hydrocarbons, have a uniform electron density distribution and no polarity. Heteroatoms such as oxygen, nitro-gen, or the halogen atoms, on the other hand, have a higher electron-attracting power, i.e., higher electronegativity, and thus introduce polarity into organic com-pounds. A simple picture expressing this notion is drawn by showing these atoms bearing a partial negative charge; consequently, the atoms to which they are bonded carry a partial positive charge (Figure 3-6a). Metal atoms, on the contrary, are less electronegative than carbon atoms and thus give electron density away, pro-viding adjacent carbon atoms with a partial negative charge. Reagents with sites of high electron density (nucleophilic agents) seek atoms with low electron density and, conversely, atoms with low electron density (electrophilic agents) bind to atoms with high electron density.

Appealing and important as this concept of a molecule consisting of partially charged atoms has been for many decades for explaining chemical reactivity and discussing reaction mechanisms, chemists have only used it in a qualitative man-ner, as they can hardly attribute a quantitative value to such partial charges. Quan-tum mechanical methods (see Section 7.4) as well as empirical procedures (see

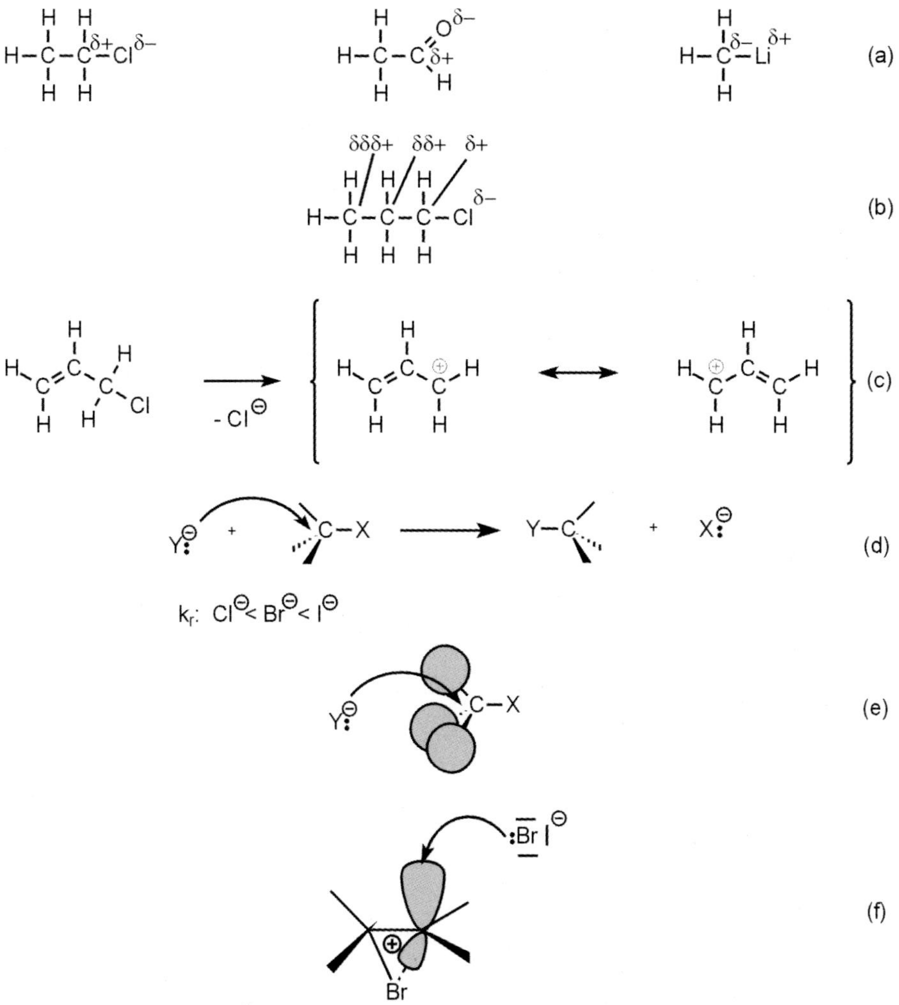

$$H-\underset{\underset{H}{|}}{\overset{\overset{H}{|}}{C}}-\overset{\overset{H}{|}}{\underset{\underset{H}{|}}{C}}\overset{\delta+}{=}Cl^{\delta-} \qquad H-\overset{\overset{H}{|}}{\underset{\underset{H}{|}}{C}}-\overset{\overset{O^{\delta-}}{\diagup}}{\underset{\diagdown_{H}}{C}}_{\delta+} \qquad H-\overset{\overset{H}{|}}{\underset{\underset{H}{|}}{C}}\overset{\delta-}{-}\overset{\delta+}{Li} \qquad \text{(a)}$$

$$\overset{\delta\delta\delta+ \quad \delta\delta+ \quad \delta+}{H-\underset{\underset{H}{|}}{\overset{\overset{H}{\diagup}}{C}}-\underset{\underset{H}{|}}{\overset{\overset{H}{\diagup}}{C}}-\underset{\underset{H}{|}}{\overset{\overset{H}{\diagup}}{C}}-Cl}^{\delta-} \qquad \text{(b)}$$

$$\text{(c)}$$

$$\text{(d)}$$

$$k_r: \; Cl^{\ominus} < Br^{\ominus} < I^{\ominus}$$

$$\text{(e)}$$

$$\text{(f)}$$

**Figure 3-6.** a) The charge distribution, b) the inductive effect, and c) the resonance effect, d) the polarizability effect, e) the steric effect, and f) the stereoelectronic effect.

Section 7.1) have been developed to assign quantitative values to the partial charges of the atoms in a molecule [1, 2]. This opens the door to defining chemical reactivity on a more quantitative basis.

### 3.4.1.2 Inductive Effect

The polarizing influence of an electronegative atom decreases with the number of intervening $\sigma$-bonds. This is called the inductive effect and is indicated in Figure 3-6b by a progression of $\delta$ symbols. It is generally accepted that the inductive effect is attenuated by a factor of 2–3 by each intervening bond. The inductive effect is not only operative in the ground state of a molecule but also exerts its influence when bonds are broken heterolytically. Then, electronegative atoms adjacent to the reacting bond will stabilize incipient negative charges; and vice versa. It has been shown that residual electronegativity as calculated by the PEOE method (see Section 7.1) can be taken as a quantitative measure of the inductive effect [3].

### 3.4.1.3 Resonance Effect

The charges that are generated on heterolysis, on polar breaking of a bond, can also be stabilized by delocalization, as observed in conjugated $\pi$-systems (Figure 3-6c). This is called the resonance or mesomeric effect and it usually has an even greater influence on chemical reactivity than the inductive effect [2]. The RAMSES data structure (see Section 2.6.2) is particularly suited for determining conjugated $\pi$-systems.

### 3.4.1.4 Polarizability Effect

An electric field induces a dipole in a molecule; the magnitude of this dipole is proportional to the polarizability of the molecule, which is measured by the mean molecular polarizability due to motion-averaging. In chemical reactions, charges within a molecule may be generated by bond breaking or bond formation. Atoms that have a high polarizability can stabilize such charges, but they do so more the closer they are to these charges. Furthermore, groups that have more polarizable electrons can better provide such electrons to the reacting complex. This explains the increase in reaction rate on going from the chloride through the bromide to the iodide ion in nucleophilic aliphatic substitution (Figure 3-6d). In Section 7.1, a method for the quantification of the polarizabilitiy effect is presented [4].

### 3.4.1.5 Steric Effect

The attack by a reagent of a molecule might be hampered by the presence of other atoms near the reaction site. The larger these atoms and the more are there, the higher is the geometric restriction, the steric hindrance, on reactivity. Figure 3-6e illustrates this for the attack of a nucleophile on the substrate in a nucleophilic aliphatic substitution reaction.

### 3.4.1.6 Stereoelectronic Effects

Some reactions require the bonds being broken or made in a reaction to be aligned with other parts ($\pi$- or free electrons) of a molecule. These requirements are called stereoelectronic effects. Figure 3-6f shows that the bromide ion has to open a bromonium ion by an *anti* attack in order that the new bond is formed concomitantly with the breaking of one bond of the three-membered ring.

3.4.2
## Simple Approaches to Quantifying Chemical Reactivity

### 3.4.2.1 Frontier Molecular Orbital Theory

With the advent of quantum mechanics, quite early attempts were made to obtain methods to predict chemical reactivity quantitatively. This endeavor has now matured to a point where details of the geometric and energetic changes in the course of a reaction can be calculated to a high degree of accuracy, albeit still with quite some demand on computational resources.

In view of this, early quantum mechanical approximations still merit interest, as they can provide quantitative data that can be correlated with observations on chemical reactivity. One of the most successful methods for explaining the course of chemical reactions is frontier molecular orbital (FMO) theory [5]. The course of a chemical reaction is rationalized on the basis of the highest occupied molecular orbital (HOMO) and the lowest unoccupied molecular orbital (LUMO), the frontier orbitals. Both the energy and the orbital coefficients of the HOMO and LUMO of the reactants are taken into account.

As an example, we shall discuss the Diels–Alder reaction of 2-methoxybuta-1,3-diene with acrylonitrile. Figure 3-7 gives the reaction equation, the correlation diagram of the HOMOs and LUMOs, and the orbital coefficients of the correlated HOMO and LUMO.

FMO theory requires that a HOMO of one reactant has to be correlated with the LUMO of the other reactant. The decision between the two alternatives – i.e., from which reactant the HOMO should be taken – is made on the basis of which is the smaller energy difference; in our case the HOMO of the electron rich diene, **3.1**, has to be correlated with the LUMO of the electron-poor dienophile, **3.2**. The smaller this HOMO–LUMO gap, the higher the reactivity will be. With the HOMO and LUMO fixed, the orbital coefficients of these two orbitals can explain the regioselectivity of the reaction, which strongly favors the formation of **3.3** over **3.4**. The new bonds will be made between those two ends of the reactants where the orbital coefficients of the frontier orbitals match best, where both have the largest coefficients.

The importance of FMO theory lies in the fact that good results may be obtained even if the frontier molecular orbitals are calculated by rather simple, approximate quantum mechanical methods such as perturbation theory. Even simple additivity schemes have been developed for estimating the energies and the orbital coefficients of frontier molecular orbitals [6].

### 3.4.2.2 Linear Free Energy Relationships (LFER)

Quite some time ago, in the 1940s, inroads were made into quantifying chemical reactivity on the basis of ordering series of reaction observations and their associated quantitative data.

Hammett [7] was the first to develop an approach that was later subsumed under Linear Free Energy Relationships (LFER). He showed that the acidity constants of a

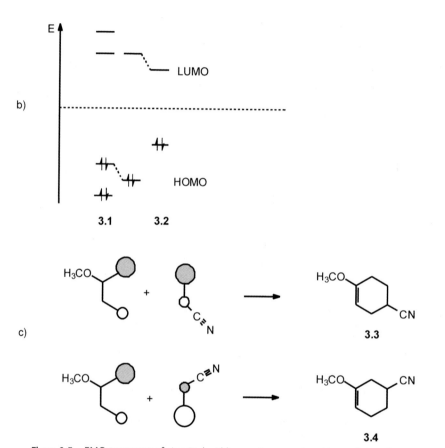

**Figure 3-7.** FMO treatment of a) a Diels–Alder reaction equation, b) correlation diagram, c) orbital coefficients.

series of benzoic acids can be broken down into contributions of the substituents and an inherent sensitivity of the reaction system.

$$\log K_x/\log K_0 = \sigma_x \cdot \rho \tag{1}$$

The logarithm of the equilibrium constant, $K_x$, for the chemical equation shown in Figure 3-8a for a substituted benzoic acid can be related to the logarithm of the

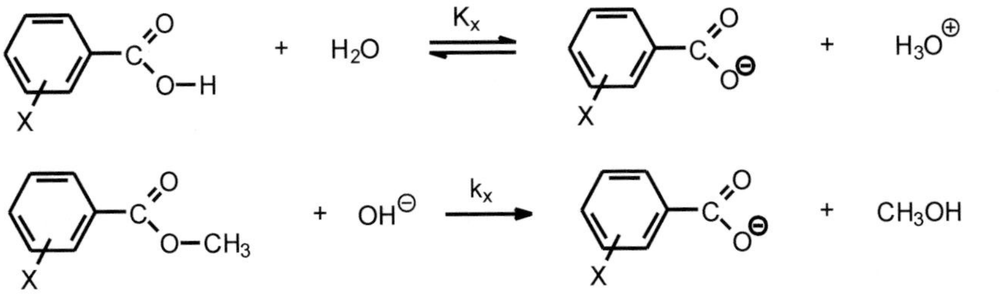

**Figure 3-8**  a) The dissociation of substituted benzoic acids (X = substituent), and b) the hydrolysis of benzoic acid methyl esters.

equilibrium constant, $K_0$, of the unsubstituted benzoic acid by Eq. (1). In this equation, $\sigma_x$ is a constant specific to the substituent X, and $\rho$ is specific to the reaction investigated, showing how sensitively the reaction responds to different substituents. For the acidity of benzoic acids $\rho$ was set equal to 1 allowing one to fix a scale for the substituent constants, $\sigma$. Similar relationships could be set up for other equilibrium constants of chemical reactions. It was also shown that in a similar manner reaction rates, $k$, such as for the hydrolysis of benzoic acid esters (Figure 3-8b) can be represented (Eq. (2)).

$$\log k_x/\log k_0 = \sigma_x \cdot \rho \tag{2}$$

Table 3-1 gives substituent constants for several groups.

Thermodynamics shows that equilibrium constants can be related to Gibbs free energies, $\Delta G$, by Eq. (3).

$$\Delta G^\circ = -RT \ln K \tag{3}$$

**Table 3-1.**  Substituent constants for various groups to be used in Eqs. (1) and (2) ($\sigma_m$ for substituents in the *meta* position, $\sigma_p$ for substituents in the *para* position).

| Substituent | $\sigma_m$ | $\sigma_p$ |
|---|---|---|
| H | ≡0.0 | ≡0.0 |
| CH$_3$ | −0.069 | −0.170 |
| C$_6$H$_5$ | 0.06 | −0.01 |
| CF$_3$ | 0.43 | 0.54 |
| CN | 0.56 | 0.66 |
| NH$_2$ | −0.16 | −0.66 |
| NO$_2$ | 0.71 | 0.778 |
| OCH$_3$ | 0.115 | −0.268 |
| OH | 0.121 | −0.37 |
| Cl | 0.373 | 0.227 |
| Br | 0.391 | 0.232 |

Thus, Eq. (1) can be written as in Eq. (4):

$$\Delta G_x - \Delta G_0 = -2.3RT\sigma_x \cdot \rho \qquad (4)$$

This shows that Eqs. (1) and (2) are basically relationships between the Gibbs free energies of the reactions under consideration, and explains why such relationships have been termed linear free energy relationships (LFER).

Taft showed that LFER can also be established for aliphatic systems [8]. Taft compared the hydrolysis of substituted aliphatic methyl esters under basic conditions with the corresponding acid-catalyzed reactions.

Taft then noted that the tetrahedral intermediates of both reactions differ by only two protons, suggesting that the steric effect in both reactions is expected to be the same. Taking the difference in these reaction rates, thus allowed the quantification of the inductive effect.

This work already showed that substituent constants of one reaction can only be transferred to another reaction when similar effects are operating and when they are operating to the same extent. In order to find a broader basis for the transferability of substituent constants, they were split into substituent constants for the resonance effect and those for the inductive effect.

Decades of work have led to a profusion of LFERs for a variety of reactions, for both equilibrium constants and reaction rates. LFERs were also established for other observations such as spectral data. Furthermore, various different scales of substituent constants have been proposed to model these different chemical systems. Attempts were then made to come up with a few fundamental substituent constants, such as those for the inductive, resonance, steric, or field effects. These fundamental constants have then to be combined linearly to different extents to model the various real-world systems. However, for each chemical system investigated, it had to be established which effects are operative and with which weighting factors the fundamental constants would have to be combined. Much of this work has been summarized in two books and has also been outlined in a more recent review [9–11].

The importance of all this work lies in the fact that it established for the first time that chemical reactivity data for a wide series of reactions can be put onto a quantitative footing. With continuing research, however, it was found that the various chemical systems required quite specific substituent constants of their own, leading to a decline in interest in LFER. Nevertheless, substituent constant scales are still in use and methods for calculating or correlating them are still of interest [12].

### 3.4.2.3 Empirical Reactivity Equations
LFER suffer from an artificial separation of a molecule into skeleton, reaction site, and substituent. The physicochemical effects mentioned in section 3.4.1 and the methods presented in section 7.1 for their calculation consider a molecule as a

whole and provide quantitative values that can be used for the correlation of reactivity data by statistical methods or neural networks.

As an example, experimental kinetic data on the hydrolysis of amides under basic conditions as well as under acid catalysis were correlated with quantitative data on charge distribution and the resonance effect [13]. Thus, the values on the free energy of activation, $\Delta G^{\neq}$, for the acid catalyzed hydrolysis of amides could be modeled quite well by Eq. (5)

$$\Delta G^{\neq} = 61.1 + 1.32 \ R^+_C + 0.33 \ R^+_N - 51.8 \ \Delta q_{NC} \text{ (in kJ / mol)} \tag{5}$$

In this equation, $R^+_C$ measures how well a positive charge can be stabilized by the resonance effect on the carbonyl carbon atom, and $R^+_N$ how well a positive charge can be stabilized on the nitrogen atom of the amide group. $\Delta q_{NC}$ gives the charge difference between the carbon and the nitrogen atom of the amide group [13]. All these effects will be considered by a chemist to be of influence on this reaction; here they could be quantified.

It was further shown that such equations could be extended to a wide variety of amides and related compounds and even allowed the prediction of the degradation products of agrochemicals of the benzoylphenylurea type [13]

## 3.5
## Reaction Classification

Since 1970 a variety of reaction classification schemes have been developed to allow a more systematic processing of the huge variety of chemical reaction instances (see Chapter III, Section 1 in the Handbook). Reaction classification serves to combine several reaction instances into one reaction type. In this way, the vast number of observed chemical reactions is reduced to a manageable number of reaction types. Application to specific starting materials of the bond and electron changes inherent in such a reaction type then generates a specific reaction instance.

From among the many reaction classification schemes, only a few are mentioned here. The first model concentrates initially on the atoms of the reaction center and the next approach looks first at the bonds involved in the reaction center. These are followed by systems that have actually been implemented, and whose performance is demonstrated.

## 3.5.1
## Model-Driven Approaches

Model-driven approaches classify reactions according to a preconceived model, a conceptual framework.

### 3.5.1.1 Hendrickson's Scheme

Hendrickson [14, 15] concentrated mainly on C–C bond-forming reactions because the construction of the carbon atom skeleton is the major task in the synthesis of complex organic compounds. Each carbon atom is classified according to which kind of atoms are bonded to it and what kind of bonds ($\sigma$ or $\pi$) are involved (Figure 3-9).

The number of bonds to R, Π, Z, and H atoms is given by the numbers $\sigma$, $\pi$, $z$, and $h$, respectively. For any uncharged carbon atom Eq. (6) must hold.

$$\sigma + \pi + z + h = 4 \tag{6}$$

These numbers carry other chemical information. For example, $z - h = x$ gives the oxidation state of a carbon atom. In effect, each carbon atom is classified according to its oxidation state, $x$, and its attachment to other carbon atoms.

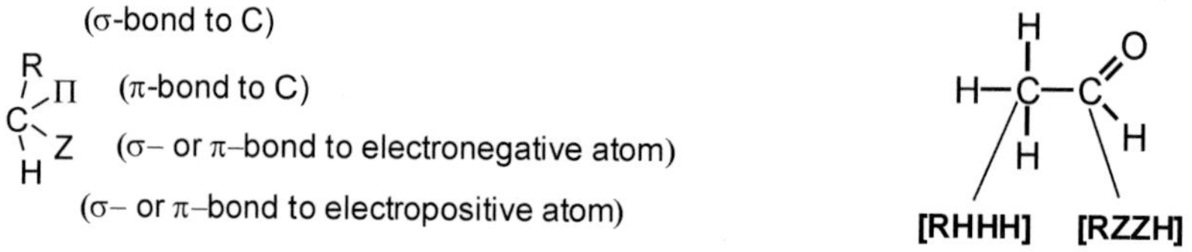

**Figure 3-9.** Hendrickson's classification of atom types, and an example.

Unit reactions at each carbon atom are then composed of unit exchanges of one bond type against another. There are 16 such exchanges possible at one carbon atom, each denoted by two letters, the first one for the bond made and the second for the bond broken. Figure 3-10 shows an example and Table 3-2 gives all the possible unit exchanges.

Skeletal changes are characterized by changes in R, with constructions having positive values (+ R) and fragmentation negative (–R); functionality changes have ±Π, ±Z, or ±H.

Complete reactions are obtained by the combination of these unit exchanges into composite reactions. Figure 3-11 gives the example of an allylic substitution.

This example again emphasizes that Hendrickson only considered the bond changes at the carbon atoms of a reaction.

On this basis Hendrickson classified organic reactions. A distinction is made between refunctionalization reactions and skeletal alteration reactions. Refunctionalizations in almost all cases have no more than four carbon atoms in the reaction center. Construction or fragmentation reactions have no more than three carbon atoms in each joining or cleaving part of the molecule. Thus, these parts are treated

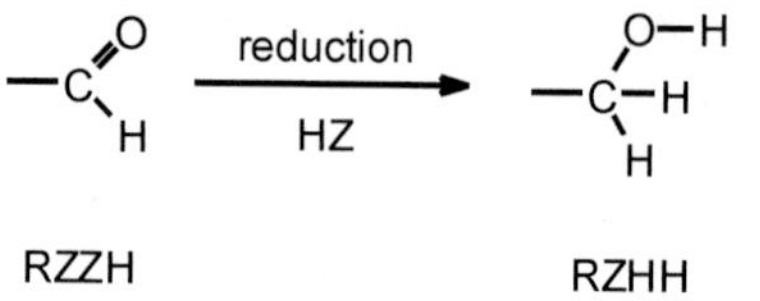

**Figure 3-10.**
Example of a unit exchange in a reaction.

**Table 3-2.** The 16 possible unit exchanges at any skeletal carbon atom.

| | | $\Delta x$ | $\Delta \pi$ | $\Delta \sigma$ |
|---|---|---|---|---|
| Substitution | HH, ZZ, RR, ΠΠ | 0 | 0 | 0 |
| Oxidation | ZH | +2 | 0 | 0 |
| Reduction | HZ | −2 | 0 | 0 |
| Elimination | ΠH | +1 | −1 | 0 |
| | ΠZ | −1 | +1 | 0 |
| Addition | HΠ | −1 | −1 | 0 |
| | ZΠ | +1 | −1 | 0 |
| Construction | RH | −1 | 0 | +1 |
| | RZ | −1 | 0 | +1 |
| | RΠ | 0 | −1 | +1 |
| Fragmentation | HR | −1 | 0 | −1 |
| | ZR | +1 | 0 | −1 |
| | ΠR | 0 | +1 | −1 |

separately as half-reactions. This classification system was used to index a database of 400 000 reactions [14].

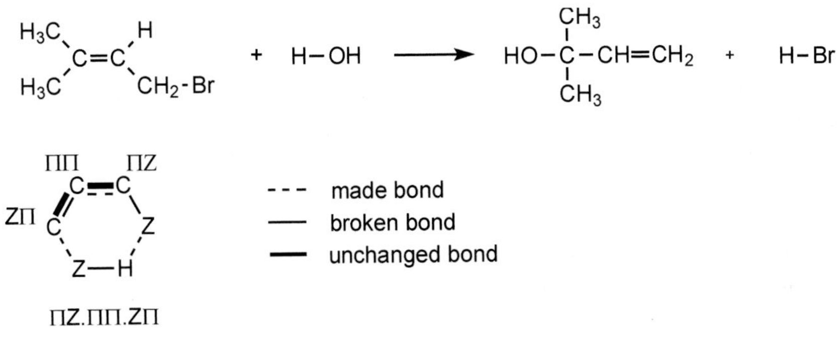

**Figure 3-11.** Allylic substitution as a composite reaction in Hendrickson's scheme.

### 3.5.1.2 Ugi's Scheme

In the mid 1970s, Ugi and co-workers developed a scheme based on treating reactions by means of matrices – reaction (R-) matrices [16, 17]. The representation of chemical structures by bond and electron (BE-) matrices was presented in Section 2.4. BE-matrices can be constructed not only for single molecules but also for ensembles of them, such as the starting materials of a reaction, e.g., formaldehyde (methanal) and hydrocyanic acid as shown with the BE-matrix, $B$, in Figure 3-12. Figure 3-12 also shows the BE-matrix, $E$, of the reaction product, the cyanohydrin of formaldehyde.

Having the BE-matrices of the beginning, $B$, and the end, $E$, of a reaction, one can calculate $E - B = R$. As can easily be seen, the entries $r_{ij}$ in the R-matrix indicate the bonds broken and made in the course of this reaction.

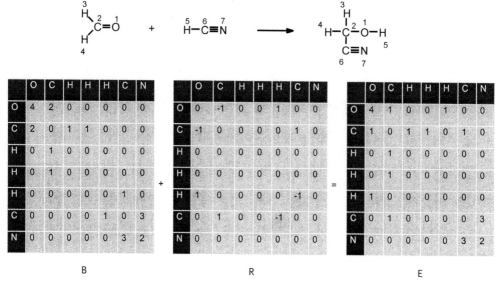

**Figure 3-12.** The reaction of formaldehyde with hydrocyanic acid to give a cyanohydrin, and the matrix representation of this reaction.

An R-matrix has a series of interesting mathematical properties that directly reflect chemical laws. Thus, the sum of all the entries in an R-matrix must be zero, as no electrons can be generated or annihilated in a chemical reaction. Furthermore, the sum of the entries in each row or column of an R-matrix must also be zero as long as there is not a change in formal charges on the corresponding atom. An elaborate mathematical model of the constitutional aspects of organic chemistry has been built on the basis of BE- and R-matrices [17].

Clearly, BE- and R-matrices have far too many entries of zero to be useful for direct computer implementation. Furthermore, the number of entries in BE- and R-matrices increases by $N^2$, N being the number of atoms in the molecule, so any implementation will try to use a representation such as a connection table where the number of entries increases linearly with the number of atoms. Using a connection table, an R-matrix will be stripped down to its non-zero elements. In the further discussion we will therefore only consider the bonds being broken and made in a reaction.

The merit of the mathematical model [17] inherent in the BE- and R-matrices lies in the fact that it emphasized two essential points:

1. The representation of chemical species should take account of all valence electrons.
2. Reactions should be represented by the shifting of bonds and electrons in the reaction center.

An R-matrix expresses the bond and electron rearrangement in a reaction. The R-matrix of Figure 3-12 reflects a reaction scheme, the breaking and the making of two bonds, that is at the foundation of the majority of all organic reactions. Figure 3-13 shows this scheme and some examples of it, such as nucleophilic aliphatic substitutions (a), additions to multiple bonds and (in reverse) elimination reactions (b), electrocyclic reactions (c), and electrophilic aromatic substitutions (d).

Concentration on the types of bonds broken or made in a reaction provides a basis for reaction classification. We first show this only for one bond (Figure 3-14). On the first level of a hierarchy, a bond can be distinguished by whether it is a single, double, or triple. Then, on the next level, a further distinction can be made on the basis of the atoms that comprise the bond.

Any more complicated reaction scheme involving several bonds can be classified accordingly. Thus, a comprehensive system for a hierarchical classification of reactions can be built.

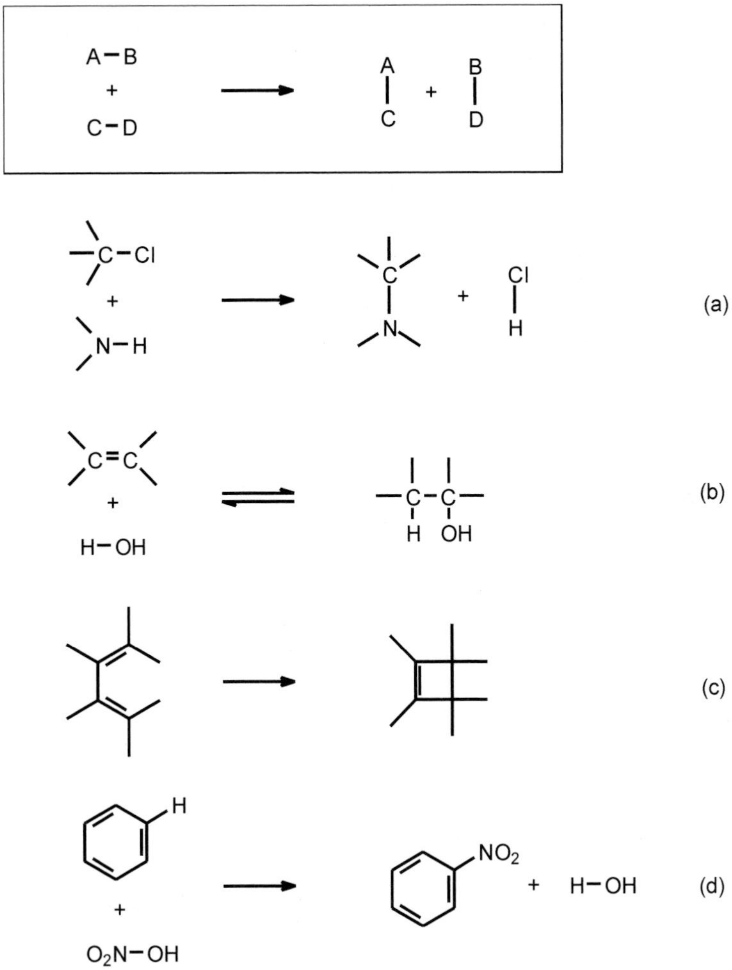

**Figure 3-13.** The reaction scheme comprising the breaking and the making of two bonds and some examples of reactions following this scheme.

A—B
A–B
A=B
A≡B
C–C
C–O
C=C
C=O
C≡C

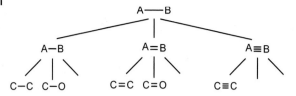

Figure 3-14. Different levels of specification for a bond participating in a reaction.

Some systematic studies on the different reaction schemes and how they are realized in organic reactions were performed some time ago [18]. Reactions used in organic synthesis were analyzed thoroughly in order to identify which reaction schemes occur. The analysis was restricted to reactions that shift electrons in pairs, as either a bonding or a free electron pair. Thus, only polar or heterolytic and concerted reactions were considered. However, it must be emphasized that the reaction schemes list only the overall change in the distribution of bonds and free electron pairs, and make no specific statements on a reaction mechanism. Thus, reactions that proceed mechanistically through homolysis might be included in the overall reaction scheme.

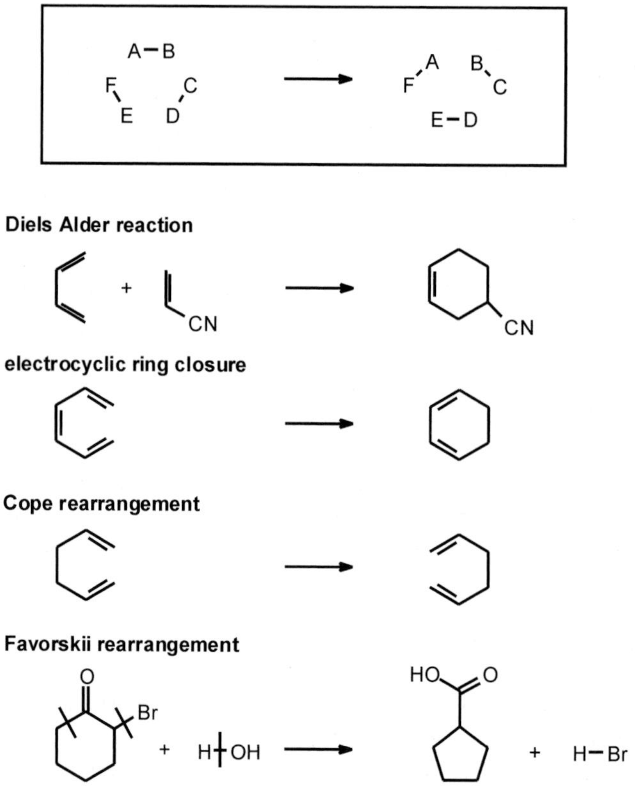

A–B
F
C
E
D
A
B
F
C
E–D
Diels Alder reaction
CN
CN
electrocyclic ring closure
Cope rearrangement
Favorskii rearrangement
O
Br
H–OH
HO
O
H–Br

Figure 3-15. The reaction scheme breaking three and making three bonds, and some of the reaction types that fall into this scheme.

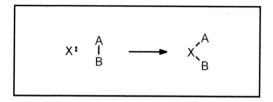

**Figure 3-16.** A reaction scheme that changes the number of bonds at one atom, and some specific examples.

Next, an attempt was made to evaluate the quantitative importance of the various reaction schemes [19]. To this effect, a printed compilation of 1900 reactions dealing with the introduction of one carbon atom bearing a functional group [20] was analyzed and each reaction assigned manually to a corresponding reaction scheme. The results are listed in Table 3-3.

Clearly, this choice of a reference set of organic reactions is arbitrary, not necessarily representative of the whole set of organic reaction types described in the literature, and therefore not free from bias. However, it does give some indication of the relative importance of the various reaction schemes. It is quite clear that the reaction scheme shown in Figure 3-13 (R1 of Table 3-3) comprises the majority of organic reactions; in most compilations of reactions it will account for more than 50 % of all reactions.

Such an analysis of the literature for assigning reaction types to different reaction schemes definitely has merits. However, it does not say anything about the importance of a reaction type, such as how frequently it is actually performed in the laboratory.

Clearly, such statistics are impossible to obtain on a worldwide basis. However, it is quite clear that organic reaction types that follow reaction scheme R1 (Table 3-3, Figure 3-13) are among the most frequently performed. This shifts the balance even further in the direction of this reaction scheme, lending overwhelming importance to it.

The second most important reaction scheme is the next higher homolog to that shown in Figure 3-13, involving the breaking and making of three bonds. Figure 3-15 shows this reaction scheme and some reaction types that follow it.

It has to be emphasized that a reaction scheme shows the overall bond change in a reaction and does not imply anything about the timing of the bond breaking and making. Thus, whereas the first three reaction types in Figure 3-15 are concerted reactions simultaneously breaking and making the three bonds, the last one, the

**Table 3-3.** Quantitative analysis of reaction schemes.

| Notation | Reaction scheme | Frequency [%] |
|---|---|---|
| R23 | A–B → A + B: | 0.60 |
| R32 | A: + :B → A=B | 0.40 |
| R11 | **A–B + C → A–C + B** | 0.47 |
| R21 | A–B + C: → A–C–B | 6.11 |
| R33 | A + :B–C → A–B + C: | 0.20 |
| R12 | A: + B–C → A–C + B: | 1.68 |
| R25 | A + B–C + D: → A–B + C–D | 0.74 |
| R1 | A–B + C–D → A–C + B–D | 51.38 |
| R3 | A–B + C–D → A + B–C + D: | 1.01 |
| R5 | A–B + :C–D → A–C–B + D: | 0.87 |
| R8 | A–B + C–D + E: → A–C + D–E + B: | 2.35 |
| R34 | A=B → A: + B: | 0.27 |
| R35 | A=B + C: → A=C + B: | 0.07 |
| R36 | A=B + C: + D: → C–A–D + :B–E | 0.20 |
| R37 | A=B + C–D + E: → C–A–D + :B–E | 0.67 |
| R31 | A=B + C–D + E–F → C–A–E + D–B–F | 3.63 |
| R2 | A–B + C–D + E–F → A–C + D–E + B–F | 19.48 |
| R15 | A–B + C–D + :E–F → A–E–D + B–C + F: | 2.55 |
| R28 | A–B + C–D + E–F + G: → A–F + B–D + E–G + C: | 0.34 |
| R17 | A–B + C–D + E–F + G–H → A–D + B–H + C–E + F–G | 0.74 |
| R38 | A–B + C–D + E–F + G–H → A–C + B–G + D–E–H + F: | 0.13 |
| R30 | A–B + C–D + E–F + G–H + I: → A–I + B–D + C–F E–G + E–G + H: | 0.13 |
| R22 | A–B–C → A–C + B: | 0.20 |
| R7 | A–B–C + D–E → D–B–E + A–C | 0.20 |
| R9 | A–B–C + D–E → A–D + E–C + B: | 1.21 |
| R39 | A–B–C + D–E + F–G + H: → A–F + C–H + E–B–G +D: | 0.27 |
| R40 | A–B–C + D–E + F–G + H–I → A–G + C–D + E–F + H–B–I | 0.07 |
| R41 | A–B–C + D–E–F → A–F + B=E + C–D | 067 |
| R19 | A–B–C + D–E–F + G–H → A–F B=E + C–H + D–G | 0.54 |
| R10 | A–B–C + D=E → A–D–C + B=E | 2.82 |

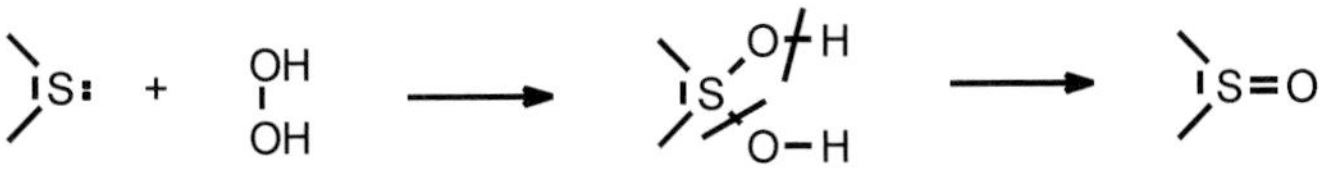

**Figure 3-17.** Consecutive application of two reaction schemes to model the oxidation of thioethers to sulfoxides.

Favorskii rearrangement, is a stepwise reaction with the breaking and making of the three bonds at different time intervals.

The two reaction schemes of Figures 3-13 and 3-15 encompass a large proportion of all organic reactions. However, these reactions do not involve a change in the number of bonds at the atoms participating in them. Therefore, when oxidation and reduction reactions that also change the valency of an atom are to be considered, an additional reaction scheme must be introduced in which free electron pairs are involved. Figure 3-16 shows such a scheme and some specific reaction types.

Clearly, for symmetry reasons, the reverse process should also be considered. In fact, early versions of our reaction prediction and synthesis design system EROS [21] contained the reaction schemes of Figures 3-13, 3-15, and 3-16 and the reverse of the scheme shown in Figure 3-16. These four reaction schemes and their combined application include the majority of reactions observed in organic chemistry. Figure 3-17 shows a consecutive application of the reaction schemes of Figures 3-16 and 3-13 to model the oxidation of thioethers to sulfoxides.

It has to be emphasized that these formal reaction schemes of Figures 3-13, 3-15, and 3-16 have the potential to discover novel reactions. Application of these bond- and electron-shifting schemes to specific molecules and bonds may correspond to a known reaction but may also model a completely novel reaction.

Herges has systematically investigated certain reaction schemes and their realization in chemistry; this led him to instances that were without precedent. He could then verify some of these experimentally and thus discover new reactions [22].

### 3.5.1.3 InfoChem's Reaction Classification

The reaction databases of MDL Information Systems, Inc., use a reaction classification method developed by InfoChem GmbH [23] (see also Chapter X, Section 3.1 of the Handbook). The algorithm considers, for each atom of the reaction center, the atom type, valence state, total number of bonded hydrogen atoms, number of $\pi$-electrons, aromaticity, and formal charge. These pieces of information are merged into a hash code [24]. This comprises the broad classification. a medium classification is obtained from the atoms of the reacting bonds and the atoms directly bonded to them (the $a$-atoms), again by hashcoding. A narrow classification is generated from the atoms of the reaction center and their $a$- and $\beta$-neighboring atoms through hashcoding.

This reaction classification allows users to browse through hits of reaction searches, thus enabling them to focus on the types of reaction in which they are interested.

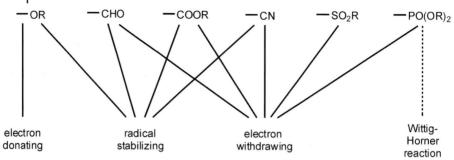

electron            radical            electron            Wittig-
donating         stabilizing      withdrawing      Horner
                                                      reaction

**Figure 3-18.** A different classification of substituents at the reaction site.

### 3.5.2
### Data-Driven Approaches

Whereas a model-driven method imposes a rigid classification scheme onto a set of reactions, the data-driven methods try to derive a classification from the data presented.

#### 3.5.2.1 HORACE

Gelernter and Rose [25] used machine learning techniques (Chapter IX, Section 1.1 of the Handbook) to analyze the reaction center. Based on the functionalities attached to the reaction center, the method of conceptual clustering derived the features a reaction needed to possess for it to be assigned to a certain reaction type. A drawback of this approach was that it only used topological features, the functional groups at the reaction center, and its immediate environment, and did not consider the physicochemical effects which are so important for determining a reaction mechanism and thus a reaction type.

A functional group can, however, have different effects, depending on the mechanism of a reaction type and depending on the electron demand on the reaction center. Thus, Figure 3-18 shows how functional groups can be considered as either electron-donating or radical-stabilizing (e.g., OR), or as either radical-stabilizing or electron-withdrawing (e.g., CN). Furthermore, whereas in many reactions a –PO(OR)$_2$ group is electron-withdrawing such as the –CHO, –COOR, –CN, or –SO$_2$R groups, yet the Wittig–Horner reaction needs a –PO(OR)$_2$ group at a carbon atom and none of the other groups in this list can initiate a Wittig–Horner reaction. Thus, neither an unequivocal classification nor a universal generalization of functional groups is possible.

This deficiency of only working with functional groups was rectified by the HORACE system (Hierarchical Organization of Reactions through Attribute and Condition Eduction) [26]. HORACE used the same set of 114 structural classification features as in Ref. [25], e.g., simple functional groups (alcohol, carbonyl, carboxylic acid, etc.) or larger groups (different heterocycles – oxazoles, thiazoles, etc.). On top of that, physicochemical features such as charge distribution and measures of the inductive and resonance effect at the reaction center were used in reaction classification.

HORACE used alternating phases of classification (which topological or physicochemical features are required for a reaction type) and generalization (which features are allowed and can be eliminated) to produce a hierarchical classification of a set of reaction instances.

### 3.5.2.2 Reaction Landscapes

The work on HORACE underlined the overwhelming importance for reaction classification of electronic effects at the reaction center. The question that then came up was whether the functional groups could be dropped altogether and the reactions could be classified only on the basis of physicochemical descriptors. Then, with only numerical values characterizing the reaction center, there would be no need to balance topological and physicochemical effects by conceptual clustering, and simpler methods could be used for reaction clustering. Clearly, it would have to be an unsupervised learning method because we wanted to have a data-driven method and did not want to impose a rigid model on classification. Furthermore, it was expected that problems might come up that needed a nonlinear method. We decided to use a self-organizing neural network (the Kohonen network) (see Section 9.4), as this is a powerful nonlinear unsupervised learning method.

The method that was developed builds on computed values of physicochemical effects and uses neural networks for classification. Therefore, for a deeper understanding of this form of reaction classification, later chapters should be consulted on topics such as methods for the calculation of physicochemical effects (Section 7.1) and artificial neural networks (Section 9.4).

Previous work in our group had shown the power of self-organizing neural networks for the projection of high-dimensional datasets into two dimensions while preserving clusters present in the high-dimensional space even after projection [27]. In effect, 2D maps of the high-dimensional data are obtained that can show clusters of similar objects.

An object, $s$, represented by a set of descriptors $x_s = (x_1, x_2, \dots x_m)$ is mapped into a 2D arrangement of neurons each containing as many weights, $w_{ji}$, as the object has descriptors. The neuron that obtains the object, and is the winning neuron, has weights most similar to the descriptors of the object. A competitive learning algorithm will then adjust the weights of the neurons to make them even more similar to the descriptor values of the object (see Section 9.4). Objects having similar descriptors will be mapped into the same or closely adjacent neurons, thus leading to clusters of similar objects.

We will show here the classification procedure with a specific dataset [28]. A reaction center, the addition of a C–H bond to a C=C double bond, was chosen that comprised a variety of different reaction types such as Michael additions, Friedel–Crafts alkylation of aromatic compounds by alkenes, or photochemical reactions. We wanted to see whether these different reaction types can be discerned by this newly developed procedure. A search in the 1994 ChemInform RX database [29] produced 120 hits for the reaction center shown in Figure 3-19; some examples are also given in this figure.

The next question is how to represent the reacting bonds of the reaction center. We wanted to develop a method for reaction classification that can be used for knowledge extraction from reaction databases for the prediction of the products of a reaction. Thus, we could only use physicochemical values of the reactants, because these should tell us what products we obtain.

Previous studies with a variety of datasets had shown the importance of charge distribution, $q_{tot}$, of $\sigma$-electronegativity, $\chi_\sigma$ (inductive effect), of $\pi$-electronegativity, $\chi_\pi$ (resonance effect), and of effective polarizability, $a_{eff,i}$ (polarizability effect) (for details on these methods see Section 7.1). All four of these descriptors on all three carbon atoms were calculated. However, in the final study, a reduced set of descriptors, shown in Table 3-4, was chosen that was obtained both by statistical methods and by chemical intuition.

Next, the architecture of the Kohonen network had to be chosen. With seven descriptors for each reaction, a network of neurons with seven weights had to be

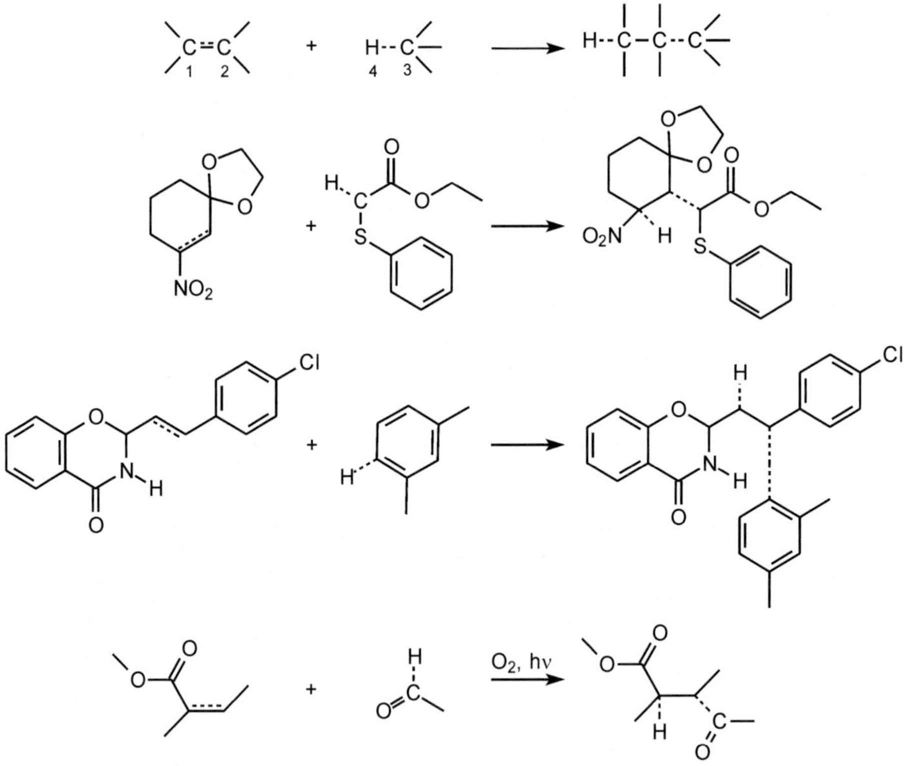

**Figure 3-19.** Reaction center of the dataset of 120 reactions (reacting bonds are indicated by broken lines), and some reaction instances of this dataset.

**Table 3-4.** Seven physiochemical property data used to characterize each reaction center.

| Electronic variable[a] | C = C | | + | H – C | | → | H – C – C – C | | | |
|---|---|---|---|---|---|---|---|---|---|---|
| | 1 | 2 | | 4 | 3 | | 4 | 1 | 2 | 3 |
| $q_{tot}$ | | X | | | X | | | | | |
| $\chi_\sigma$ | X | | | | X | | | | | |
| $\chi_\pi$ | X | | | | X | | | | | |
| $\alpha_i$ | | | | | X | | | | | |

[a] $q_{tot}$ = total charge; $\chi_\sigma$ = σ-electronegativity; $\chi_\pi$ = π-electronegativity; $\alpha_i$ = effective atom polarizability.

taken. A reasonable start for the investigation of a dataset by a Kohonen network is always to start with a network having about as many neurons as there are objects in the dataset. In this investigation of a dataset of 120 reactions we selected a network with 12 x 12 = 144 neurons.

Sending the entire dataset through this network leads to a distribution of the 120 reactions across the 2D arrangement of neurons. The question is now, does this distribution make sense? Remember we have used an unsupervised learning method and therefore have not said anything about the membership of a reaction in a certain reaction type. In order to analyze the mapping, these 120 reactions were classified intellectually by a chemist and the neurons were patterned according to the assignment of a reaction to a certain type (this was done a posteriori, after training of the network; we still have unsupervised learning!). Figure 3-20a shows the map thus patterned. It can be seen that reactions considered by a chemist to belong to one and the same reaction type are to be found in contiguous parts of the Kohonen map. This becomes even clearer when we pattern the empty neurons, those neurons that did not obtain a reaction, on the basis of their $k$ nearest neighbors: a neuron obtains a pattern by a majority decision of its nearest neighbors (Figure 3-20b).

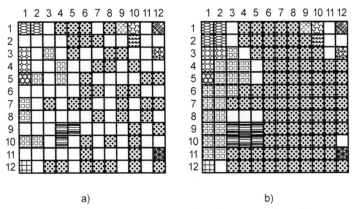

a)                          b)

**Figure 3-20.** Distribution of the dataset of 120 reactions in the Kohonen network. a) The neurons were patterned on the basis of intellectually assigned reaction types; b) in addition, empty neurons were patterned on the basis of their $k$ nearest neighbors.

Reactions belonging to the same reaction type are projected into coherent areas on the Kohonen map; this shows that the assignment of reaction types by a chemist is also perceived by the Kohonen network on the basis of the electronic descriptors. This attests to the power of this approach.

There are finer details to be extracted from such Kohonen maps that directly reflect chemical information, and have chemical significance. A more extensive discussion of the chemical implications of the mapping of the entire dataset can be found in the original publication [28]. Clearly, such a map can now be used for the assignment of a reaction to a certain reaction type. Calculating the physicochemical descriptors of a reaction allows it to be input into this trained Kohonen network. If this reaction is mapped, say, in the area of Friedel–Crafts reactions, it can safely be classified as a feasible Friedel–Crafts reaction.

A wider variety of reaction types involving reactions at bonds to oxygen atom bearing functional groups was investigated by the same kind of methodology [30]. Reaction classification is an essential step in knowledge extraction from reaction databases. This topic is discussed in Section 10.3.1 of this book.

## 3.6
## Stereochemistry of Reactions

Many chemical reactions proceed with a clearly defined stereochemistry, requiring the bonds to be broken and made in the reaction to have a specific geometrical arrangement. This is particularly true for reactions that are controlled by enzymes.

Thus, to name just a few examples, a nucleophilic aliphatic substitution such as the reaction of the bromide **3.5** with sodium iodide (Figure 3-21a) can lead to a range of stereochemical products, from a 1:1 mixture of **3.6** and **3.7** (racemization) to only **3.7** (inversion) depending on the groups a, b, and c that are bonded to the central carbon atom. The ring closure of the 1,3-butadiene, **3.8**, to cyclobutene

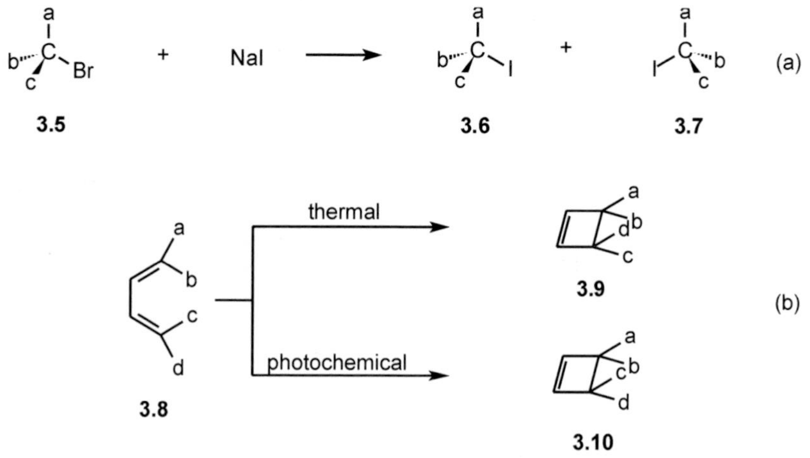

**Figure 3-21.** Some reactions that proceed under stereochemical control.

(electrocyclic reaction) leads either to the product **3.9** in a thermal reaction by disrotatory movements of the substituents at the end of the butadiene system or to the product **3.10** in a photochemical reaction by conrotatory ring closure (Figure 3-21b).

The stereochemistry of reactions has to be handled in any detailed modeling of chemical reactions. Section 2.7 showed how permutation group theory can be used to represent the stereochemistry of molecular structures. We will now extend this approach to handle the stereochemistry of reactions also [31].

## 3.7 Tutorial: Stereochemistry of Reactions

Let us first repeat the essential features of handling the stereochemistry of molecular structures by permutation group theory:

1. A molecule is split into a skeleton and ligands.
2. Both the skeleton and the ligands are independently numbered.
3. The numbering of the skeleton can be fixed arbitrarily, but thereafter it must always be kept the same.
4. The ligands can be numbered by any algorithm. For our purposes here we will base the numbering on the atomic number of the $\alpha$-atoms, then of the $\beta$-atoms, in analogy to the Cahn–Ingold–Prelog rules [32].
5. A reference isomer is defined as the structure where ligand number *1* is on skeleton position **1**, ligand no. *2* is on skeleton position **2**, etc. This reference isomer obtains the descriptor (+1).

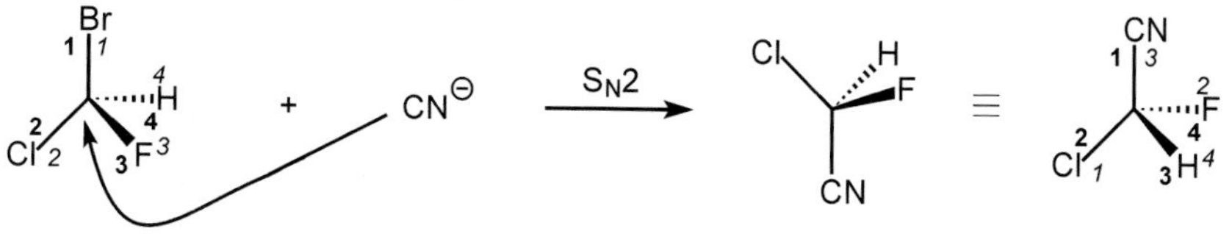


educt

$\begin{pmatrix} 1\,2\,3\,4 \\ \mathbf{1\,2\,3\,4} \end{pmatrix}$ **(+1)**

product

$\begin{pmatrix} 3\,1\,4\,2 \\ \mathbf{1\,2\,3\,4} \end{pmatrix}$ **(-1)**

$$\underset{\mathbf{(+1)}}{\begin{pmatrix} 1\,2\,3\,4 \\ \mathbf{1\,2\,3\,4} \end{pmatrix}} = (23)(31)(24)\underset{\mathbf{(-1)}}{\begin{pmatrix} 3\,1\,4\,2 \\ \mathbf{1\,2\,3\,4} \end{pmatrix}}$$

$$\underset{p_{ed}}{(+1)} \qquad p_m = (-1)^3 = (-1) \qquad p_{prod}$$

**Figure 3-22.** The treatment of the stereochemistry of an $S_N2$ reaction by permutation group theory.

Figure 3-22 shows a nucleophilic aliphatic substitution with cyanide ion as a nucleophile. This reaction is assumed to proceed according to the $S_N2$ mechanism with an inversion in the stereochemistry at the carbon atom of the reaction center. We have to assign a stereochemical mechanistic factor to this reaction, and, clearly, it is desirable to assign a mechanistic factor of $(+1)$ to a reaction with retention of configuration and $(-1)$ to a reaction with inversion of configuration. Thus, we want to calculate the parity of the product, $p_{prod}$, of a reaction from the parity of the educt, $p_{ed}$, and the mechanistic factor, $m$, of the reaction according to Eq. (7).

$$p_{prod} = m \cdot p_{ed} \tag{7}$$

We see from Figure 3-22 that we need three transpositions to transform the isomer of the product of this reaction into the reference isomer. Thus, for Eq. (7) we obtain: $p_{prod} = (-1) \cdot (+1) = (-1)$ and everything seems fine as the descriptor of the product is inverted from the descriptor of the educt in this reaction with a stereochemical course of inversion.

Now we carry out the same reaction with thiolate as nucleophile, as shown in Figure 3-23. However, we must now realize that the product has a parity of $(+1)$ although we have the same mechanism as in Figure 3-22 with inversion of configuration.

What has happened? The parity of the product in the reaction shown in Figure 3-23 is different from that in Figure 3-22 because the position in the numbering scheme taken by the incoming ligand, SH, with respect to the three remaining ligands, Cl, F, H, differs from the one the ligand CN has taken. In other words, we have to extend Eq. (7) by a nomenclature factor, $n$, to obtain Eq. (8), which takes account of the positions of the incoming and of the leaving ligand in the numbering scheme.

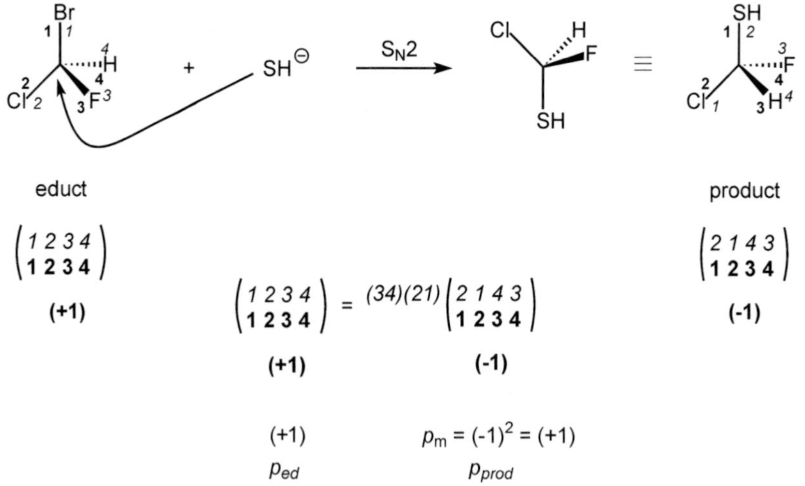

**Figure 3-23.** The treatment of the stereochemistry of a further $S_N2$ reaction by permutation group theory.

|        | Br | Cl | F | CN | H |
|--------|----|----|---|----|---|
| educt  | 1  | 2  | 3 | -  | 4 |
| product| -  | 1  | 2 | 3  | 4 |

$$k = 1 + 3 = 4$$
$$n = (-1)^4 = (+1)$$

|        | Br | Cl | SH | F | H |
|--------|----|----|----|---|---|
| educt  | 1  | 2  | -  | 3 | 4 |
| product| -  | 1  | 2  | 3 | 4 |

$$k = 1 + 2 = 3$$
$$n = (-1)^3 = (-1)$$

**Figure 3-24.** The ranking of the leaving and the entering groups in the reactions of Figures 3-22 and 3-23.

$$p_{prod} = n \cdot m \cdot p_{ed} \tag{8}$$

It can easily be verified that the nomenclature factor, $n$, can be calculated according to Eq. (9)

$$n = (-1)^k \tag{9}$$

with $k$ = position of leaving group + position of entering group.

Figure 3-24 shows how the nomenclature factor is calculated for the two reactions shown in Figures 3-22 and 3-23. The position of the leaving group has to be taken from the set of ligands in the educt, whereas the position of the entering group has to be seen with respect to the set of ligands of the product.

Thus, we see that we have not felt the need for a nomenclature factor in the first reaction shown in Figure 3-22, for the sole reason that the nomenclature factor has a value of (+1).

The stereochemistry of reactions can also be treated by permutation group theory for reactions that involve the transformation of an $sp^2$ carbon atom center into an $sp^3$ carbon atom center, as in additions to C=C bonds, in elimination reactions, or in electrocyclic reactions such as the one shown in Figure 3-21. Details have been published [31].

## Essentials

- The representation of a chemical reaction should include the connection table of all participating species (starting materials, reagents, solvents, catalysts, products) as well as information on reaction conditions (temperature, concentration, time, etc.) and observations (yield, reaction rates, heat of reaction, etc.).
- However, reactions are only insufficiently represented by the structure of their starting materials and products.
- It is essential to indicate also the reaction center and the bonds broken and made in a reaction – in essence, to specify how electrons are shifted during a reaction.
- In this sense, the representation of chemical reactions should consider some essential features of a reaction mechanism.

- Further insight into the driving forces of chemical reactions can be gained by considering major physicochemical effects at the reaction center.
- Specification of the reaction center is important for many queries to reaction databases.
- Reaction types can be derived automatically through classification of reaction instances.
- Reaction classification is an essential step in knowledge acquisition from reaction databases.
- There are two fundamental approaches to automatic reaction classification: model-driven and data-driven methods.
- The stereochemistry of reactions can be treated by means of permutation group theory.

### Selected Reading

- L. Chen, Reaction classification and knowledge acquisition, in *Handbook of Chemoinformatics – From Data to Knowledge*, J. Gasteiger (Ed.), Wiley-VCH, Weinheim, **2003**.
- G. Grethe, Analysis of reaction information, in *Handbook of Chemoinformatics – From Data to Knowledge*, J. Gasteiger (Ed.), Wiley-VCH, Weinheim, **2003**.
- J. Dugundji, I. Ugi, *Topics Curr. Chem.* **1973**, *39*, 19.
- J. R. Rose, J. Gasteiger, *J. Chem. Inf. Comput. Sci.* **1994**, *34*, 74–90.
- L. Chen, J. Gasteiger, *J. Am. Chem. Soc.,* **1997**, *119*, 4033–4042.
- J. Blair, J. Gasteiger, C. Gillespie, P. D. Gillespie, I. Ugi, *Tetrahedron*, **1974**, *30*, 1845–1859.

# References

[1] J. Gasteiger, M. Marsili, *Tetrahedron* **1980**, *36*, 3219–3228.

[2] J. Gasteiger, H. Saller, *Angew. Chem.* **1985**, *97*, 699–701; *Angew. Chem. Int. Ed. Engl.* **1985**, *24*, 687–689.

[3] M. G. Hutchings, J. Gasteiger, *Tetrahedron Lett.* **1983**, *24*, 2541–2544.

[4] J. Gasteiger, M.G. Hutchings, *J. Chem. Soc. Perkin 2*, **1984**, 559–564.

[5] I. Fleming, *Frontier Orbitals and Organic Chemical Reactions*, Wiley, New York, **1976**.

[6] J.S. Burnier, W.L. Jorgensen, *J. Org. Chem.* **1983**, *48*, 3923–3941.

[7] L.P. Hammett, *Physical Organic Chemistry*, McGraw-Hill, New York, **1940**.

[8] R.W. Taft, *Steric Effects in Organic Chemistry*, M.S. Newman (Ed.), Wiley, New York, **1956**.

[9] N.B. Chapman, J. Shorter (Eds.), *Advances in Linear Free Energy Relationships*, Plenum Press, London, **1972**.

[10] N.B. Chapman, J. Shorter (Eds.), *Correlation Analysis in Chemistry*, Plenum Press, London, **1978**.

[11] J. Shorter, Linear Free Energy Relationships (LFER), in *Encyclopedia of Computational Chemistry, Vol. 2*, P.v.R. Schleyer, N.L. Allinger, T. Clark, J. Gasteiger, P.A. Kollman, H.F. Schaefer III, P.R. Schreiner (Eds.), Wiley, Chichester, UK, **1998**, pp. 1487–1496; *http://www.mrw.interscience.wiley. com/ecc*

[12] P. Ertl, *Quant. Struct.–Act. Relat.* **1997**, *16*, 377–382.

[13] J. Gasteiger, U. Hondelmann, P. Röse, W. Witzenbichler, *J. Chem. Soc. Perkin 2*, **1995**, 193–204.

[14] J.B. Hendrickson, T. Sander, *J. Chem. Inf. Comput. Sci.* **1995**, *35*, 251–260.

[15] J.B. Hendrickson, L. Chen, Reaction classification, in *Encyclopedia of Computational Chemistry, Vol. 4*, P.v.R. Schleyer, N.L. Allinger, T. Clark, J. Gasteiger, P.A. Kollman, H.F. Schaefer III, P.R. Schreiner (Eds.), Wiley, Chichester, **1998**, pp. 2381–2402; *http://www.mrw.interscience.wiley. com/ecc*

[16] I. Ugi, P.D. Gillespie, *Angew. Chem.* **1971**, *83*, 982; *Angew. Chem. Int. Ed. Engl.* **1971**, 10, 915.

[17] J. Dugundji, I. Ugi, *Topics Curr. Chem.* **1973**, *39*, 19.

[18] J.C.J. Bart, E. Garagnani, *Z. Naturforsch.* **1977**, *32b*, 455–464.

[19] E. Garagnani, J.C.J. Bart, *Z. Naturforsch.* **1977**, *32b*, 465–648.

[20] J. Matthieu, J. Weill-Raynal, *Formation of C–C Bonds, Vol. 1, Introduction of a Functional Carbon Atom*, G. Thieme, Stuttgart, **1973**.

[21] J. Gasteiger, M.G. Hutchings, B. Christoph, L. Gann, C. Hiller, P. Löw, M. Marsili, H. Saller, K. Yuki, *Topics Curr. Chem.* **1987**, *137*, 19–73.

[22] R. Herges, C. Hoock, *Science*, **1992**, *225*, 711–713.

[23] *http://www.infochem.de/classify.htm*

[24] W.-D. Ihlenfeldt, J. Gasteiger, *J. Comput. Chem.* **1994**, *15*, 793–813.

[25] H. Gelernter, J.R. Rose, C. Chen, *J. Chem. Inf. Comput. Sci.* **1990**, *30*, 492–504.

[26] J.R. Rose, J. Gasteiger, *J. Chem. Inf. Comput. Sci.* **1994**, *34*, 74–90.

[27] J. Zupan, J. Gasteiger, *Neural Networks in Chemistry and Drug Design*, 2nd edn., Wiley-VCH, Weinheim, **1999**.

[28] L. Chen, J. Gasteiger, *J. Am. Chem. Soc.* **1997**, *119*, 4033–4042.

[29] *The ChemInform RX reaction* database is produced by FIZ Chemie, Berlin, Germany, and marketed by MDL Information Systems, Inc., San Leandro, CA, USA.

[30] H. Satoh, O. Sacher, T. Nakata, L. Chen, J. Gasteiger, K. Funatsu, *J. Chem. Inf. Comput. Sci.*, **1998**, *38*, 210–219.

[31] J. Blair, J. Gasteiger, C. Gillespie, P.D. Gillespie, I. Ugi, *Tetrahedron*, **1974**, *30*, 1845–1859.

[32] R.S. Cahn, C.K. Ingold, V. Prelog, *Angew. Chem.* **1966**, *78*, 413–447; V. Prelog, G. Helmchen, *Angew. Chem.* **1982**, *94*, 614–631; *Angew. Chem. Int. Ed. Engl.* **1982**, *21*, 567–583.

# 4
# The Data

*Giorgi Lekishvili*

## Learning Objectives

- To gain a general overview on data and its pre-processing for learning
- To know, in outline, the pathways for data acquisition
- To understand what datasets are and how to estimate their quality
- To be able to deal with outliers and redundancy
- To know how to carry out scaling, mean-centering, and auto-scaling
- To understand data transformations and their applicability
- To know how to select an optimal subset of descriptors
- To become familiar with dataset optimization techniques
- To know how to validate results
- To understand what training and test sets are, and how to make use of them

## 4.1
## Introduction

### 4.1.1
### Data, Information, Knowledge

The real world is one of uncertainty. Suppose we are carrying out a reaction. We have obtained a product. In the beginning we observe a total uncertainty regarding the molecule. We have no information about its composition, the constitution of the skeleton, its stereochemical features, its physical properties, its biological activities, etc. Step by step, by routine experiments, we collect data. When the acquisition of the structural information is complete there is no uncertainty, at least about its structure. Well, we may not have perfect experiments, so this will require us to reserve space for the missing relevant information. However, it is rather more noise than genuine uncertainty, which, by the way, will never be eliminated.

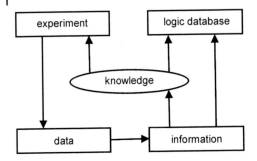

**Figure 4-1.** Research pathways.

We are now obtaining an initial idea of what information could be. The information about a real physical system is a measure of decreasing uncertainty of the system by means of some physical (including mental) activities [1].

Information itself is hardly ever just the end of research – it stands at the beginning of knowledge. If information can be more or less defined in terms of formal logic and even in a discrete-algebraic manner, the very essence of knowledge is truly philosophical and therefore is far beyond the scope of this textbook in general, and this chapter in particular. However, we may speak about the relation of knowledge to data in an intuitive way. It can be clear to us that retrieval and analysis of data can bring us new knowledge.

Let us define knowledge as the perception of the logical relations among the structures of the information. One thing we have to bear in mind is that any systematic treatment of information needs some previous knowledge. Therefore, research, in the long run, is always an iterative process, as depicted on Figure 4-1.

The major task of chemoinformatics is to find these relationships between the data on the molecular structure and the data on the physical, chemical, or biological properties of the molecules.

Two factors matter most in gaining knowledge from data: first, the quality of the data; and secondly, the method one applies to the data, and by which one learns from them.

The aim of this chapter is to deal with the first task, analysis of the quality of data. Very quickly we are going to demonstrate how to get high quality data, before we move on to show how they become useful for learning.

### 4.1.2
### The Data Acquisition Pathway

We learn from data. Therefore, the way we prepare the data for the learning process will crucially condition the quality of learning and the reliability of the extracted knowledge.

The first stage in data acquisition is the identification of the task; that is, we have to know what kind of physical properties/biological activities we are going to model.

Once we have defined this, we have to compile the initial dataset. First of all, we decide upon the composition of the dataset. Usually, we take initially as many compounds as possible.

The next and very important step is to make a decision about the descriptors we shall use to represent the molecular structures. In general, modeling means assignment of an abstract mathematical object to a real-world physical system and subsequent revelation of some relationship between the characteristics of the object on the one side, and the properties of the system on the other.

Perhaps the best idea is to compute as many descriptors as possible and then to select an optimal subset by applying sophisticated techniques, discussed below. First, we have an initial, and probably utterly crude, dataset. Genuine data pre-processing has only just started. The task is to assess the quality of the data. One of the topics for discussion in this chapter is the methods by which one finds out the potential drawbacks of the dataset.

The quality may suffer from the presence of so-called outliers, i.e., compounds that have low similarity to the rest of the dataset. Another negative feature may be just the contrary: the dataset may contain too many too highly similar objects.

Another problem is to determine the optimal number of descriptors for the objects (patterns), such as for the structure of the molecule. A widespread observation is that one has to keep the number of descriptors as low as 20 % of the number of the objects in the dataset. However, this is correct only in case of ordinary Multi-Linear Regression Analysis. Some more advanced methods, such as Projection of Latent Structures (or, Partial Least Squares, PLS), use so-called latent variables to achieve both modeling and predictions.

Once the quality of the dataset is defined, the next task is to improve it. Again, one has to remove outliers, find out and remove redundant objects (as they deliver no additional information), and finally, select the optimal subset of descriptors.

The final stage of compiling a maximally refined dataset is to split it into a training and test dataset. The definition of a test dataset is an absolute must during learning, as, in fact, it is the best way to validate the results of that learning.

And, last but by far not least; we must mention a very important part of data pre-processing. It is up to a researcher to decide when to employ these techniques. Figure 4-2 displays a step-by-step preparation of a dataset.

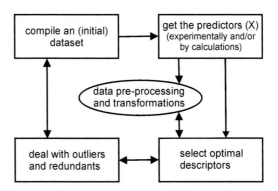

**Figure 4-2.** A high-quality dataset has to be prepared step-by-step, and often iteratively.

The two main ways of data pre-processing are mean-centering and scaling. Mean-centering is a procedure by which one computes the means for each column (variable), and then subtracts them from each element of the column. One can do the same with the rows (i.e., for each object). Scaling is a a slightly more sophisticated procedure. Let us consider unit-variance scaling. First we calculate the standard deviation of each column, and then we divide each element of the column by the deviation.

Furthermore, one may need to employ data transformation. For example, sometimes it might be a good idea to use the logarithms of variables instead of the variables themselves. Alternatively, one may take the square roots, or, in contrast, raise variables to the $n$th power. However, genuine data transformation techniques involve far more sophisticated algorithms. As examples, we shall later consider Fast Fourier Transform (FFT), Wavelet Transform and Singular Value Decomposition (SVD).

All these topics are discussed in greater detail later in this chapter.

## 4.2
## Data Acquisition

### 4.2.1
### Why Does the Quality of Data Matter?

The question is rhetorical. Once we learn from data, the quality of learning and the reliability of the obtained knowledge are conditioned by the quality of the data. However, some examples of how data affect the results and the conclusions we draw from them may be of interest.

Let us start with a classic example. We had a dataset of 31 steroids. The spatial autocorrelation vector (more about autocorrelation vectors can be found in Chapter 8) stood as the set of molecular descriptors. The task was to model the Corticosteroid Binding Globulin (CBG) affinity of the steroids. A feed-forward multilayer neural network trained with the back-propagation learning rule was employed as the learning method. The dataset itself was available in electronic form. More details can be found in Ref. [2].

An observation of the results of cross-validation revealed that all but one of the compounds in the dataset had been modeled pretty well. The last (31st) compound behaved weirdly. When we looked at its chemical structure, we saw that it was the only compound in the dataset which contained a fluorine atom. What would happen if we removed the compound from the dataset? The quality of learning became essentially improved. It is sufficient to say that the cross-validation coefficient increased from 0.82 to 0.92, while the error decreased from 0.65 to 0.44. Another learning method, the Kohonen's Self-Organizing Map, also failed to classify this 31st compound correctly. Hence, we had to conclude that the compound containing a fluorine atom was an obvious outlier of the dataset.

Another misleading feature of a dataset, as mentioned above, is redundancy. This means that the dataset contains too many similar objects contributing no new information. Indeed, if some two objects are highly similar, then the information content of the second object can always be more or less successfully reconstructed from the information delivered by the first object.

An example of how redundancy in a dataset influences the quality of learning follows. The problem implied classification of objects in a dataset through their biological activities with the help of Kohonen Self-Organizing Maps. Three subgroups were detected. The first one contained highly active compounds, the second subgroup comprised compounds with low activity, and the rest, the intermediately active compounds, fell into the third subgroup. There were 91 highly active, 540 intermediately active, and 492 inactive compounds. As one can see, the dataset was not balanced in the sense that the intermediately active and the inactive compounds outnumbered the active compounds. A first attempt of balancing the dataset by means of a mechanical (i.e., by chance) removal of the intermediately active and the inactive compounds decreased the quality of learning. To address this problem, an algorithm for finding and removing redundant compounds was elaborated. One can find more details on the algorithm in Section 4.3.4. This time the learning yielded essentially improved results. It is sufficient to say that if in the case of the primary dataset, only 21 compounds from 91 were classified correctly, whereas in the optimized dataset (i.e., that with no redundancy) the correctness of classification was increased to 65 out of 91.

As another example, we shall consider the influence of the number of descriptors on the quality of learning. Lucic et. al. [3] performed a study on QSPR models employing connectivity indices as descriptors. The dataset contained 18 isomers of octane. The physical property for modeling was boiling points. The authors were among those who introduced the technique of orthogonalization of descriptors. By applying this method, they demonstrated that the removal of insignificant variables increases the quality and reliability of the models despite the fact that the correlation coefficient, $r$, always decreases, although only slightly. For example, the characteristics of a model with six orthogonalized descriptors were: $r = 0.99288$, $s = 0.9062$, $F = 127.4$; and the quality of this model was sufficiently improved after removal of the two least significant descriptors, to: $r = 0.9925$, $s = 0.8553$, $F = 214.4$. Here, $s$ is the standard deviation and $F$ is the Fisher ratio.

## 4.2.2
## Data Complexity

Systems can possess different extents of complexity. To measure complexity, the information content of the system can be used. Application of information theory is increasingly fruitful for modeling biological activities with regard to the symmetry of molecules.

The reason why complexity and symmetry are linked together is quite straightforward. Indeed, a representation of highly symmetrical systems requires fewer characteristics than that of objects having low symmetry because, if we know the characteristics of one object, we can employ them to represent all those which are symmetrical with the given one.

Let us consider a system $S$ with $n$ objects. Suppose we have a criterion which enables us to distribute the objects into different subsets of $S$. One condition is that no object can belong to any two different subsets. Once the distribution is complete, we may have $m$ subsets containing $n_i$ objects, correspondingly, so that $\Sigma_1 n_i = n$ and $I = 1, 2, ..., m$.

The Shannon Equation (Eq. (1)) [4] enables one to evaluate the information content, $I$ (also known as the Shannon entropy). of the system.

$$I = -\sum_i^m \frac{n_i}{n} \log_2 \frac{n_i}{n} \tag{1}$$

Whatever the criterion is, we may have the following two extreme situations. The first one occurs when all the objects fall into the same subset (such subsets are known in discrete algebra as classes of equivalence). The second is when each subset contains one, and only one, object.

The first case corresponds to zero information content.

$m = 1$, $n_i = n$, and thus: $I = -(n/n) \times \log_2 (n/n) = (-1) \times \log_2 (1) = 0$

The second case is completely different.

$m = n$, $n_i = 1$, for all $i$: $I = -n \times (1/n) \times \log_2 (1/n) = -(-\log_2 n) = \log_2 n$

Therefore, the maximal complexity of the system cannot exceed this number ($\log_2 n$, where, as mentioned above, $n$ is the number of system elements). Needless to say, real-world systems reveal what can be called medium extents of complexity, i.e., $0 < I < \log_2 n$.

Returning to datasets from chemoinformatics, we may conclude that the first case stands for the complete degeneracy of the dataset, when all but one data object are redundant. The second case corresponds to the weird situation in which all the objects of the dataset are outliers. We have thus arrived at the extreme extents of the dataset complexity.

As we should remember now, we distribute the objects into subsets in accordance with some criterion, not having known even the number of subsets themselves. That is why the evaluation of data complexity is still a challenging problem.

As oversimplified cases of the criterion to be used for the clustering of datasets, we may consider some high-quality Kohonen maps, or PCA plots, or hierarchical clustering.

### 4.2.3
### Experimental Data

Data can be derived in several ways, but an experiment is the process we intuitively link with deriving data. Even mental activities are often called mental experiments, especially in quantum mechanics. The better the experiment, the less noisy are the data. Experimental design is a discipline which teaches one to plan and carry out experiments in such a way that the maximum possible amount of relevant information is gained.

The key tasks are:

- determination of the absolute importance that variables have in generating responses in complex systems;
- identification and elimination of irrelevant variables;
- creation of simple mathematical models for experimental optimization;
- reduction of the costs of experiments;
- enhancement of the quality of products and of processes;
- abatement of noise effects.

More details can be found in Chapter IV, Section 2.2 of the Handbook.

### 4.2.4
### Data Exchange

According to an elegant remark by Davies [5], "Modern scientific data handling is multitechnique, multisystem, and manufacturer-independent, with results being processed remotely from the measuring apparatus." Indeed, data exchange and storage are steps of the utmost importance in the data acquisition pathway. The simplest way to store data is to define some special format (i.e., collection of rules) of a flat file. Naturally, one cannot overestimate the importance of databases, which are the subject of Chapter 5 in this book. Below we discuss three simple, yet efficient, data formats.

#### 4.2.4.1  DAT files
This format was developed in our group and is used fruitfully in SONNIA, software for producing Kohonen Self-Organizing Maps (KSOM) and Counter-Propagation (CPG) neural networks for chemical application [6]. This file format is ASCII-based, contains the entire information about patterns and usually comes with the extension "dat".

As an example, the file in Figure 4-3 has been taken from Ref. [6].

```
!!Name: steroids_s.ctx
! Input vector created by rcode
! Parameters:
! lower distance border: 0.00000000, Upper distance border 12.80000000
! input dimension: 128, non weighted
!!Property: radial distribution function A1
.... 8.830693e-02 8.825346e-02 8.786038e-02 .... -6.279 2 !aldosterone
.... 1.721332e+00 1.721456e+00 1.722374e+00 .... -5.000 3 !androstanediol
.... 1.721336e+00 1.721485e+00 1.722592e+00 .... -5.000 3 !5-androstenediol
.... -5.563186e-01 -5.564730e-01 -5.576087e-01 .... -5.763 3 !4-androstenedione
....
....
....
```

**Figure 4-3.** A sample file in the "dat" format.

The header delivers any relevant additional information helpful to define the nature of the task and its peculiarities. The body contains the variables, which are usually floats. Next, there are the integers defining the class to which the patterns belong (indispensable for classification tasks) and pattern names.

### 4.2.4.2 JCAMP-DX

This format considers so-called labelled data records (LDR) all having the same basic structure of the form: ##descriptor = "something", where the descriptor is the name of the LDR and the "=" signifies the end of the label. The lines can be up to 80 characters long, terminating in a carriage return or linefeed. However, LDRs can run over more than one line.

The JCAMP-DX file format is split into the sections CORE and NOTES with the intention of keeping less important data separated from the essential content. The CORE itself contains CORE HEADER and CORE DATA. NOTES are just between HEADER and DATA (see Figure 4-4 for an example).

More details are given in Chapter IV, Section 3 of the Handbook.

```
##TITLE=ATRAZIN
##JCAMP-DX=4.24
##DATA TYPE=INFRARED SPECTRUM
##DATE=24/2/1998
##TIME=11:29:57
##SAMPLING PROCEDURE=Pre˙ling
##ORIGIN=K˙
##DATA PROCESSING=no operation
##XUNITS=POINTS
##YUNITS=ABSORBANCE
##RESOLUTION=4
##FIRSTX=3999.6401
##LASTX=399.19263
##DELTAX=-1.9284668
##MAXY=0.66135877
##MINY=0.11461937
##XFACTOR=1
##YFACTOR=6.1593836e-010
##NPOINTS=1868
##FIRSTY=0.14059831
##XYDATA=(X++(Y..Y))
4000+228266856+228266864+228266864+228266864+228266864+228266864+228266864
3986+228266864+228266864+228247920+228207360+228135104+228066688+228054928
3973+228093040+228131792+228129872+228066496+227984528+227996192+228121216
3959+228285648+228561040+229026560+229534592+229861408+229881376+229739104
3946+229775680+230029728+230218544+230250416+230228560+230130224+229999696
3932+230141328+230693088+231415600+232074992+232432432+232071104+230917200
3919+229677872+229325744+230287504+232224432+234300304+235796224+236741136
3905+237632960+238485968+239225360+240139456+240898432+241052560+240745552
3892+239793760+238262160+236940656+236445264+237130672+238566496+239652464
3878+240238768+241006512+241866912+241581504+239823072+238816000+240719888
```

**Figure 4-4.** An example of the JCAMP-DX file format.

```
<?xml version="1.0"?>
<!DOCTYPE PMML PUBLIC "PMML 2.0" "http://www.dmg.org/v2-0/pmml_v2_0.dtd">
<PMML  version="2.0">
    <Header copyright="dmg.org"/>
```

```
<DataDictionary numberOfFields="3">
     <DataField name="marital status" optype="categorical">
         ...
     </DataField>
     ...
</DataDictionary>
```
data dictionary

```
<ClusteringModel modelName="Mini Clustering" functionName="clustering"
  modelClass="centerBased" numberOfClusters="2">
     <MiningSchema>
         ...
     </MiningSchema>
```
mining schema

```
<ClusteringField field="marital status"
   compareFunction="squaredEuclidean"/>
...
<CenterFields>
  <DerivedField name="c1">
     <NormContinuous field="age">
         <LinearNorm orig="45" norm="0"/>
         ...
     </NormContinuous>
  </DerivedField>
```
clustering fields

```
    ...
</CenterFields>
<Cluster name="marital status is d or s">
     <Array n="5" type="real">
         0.524561 0.486321 0.128427 0.459188 0.412384</Array>
</Cluster>
...
</ClusteringModel>
```
clustering results

```
</PMML>
```

**Figure 4-5.** A sample PMML file.

### 4.2.4.3 **PMML**

Predictive Model Markup Language (PMML) is far more than just another format of a data container flat file [7]. As is clear from the name, it is an XML-based markup language delivering all the power of XML. Readers are recommended to consult Section 2.4.5 and the website *www.xml.org* for more details on XML and its applications in chemistry.

An example of a PMML document is shown in Figure 4-5.

PMML is powerful enough to provide information about:

- the data dictionary;
- the mining schema;
- data flow;
- transformations;
- statistics;

- conformance;
- taxonomy;
- trees;
- regression;
- general regression;

- cluster models;
- association rules;
- the neural network;
- naive Bayes classifiers;
- sequences.

More details can be found in Ref. [7] and in the documentation mentioned and/or given therein.

4.2.5
### Real-World Data and Their Potential Drawbacks

We have already mentioned that real-world data have drawbacks which must be detected and removed. We have also mentioned outliers and redundancy. So far, only intuitive definitions have been given. Now, armed with information theory, we are going from the verbal model to an algebraic one.

Let us first define the information content per object. A (discrete) system can be split into classes of equivalence, whose number can vary from 1 to $n$, where $n$ is the number of the elements (objects) in the system. No element can belong simultaneously to more than one class. Therefore, the information content ($IC$) of the system is additive, at least class-wise. This means that the information content of the system is the sum of the information contents of the classes. The $IC$ of a class can be given by Eq. (2), where $n_i$ is the number of elements in the $i$th class. (Recall, also, that log $(x) = -\log (1/x)$.)

$$IC_i = \frac{n_i}{n} \log_2 \frac{n}{n_i}, \tag{2}$$

Now we calculate the information content per object ($ICO$). The $ICO$ of the $j$th object, which belongs to the class of equivalence $i$, can be evaluated via Eq. (3).

$$ICO_j(j \in i) = \frac{IC_i}{n_i} = \frac{1}{n} \log_2 \frac{n}{n_i}. \tag{3}$$

As we can see, all objects inside a given class must have equal $ICO$s.

The average information content per object, $AICO$, is given by Eq. (4), in which $m$ is the number of equivalence classes.

$$AICO = \frac{\sum_{i}^{m} ICO(j \in i)}{m}, \tag{4}$$

Let us illustrate the approach with examples. Suppose we have a dataset with some 500 compounds. First, we apply a set of descriptors. Employing some procedure or other, we obtain, say, ten classes of equivalence in which the elements are distributed as given in Table 4-1.

Equation (4) provides a value of $AICO = 0.00852$. As we see, the populations of classes 3 and 8 bear redundancy in information, as their $ICO$s are quite low here. In contrast, classes 6 and 7 are clearly outliers. Their $ICO$s are too high in compar-

**Table 4-1.** The distribution yielded by the first set of descriptors.

|  | Class ID | | | | | | | | | |
|---|---|---|---|---|---|---|---|---|---|---|
|  | 1 | 2 | 3 | 4 | 5 | 6 | 7 | 8 | 9 | 10 |
| Population | 22 | 45 | 120 | 15 | 58 | 2 | 5 | 178 | 34 | 21 |
| $IC_i$ | 0.198 | 0.313 | 0.494 | 0.152 | 0.360 | 0.032 | 0.066 | 0.530 | 0.264 | 0.192 |
| $ICO$ | 0.009 | 0.007 | 0.004 | 0.010 | 0.006 | 0.016 | 0.0132 | 0.003 | 0.008 | 0.009 |

ison with the average. The question is whether outliers have to be removed immediately. Their *ICOs* are really high, and thus they may deliver valuable and original information. Therefore, before having them removed, one must also test different sets of descriptors.

We must now mention, that traditionally it is the custom, especially in chemometrics, for outliers to have a different definition, and even a different interpretation. Suppose that we have a *k*-dimensional characteristic vector, i.e., *k* different molecular descriptors are used. If we imagine a *k*-dimensional hyperspace, then the dataset objects will find different places. Some of them will tend to group together, while others will be allocated to more remote regions. One can by convention define a margin beyond which there starts the realm of "strong" outliers. "Moderate" outliers stay near this margin.

We have to apply projection techniques which allow us to plot the hyperspaces onto two- or three-dimensional space. Principal Component Analysis (PCA) is a method that is fit for performing this task; it is described in Section 9.4.4. PCA operates with latent variables, which are linear combinations of the original variables. The first few principal components store most of the relevant information, the rest being merely the noise. This means that one can use two or three principal components and plot the objects in two or three-dimensional space without losing information.

## 4.3
## Data Pre-processing

### 4.3.1
### Mean-Centering, Scaling, and Autoscaling

Mean-centering, as is shown by experience, can be successfully employed in combination with another data pre-processing technique, namely scaling, which is discussed later.

Let $a_j$ be the average of the *j*th column vector of a data matrix (Eq. (5)).

$$a_j = \frac{\sum_{j=1}^{N} x_{ij}}{n}. \tag{5}$$

Here, $x_{ij}$ is the *i*th entry of the *j*th column vector and *n* is the number of objects (rows in the matrix). The essence of mean-centering is to subtract this average from the entries of the vector (Eq. (6)).

$$x_{ij} \leftarrow x_{ij} - a_j \tag{6}$$

What we have now is the resultant data matrix containing the column vectors with the average values each being zero.

Scaling is quite often applied in chemometrics. However, before we start to examine it, we have to consider the following. Classical chemometrics often deals with tasks which differ from those in molecular design. An important chemometrical task is to model experimental data, i.e., the independent variables are measured experimentally in just the same way as the responses. On the other hand, in molecular design and QSAR these independent variables are mostly calculated by some method or other. Even the physical properties of molecules, which are applied as descriptors, are often evaluated theoretically. It leads to a situation where the variables have more or less similar ranges in their numerical values. Quite a different picture occurs when a model is based on experimental data. The latter can differ widely from each another, and this difference can result in misleading models, especially when PCA or PLS is used as a mapping device. Therefore, the major task of molecular design is to establish better descriptors and to select the optimal subset of them. It is noteworthy that only very few papers were published in the field of molecular design where data pre-processing techniques were used. One area where the optimal subset selection problem is crucial in chemometrics is the case where spectral data (wavelengths) are used as independent variables. Robust wavelength selection is an exciting problem.

In general, scaling of a variable in the data matrix can be viewed as a multiplication of the corresponding column vector entries with some number. If the significances of the variables to the model are known prior to modeling, then it might be a good idea to upscale the highly relevant variables. In contrast, if a variable is supposed to bear merely noise, then its significance must be downscaled. However, this is a rare case in reality. Therefore, unit-variance scaling (UV-scaling) is most often used. Moreover, scaling itself is sometimes associated with UV-scaling.

Let $s_j$ be the standard deviation of the $j$th variable (i.e., column vector in the data matrix), as defined in Eq. (7), where $x_{ij}$, $n$, and $a$ are the same as in the previous equations.

$$s_j^2 = \frac{\sum\limits_{i=1}^{N}(x_{ij} - a_j)^2}{n - 1} \tag{7}$$

UV-scaling functions as set out in Eq. (8).

$$x_{ij} \leftarrow \frac{x_{ij}}{s_j} \tag{8}$$

The result is that each column vector has unit standard deviation.

What is the objective of using UV-scaling? As was often mentioned above, the major task during modeling is to separate the relevant information in the data from the noise. Which variables are more informative? The answer to this question is obvious: those which are more diverse. The standard deviation comes to help: the higher it is, the more diverse the variables. In fact, the variables appear more diverse – only *appear*, and nothing more – because, if different column vectors have different ranges, then even fewer diverse variables with high ranges may appear significant, as they will possess high standard deviations! This is particu-

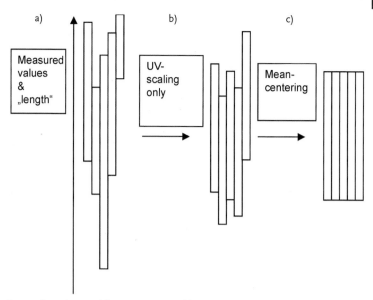

**Figure 4-6.** Autoscaling. The variables are represented by variance bars. a) Raw data; b) the data after UV-scaling only; c) the autoscaled data [8].

larly true when one wants to apply PCA-like methods. Let us follow Erikson et. al. [8]:

"A variable with a large numerical range automatically gets a large initial variance, whereas a variable with a small numerical range will get a lower initial variance. Then, since PCA is a maximum variance projection method, it follows that a variable with a large variance will have a better chance to be expressed in the modeling than a low-variance variable."

What UV-scaling does is to "concentrate" the relevant information into the same range for all the variables (or, at least, for those subjected to this method). Then, the loading matrix yielded by PCA will show the importance of the initial variables.

As mentioned above, one can use UV-scaling together with mean-centering. This is called autoscaling (Eq. (9)).

$$x_{ij} \leftarrow \frac{x_{ij} - a_j}{s_j} \qquad (9)$$

The process can be illustrated by Figure 4-6.

If the task is multivariate calibration, for example, the proper choice of a pre-processing method will essentially affect the quality of the resultant model. For more details about the use of these techniques together with PCA and PLS, readers are advised to consider the fundamental monograph by Erikson et al [8].

4.3.2
**Advanced Methods**

4.3.2.1 **Fast Fourier Transformation**

Fast Fourier Transformation is widely used in many fields of science, among them chemometrics. The Fast Fourier Transformation (FFT) algorithm transforms the data from the "wavelength" domain into the "frequency" domain. The method is almost compulsorily used in spectral analysis, e. g., when near-infrared spectroscopy data are employed as independent variables. Next, the spectral model is built between the responses and the Fourier coefficients of the transformation, which substitute the original X-matrix.

As stated in Ref. [9],

*"The most important feature of the Fourier analysis is the reduction of the multicollinearity and the dimension of the original spectra. However, the Fourier coefficients bear no simple relationship to individual features of the spectrum so that it will not be clear what information is being used in calibration."*

4.3.2.2 **Wavelet Transformation**

Wavelet transformation (analysis) is considered as another and maybe even more powerful tool than FFT for data transformation in chemometrics, as well as in other fields. The core idea is to use a basis function ("mother wavelet") and investigate the time-scale properties of the incoming signal [8]. As in the case of FFT, the Wavelet transformation coefficients can be used in subsequent modeling instead of the original data matrix (Figure 4-7).

The method has many applications; among them are Denoising Smoothing (DS), compression, and Feature Extraction (FE), which are powerful tools for data transformations. See the "Selected Reading" section at the end of this chapter for further details.

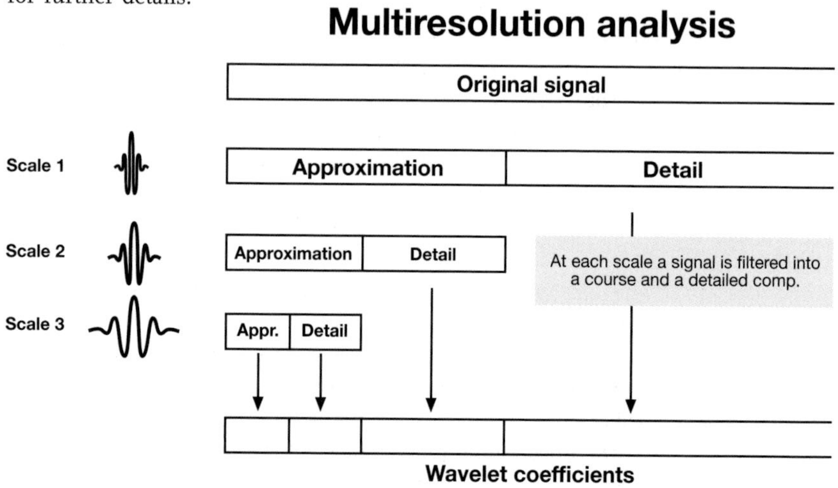

**Figure 4-7.** Overview of the wavelet transform and multi-resolution analysis scheme [10].

### 4.3.2.3  Singular Value Decomposition

It may look weird to treat the Singular Value Decomposition (SVD) technique as a tool for data transformation, simply because SVD is the same as PCA. However, if we recall how PCR (Principal Component Regression) works, then we are really allowed to handle SVD in the way mentioned above. Indeed, what we do with PCR is, first of all, to transform the initial data matrix $X$ in the way described by Eqs. (10) and (11).

$$X = UVW^T \tag{10}$$

$$X^+ = V(diag\ (1/w_{ij})\ U^T) \tag{11}$$

Here $W$ is diagonal matrix of singular values, $V^T$ is the transpose of the second resultant matrix, being actually the same as the loading matrix in PCA, and $X^+$ is the matrix, which is applied for further modeling.

Anyway, we do not use the whole resultant matrix; instead, we need only the first few columns of $X^+$, as SVD guarantees that all of the relevant information is concentrated there, the rest being noise.

Now, one may ask, what if we are going to use Feed-Forward Neural Networks with the Back-Propagation learning rule? Then, obviously, SVD can be used as a data transformation technique. PCA and SVD are often used as synonyms. Below we shall use PCA in the classical context and SVD in the case when it is applied to the data matrix before training any neural network, i.e., Kohonen's Self-Organizing Maps, or Counter-Propagation Neural Networks.

The profits from using this approach are clear. Any neural network applied as a mapping device between independent variables and responses requires more computational time and resources than PCR or PLS. Therefore, an increase in the dimensionality of the input (characteristic) vector results in a significant increase in computation time. As our observations have shown, the same is not the case with PLS. Therefore, SVD as a data transformation technique enables one to apply as many molecular descriptors as are at one's disposal, but finally to use latent variables as an input vector of much lower dimensionality for training neural networks. Again, SVD concentrates most of the relevant information (very often about 95 %) in a few initial columns of the scores matrix.

### 4.3.3
### Variable Selection

There was a time when one could use only a few molecular descriptors, which were simple topological indices. The 1990s brought myriads of new descriptors [11]. Now it is difficult even to have an idea of how many molecular descriptors are at one's disposal. Therefore, the crucial problem is the choice of the optimal subset among those available.

Much labor has been dedicated to establishing a common technique enabling one to solve the problem of choice. Some solutions suggested are useful, others are less efficient. Below we shall examine the most prominent ones.

### 4.3.3.1 **Genetic Algorithm (GA)-Based Solutions**

The idea behind this approach is simple. First, we compose the characteristic vector from all the descriptors we can compute. Then, we define the maximum length of the optimal subset, i.e., the input vector we shall actually use during modeling. As is mentioned in Section 9.7, there is always some threshold beyond which an increase in the dimensionality of the input vector decreases the predictive power of the model. Note that the correlation coefficient will always be improved with an increase in the input vector dimensionality.

Let us see how the approach works in practice. One of the first studies dedicated to the applications of GA with regard to this task was that by Rogers and Hopfinger [12]. However, the pioneering efforts are due to the Nijmegen chemometrics research group led by Buydens [13, 14].

The simplest algorithm can be formulated as follows. After the computation of all available descriptors and setting the length of the input vector, which would be much shorter than the initial one, GA comes in action. The selection is encoded with bit-strings having the same number of entries as the dimensionality of the initial vector. Usually each of the chosen variables is represented by a one, whereas each of the dropped ones is represented by a zero. Therefore, the bit-strings can be treated in a unique GA-like way. The model is built with the selected variables and its robustness has to be evaluated. The key problem is to find a fitness or cost function. The worst choice would probably be the correlation coefficient *r*; the Standard Error (SE) and cross-validation correlation coefficient *q* would be slightly better. However, application of test set-based evaluations is preferred. Our recommendations are *LOF* (Lack of Fitting) [15] and *CoSE* (Compound Standard Error), which are defined in Eqs. (12) and (13) [16].

$$LOF = \frac{SE}{\left(1 - \dfrac{c + dp}{n}\right)^2} \tag{12}$$

In Eq. (12), *SE* is the standard error, *c* is the number of selected variables, *p* is the total number of variables (which can differ from *c*), and *d* is a smoothing parameter to be set by the user. As was mentioned above, there is a certain threshold beyond which an increase in the number of variables results in some decrease in the quality of modeling. In fact, the smoothing parameter reflects the user's guess of how much detail is to be modeled in the training set.

$$CoSE = 0.5 \sqrt{\frac{\sum\limits_{i=1}^{n} (y_{\exp,i} - y_{est,i})^2}{n-1}} + \sqrt{\frac{\sum\limits_{j=1}^{m} (y_{\exp,j} - y_{est,j})^2}{m-1}} \tag{13}$$

In Eq. (14) "est" stands for the calculated (estimated) response, "exp" for the experimental one, and *n* and *m* are the numbers of objects in the training set and the test set respectively. "CoSE" stands for COmpound Standard Error. As an option, one can employ several test sets, if needed.

Once the cost of the solution is calculated, the relevant models are probabilistically kept; new solutions are generated via crossover and mutation (it is a good idea to allow a change in the length of the input vector, i.e., making it longer or shorter by one or two descriptors during mutation), and the process goes on until convergence.

### 4.3.3.2 Orthogonalization-Based Solutions

Real-world molecular descriptors are highly inter-correlated. This means that most of the relevant information content of a descriptor $k$ can be successfully reconstructed from some other descriptor $l$, which is highly (say, >75%) correlated with $k$. Among other drawbacks, this phenomenon makes it even more difficult to decide which particular descriptors of the highly inter-correlated pairs, or even sets, have to be picked up. Similarly to PCA, orthogonalization yields new variables, which are orthogonal to each another pair-wise, i.e., the correlation coefficients are zero each time. However, unlike PCA, the new variables are by no means latent, at least in the sense that they have a clear interpretation. Each orthogonalized variable delivers only that part of the original counterpart which has no analog in other descriptors. What this means is that each new descriptor contributes independently to the model. Therefore, the higher the individual correlation coefficient of this variable with a response, the higher is the significance of the variable to the model! Just as in PCA, one can (and must) quit with the first few orthogonalized variables, as they explain the bulk of the response. Again, we emphasize that the orthogonalized variables bear a clear meaning and can be interpreted in a straightforward and simple way.

More details can be found in the contributions from Randic [17] and Lucic & Trinajstic [18].

### 4.3.3.3 Simulated Annealing (SA)-Based Solutions

Simulated Annealing-based solutions [19] are conceptually the same as Genetic Algorithm-based approaches. However, the SA-based techniques, in our experience, are more sensitive to the initial settings of the parameters. Nevertheless, once the correct ones are found, the method can achieve the efficiency of GA-based solutions. We must point out that SA-based solutions have never outperformed the GA-based ones in our studies. Much of what has been mentioned regarding the GA-based solutions is also relevant for the SA technique, particularly, with respect to the cost functions.

It may be of interest to readers that all three methods mentioned above resulted in the same optimal subset of descriptors for the well-known Selwood dataset, which has become a *de-facto* standard in testing new approaches in this field [20].

See Kalivas [21] for more details about SA.

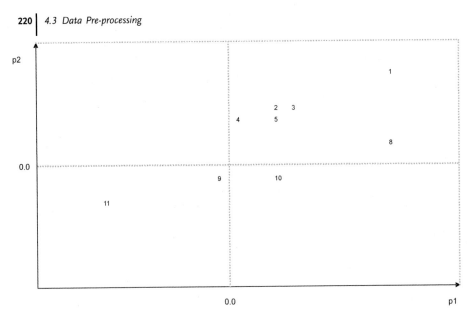

**Figure 4-8.** Plot of the first two column vectors of the loadings matrix of PCA.

### 4.3.3.4 **PCA-Based Solutions**

PCA loadings deliver very useful information about the impacts of particular variables on the models. The plot of the loadings reveals which variables contain similar information (not necessarily pair-wise) and are highly correlated (those grouped together). Furthermore, one can see which variables are inversely correlated with each other (those in diagonally opposite quadrants from the origin). Finally, one can arrive at the most important question: which variables have the highest influence? The diagnostic criterion is that the more remote the variable is from the origin, the more impact it has on the model. As an illustration, we may consider the hypothetical loading plot in Figure 4-8.

As we can see from Figure 4-8. variables 1, 8, and 11 are the most significant ones, variables 4, 9, and 10 much less so, variables 2 and 5 are highly and positively correlated, the whole group of variables 2, 3, 4, and 5 are more or less similar, and variables 9 and 4 are inversely correlated.

### 4.3.4
**Object Selection**

Although the problem of compilation of training and test datasets is crucial, unfortunately no *de-facto* standard technique has been introduced. Nevertheless, we discuss here a method that was designed within our group, and that is used quite successfully in our studies. The method is mainly addressed to the task of finding and removing redundancy.

It has often been mentioned in this chapter that many molecular descriptors can well be highly inter-correlated. Therefore, any significant information content of a descriptor can be reconstructed from another one which is highly correlated with it. This is a crucial problem. Similarly, a dataset can contain such objects which are highly similar to others. Therefore, the redundancy can be referred to as a "horizontal" inter-correlation. The biggest difference, however, is that usually there are many fewer compounds than molecular descriptors available for compiling a training set, and the relevance of the choice of a suitable descriptor is utterly dependent on the structure of the training set. Hence, the way to solve the problem has to be different.

Let us outline one of our approaches with the following simple example. Suppose we have a dataset of compounds and two experimental biological activities, of which one is a target activity (TA) and the other is an undesirable side effect (USE). Naturally, those with high TA and low USE form the first subclass, those with low TA and high USE the second, and the rest go into the third, intermediate subclass.

Next, we select some "pillar" compounds inside each or some of those subclasses, i.e., those having the highest norm of the characteristic vector. We can employ two pillars, the "lowest" (that with the lowest norm) along with the "highest", and keep only those compounds which are reasonably dissimilar to the pillar (or to both pillars). The threshold of "reasonability" is to be set by the user.

The functionality of the algorithm can be exemplified with the help of a real-world dataset.

Initially the dataset contained 818 compounds, among which 31 were active (high TA, low USE), 157 inactive (low TA, high USE), and the rest intermediate. When the complete dataset was employed, none of the active compounds and 47 of the inactives were correctly classified by using Kohonen self-organizing maps (KSOM).

Next, the technique described above (our method) was applied to the subclasses of inactive and intermediate compounds. This time 25 active compounds and 35 inactive compounds (from the remaining 68) were correctly classified by the same method, namely, KSOM.

It is interesting to note that in a case where the same numbers of inactive and of intermediate compounds were mechanically removed (i.e., without consideration of similarity), no improvement was observed – moreover, the model quality became even worse.

In our experience, another important advantage of the method is that one can use all the available descriptors without taking care over their choice. The method does not require any significant CPU resources, even when applied to a large dataset.

**4.4**
**Preparation of Datasets for Validation of the Model Quality**

4.4.1
**Training and Test Datasets**

The most important task of modeling is prediction. The model itself is needed for evaluating the biological activities (and/or physical properties) of compounds, where it is either difficult or costly to measure the activities experimentally.

The danger is overfitting/overtraining, which would lead to the model obtained being too tightly linked to the dataset that has been used to build it. Thereafter, it would be useless to make predictions regarding any other related dataset. The problem is serious. Before the model is applied to a dataset containing compounds with unknown activities, it has to be tested with the help of another dataset, where the activities have been measured. In other words, it is an absolute necessity to split the initial dataset into training and test datasets.

The purpose of a training dataset is to build a model, whereas the test set enables one to check the quality of the model against, e.g., overfitting. Sometimes one needs more than two datasets for creating a valid model. There are learning methods which have to come to convergence, and only then is learning stopped. Usually, the goodness of a model, evaluated in some way, is applied as a convergence criterion. Again, the same question arises: how can one evaluate the quality of a model during the learning process? The solution (Figure 4-9) is to apply another dataset, commonly known as the control dataset [22].

How should the initial dataset be split into two or three parts? The answer is not so trivial as it might appear. Suppose we have detected and removed neither the outliers nor the redundancy. Then, there is a danger that the last two datasets (control and test) may contain much less relevant information than the training set. Now, even if the model built via the training set is good enough, when the control set contains too many outliers the diagnostics will be invalid. This will lead to the loss of a good model and a wrong assignment of convergence.

Therefore, the initial dataset and the resultant sets must ideally have the same information content.

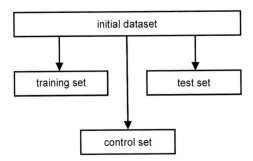

**Figure 4-9.** Splitting of the initial dataset into two or three parts for evaluation purposes.

### 4.4.2
### Compilation of Test Sets

The easiest way to extract a set of objects from the basic dataset, in order to compile a test set, is to do so randomly. This means that one selects a certain number of compounds from the initial (primary) dataset without considering the nature of these compounds. As mentioned above, this approach can lead to errors.

Another method of detection of overfitting/overtraining is cross-validation. Here, test sets are compiled at run-time, i.e., some predefined number, $n$, of the compounds is removed, the rest are used to build a model, and the objects that have been removed serve as a test set. Usually, the procedure is repeated several times. The number of iterations, $m$, is also predefined. The most popular values set for $n$ and $m$ are, respectively, 1 and $N$, where $N$ is the number of the objects in the primary dataset. This is called one-leave-out cross-validation.

Our recommendation is that one should use $n$-leave-out cross-validation, rather than one-leave-out. Nevertheless, there is a possibility that test sets derived thus would be incompatible with the training sets with respect to information content, i.e., the test sets could well be outside the modeling space [8].

Hence, the main danger in the process of compiling test sets remains. Fortunately, there are some other approaches which, although they may look more sophisticated, do diminish the possibility of such incompatibilities.

First, one can check whether a randomly compiled test set is within the modeling space, before employing it for PCA/PLS applications. Suppose one has calculated the scores matrix $T$ and the loading matrix $P$ with the help of a training set. Let $z$ be the characteristic vector (that is, the set of independent variables) of an object in a test set. Then, we first must calculate the scores vector of the object (Eq. (14)).

$$t = zP \tag{14}$$

Next the error is calculated (Eq. (15), where $I$ is the identity matrix).

$$e = z - tP' = z(I - PP') \tag{15}$$

Now, if the error, $e$, is greater than that of the training stages, the object is outside the modeling space. Further details can be found in Chapter 9 and also in Ref. [8].

The Kohonen Self-Organizing Maps can be used in a similar manner. Suppose $x_k$, $k = 1, ..., N$ is the set of input (characteristic) vectors, $w_{ij}$, $I = 1, ..., I, j = 1, ..., J$ is that of the *trained* network, for each $(i, j)$ cell of the map; $N$ is the number of objects in the training set, and $I$ and $J$ are the dimensionalities of the map. Now, we can compare each $x_k$ with the $w_{ij}$ of the particular cell to which the object was allocated. This procedure will enable us to detect the maximal $(e^2_{max})$ and minimal $(e^2_{min})$ errors of fitting. Hence, if the error calculated in the way just mentioned above is beyond the range between $e^2_{max}$ and $e^2_{min}$, the object probably does not belong to the training population.

The methods discussed above enable one to examine carefully the content of test sets, thus improving the quality of modeling.

**Essentials**

- The quality and the relationship of the data to the information and knowledge are crucial points in the learning process
- The data has to be prepared (pre-processed) to learn from it
- The complexity of a system has to be evaluated
- Data handling (management and storage) needs a good performance to exchanging the data
- It is necessary to pre-process data by mean-centering, scaling or autoscaling
- Widely used methods of data transformation are Fast Fourier and Wavelet Transformations or Singular Value Decomposition
- Variable and pattern selection in a dataset can be done by genetic algorithm, simulated annealing or PCA
- The quality of a model should be validated by compilation of training, test and control datasets

**Selected Reading**

Application of Statistical Methods

- M. Otto, *Chemometrics*. Wiley-VCH, Weinheim, **1999**.
- D.L. Massart, B.G.M. Vandeginste, L.M.C. Buydens, S.De Jong, *Handbook of Chemometrics and Qualimetrics*: 2-Vol. set, Elsevier, Amsterdam, **1998**.

Neural Networks and Machine Learning

- M. Smith, *Neural Networks in Statistical Modelling*. Van Nostrand Reinhold, New York, **1993**.
- T.M. Mitchell, *Machine Learning*. McGraw-Hill, New York, **1997**.
- J. Zupan, J. Gasteiger, *Neural Networks in Chemistry and Drug Design*. Wiley-VCH, Weinheim, **1999**.
- N. Christiani, J. Shawe-Taylor, *Support Vector Machines*. Cambridge University Press, Cambridge, UK, **2000**.

Digital Signal Processing and Chemistry

- L. Eriksson, J. Trygg, E. Johansson, R. Bro, S. Wold, Orthogonal signal correction, wavelet analysis, and multivariate calibration of complicated process fluorescence data. *Anal. Chim. Acta,* **2000**; *420,* 181–195.

**Available Software**

- SPSS – applicable in a wide range of engineering tasks. More details are available at *http://www.spssscience.com/sigmastat/index.cfm*
- The Unscrambler family from CAMO is specially designed by chemometricians. However, this package is also applicable in a wide range of engineering tasks. *http://www.camo.no/p2_tuf.htm*
- The SIMCA family from Umetrics – *http://www.umetrics.com*
- ELECTRAS – web-based data analysis system. The software supports 2 x 2 different modes of action: the modes for expert and novice engineers and the modes for expert and novice computational chemists. *http://www2.chemie.uni-erlangen.de/projects/eDAS/index.html*
- SONNIA – KSOM and CPG neural networks. The key features are robustness of training and excellent visualization capabilities. *http://www.mul-net.de*

# References

[1] L. Brillouin, *Science and Information Theory*, 2nd ed. Academic Press, New York, **1962**.

[2] J. Sadowski, M. Wagener, J. Gasteiger, *J. Am. Chem. Soc.*, **1995**, *117*, 5569–7775.

[3] B. Lucic, S. Nikolic, N. Trinajstic, D. Juretic, A. Juric, *J. Chem. Inf. Comput. Sci.*, **1995**, *35*, 532–538.

[4] C.E. Shannon, W.W. Weaver, *Mathematical Theory of Communication*. University of Illinois Press, Chicago, **1949**.

[5] A.J. Davies, *Spectral Data: Standard Exchange Formats*. CSA23.

[6] A. Teckentrup, L. Terfloth, J. Gasteiger, *SONNIA 4.10.* Manual, Erlangen, **2002**.

[7] *http://www.dmg.org*

[8] L. Eriksson, E. Johansson, N. Kettaneh-Wold, S. Wold. *Introduction to Multi- and Megavariate Data Analysis using Projection Methods (PCA & PLS)*. Umetrics AB, Umea, **1999**.

[9] J. Chen, X.Z. Wang, *J. Chem. Inf. Comput. Sci.*, **2001**, *41*, 992–1001.

[10] Taken from *http://www.acc.umu.se/ ~tnkjtg/chemometrics/editorial/ may2002.html#wavelet* – see references therein.

[11] R. Todeschini, V. Consonni, *Handbook of Molecular Descriptors*. Wiley-VCH, Weinheim, **2002**.

[12] D. Rogers, A.J. Hopfinger, *J. Chem. Inf. Comp. Sci.*, **1994**, *34*, 854–866.

[13] C.B. Lukasius, G. Kateman, *Chemom. Intell. Labor. Syst.* **1993**, *19*, 1–33.

[14] R. Wehrens, L.M.C. Buydens, *TrAC*, **1998**, *17*, 193–203.

[15] J. Friedman, *Multivariate Adaptive Regression Splines*, Technical Report No. 102. Stanford University, CA, Nov. **1988**, Aug. **1990**.

[16] H. Swierenga, P.J. de Groot, P.J. de Weijer, M.W.J. Derksen, L.M.C. Buydens, *Chemom. Int. Labor. Syst.*, **1998**, *41*, 237–248.

[17] M. Randic, *New J. Chem.* **1991**, *15*, 517–525.

[18] B. Lucic, N. Trinajstic, S. Sild, M. Karelsen, A.R. Katritzky, *J. Chem. Inf. Comput. Sci.*, **1999**, *39*, 620–621 (and references therein).

[19] J. Korst, E.H. Aarts, A. Korst, *Simulated Annealing and Boltzmann Machines: A Stochastic Approach to Combinatorial Optimization and Neural Computing*. John Wiley & Sons, New York, **1989**.

[20] *http://www.ndsu.nodak.edu/qsar_soc/ resource/datasets/selwood.htm*

[21] J. Kalivas, *Adaption of Simulated Annealing to Chemical Optimization Problems*. Elsevier Science, New York, **1995**.

[22] J. Zupan, J. Gasteiger, *Neural Networks in Chemistry and Drug Design*. Wiley-VCH, Weinheim, **1999**.

# 5
# Databases and Data Sources in Chemistry

*T. Engel*

## Learning Objectives

- To understand introductory basic database theory
- To become familiar with the classification of chemical databases according to their data content
- To get to know various databases covering the topics of bibliographic data, physicochemical properties, and spectroscopic, crystallographic, biological, structural, reaction, and patent data
- To be able to access chemical information available on the Internet

## 5.1
## Introduction

The preceding chapters of this book deal with methods for representing chemical structures and reactions. As there is a huge and continuously increasing amount of data associated with chemical compounds, it is impossible to handle mountains of data by conventional techniques. The multi-faceted information on compounds, such as literature, physicochemical properties, spectra, etc., and on reactions can be handled in a comprehensive manner only by electronic methods. Such a system for storing and retrieving these data is generally called an information system (Figure 5-1), and comprises application programs (e.g., search engines) and a data stock or database, which can also be part of a database system. In chemistry the term "database" is often used for an entire information system, a database system, or a data-file itself. In this chapter, "database" is employed as a synonym for the entire information system. As models, languages, and management systems of databases fill volumes of books, this chapter can only give a flavor of this topic. For further literature, see Refs. [1–3].

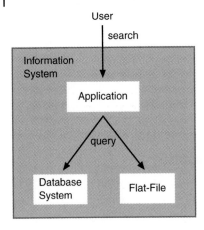

**Figure 5-1.**
The organization of an information system.

## 5.2
## Basic Database Theory

### 5.2.1
### Databases in the Information System

There are basically two different approaches in an information system for providing information with a database (see Figures 5-1 and 5-2).

In one method, the data are made available by means of a database system (DBS) (Figure 5-2). The DBS is a tool for efficient computer-supported organization, generation, manipulation and management of huge data collections. There, the database is only one part of this system where data can be stored easily, quickly, and reliably. If the data are organized in a DBS, the data store is called a container. A second important part of the software is the DataBase Management System (DBMS) (e.g. Oracle, DBase, MSAccess). This software allows the storage of data according to a database structure, which facilitates the retrieval or manipulation of data, user management, security, backup, load balancing, etc.

If the database is not integrated in a database system, the database is called a flat-file. As the name indicates, the data are stored in a file that can be used directly by the user.

The database is defined as a self-describing collection of integrated records, mainly stored on hard disk or a CD-ROM. The structure of the database (tables, objects, indices, etc.) is described by metadata (data about data) and is stored within the database as a data dictionary (system catalog). Figure 5-3a presents the units for organizing data in a database. The smallest unit is a bit (0 or 1), which is a component of a byte (8 bits = 1 byte). In a database, the bytes express fields of one or more records. These variable-lenght subsets of data of a particular entity consist of fields with unique information or characters (numeric, graphical, etc.). Records or data sets include different attributes, which describe corresponding object proper-

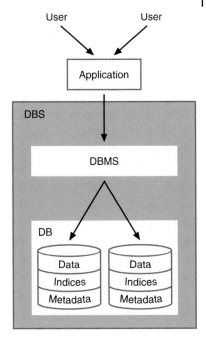

Figure 5-2. The database(s) (DB) with organized data and metadata are part of the Database System (DBS), which is managed by the Database Management System (DBMS).

ties (e.g., name, CAS Registry Number, etc.). Some or all of these records are put into a relationship with each other and are stored in a file (see also Figure 5-9). In order to transfer records optimally between the hard disk and the memory, one or more records are compiled on a page. A file contains various pages with congeneric data and is called a container [1].

In a flat-file system the database is called a file.

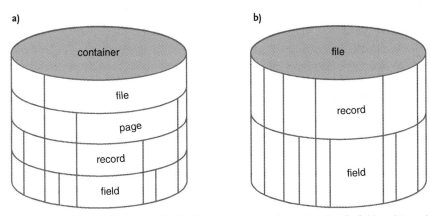

Figure 5-3. a) Main organization of a database or container; the basic units of a field are bits and bytes. b) Example of data organization in a flat-file.

## 5.2.2
### Search Engine

Most database users do not know how the data are organized in a database system (DBS); they depend solely on the application programs. This is sufficient for most database searches where users can receive large amounts of results quickly and easily, e.g., on literature or other information. Nevertheless, a basic knowledge on where and how to find deeper or more detailed information is quite useful. Due to their complex nature, comprehensive searches (e.g., for processes or patents) are not recommended for beginners. However, most local (in-house), online, and CD-ROM databases provide extensive tutorials and help functions that are specific to the database, and that give a substantial introduction into database searching.

The initial step is to identify which database, from a few thousands worldwide (about 10 000 in 2002), provides the requested information. The next step is to determine which subsection of the topic is of interest, and to identify typical search terms or keywords (synonyms, homonyms, different languages, or abbreviations) (Table 5-1). During the search in a database, this strategy is then executed (money is charged for spending time on some chemical databases). The resulting hits may be further refined by combining keywords or database fields, respectively, with Boolean operators (Table 5-2). The final results should be saved in electronic or printed form.

**Table 5-1.** Steps involved in searching a database.

- definition of the main focus of the search
- collection of search terms and keywords
- planning of the search strategy by assigning interconnecting terms (Boolean operators)
- choice of the database(s)
- login
- execution of the strategy, including refinement
- display, printing, or saving of the results
- logout

## 5.2.3
### Access to Databases

More than 10 000 databases exist that provide a small or large amount of data on various topics (including chemistry). The contents in databases are supplied by approximately 3500 database developers (e.g., the Chemical Abstracts Service, MDL Information Systems, etc.). Since there is a variety of topics from economics to science, as well as a variety of structures of the database, only some of the vendors (~2000) offer one or more databases as either local or as online databases (Figure 5-4) [4]. Usually, databases are provided by hosts that permit direct access to more than one database. The search occurs primarily through different individual soft-

**Table 5-2.** Basic search tools of Boolean operators and truncation.

*Boolean operators*

The Boolean search operators are used to specify logical relationships among the terms being searched.

| | | |
|---|---|---|
| AND | two or more search terms have to be in the same record (document) (e.g., acetylsalicylic AND headache) | |
| OR | either search term has to be in the same record (e.g., acetylsalicylic OR headache) | |
| NOT | excludes the following term (retrieves documents with the first term and not the second) (e.g., acetylsalicylic NOT synthesis); however, this often has the danger that some important documents are not found! | |

*Truncation (wildcards)*

Truncation allows one to search a term that is not exactly defined, such as singulars, plurals, declarations, or different spellings. The types of truncation that could be used depend on the database retrieval technique.

| | |
|---|---|
| ! | substitutes one variable character in the search term (e.g., analys!s matches analysis and analyses) |
| # | substitutes zero or one character in the search term (e.g., acetylsalicylic#acid matches acetylsalicylic acid and acetylsalicylicacid) |
| ? or * | replaces any number of characters (e.g., acetylsali* matches acetylsalicylic, acetylsali-cylsäure, acetylsalicylique, etc.) |

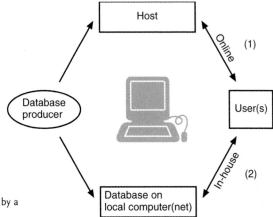

**Figure 5-4.** Databases can be classified as online (1), provided by a host, and in-house (local) (2).

ware or application programs, due to the existence of distinct types of data (bibliographic, factual, etc.; see Section 5.3). One of the most famous (online) "tools" in chemistry is STN Express [5].

5.2.4
## Types of Database Systems

Each database models data differing in origin, nature, or designation, in a structured and organized manner. Therefore the most important task in conceiving an effective database is to structure the import data. Different conceptual models exist for organizing data in a structured manner [6]. In the 1950s, the file system as a precursor of databases was developed, the file-system. The first genuine database systems became available in the 1970s as hierarchical and network systems. Then, in the 1980s, relational systems emerged. In the 1990s, object-based databases became available.

### 5.2.4.1  Hierarchical Database System
A hierarchical system is the simplest type of database system. In this form, the various data types also called entities (see Figure 5-3) are assigned systematically to various levels (Figure 5-5). The hierarchical system is represented as an upside-down tree with one root segment and ordered nodes. Each parent object can have one or *more children* (objects) but each child has only *one parent*. If an object should have more than one parent, this entity has to be placed a second time at another place in the database system.

In order to trace (find, change, add, or delete) a segment in the database, the sequence in which the data are read is important. Thus, the sequence of the hierarchical path is: parent > child > siblings. The assignment of the data entities uses pointers. In our example, the hierarchical path to **K** is traced in Figure 5-6.

Typical examples of hierarchical database systems are the file system of personal computers or the organization of parts (e.g., a construction plan). In the case of car parts, the objects (e.g. **B** = rear suspension, **E** = right wheel, **J** = rim, **K** = screw) are

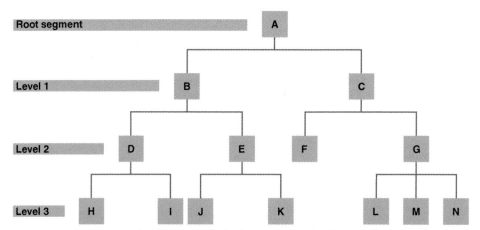

**Figure 5-5.**  Hierarchical structure of a database. For example, object **E** on level 2 is the parent of the child objects **J** and **K**.

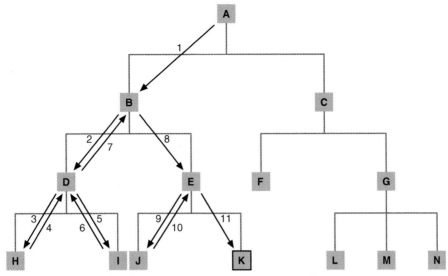

**Figure 5-6.** Hierarchical path to trace the sequence to access object **K**.

very different and exist independently of each other but the relationship between them is fixed (if you want to change a wheel it is obvious what else has to be changed). The first version of ISIS databases were hierarchical databases (seen in the RD-file format).

The primary advantage of hierarchical databases is that the relationship between the data at the different levels is easy. The simplicity and efficiency of the data model is a great advantage of the hierarchical DBS. Large data sets (series of measurements where the data values are dependent on different parameters such as boiling point, temperature, or pressure) could be implemented with an acceptable response time.

The disadvantage is that implementation and management of the database requires a good knowledge of the organization (physical level) of the data storage. Additionally, it is difficult to manipulate (edit) the structure of the database. New relationships or nodes result in a complex system of management tasks. Therefore, a modification of the logical data-independent data structure in this DBS with limited flexibility may lead to significant modifications to the application programs. Furthermore, the hierarchical model has the problem that a child cannot be related to multiple parents (in our example of the car, the object "rim" cannot be part of different wheels).

### 5.2.4.2 Network Model

The network model of a database system is an improvement over the hierarchical model. This model was developed in 1969 by the Data Base Task Group (DBTG) of CODASYL (Conference on Data System Languages) [8], because sometimes the re-

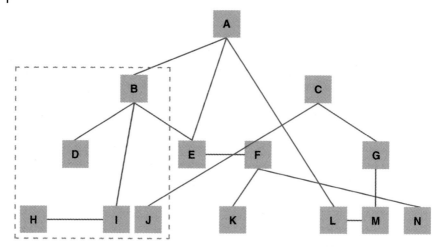

**Figure 5-7.** Network model of a database.

lationships between the records in real-world data are more complex than one can model by a hierarchical system. In the network database model, a single object can point to many other objects, and vice versa (e.g., an object (child) **E** can have several parents – **A** and **B**) (Figure 5-7). While the complex relationships improve the access to desired records, the clarity of the hierarchical system is lost through the complexity of the database organization. This has the consequence that database design and navigational data access are more complicated. Whereas, the objects in the hierarchical model have a clear top-down relationship: they are interconnected to each other in the network model. Thus, it is possible that a linked object may be deleted during database management, resulting in other objects that are still available but no longer have any relationships (Figure 5-8).

This database system is implemented in only a few instances because of its complexity and its liability to errors, although it is a model for the World Wide Web.

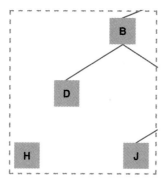

**Figure 5-8.** Detail of Figure 5-7 after deletion of object I. Other important objects such as **H** can lose their relationships.

### 5.2.4.3 Relational Model

The characteristic of a relational database model is the organization of data in different tables that have relationships with each other. A table is a two-dimensional construction of rows and columns. All the entries in one column have an equivalent meaning (e.g., name, molecular weight, etc.) and represent a particular attribute of the objects (records) of the table (file) (Figure 5-9). The sequence of rows and columns in the table is irrelevant. Different tables (e.g., different objects with different attributes) in the same database can be related through at least one common attribute. Thus, it is possible to relate objects within tables indirectly by using a key. The range of values of an attribute is called the domain, which is defined by constraints. Schemas define and store the metadata of the database and the tables.

Relational database models utilize memory very efficiently, avoiding repetition of data. It is possible to extract both individual data elements and combinations of them from a table. The main advantage of this structure is that it offers the possibility of changing the structure of the database (adding or deleting tables) without

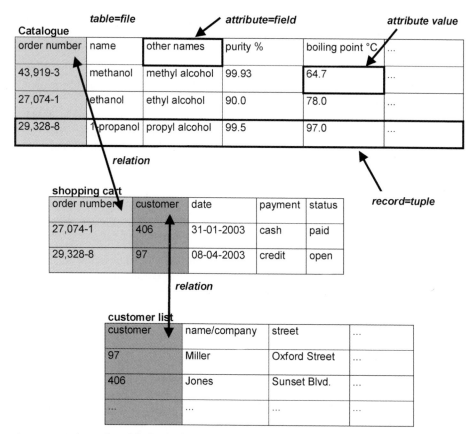

**Figure 5-9.** Relational model of a database. The records of each individual table, with different attributes, are related through at least one common attribute.

changing the application programs, which were based on prior database structures (data independence). Furthermore, a relational database allows the creation of many virtual tables (and consequently views) of the data with different logical structures, combining different tables or parts of them. For this purpose, the structure of the database does not have to be changed.

A disadvantage of the relational database management system (RDBMS) might be the overload of hardware and operating systems, which make the system slower.

Among many approaches to manipulating a relational database, the most prevalent one is a language called SQL (Structured Query Language) [2].

The relational database model was developed by Codd at IBM in 1970 [9]. Oracle provided the first implementation in 1979. The hierarchical database IMS was replaced by DB2, which is also an RDBMS. There exist hundreds of other DBMSs, such as SQL/DS, XDB, My SQL, and Ingres.

### 5.2.4.4 Object-Based Model

For a variety of applications such as computer-aided engineering systems, software development, or hypermedia, the relational database model is insufficient. In an RDBMS, it is difficult to model complex objects and environments; the various extensive tables become complicated, the integrity is problematic to observe, and the performance of the system is reduced. This led to two sophisticated object-based models, the object-oriented and the object-relational model, which are mentioned only briefly here. For further details see Refs. [10] and [11].

The main difference from the relational DBS is that the data are now stored in object types with a unique identity number (ID), attributes, and operations. Therefore, the relationship between objects is completely different from that in an RDBMS.

### 5.3
## Classification of Databases

The user is often more interested in the contents than in the technical organization of databases. The wide variety of data allows the classification of databases in chemistry into literature, factual (alphanumeric), and structural types (Figure 5-10) [12, 13].

A strict separation of these three types of databases is difficult; hence most databases contain a mixture of data types. Therefore the classification given here is based on the predominating data type. For example, the major emphasis of a patent database is on literature, whereas it also comprises numeric and structural data. Another type is the integrated database, which provides a supplement of additional information, especially bibliographic data. Thus, different database types are merged, a textual database and one or more factual databases.

One of the more prominent integrated databases is provided by CAS – the Registry CAPLUS, CA, CAOLD. All of these diverse databases are based on the

Evaluation of database models.

| Advantages | Disadvantages |
|---|---|
| *Hierarchical database system* | |
| • simple model | • requires knowledge of data storage |
| • fast performance | • each relationship must be defined |
| • data independence and data integrity strongly maintained | • each record has *only* one field and *one* relationship between two fields, which results in complex management tasks |
| • useful for large databases | • modification of data structure leads to modification of application programs |
| *Network model* | |
| • more efficient than relational model | • more difficult to design and to use than relational model |
| • better integrity support than relational model | • requires knowledge of data structure less data independence than relational model |
| *Relational model* | |
| • ease of use without in-depth knowledge | • less efficient than other models |
| • high degree of data independence | • integrity problems |
| • SQL capability | • possible to misuse |
| *Object-based model* | |
| • data representation in many different data types | • difficult to design DBMS |
| • supports temporal and higher-dimensional data | • slow performance |
| • objects can be reused | |

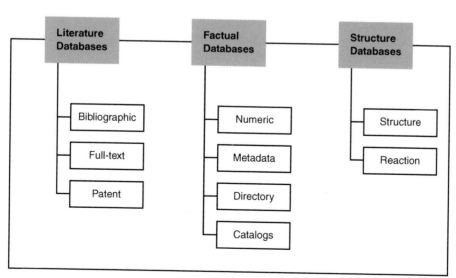

**Figure 5-10.** Classification of databases.

chemical substance, and are linked through the CAS Registry Number. Each CAS Registry Number is uniquely linked with a compound.

Therefore, it is preferable to separate chemical databases into categories by subject: bibliography, biology, environment, patents, chemical and physical properties, spectroscopy, structures, toxicology, or other special databases. Each type contains particular information, which can be managed individually. In Sections 5.4–5.15 the most important databases, grouped according to the above classification, are briefly introduced. Some of them are treated in detail in tutorials. An overview of all the databases is given in the form of a table in Section 5.18.

### 5.3.1
### Literature Databases

In literature databases, the fields describe objects, using characters as strings, e.g., letters, numbers, and special characters (including a blank). In this manner, author names, titles, journals or books, publication year, keywords, or abstracts are stored and retrieved from these databases. The text-based databases are divided into bibliographic and full-text databases. In contrast to full-text databases, bibliographic databases do not contain the text proper, but only reference the information. Thus, a retrieval in a bibliographic database results in a literature citation whereas in full-text databases the complete article is found. Full-text databases do not only contain the complete text (including references), but may also store facts and figures in the text. Due to the comprehensive information stored in literature databases, the search time is substantially greater than that for bibliographic databases, rendering this system more expensive to search.

Typical bibliographic databases are the CA File of Chemical Abstracts Service (CAS) or Medline of the US National Library of Medicine. Most electronic journals provide articles as full-text files, e.g., the *Journal of the American Chemical Society* (JACS).

### 5.3.2
### Factual Databases

Factual databases mainly contain alphanumeric data on chemical compounds. In contrast to bibliographic databases, factual databases directly describe the objects (primary data on chemical compounds) and provide the required information on them. Factual databases can be divided into numeric databases, metadatabases, research project databases, and catalogs of chemical compounds.

#### 5.3.2.1   Numeric Databases
Numeric databases primarily contain numeric data on chemical compounds, such as physicochemical values and the results of series of measurements. Therefore, the files correspond to printed tables of numeric property data. Since the attributes of numeric data are different from those of text data, the search has to be managed

---

**Excursion: Classification of Scientific Literature**

*Primary Literature*
This comprises original publications in scientific journals or serials (proceedings), in which the latest information and data are published for the first time. Primary literature includes dissertations and theses, journals, patents, conference proceedings, research reports, and preprints (the latter three are often called gray literature).

*Secondary Literature*
Publications of this kind are described as non-original. They are abstracting services and handbooks that catch the primary literature, condense the important contents, and make this information available (searchable). Secondary literature is not evaluated and is provided in both printable and electronic forms. Examples are *Gmelin, Beilstein, Citations: Chemisches Zentralblatt, Chemical Abstracts*, or *Science Citation Index*; handbooks include *Houben-Weyl*, and *Landolt–Börnstein*.

*Tertiary Literature*
Monographs, reference books, and encyclopedias, e.g., *Ullmann's Encyclopedia of Industrial Chemistry*, the *Kirk-Othmer Encyclopedia of Chemical Technology*, or the *Encyclopedia of Computational Chemistry* are included in this type of literature, which is furthest from the primary literature as concerns time and content. In most cases, tertiary literature summarizes a topic with information from different sources, and additionally evaluates the contents.

Handbooks such as *Gmelin* and *Beilstein* are often added to the tertiary literature, due to the high degree of processing their information has undergone. However, they are used as citations, and this is why they are integrated within the secondary literature.

---

differently. Specifically, many of the physicochemical terms are associated with units and are dependent on other parameters (e.g., boiling point and pressure, or spectra and solvent). The search is then specified by a range with two endpoints (e.g., molecular weight: 50–70) or is open-ended (e.g., $<50$, $\geq70$). Additionally, some numeric databases also allow calculations to be made with the data.

In addition to the numeric data (color, solubility, refraction index, spectra, etc.), these factual databases also include a bibliographic section with references or sources and a section with information for the identification of a compound (e.g., name, CAS Registry Number, molecular weight).

Typical numeric databases are Beilstein, SpecInfo, DETHERM, and the Cambridge Structural Database.

### 5.3.2.2  Catalogs of Chemical Compounds

Factual databases may provide the electronic version of printed catalogs on chemical compounds. The catalogs of different suppliers of chemicals serve to identify chemical compounds with their appropriate synonyms, molecular formulas, molecular weight, structure diagrams, and – of course – the price. Sometimes the data are linked to other databases that contain additional information. Structure and substructure search possibilities have now been included in most of the databases of chemical suppliers.

Typical catalogs are Chemline and MRCK.

### 5.3.2.3  Research Project Databases

Research project databases include information on abstracts and reports categorized by research projects. Such factual databases allow one to search for projects in various fields of science and technology with numeric and textual queries.

Typical research project databases are UFORDAT (Environment Research in Progress) or Federal Research in Progress (FEDRIP).

### 5.3.2.4  Meta-databases

These are databases that provide links to other databases or data sources. In this case, records describe objects that are other databases. The "Gale Directory of Databases" [14] is one of them. The connection between the databases flows through the meta-data of each database.

### 5.3.3
### Structure Databases

Structure databases are databases that contain information on chemical structures and compounds. The compounds or structure diagrams are not stored as graphics but are represented as connection tables (see Section 2.4). The information about the structure includes the topological arrangement of atoms and the connection between these atoms. This strategy of storage is different from text files and allows one to search chemical structures in several ways.

Examples of structure databases are Beilstein, Gmelin, and CAS Registry.

### 5.3.4
### Reaction Databases

Reaction databases additionally contain information on chemical reactions, giving the reaction participants and reaction conditions of both single- and multi-step reactions.

ChemInform is one of the reaction databases.

**5.4**
**Literature Databases**

**5.4.1**
**Chemical Abstracts File**

The Chemical Abstracts (CA) File of the Chemical Abstracts Service (CAS) [15] is the main abstracting and indexing service for chemistry, chemical engineering, and biochemistry. It includes conference proceedings, technical reports, books, dissertations (from 1967), reviews, meeting abstracts, electronic journals, web reports, international journals, and patents. The database has the broadest coverage of all the chemistry databases and is provided by different hosts: DIALOG [16], DataStar, Questel-Orbit [17], STN International, and particularly SciFinder. The bibliographic database comprises more than 22 million records (March, 2003) from 1907 to the present, and is updated each week with about 14 000 new citations.

An introduction to using the Chemical Abstracts System can be found in the tutorial in Section 5.5.

**5.4.2**
**SCISEARCH**

SCISEARCH contains bibliographic citations (links) to publications in science and technology. The database represents the electronic online version of the expanded Science Citation Index (SCI) and parts from the Current Contents of the Institute for Scientific Information (ISI). More than 5900 science and technical journals are included in the database with more than 20 million records (October, 2002). Searches can be performed on the bibliographic data, along with where, and how often, an author or publication is cited.

**5.4.3**
**Medline (Medical Literature, Analysis, and Retrieval System Online)**

Medline covers primarily biomedical literature, containing more than 13 million citations (October, 2002) of articles from more than 4600 journals published since 1958 [18]. The database covers basic biomedical research, clinical sciences, dentistry, pharmacy, veterinary medicine, pre-clinical sciences, and life science. Medline, a subset of PubMed, is a bibliographic database produced by the US National Library of Medicine (NLM). The database is available free of charge via SciFinder Scholar or PubMed [19].

**5.5**
**Tutorial: Using the Chemical Abstracts System**

The Chemical Abstracts System (CAS) produces a set of various databases ranging from bibliographic to chemical structure and reaction databases. All the databases originate from the printed media of *Chemical Abstracts*, which was first published in 1907 and is divided into different topics. Author index, general index, chemical structure index, formula index, and index guide are entries to the corresponding database (Table 5-3).

**5.5.1**
**Online Access**

The databases of CAS are accessible through two major tools – STN software and SciFinder. STN Information was formed by the collaboration of CAS with FIZ Karlsruhe, Germany and the Japan Information Center for Science and Technology (JICST) in order to meet customer needs more effectively for access to the rapidly growing resources of scientific and technical information. For a long time searching in CA databases had to be performed in alphanumeric form with the Messenger language. Because of the complexity of the task searching was mostly performed by experts. Then, the STN Express software was introduced for searching in CAS files. Next, STN Easy, a browser-based interface for novice users, was developed along with STN on the Web, which provided the STN command line capabilities within a web browser.

Then, in the early and mid-1990s, CAS developed SciFinder and SciFinder Scholar to address the needs of professional chemists and other scientists. SciFinder was developed to allow more intelligence in data access, such as smart structure searching, research topic exploration, advanced author searching, and powerful refine and analysis capabilities including "categorize" and "panorama".

In the mid and late 1990s CAS developed the ChemPort module as a linked gateway to primary literature. The CAS online delivery clients are able to move into ChemPort to display primary literature by presenting stored or dynamically generated URLs to ChemPort. Conversely, ChemPort will link a user to SciFinder to provide access to the CAS databases.

**5.5.2**
**Access to CAS with SciFinder Scholar 2002**

**5.5.2.1 Getting Started**
Access to CAS databases is only possible on computers on which the SciFinder software has been installed. It is directly available at CAS, computational service centers, or library services with online access. The database is not free of charge; access can be obtained only via these services. After the licensed software has been installed and online access is obtained, the program can be started.

**Table 5-3.** Databases of the Chemical Abstracts Service

| | |
|---|---|
| CAPLUS | The CAPLUS file is the most current and most comprehensive chemistry bibliographic database available from CAS, covering international journals, patents, patent families, technical reports, books, conference proceedings, and dissertations from all areas of chemistry, biochemistry, chemical engineering, and related sciences. CAPLUS includes all the data in the CA File (from 1907 to the present) along with a significant amount of information from additional sources, updated daily. There are more than 22 million records (February, 2003) from 1947 to the present. |
| CAOLD | The CAOLD File (pre-1967, Chemical Abstracts File) contains records for *CA* references from 1907 through 1966. |
| REGISTRY | The REGISTRY File is a chemical structure and dictionary database containing unique substance records that are produced when new substances are identified by the CAS Registry System. The REGISTRY File contains records cited in CAPLUS, CA, and CAOLD online files, and special registrations. There are more than 20 million organic and inorganic substances and more than 24 million biosequences (October, 2002) in the Registry database, which is updated daily. |
| CASREACT® | The CASREACT File (The Chemical Abstracts Reaction Search Service) is a chemical reaction database with reaction information derived from journal documents from 1974 to the present and from patent documents from 1982 to date. The document-based file contains both 3 million single-step and 3.6 million multi-step reactions (February, 2003). |
| MARPAT® | The MARPAT File is a Markush structure search service. It contains the Markush structure records for patents found in the CA File with the patent publication year from 1988 to the present. The records contain the Markush structures found in the claims and often the disclosure of the patent, the bibliographic information, in-depth substance and subject indexing including CAS Registry Numbers, and an abstract. All of these data can be displayed. |
| CHEMCATS® | CHEMCATS is a catalog file containing information on about 6 million commercially available chemicals and their worldwide suppliers (702) (March, 2003). |
| CHEMLIST® | The CHEMLIST File contains chemical substances on national inventories, registered by the US Environmental Protection Agency (EPA). The data in CHEMLIST are from 1979 to the present. There are more than 228 380 records (September, 2002). CHEMLIST is updated weekly with more than 50 additions to existing records or new substances. |
| CIN® | CIN (Chemical Industry Notes) is a bibliographic database covering worldwide business events in the chemical industry since 1974. CIN monitors about 80 worldwide periodicals including: journals, trade magazines, newsletters, government publications, and special reports. There are more than 1 488 320 records (September, 2002). CIN is updated weekly with about 1000 records per week. |
| TOXCENTER | TOXCENTER on STN is a bibliographic database that covers the pharmacological, biochemical, physiological, and toxicological effects of drugs and other chemicals. The data in TOXCENTER are from 1907 to the present. There are more than 5.7 million records (December, 2002). It is updated weekly. |

### 5.5.2.2 **Searching within Various Topics**

The first window that appears after starting the program and accepting the license agreement is the SciFinder Scholar window with the Explore dialog box, which offers six basic search topics (Figure 5-11): Chemical Substance or Reaction, Research Topic, Author Name, Document Identifier, Company Name/Organization, and Browse Table of Contents. At this point the user can choose the search topic that is most relevant according to the information that is available.

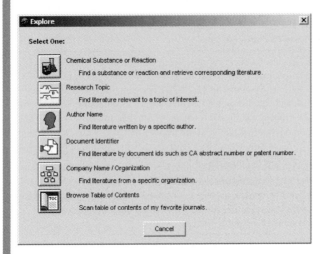

**Figure 5-11.**
The Explore dialog box.

Thus, if the user wants to look for literature including requested chemicals or reactions, it is possible to query the database by the first option: "Chemical Substance or Reaction". The compound can be entered as a query in three different ways: drawing the chemical structure in a molecule editor (Chemical Structure); searching by names or identification number, such as the CAS Number (Structure Identifier); and searching by molecular formula (Figure 5-12).

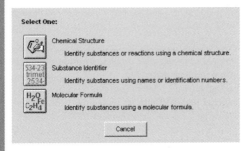

**Figure 5-12.**
The Explore by
Chemical Substance
window.

Once the query has been entered, the search for the compound can be executed with an exact match or a substructure search. All resulting matches, up to 10 000, are then displayed in a tabular form with brief information on the compound, e.g., the CAS Number (Figure 5-13). More information relating to the substance is available via the microscope button.

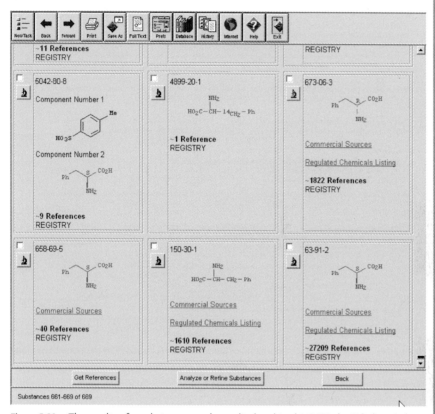

**Figure 5-13.** The results of a substance search are displayed in the SciFinder Scholar window.

One or more interesting hits can be selected by clicking the small white box next to the structure. The resulting list of literature can be analyzed or refined (Figure 5-14).

If the query provides thousands of hits, the analyze features are particularly advantageous. One method is to analyze the results by any of the criteria that are listed, e.g., by language (default), author names, journals, publication year, and so on. If one specification is selected and the choice is modified, the hit list will be updated. A more specific analysis is available with the "Refine" option, where the user has the opportunity to choose one of eight criteria (including the search topics above) with further individual input. Several refinements of the hit list can reduce the result to a concise list of literature. To read the abstract of an article, the microscope button (to the right of the citation) has to be pressed (Figure 5-15).

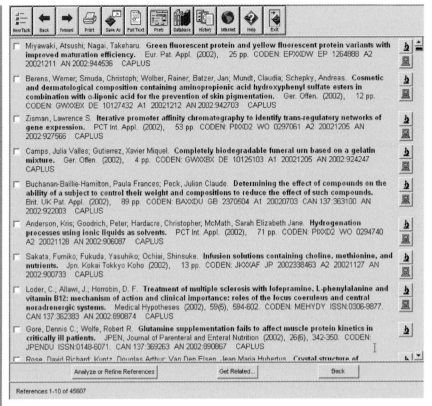

**Figure 5-14.** The retrieval result of a name or literature search displayed in the SciFinder Scholar window.

If there is a monitor-icon under the microscope, the full-text article is additionally available via ChemPort.

The remaining five search topics (Research Topic, Author Name, Document Identifier, Company Name/Organization, and Browse Table of Contents) are conducted in a similar fashion, with the input being the only difference between the criteria. Thus, in "Research Topic" the entry can be any, or even several, keywords or phrases. In "Author Name", literature written by a specific author will be found, including alternative spelling. "Document Identifier" can also be entered directly in the query. Document identifiers are CA abstract numbers, patent numbers, patent application numbers, or priority application numbers. The last two search topics (Company Name/Organization, and Browse Table of Contents) allow one to search for literature from specific companies or to view the list of journals which are available in the database.

All results can be saved or printed at any stage of the retrieval.

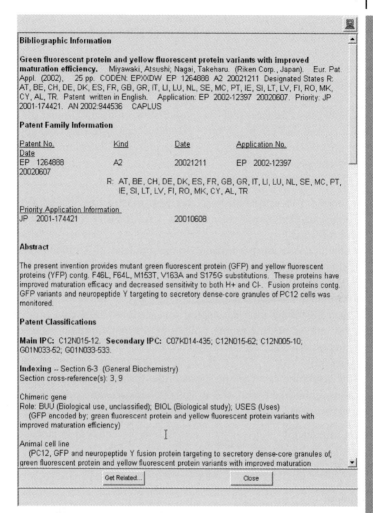

**Figure 5-15.** Detail window of a selected reference.

In every step of the query, users can see examples given by the program, in order to guide them through the literature search. Thus, a novice user can operate this system immediately.

## 5.6
## Property (Numeric) Databases

Beilstein and Gmelin are the world's largest factual databases in chemistry. Beilstein contains facts and structures relating to organic chemistry, whereas Gmelin provides information on inorganic, coordination, and organometallic compounds.

Both database sources contain evaluated data on millions of compounds, and allow the retrieval of bibliographic information and of structures.

### 5.6.1
### Beilstein Database

The Beilstein database [20] has more than 8.3 million (October, 2002) organic substance records from the *Beilstein Handbook* and abstracted from about 180 journals in organic chemistry from 1779 to the present. All documents are critically evaluated and peer-reviewed.

The Beilstein File includes information on:

- substance identification (e. g. structure/substructure, chemical name, chemical name segments, molecular formula, CAS Registry Number, physical properties, or keywords);
- chemical data on 8.3 million compounds and of 5 million chemical reactions (preparation, reaction, isolation from natural products, chemical derivatives, purification);
- 35 million associated physicochemical property and bioactivity records, including data describing pharmacological and environmental data (numeric fields: e.g., 3.8 million melting points, boiling points; non-numeric fields: e.g., 6.2 million preparations, 1.5 million IR spectra, 2.3 million NMR spectra, 500 000 absorption spectra);
- bibliographic data (author, journal title, Beilstein citation, CODEN, patent numbers, publication year) of over 750 000 abstracts and titles indexed from the primary organic chemical literature since 1980.

An introductory section about searching in the Beilstein Database can be found in the tutorial in Section 5.7.

### 5.6.2
### Gmelin

Gmelin [21] is also a structure and factual database. It is a comprehensive, electronically searchable source of structures and properties in inorganic and organometallic chemistry. The database contains substance records from the *Gmelin Handbook of Inorganic and Organometallic Chemistry* (1772–1975) and also from 110 of the most important inorganic, organometallic, and materials science journals from 1975 to the present. Gmelin currently comprises 1.4 million compounds including, for instance, coordination compounds, alloys, glasses and ceramics, polymers, and minerals.

The database is produced by the German Chemical Society (GDCh) and provided by MDL Information Systems Inc. [22].

Gmelin contains over 800 different chemical and physical property fields, and a detailed index of the original literature. Broad categories of data found in the database include:

- identification, including material composition and structural data;
- chemical properties, including behavior, preparation, and reaction details;
- electrochemical data;
- electrical, magnetic, mechanical, molecular, and optical properties;
- solubility and vapor pressure in solution;
- spectroscopic data;
- thermodynamic data;
- quantum chemical calculations.

Besides structure and substructure searches, Gmelin provides a special search strategy for coordination compounds which is found in no other database: the ligand search system. This superior search method gives access to coordination compounds from a completely different point of view: it is possible to retrieve all coordination compounds with the same ligand environment, independently of the central atom or the empirical formula of the compound.

### 5.6.3
### DETHERM

Another numeric database including bibliographic information is DETHERM. This database provides thermophysical property data (phase equilibrium data, critical data, transport properties, surface tensions, electrolyte data) for about 21 000 pure compounds and 101 000 mixtures. DETHERM, with its 4.2 million data sets, is produced by Dechema, FIZ Chemie (Berlin, Germany) and DDBST GmbH (Oldenburg, Germany). Definitions of the more than 500 properties available in the database can be found in NUMERIGUIDE (see Section 5.18).

### 5.7
### Tutorial: Searching in the Beilstein Database [23]

This tutorial, which is based on the Beilstein update BS0202PR (May, 2002) and on the retrieval program CrossFire Commander V6, shows some typical advanced search examples in the Beilstein database. It is assumed that the user already knows some of the basic features of the retrieval program. Moreover, in this tutorial the CrossFire Structure Editor is used instead of the ISIS/Draw Structure Editor. The first example is a combined application of structure and fact retrieval, whereas the second example demonstrates reaction retrieval.

### 5.7.1
### Example 1: Combined Structure and Fact Retrieval

The aim of the first example is to look for polychlorinated biphenyls (PCB) for which $^{13}$C-NMR spectra, measured in deuterochloroform, as well as the partition coefficients between 1-octanol and water are known. Since it is not reliable to perform the search by compound names, the polychlorinated biphenyls are retrieved by means of a generic structural formula. For this purpose, the structural formula of the parent compound has to be drawn with the structure editor in the usual way.

In order to allow any multiple chlorination of the biphenyl skeleton, the user may define an atom list (consisting of hydrogen and chlorine atoms) and substitute all H-atoms by this list. One may click on the drop-down selection box behind the element icons, select the options "*Generics ...*", set the user-defined atom to A1 and quit by the OK button. As a result this atom selection is active for the subsequent drawing steps. After this atom list is drawn ten times as the ten substituents, its composition has to be defined by clicking the $A_n$ icon on the left-hand side of the structure editor and by selecting H and Cl in the periodic table (Figure 5-16).

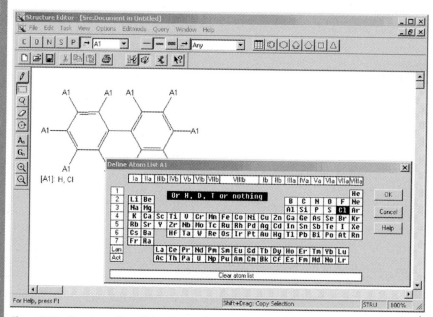

**Figure 5-16.** Structure editor of the CrossFire Commander V6, showing the definition of an atom list (consisting of hydrogen and chlorine atoms) for polychlorinated biphenyls.

After completion of the input of the generic structural formula, the user has to return to the Commander window. The remaining requirements have to be specified in the fact editor. After this editor has been opened, a tabular form with three columns is presented. The columns are assigned to the logical operators, to the search field codes, and to the applied search strings, respectively. At first, the input for the spectroscopic method may be done in the parameter field codes of NMR for the nucleus and the solvent, respectively. Browsing of the corresponding index is recommended to see how the information is stored there. In order to assure that the search strings are closely related (i.e., that

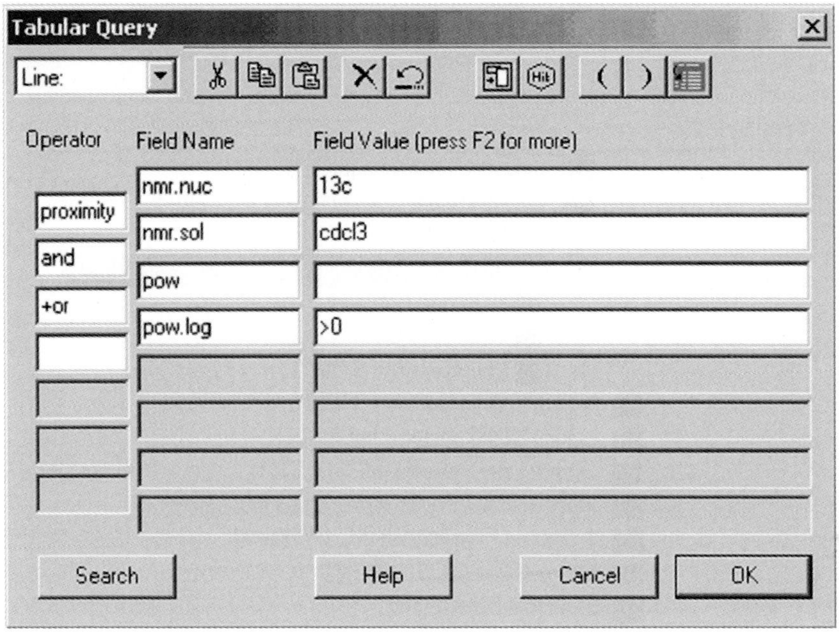

**Tabular Query**  ⊠

Line: ▾  ✂ 🗈 🗒 ✕ ↺   🗐 🕮  ( )  🗒

| Operator | Field Name | Field Value (press F2 for more) |
|---|---|---|
|  | nmr.nuc | 13c |
| proximity | nmr.sol | cdcl3 |
| and | pow |  |
| +or | pow.log | >0 |
|  |  |  |
|  |  |  |
|  |  |  |
|  |  |  |

| Search | | Help | | Cancel | | OK |

**Figure 5-17.** Fact editor of the CrossFire Commander V6; it is emphasized that the input can be provided in lower-case as well as in upper-case letters.

they belong to the same experiment), the neighborhood operator PROXIMITY must be applied (Figure 5-17). If users do not know the search field code for the partition coefficient, they may position the cursor in the second column *(Field Name)* and apply the *List Value* option (alternatively, the F2 function key). A new window is opened where users can either browse the database struc-ture or apply the *Find* button with subsequent specification of the search field (Figure 5-18). Because the partition coefficients and their decadic logarithms are stored alternatively, the corresponding search field codes are selected and combined by the Boolean operator OR. The spectroscopic and partition behavior topics are combined by the Boolean operator AND. In order to take into account the correct hierarchy of the Boolean operators, the logical expression has to be nested by means of the parenthesis icons; the result is indicated in the fact editor by a plus sign in front of the Boolean operator, whereas the parentheses can be seen in the CrossFire Commander in the query box (Figure 5-19).

As can be seen in Figure 5-17, some search fields (e.g., *POW* [= Power]) do not need any input in the search mask; this means that all entries with any content of those fields are retrieved. However, other fields always demand an input. In case the input is omitted (for example for the decadic logarithm of the partition coefficient), a corresponding error message results. Since the PCB are more soluble in the organic phase, the input of that field is restricted to positive values.

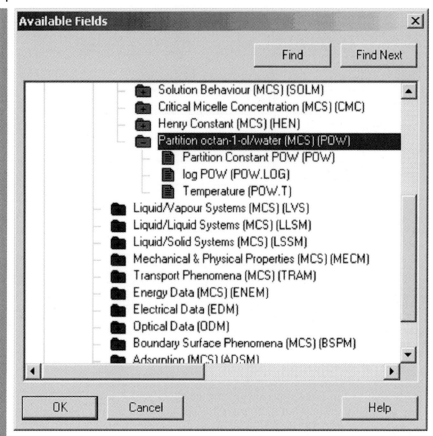

**Figure 5-18.** Interactive display of the detailed Beilstein database structure.

After returning to the CrossFire Commander the combined search can be started. First the fact retrieval is performed. Then the given generic formula is searched. Hereby the specified query options, which can be recognized as check boxes in Figure 5-19 in the right-hand column besides the display of the retrieved structure, neglect multicomponent systems, isotopic and charged species, radicals, and ring annelations. Finally, both intermediate results are automatically combined by means of the Boolean AND operator and the final result is offered for display. In the present update 28 substances result after about 30 seconds. Because, in general, the entries can be very extensive, it is recommended to use the display format "*Hit only*", which displays only the identification data together with that information which fulfills the given logical requirement. How to select display formats is described at the end of Example 2 (Section 5.7.2).

An alternative to retrieval by means of an atom list is a structure search performed by means of a user-defined generic group which contains only chlorine

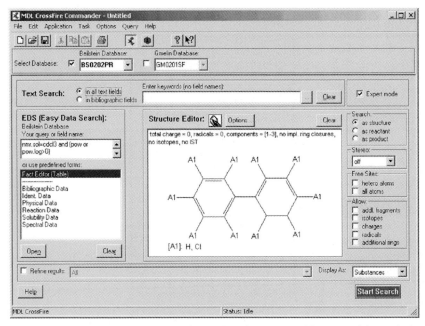

**Figure 5-19.** CrossFire Commander V6 with a combined structure and fact retrieval; here only the end of the fact requirement can be seen.

atoms. For this purpose users may click the drop-down selection box behind the element icons, select the option *"Generics ..."*, set the user-defined generic group to G1 and quit by the OK button. As a result this generic group is active for the subsequent drawing steps. After the G1 group is drawn ten times, the composition of this generic group has to be defined. Therefore, a non-bonded chlorine atom has to be drawn and selected, followed by the clicking of the $G_n$ icon on the left-hand side of the structure editor (see also Figure 5-16). In the final step, the Markush frequency (the maximal occurrence of the generic group in a retrieved fragment) must be set to 10 by clicking the G1 symbol in front of the definition by means of the pencil tool.

It is emphasized that both versions of the generic structure query may include the parent compound and monochlorinated derivatives.

Because the field code *POW* is also a hierarchical group code (see also Figure 5-17), the solution behavior retrieval *POW* OR *POW.LOG* can be abbreviated in the fact editor to simply *POW* without further specification.

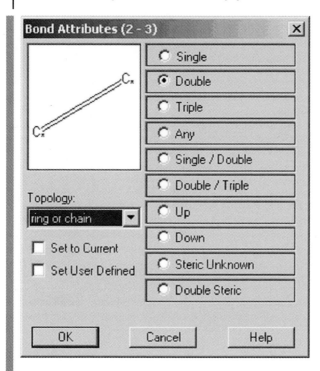

Figure 5-20. Dialog box to specify the bond attributes.

### 5.7.2
### Example 2: Reaction Retrieval

In the second example, the aim is to look for Diels–Alder reactions between aliphatic dienes and cyclic dienophiles with yields larger than 50 % at room temperature, and with the electron-attracting groups aldehyde, keto, carboxylic, cyano, and nitro groups.

For this purpose the 1,3-butadiene, ethene, and cyclohexene are drawn as separate fragments on the same screen of the structure editor. Since the bonds in open-chain fragments are set by default to *"ring or chain"*, the user has to adjust the attributes of the bonds. After clicking the bond with the pencil tool, a new window with the bond attributes is opened where users can select the desired topology in the drop-down menu (Figure 5-20). Since numerous derivatives are involved in Diels–Alder reactions, all atoms are allowed to be substituted after selecting all fragments and choosing the option *"Set Free Sites"* in the Query menu of the structure editor (see also Figure 5-16).

The $C_2$ and cyclohexene fragments are each substituted by the same user-defined generic group. In order to define this group users have to draw carbonyl, cyano, and nitro groups, select them and to click the $G_n$ icon. For generic groups consisting of polyatomic fragments the atoms which are bound to the parent structure must be specified. After the bound atom has been clicked, a window with the corresponding

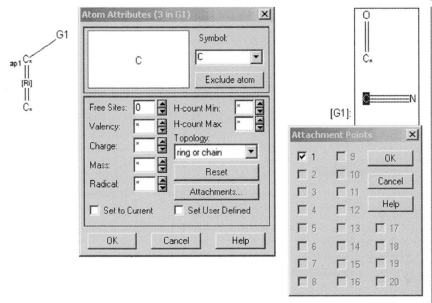

**Figure 5-21.** Part of the structure editor screen with dialog windows to specify the attachment points in user-defined generic groups.

atom attributes is opened, which allows one to define the attachment points (Figure 5-21). This procedure has to be repeated for all fragments of the user-defined generic group.

After defining the structure attributes, users have to switch within the structure editor to the reaction tools. By selecting each fragment and clicking the corresponding *Reactant* or *Product* button, respectively, the role of the reaction partners is assigned. The reaction attributes of bonds and atoms can be specified by clicking the bonds and atoms with the pencil tool. For example, both of the bonds in the cyclohexene fragment which are formed during the reaction can be tagged by the option "*Break/Make Bond*". To ensure that corresponding atoms in the reactants and the product are the same, a so-called mapping is performed: Pseudo-bonds are drawn, using the pencil tool, as dashed lines between corresponding atoms in the reactant and product fragments.

After return to the Commander window, the reaction retrieval may be executed separately: 629 Diels–Alder reactions between aliphatic dienes and cyclic dienophiles are found. This partial result can be narrowed down by restricting the reaction conditions by means of the fact editor. The search field codes for the yield and the temperature can be found to be *RX.NYD* and *RX.T*, respectively, either by browsing the database structure or by applying the *Find* option, as described in the first example. To ensure that the retrieved reaction conditions belong to the same experiment, both search terms must be connected by means of the PROXIMITY operator. Before the retrieval is started, the option "*Refine results*" in

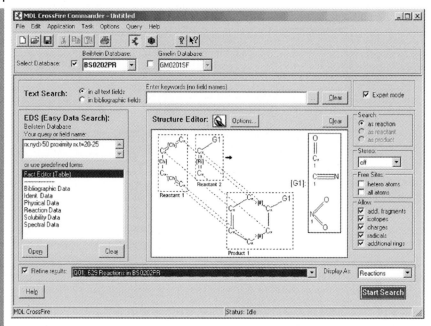

**Figure 5-22.** CrossFire Commander V6 for the refinement of a previously performed reaction retrieval by means of a fact search.

the CrossFire Commander (Figure 5-22) has to be chosen; the query number of the reaction search should be displayed. At last, the final result is achieved, which fulfills the given logical requirement.

Figure 5-23 shows the first hit in the display format "*Hit only*". After reaction retrievals, this format only displays the identification of the reaction (names of the educts and products, their Beilstein registry numbers and the reaction equation) together with those reaction conditions which fulfill the given retrieval requirement. Other formats can be chosen in the display window by clicking the *View* button or selecting one of the options in the drop-down menu *View*. The format "*All*" would inform that there is another reaction of these educts, however, for which no yield or reaction temperature is specified.

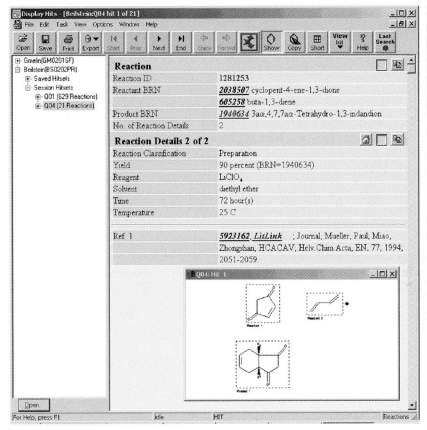

**Figure 5-23.** Display of the first hit (in "*Hit only*" format) of the reaction retrieval shown in Figure 5-22.

## 5.8
## Spectroscopic Databases

The large databases CA, Beilstein, and Gmelin do not provide methods for directly searching spectroscopic data. Detailed retrieval of spectroscopic information is provided in databases that contain one or more types of spectra of chemical compounds. Section 5.18 gives an overview of the contents of larger databases including IR, NMR, and mass spectra.

One of the largest spectra collections – SpecInfo – is introduced in more detail below.

### 5.8.1
### SpecInfo

SpecInfo, from Chemical Concepts, is a factual database information system for spectroscopic data with more than 660 000 digital spectra of 150 000 associated structures [24]. The database covers nuclear magnetic resonance spectra ($^1$H-, $^{13}$C-, $^{15}$N-, $^{17}$O-, $^{19}$F-, $^{31}$P-NMR), infrared spectra (IR), and mass spectra (MS). In addition, experimental conditions (instrument, solvent, temperature), coupling constants, relaxation time, and bibliographic data are included. The data is cross-linked to CAS Registry, Beilstein, and NUMERIGUIDE.

The features of SpecInfo include:

- a structure editor and spectra display;
- search for exact structure and substructure, name, molecular formula, or weight ranges;
- spectrum similarity search;
- prediction of NMR shifts for $^1$H, $^{13}$C, $^{19}$F, and $^{31}$P NMR;
- display of physical data, CAS Registry Number, experimental conditions;
- zooming and printing of spectra.

This database has some additional commands and search fields, which are tailored to the specific requirements of retrieving spectroscopic data, e.g., peak or multiplicity searches.

SpecInfo has an additional tool for calculating NMR spectra that is based on the data sets of the compounds contained in the database. This leads to quite reliable calculated spectral parameters for the compound classes which are registered in the database.

### 5.9
### Crystallographic Databases

A crystal is a solid with a periodic lattice of microscopic components. This arrangement of atoms is determined primarily by X-ray structure analysis. The smallest unit, called the unit cell, defines the complete crystal, including its symmetry. Characteristic crystallographic 3D structures are available in the fields of inorganic, organic, and organometallic compounds, macromolecules, such as proteins and nucleic acids.

The two major databases containing information obtained from X-ray structure analysis of small molecules are the Cambridge Structural Database (CSD) [25] and the Inorganic Crystal Structure Database (ICSD) [26]; both are available as in-house versions. CSD provides access to organic and organometallic structures (mainly X-ray structures, with some structures from neutron diffraction), data which are mostly unpublished. The ICSD contains inorganic structures.

3D structures of macromolecules, especially proteins and nucleic acids, are found in the Protein Data Bank (PDB) [27].

## 5.9.1
## Inorganic Crystal Structure Database (ICSD)

ICSD is a numeric database with 65 000 inorganic crystal structures (December, 2002) updated twice a year. All the atomic coordinates of the non-carbon structures which have been completely determined are included. ICSD is produced by FIZ Karlsruhe and the National Institute of Standards and Technology (NIST).

The database offers the following information:

- compound name, mineral name and source, and CAS Registry Number;
- molecular formula;
- crystal lattice parameters (number of formula units of each elementary cell and cell volume);
- Hermann–Mauguin space group symbols;
- atomic parameters;
- oxidation state of the elements;
- physicochemical factors: temperature, pressure of the measurement;
- experimental methods;
- bibliographic citations.

## 5.9.2
## Cambridge Structural Database (CSD)

The Cambridge Structural Database (CSD) contains crystal structure information for over 250 000 organic and organometallic compounds with up to 1000 atoms. All of these crystal structures have been analyzed using X-ray or neutron diffraction techniques. The database is produced and provided by the Cambridge Crystallographic Data Centre, and contains information dating back to1930.

For each crystallographic entry in the CSD, the following information is stored:

- bibliographic information for the particular entry, including authors' names and the full journal reference;
- a connection table for the molecule;
- structural and experimental information on the crystal structure, mainly numeric data including atomic coordinates, space group symmetry, covalent radii, and crystallographic connectivity.

## 5.9.3
## Protein Data Bank (PDB)

The PDB contains 20 254 experimentally determined 3D structures (November, 2002) of macromolecules (nucleic acids, proteins, and viruses). In addition, it contains data on complexes of proteins with small-molecule ligands. Besides information on the structure, e.g., sequence details (primary and secondary structure information, etc.), atomic coordinates, crystallization conditions, structure factors,

3D images, and a variety of links to other resources (bibliographic citations), the data entries are annotated by RCSB (Research Collaboratory for Structural Bioinformatics) with additional information.

The PDB is updated irregularly and is freely available [27].


## 5.10
## Molecular Biology Databases

Many biochemical databases with sophisticated topics have been developed for solving various problems. Since 1996 the first issue of each journal volume of *Nucleic Acid Research* has been reserved for the presentation of molecular biology databases [28]. A comprehensive catalog on the Internet is DBCAT, currently listing 511 databases [29, 30].

The largest bibliographic database in the biological sciences is BIOSIS (contents of *Biological Abstracts*) [31], which covers biological and medical literature on biology, microbiology, clinical and experimental medicine, biochemistry, biophysics, and instrumentation and methods, comprising information from 9000 life science journals since 1969. The database has more than 13 million records (September, 2002) and is produced by Biological Abstracts and updated weekly.

Besides such textual databases that provide bibliographic information, sequence databases have attained an even more important role in biochemistry. Sequence databases are composed of amino acid sequences of peptides or proteins as well as nucleotide sequences of nucleic acids. The 20 amino acids are mostly represented by a three-letter code or by one letter (according to the biochemical conventions); the four nucleic acids are defined by a one-letter code. Thus the composition of a biochemical compound is searchable by text retrieval methods.

The three largest primary sequence databases are GenBank (USA) [32], EMBL (Germany/UK) [33], and DDBJ (DNA Data Bank of Japan) [34]. The three institutions providing these databases established a collaborative project known as the "International Nucleotide Sequence Database Collaboration". Newly registered entries are maintained by daily data exchange and synchronization to ensure that all updated records are shared between the three groups. This tripartite organization has been playing a key role in acquisition, storage, and distribution of the human genome sequence data.

### 5.10.1
### GenBank (Genetic Sequence Bank)

GenBank [32] is a text–numeric database of genetic sequences with more than 28 billion bases in 22 million sequences (January, 2003) from genetic research. The collection of all publicly available sequences is annotated with information such as sequence description, source organism, sequence length, or references. The database, established in 1967, is updated daily and produced by the National Center for Biotechnology Information (USA).

## 5.10.2
## EMBL

EMBL (European Molecular Biology Laboratory) [33] is a nucleotide sequence database provided from the online host EBI. Release 73 (December, 2002) consists of over 20 million nucleotide sequences with more than 28 billion nucleotides. The information includes sequence name, species, sequence length, promoter, taxonomy, and nucleic acid sequence.

## 5.10.3
## PIR (Protein Information Resource)

The protein sequence database is also a text–numeric database with bibliographic links. It is the largest public domain protein sequence database. The current PIR-PSD release 75.04 (March, 2003) contains more than 280 000 entries of partial or complete protein sequences with information on functionalities of the protein, taxonomy (description of the biological source of the protein), sequence properties, experimental analyses, and bibliographic references. Queries can be started as a text-based search or a sequence similarity search. PIR-PSD contains annotated protein sequences with a superfamily/family classification.

PIR is produced by the National Biomedical Research Foundation (NBRF) [35].

## 5.10.4
## SWISS-PROT

The SWISS-PROT database [36] release 40.44 (February, 2003) contains over 120 000 sequences of proteins with more than 44 million amino acids abstracted from about 100 000 references. Besides sequence data, bibliographical references, and taxonomy data, there are highly valuable annotations of information (e.g., protein function), a minimal level of redundancy, and a high level of integration with other databases (EMBL, PDB, PIR, etc.). The database was initiated in 1987 by a partnership between the Department of Medicinal Biochemistry of the University of Geneva, Switzerland, and the EMBL. Now SWISS-PROT is driven as a joint project of the EMBL and the Swiss Institute of Bioinformatics (SIB).

## 5.10.5
## CA Registry

The CA Registry of CAS also comprises 1.6 million protein and peptide sequences and over 21.6 million nucleic acid sequences from literature sources and from GenBank.

## 5.11
## Structure Databases

As already mentioned (Section 5.3), the stored structure information in this type of database makes it possible to search for chemical structures in several ways. One method is to draw a structure (via a molecule editor) and to perform either a precise structure search (full structure search) or a search containing part of the input structure (substructure search) (see Sections 6.2–6.4). The databases also allow the searching of chemical names and molecular formulas (see Section 6.1). The search results are in most cases displayed in a graphical manner.

Examples of structure databases are Beilstein, Gmelin, and the CAS Registry.

### 5.11.1
### CAS Registry

The CAS Registry contains information on all the chemical compounds published in the literature since 1957. The sources of these 21 million compounds are 9000 international journals containing chemical information. The database includes CAS Registry Numbers, the CAS name (not conforming to the IUPAC convention) with synonyms, and molecular and structural formulas. The CAS provides a weekly update (see Section 5.4).

### 5.11.2
### National Cancer Institute (NCI) Database

The National Cancer Institute (NCI) database is a collection of more than half a million structures, assembled by NCI's Developmental Therapeutics Program (DTP) or its predecessors in the course of NCI's anti-cancer screening efforts that started in the late 1950s (plus the more recent anti-HIV screening) [37–39]. Approximately half of this database is publicly available without any usage restrictions, and is therefore called the "Open NCI Database." For each of these structures (more than 250 000) the DTP record contains at least the chemical structure as a connection table and an NCI accession number, the NSC number.

For various subsets of this database, additional experimental data have been made available by DTP. These are primarily the 60-cell-line cancer screening results, i.e., GI50, LC50, and TGI values for about 41 000 structures, and about 44 000 AIDS antiviral screening results. Also available are approximately 127 000 CAS numbers and 3600 measured $\log P$ values. Several groups, both inside and outside NCI, as well as various companies, have taken these data and analyzed, processed, reformatted, and repackaged them, often with the goal of making the information contained in them more accessible. One of the most comprehensive of these efforts is the "Enhanced NCI Database Browser" (*http://cactus.nci.nih.gov*, mirrored at *http://www2.ccc.uni-erlangen.de/ncidb2*). In this service, the DTP dataset has been augmented by a large amount of additional, mostly computed, data, such as calculated log*P* values, predicted biological activities, systematically determined names, 3D structures calculated by CORINA, and others.

The Web-based graphical user interface permits a choice from numerous criteria and the performance of rapid searches. This service, based on the chemistry information toolkit CACTVS, provides complex Boolean searches. Flexible substructure searches have also been implemented. Users can conduct 3D pharmacophore queries in up to 25 conformations pre-calculated for each compound. Numerous output formats as well as 2D and 3D visualization options are supplied. It is possible to export search results in various forms and with choices for data contents in the exported files, for structure sets ranging in size from a single compound to the entire database. Additional information and down-loadable files (in various formats) can be obtained from this service.

## 5.12
### Chemical Reaction Databases

Compounds are stored in reaction databases as connection tables (CT) in the same manner as in structure databases (see Section 5.11). Additionally, each compound is assigned information on the reaction center and the role of each compound in the specific reaction scheme (educt, product, etc.) (see Chapter 3). In addition to reaction data, the reaction database also includes bibliographic and factual information (solvent, yield, etc.). All these different data types render the integrated databases quite complex. The retrieval software must be able to recall all these different types of information.

A review on reaction databases can be found in Ref. [40] and an introductory section on searching in the ChemInform Reaction Database is contained in the tutorial in Section 5.13.

### 5.12.1
### CASREACT

CASREACT (Chemical Abstracts Reaction Search Service) is a reaction database started in 1985 with more than 6.7 million reactions (3 million single-step and 3.7 million multi-step reactions) (March, 2003) derived from 400 000 documents (journals, patents, etc.). The records contain the following information:

- connection tables of reactants, reagents, and products (structure-searchable);
- CAS Registry Number (of all reaction participants);
- reaction sites;
- yields;
- bibliographic data.

5.12.2
**ChemInform RX**

The ChemInform Reaction database is based on ChemInform, a weekly service bringing abstracts on new publications in organic and inorganic chemistry. The abstracting service is produced by FIZ Chemie, Berlin, and has been published for more than 30 years [41]. Each printed journal contains about 350 abstracts collected from approximately 250 primary journals. This abstracting service is focused on the information needs of organic and organometallic synthesis chemists. The ChemInform Reaction database (abbreviated ChemInform RX or CIRX) is the electronic version of this highly respected compendium [42]. It was established in 1991 by FIZ Chemie, Berlin, and has become a unique worldwide information system [43]. Currently, the ChemInform Reaction database is the most widely used in-house reaction database; it covers literature references from 1900 to the present. More than 1 million reactions and 1.1 million molecules are now stored in CIRX and it is increasing by approximately 50 000 new reactions per year.

ChemInform RX, nowadays called the ChemInform Reaction Library, is distributed as an in-house version in a client-server architecture by MDL Information Systems, Inc., San Leandro, CA, USA [22].

**5.13**
**Tutorial: Searching in the ChemInform Reaction Database [44]**

5.13.1
**Introduction**

This tutorial describes briefly some of the search capabilities possible with ChemInform RX and MDL® ISIS used as the retrieval system. In this tutorial, the CIRX databases of the years 1992–1996 are used, containing altogether 334 855 reactions.

5.13.2
**Example 1: Reaction Retrieval**

In this section, the basic concepts of reaction retrieval are explained. The first example is concerned with finding an efficient way to reduce a 3-methylcyclohex-2-enone derivative to the corresponding 3-methylcyclohex-2-enol compound (see Figure 5-24). As this is a conventional organic reaction, the CIRX database should contain valuable information on how to synthesize this product easily.

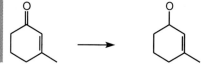

**Figure 5-24.** Query drawn for guiding examples for the reduction of 3-methylcyclohex-2-enone to 3-methylcyclohex-2-enol.

First, a query must be drawn using the MDL® ISIS/Draw program. By using this reaction query, a "current reaction search" can be performed. This type of reaction retrieval compares the starting material and the product of the reaction query with all the reactions in the CIRX database. Both query structures must match exactly, including the implicit hydrogen atoms not shown in the reaction query. In this case, one hit is found in the CIRX databases.

The next abstraction level of reaction retrieval is a so-called "reaction substructure search" in which both query structures are considered as substructures. In the case of a reaction substructure search, no hydrogen atoms are added internally during the execution of the search. Atoms which have their valencies not completely saturated are considered as open sites, where any kind of element could be bonded.

With this query 116 hits were found in the CIRX databases. The first hit of this hit list is shown in Figure 5-25.

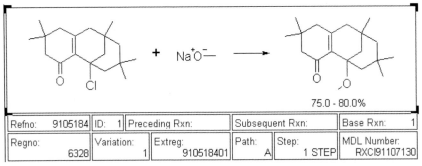

| Refno: | 9105184 | ID: | 1 | Preceding Rxn: | | Subsequent Rxn: | | Base Rxn: | 1 |
|--------|---------|-----|---|----------------|--|-----------------|--|-----------|---|
| Regno: | 6328 | Variation: | 1 | Extreg: 910518401 | | Path: A | Step: 1 STEP | MDL Number: RXCl91107130 | |

**Figure 5-25.** First record of the hit list of the reaction substructure search (MDL number RXCl91107130).

No chemist will be satisfied with this first hit, as the carbonyl group is still present in the product molecule. Although both molecules of the hit that was found fulfill the reaction query, the chemist has tacitly assumed that a reaction takes place at the carbonyl group. The declaration of atom–atom mapping numbers will solve this problem (cf. Section 3.3). Corresponding atoms in the starting material and in the product are labeled with the same mapping number (see Figure 5-26). The refinement of adding atom–atom mapping numbers leads to 105 hits. Furthermore, there are some other search criteria that can be set within a reaction query to specify the arrangement of bonds, such as setting a "Make/Break" or "Change bond order" flag. In our case, the declaration of atom–atom mapping numbers is sufficient.

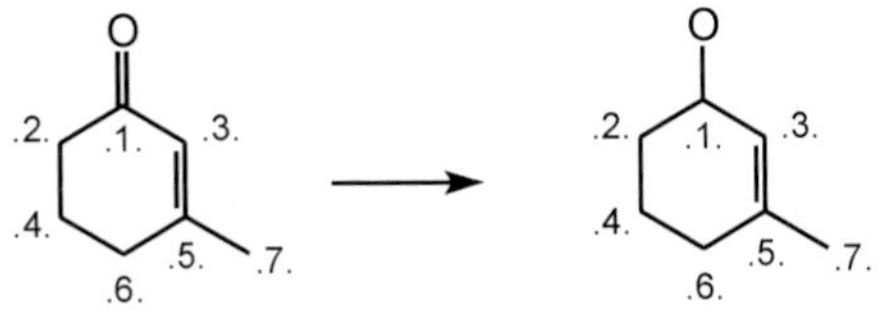

**Figure 5-26.** Reaction query with atom–atom mapping numbers.

As all unsaturated atoms are considered as open sites, the oxygen atom having one open site in the product molecule could now be attached to two carbon atoms, building an ether derivate. The same applies to the carbon atom attached to the oxygen atom in the product molecule of our query. But this carbon atom should have four neighbors altogether: one oxygen atom, two carbon atoms, and one hydrogen atom. Thus, it is necessary to add two hydrogen atoms to the oxygen and the adjacent carbon atom of the reaction query. After applying all these refinements to the initial query, the final reaction query is shown in Figure 5-27. Using this reaction query, 41 hits are found.

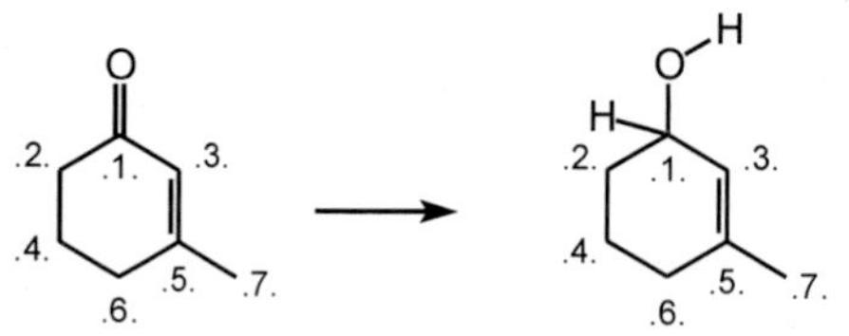

**Figure 5-27.** Final reaction query for Example 1.

To obtain more information on each reaction of the hit list, it is possible to browse through the hit list, classify it (see Section 5.13.4), print it, and export all the references on the hit list or save the reactions in different file formats for documentation purposes.

5.13.3
**Example 2: Advanced Reaction Retrieval**

The aim of the second example is to find suitable reaction conditions for running the same reduction reaction as in the first example, but in the presence of another carbonyl group which should not react. Furthermore, the reaction should lead to a product with a yield of 80 % or more and a specific stereochemical configuration.

This is achieved by making the reaction query of the first example more general, since for this example the reaction may also occur in a five-membered ring system (see Figure 5-28). For that reason only the $\alpha,\beta$-unsaturated carbonyl fragment must be drawn, which must be part of a ring system indicated by an "$R_n$" label. Furthermore, another atom–atom-mapped carbonyl fragment must be added, attached to two carbon atoms, one on each side of the reaction query. This group is not bonded

**Figure 5-28.** Reaction query for Example 2: "$R_n$," indicates that the bond must be part of a ring system, and "s3" represents an atom with three non-hydrogen attachments. The "Chiral" flag is necessary to retrieve only molecules with the identical absolute stereoconfiguration.

to the carbonyl fragment of the conjugated system, in order to indicate that the second carbonyl group could be located anywhere in the molecule.

To obtain the required product yield, the reaction substructure search is combined with a textual search using the "query builder" of the MDL® ISIS program.

One reaction is found after performing this search; it is shown in Figure 5-29. Analysis of the reaction conditions by retrieving the catalyst/solvent and conditions entry, or by reading the given literature for more information, solves the problem described at the beginning of this section.

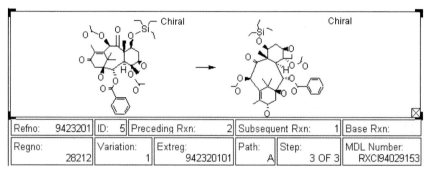

| Refno: | 9423201 | ID: | 5 | Preceding Rxn: | 2 | Subsequent Rxn: | 1 | Base Rxn: | |
|--------|---------|-----|---|----------------|---|-----------------|---|-----------|---|
| Regno: | 28212 | Variation: | 1 | Extreg: 942320101 | | Path: A | Step: 3 OF 3 | MDL Number: RXCI94029153 | |

**Figure 5-29.** Found record (MDL number: RXCI94029153) for Example 2.

## 5.13.4
### Classifying Reactions on a Hit List

An important advantage of the CIRX database (and other reaction databases provided by MDL) is the option of classifying reaction results. This feature was implemented because, as the number of reactions in the CIRX database increases, the hit list may become very large. In this case, reaction classification (see Section 3.5) is a valuable method for minimizing the work necessary to analyze the hit list. This method allows the grouping of the reaction results by data value, such as author, journal, journal year, reagent, or solvent, and on the basis of the reaction center and its environment. Three classification values are stored for each reaction in the CIRX databases by an algorithm provided by InfoChem [45]. These three values, BROAD, MEDIUM, and NARROW, represent different levels of similarity considering the structural information at the reaction center. After completing a reaction search in the CIRX databases, the reactions of the hit list can be grouped automatically using these classification values. This enables users to browse quickly through the classified hit list, as having only a small number of reaction instances in each cluster is sufficient to permit analysis. Only the reactions of the most interesting clusters need to be analyzed to solve the problem being considered.

By applying the BROAD classification feature to the hit list of the first example, described in Section 5.13.2, 18 clusters are found. The second reaction instance of the first cluster is shown in Figure 5-30.

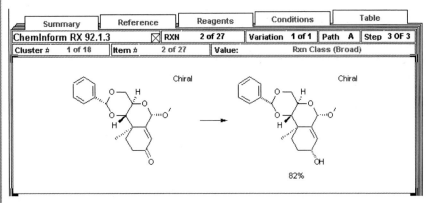

| Summary | Reference | Reagents | Conditions | Table |
|---|---|---|---|---|

| ChemInform RX 92.1.3 | ☒ RXN | 2 of 27 | Variation 1 of 1 | Path A | Step 3 OF 3 |
|---|---|---|---|---|---|

| Cluster # | 1 of 18 | Item # | 2 of 27 | Value: | Rxn Class (Broad) |
|---|---|---|---|---|---|

**Figure 5-30.** Second reaction instance of the first cluster for Example 1, obtained by applying the BROAD classification method.

## 5.14
## Patent Databases

Patent databases are databases containing information gathered from patent documents [46]. These describe both the technical aspects and the scope of a patent (patent protection) and are specific to a country or a group of countries (Table 5-4).

The information included in patent documents is structured in the following way:

- bibliographic data;
- review, sometimes supplemented by a schematic drawing;
- description;
- patent claims;
- examples.

**Table 5-4.** Definition of a patent and its functions.

| Patent | • legal document, securing monopolistic rights to the owner, preventing others from making, using, or selling the claimed invention for a certain period of time |
|---|---|
| Functions | • legal protection for novel products and processes |
| | • advances technological progress by making information available to all |

**Table 5-5.** Examples of patent databases.

| | |
|---|---|
| EUROPATFULL | full-text European patents and applications |
| INPADOC | international patent database covering more than 60 countries |
| JAPIO | bibliographic data of Japanese patent applications |
| MARPAT | CAS Patent Markush File |
| PATDPA | patents and utility models in Germany |
| PATIPC | the international patent classification |
| USPATFULL | full-text database of US patents |
| WPINDEX | database of international patent publications |

Patent databases are therefore integrated databases because facts, text, tables, graphics, and structures are combined. In patents that include chemical aspects (mostly synthesis or processing), the chemical compounds are often represented by Markush structures (see Chapter 2, Section 2.7.1). These generic structures cover many compound families in a very compact manner. A Markush structure has a core structure diagram with specific atoms and with variable parts (R-groups), which are defined in a text caption. The retrieval of chemical compounds from Markush structures is a complicated task that is not yet solved completely satisfactorily.

Patents are important for companies to protect their research. In industry, novelties are first published in patents and nowhere else. This means that only up to 10 % of the information contained in patents is available through other information systems. In addition, to avoid redundant investigations, companies can monitor the research of competitors and can claim new developments (products, compounds, etc.) on their own.

Some examples of patent databases are given in Table 5-5 and in Section 5.18.

Most databases, e.g. the World Patents Index, provide a "Learn database" function (LWPI) to give the opportunity for training in the specific retrieval strategies for patents.

## 5.14.1
### INPADOC

INPADOC (International Patent Documentation Center) is the most comprehensive bibliographic database of scientific and technological patent documents in the world. The stock encompasses more than 26 million patent documents, more than 59 million legal status data, and about 10 million patent families (January, 2003). The database contains more than 35 million patent citations from 71 patent-issuing organizations (European Patent Office, World Intellectual Property Organization (WIPO)) and is updated weekly with about 40 000 new citations.

The database is produced by the European Patent Office and is provided by the host FIZ Karlsruhe (Germany).

### 5.14.2
### World Patent Index (WPINDEX)

The World Patent Index of Derwent Information Ltd. is a broad collection of international value-added patent documents from 40 patent-issuing authorities. The bibliographic database contains 11.6 million patent records with 5.5 million images (October, 2002) and grows by 1.5 million patent documents each year. The classified and indexed documents (since 1963) are sometimes provided with additional abstracts or significant titles.

### 5.14.3
### MARPAT

The CAS Patent Markush File contains about 180 000 structure records (November, 2002) of patents and of all relevant chemical literature covered by the CA File. These documents can also be retrieved by generic structures. The database covers Markush structures of organic and organometallic molecules, but not alloys; metal oxides, inorganic salts, intermetallics and polymers are included. MARPAT, established in 1988, has more than 500 000 (December, 2002) searchable Markush structures from 38 patent-issuing organizations. The retrieved information includes bibliographic data, abstract, and CAS indexing. Approximately 250 new citations and 750 new Markush structures enter weekly in the update. MARPAT is available on STN International and STN on the Web.

### 5.15
### Chemical Information on the Internet

Scientific, and especially chemical, information is becoming increasingly available on the Internet. This has the advantage that the information is accessible to other users. Additionally, the data formats (HTML, PDF, GIF, etc.) that are used are restricted to the most important ones to provide standards that are readable worldwide. Unfortunately, however, much of the information on the Internet is not reviewed or verified by other organizations as is the case in the primary literature. Thus, the quality of the information is extremely variable.

Altogether, the Internet provides the following types of scientific information:

- teaching material, and student research projects;
- reprints of publications, proceedings, reports, and dissertations;
- (full-text) electronic journals;
- access to databases and meta-databases (public domain, commercial, library).

The huge number of websites on the Internet (about 2.5 billion web pages; December, 2002) containing chemical information is a great challenge when one is attempting to find specific information on a topic. Therefore numerous search engines have been developed and offered that provide fast access to the data.

In an evaluation of search engines for providing information on environmental topics, seven leading international search services (AltaVista, Excite, Fast, Google, HotBot, Northern Light, Yahoo!) and seven comparatively small specialty search services for chemistry (Anavista.de, ChemGuide, ChemFinder, Chemie.de, ChemIndustry.com, MetaXChem, NIOSH) were investigated [47]. Of course, none of the major search engines offers special tools for retrieving chemical information such as a CAS-Number, molecular formula, or structure search. These queries have to be transformed into proximity expressions, for example. Surprisingly, these features are not even common with the specialty search engines. Here, users are able to enter text words or substance names but cannot take advantage of Boolean logic or draw chemical structures. Often, even a CAS Number search is not provided. An adequate help feature is rarely available. Structure diagrams can be entered only with the help of applets or plug-ins. Software either has to be downloaded on the user side or has to be made available on the provider side. ChemFinder offers a structure plug-in and MetaXchem a structure and substructure search via SMILES strings which encode graphical information as text strings. None of the other chemistry-related services offered structure queries in mid-2001 [47].

Recapitulating, large search engines are useful to obtain a fast overview of search terms. To get more detailed information one has to enter qualified, specialized portals or even utilize databases (see above).

A list of large and smaller search services can be found in Table 5-6 in the tutorial in Section 5.16.

## 5.16
### Tutorial: Searching the Internet for Chemical Information

Given the enormous number of resources for chemical information available, many researchers do not have the time to learn the details of the various systems, and they end up searching in only a few resources with which they are familiar. This is a dangerous approach! Knowing that both fee and non-fee resources are available on the Internet and both hold the desired information, it is prudent to search non-fee systems first and then use proprietary databases to fill data gaps [49].

If users are inexperienced in searching for information, they should first consult search engines, meta-databases or portals (Table 5-6). Searchers who are familiar with databases may consult known databases (numeric databases, bibliographic databases, etc.) directly, being aware that they might miss new data sources (see Section 5.18). The reliability and quality of data are only given in peer-reviewed data sources.

**Table 5-6.** Free-of-charge search engines, metadatabases, and (chemical) portals.

| Large search engines | *www.altavista.com* |
|---|---|
| | *www.alltheweb.com* |
| | *www.excite.com* |
| | *www.fireball.de* |
| | *www.google.com* |
| | *www.hotbot.com* |
| | *www.lycos.com* |
| | *www.metacrawler.com* |
| | *www.northernlight.com* |
| | *www.web.de* |
| | *www.webcrawler.com* |
| | *www.yahoo.com* |
| Small search engines | *www.analytik-news.de* |
| | *www.chemfinder.cambridgesoft.com* |
| | *www.chemie.de/metaxchem* |
| | *www.chemindustry.com* |
| | *www.fiz-chemie.de/en/datenbanken/chemguide* |

The large search engines (Table 5-6) generally provide a larger number of hits, but often from commercial if not even dubious sources. Yet if information on a new or rare compound is needed, they can be recommended as a first choice. The smaller subject engines provide more reliable data, but vary considerably in their results [47].

Another approach to obtain an overview on chemical information or on information related to specified topics in chemistry, is to use websites that contain link lists. These link lists are usually provided by universities and private persons and are classified into subject areas. Table 5-7 gives an sample of the thousands of link lists in chemistry, and in addition some other valuable URLs that deal with chemoinformatics.

All the methods of obtaining information via the Internet presented above carry one risk – dead links. Although a search term may be found by a search engine in its own website-metadata database, the original link to the website could be broken and the information is lost. In this book a conscious effort has been made to limit the URLs and to reduce the web address to the index page of the server, to avoid this sometimes annoying problem.

A more stable source of information is represented by databases.

**Table 5-7.** Small sample of interesting links in chemistry and chemoinformatics.

| Short description | URL |
|---|---|
| Global Instructional Chemistry, Imperial College of Science, Technology and Medicine | *www.ch.ic.ac.uk/GIC/* |
| Links for Chemists, University of Liverpool | *www.liv.ac.uk/Chemistry/Links/links.html* |
| Links, Chemie.DE Information Service GmbH | *www.chemie.de/* |
| The Sheffield Chemdex: Directory of chemistry on the WWW since 1993 | *www.shef.ac.uk/chemistry/chemdex/* |
| Rolf Claessen's Chemistry Index | *www.claessen.net/chemistry/* |
| ChemFinder.Com, a portal of free and subscription scientific databases | *http://chemfinder.cambridgesoft.com/* |
| Forum with free access to databases and journals | *www.chemweb.com* |
| Chemistry Index of the German Virtual Chemistry and Biochemistry Library | *www.chemlin.de* |
| RSC's chemistry societies' electronic network | *http://www.chemsoc.org/* |
| Infochem's guide to chemistry software sources on the Internet | *http://www.chemistry-software.com/* |
| Links to chemistry Journals on the World Wide Web | *http://www.chem.usyd.edu.au/~haymet/paper/chem-paper.html* |
| Full-text documents on the Web | *http://chemport.fiz-karlsruhe.de/* |
| Web of Knowledge portal | *http://www.isiknowledge.com* |
| Database index for chemists | *www.chemie-datenbanken.de/* |

*Links to educational chemoinformatics*

| | |
|---|---|
| CHEMINFO of Indiana University | *www.indiana.edu/~cheminfo/* |
| University of Sheffield – Department of Information Studies | *www.shef.ac.uk/uni/academic/I-M/is/home.html* |
| University of Manchester Institute of Science and Technology (UMIST) | *www2.umist.ac.uk/chemistry/PG/msc/MScChem-inf.htm* |
| Computer-Chemie-Centrum, University of Erlangen-Nürnberg | *http://www2.chemie.uni-erlangen.de/* |

**5.17**
**Tutorial: Searching Environmental Information in the Internet [49]**

**5.17.1**
**Introduction: Difficulties in Extracting Scientific Environmental Information from the Internet**

Environmental information is available extensively on the free Internet. Two main reasons account for this. First, the freedom of environmental information law and, secondly, the fact that many environmental databases which used to be available only commercially are now accessible on the free Internet. This section focuses on environmental information which is available free of charge on the Internet. Three main paths are distinguished for searching environmental information effectively. In general, all three routes should be taken into consideration for achieving a sound search result for environmental information on chemical substances.

Thorough searches exploring environmental information on chemical substances, such as ecotoxicity parameters on daphnia, fish or algae toxicity, or biodegradation, bioaccumulation, photodegradation, or detection of chemicals in environmental media – soil, air, water, etc. – first aim at identifying the substance, then locate the data sources that contain information. Both numeric and bibliographic databases are used to search environmental parameters. Sufficient information may be found in numeric databases with data on these environmental parameters. Generally, such databases are well documented, in readable format, and often even peer-reviewed. Comprehensive searches, however, cover both numeric and bibliographic and full-text databases. The most recent research results are provided in bibliographic and full-text databases.

The prerequisite for using this approach is that the user has an intimate knowledge of the environmental databases, i.e., numeric, bibliographic, and full-text ones, which are available on the Internet. This might not be the case all the time. A different way to proceed is to consult metadatabases and portals, databases which provide information on resources available on the Internet which contain environmental and chemical information. A working knowledge of the availability and functionality of these metadatabases and portals is the basis for taking this route. The third path uses search engines. First, the chemical compound has to be identified, then a search engine must be chosen and the documents comprising environmental information retrieved. This procedure often gives reasonable results [47]. Additionally, online journals can be consulted to find information on environmental chemicals. This aspect is not discussed in this section.

## 5.17.2
## Ways of Searching for Environmental Information on the Internet

In this section, a search strategy is demonstrated by means of a search for daphnia toxicity for atrazine (1912-24-9). Some advices concerning metadatabases, portals, and search engines, as well as examples of recommended databases, are given in the following sections. Further details on finding environmental information are presented in Ref. [50].

### 5.17.2.1 Metadatabases and Portals

Metadatabases are databases which describe other databases in a comprehensive and structured way [51]. A prominent example is the DAIN Metadatabase of Internet Resources for Environmental Chemicals (*http://www.wiz.uni-kassel.de/dain*) which in 2002 comprised more than 700 entries. This metadatabase is documented in Chapter V, Section 10 of the Handbook, on "Databases on Environmental Information". It supports the user in finding the right database(s) for their purposes.

The word "portal" is often used for web sites that function as an entry to a repository of information on almost any topic on the Internet. The following portal categories for chemical issues are distinguished: general chemistry sites; substance information sites; analytical chemistry sites; environmental chemistry sites; toxicology and occupational health sites; as well as patent services [47].

Figure 5-31 gives an excerpt from the search results for the daphnia toxicity of atrazine in the large environmental portal German Environmental Information Network, called GEIN. GEIN functions as a unique web interface to data collections of 70 suppliers of environmental information, mostly governmental authorities and other German public institutions. It offers access to more than 140 000 individual web pages and nine databases, and even to hidden (dynamic) web pages such as environmental data catalogs of German federal states and the German Government. For this search example, performed on October 21, 2002, the number of hits was 100. Not only German but also English and French hits are obtained in this search. Links to publications and databases are given.

### 5.17.2.2 Search Engines

Ten recommendations for searching data on environmental chemicals on the Internet can be given:

- Choose relevant sites or databases directly (metadatabases), and always use more than one search service.
- Search by CAS Number – if known.
- Consider different spellings or synonyms of compound names.
- Make use of Advanced, Power or Precision Search.
- Reduce the number of search results by using AND or NEAR.

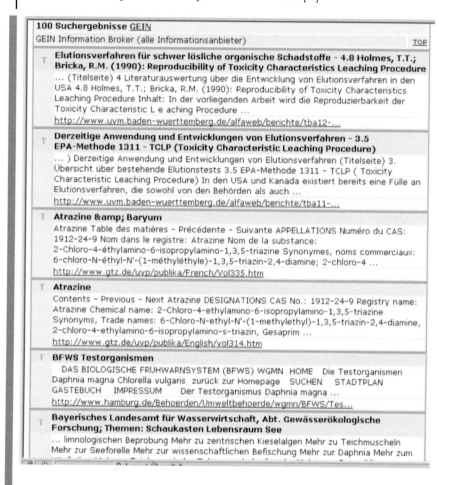

**Figure 5-31.** Search for atrazine and daphnia toxicity in the German Environmental Information Network GEIN portal (excerpt from search, October 21, 2002) *(http://www.gein.de/index_en.html).*

- Choose large search engines or meta-search engines in the case of few or no results; if applicable, try OR.
- Specialty search engines are to be preferred if a database or a detailed menu is available; in any other case make use of their directory; scan more than one to three result pages. The relevance ranking of a search engine may not correspond to your preferences.
- In general: the more well-known a substance is, the more likely it is that information can be found using specialty search engines. The less well-known a substance is, the more likely it is that information can be found using major search engines.

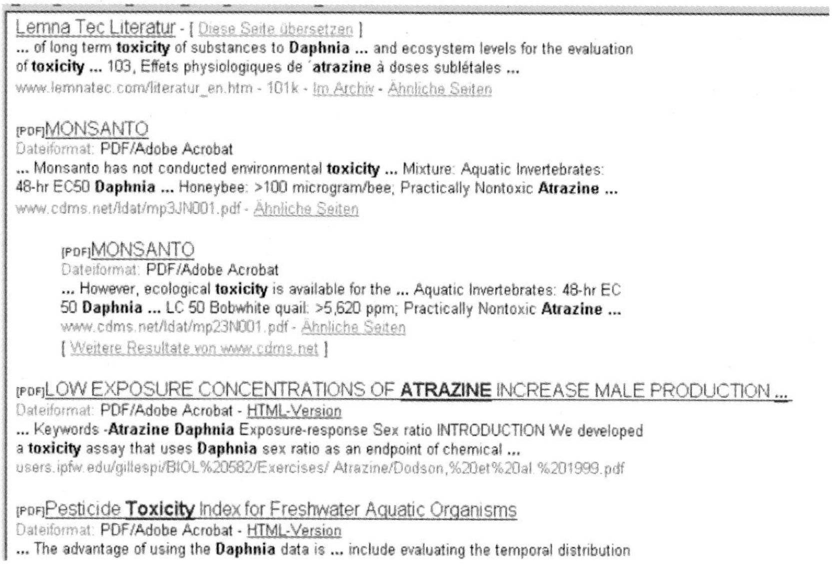

**Figure 5-32.** Search for atrazine and daphnia toxicity in Google (excerpt from search, October 21, 2002) (*http://www.google.com*).

The search example (for daphnia toxicity of atrazine) is performed using the large common search engine Google (see Figure 5-32). The search entry is the CAS Number that can be found in Chemical Name Directories such as ChemFinder [52]. It can be demonstrated by the result that not only commercial sites such as the Monsanto sites, but also university sites, governmental sites, and organizations which work on the topic are retrieved. This indicates a quite high quality of the information found by this search engine.

The search results from Google also give links to environmental databases.

### 5.17.2.3 **Databases**

Environmental databases focusing on chemical substances are introduced and discussed in this section.

A review which puts emphasis on currently available free Web resources related to environmental toxicology was published recently by Russom [53]. In this section, US sites, specifically provide those which address available empirical data sources, predictive tools, and publications of interest such as standard test methods, guidance documents, and government regulations. Fee-based and non-fee-based databases are listed and explained by Wright [48]. Again, only US resources are covered. Russom explains the ECOTOX System, which offers numeric information on chemical toxicity values for aquatic and terrestrial life (see Figure 5-33). ECOTOX integrates searching among three databases: AQUIRE (aquatic), PHYTOTOX (terrestrial plants), and TERRETOX (terrestrial wildlife). Options offer searchers

Your Report for Aquatic Data is Ready! Click Here!

Record Count : 1459

| Scientific name, Common name | Endpoint | Effect | Effect Measurement | Trend / Effect % | Media Type | Duration / Exp Typ | Conc (ug/L) | Signif / Level | Response Site / BCF | Ref # |
|---|---|---|---|---|---|---|---|---|---|---|
| Test Loc: NR | | | | | | | | | | |
| CAS #/Chemical: 1912249, 6-Chloro-N-ethyl-N'-(1-methylethyl)-1,3,5-triazine-2,4-diamine | | | | | | | | | | |
| Daphnia magna Water flea | EC50 | ITX | IMBL | -------- | FW | 24 H / S | F >39000 | | ------ | 13154 |
| Daphnia magna Water flea | EC50 | ITX | IMBL | -------- | FW | 48 H / S | F >39000 | | ------ | 13154 |
| Daphnia magna Water flea | EC50 | ITX | IMBL | INC / -------- | FW | 24 H / | A >30000 | | ------ | 17289 |
| Daphnia magna Water flea | EC50 | ITX | IMBL | INC / -------- | FW | 48 H / | A >30000 | | ------ | 17289 |
| Daphnia magna Water flea | EC50 | ITX | IMBL | -------- | FW | 48 H / S | F 6.9, 5.2 - 8.1 ppm | | ------ | 344 |
| Daphnia magna Water flea | EC50 | ITX | IMBL | -------- | FW | 48 H / S | F 115, 76 - 184 ppm | | ------ | 344 |
| Daphnia magna Water flea | EC50 | ITX | IMBL | -------- | FW | 48 H / F | F 49 ppm | | ------ | 344 |
| Daphnia magna Water flea | LC50 | MOR | MORT | -------- | FW | 48 H / S | F 6900, 5200 - 8100 | | ------ | 631 |
| Daphnia magna Water flea | LC50 | MOR | MORT | INC / -------- | | 48 H / | F 54000 | | ------ | 8970 |

**Figure 5-33.** ECOTOX database (excerpt from search, October 21, 2002) (*http://www.epa.gov/ecotox/*).

opportunities to search and limit by habitat (aquatic, terrestrial, both), kingdom (plant, animal), chemical name, CAS Number, species name, and endpoints (accumulation, biochemical, reproduction, behavior, cellular, mortality, ecosystem, etc.).

For the sample search on aquatic toxicity of atrazine, 1459 results were found. An excerpt on the daphnia acute aquatic toxicity tests (EC50, LC50) is presented in Figure 5-33.

## 5.18
## Tool: The Internet (Online Databases in Chemistry)

The following table can only give a short overview of the most important or most used databases in Chemistry and Biochemistry. It has no intention to be comprehensive or complete since the scenery of databases is changing very fast.

**Table 5-8.** Online databases in Chemistry.

| Database | Producer | Content | Type | Size (records) | Source | Access (host) | Availability (Price) | Update | URL |
|---|---|---|---|---|---|---|---|---|---|
| CA File | Chemical Abstracts Service | biochemistry, chemistry, and chemical engineering | biblio. | >21 mio substances >25 mio sequences | 9000 journals, 37 patent offices, proceedings, books | STN | commercial CD-ROM, online | weekly | www.cas.org |
| Kirk-Othmer Encyclopedia of Chemical Technology | Wiley Electronic Publishing | chemical technology | biblio. | – | – | – | CD-ROM, online | irregular | www.mrw.interscience.wiley.com/kirk |
| Medline | National Library of Medicine, USA | medicine, life science | biblio. | >13 mio | 4600 journals | STN, SciFinder, MDL | free | 4-weekly | www.nlm.nih.gov |
| SCISEARCH | Institute of Scientific Information, Thomson Scientific | science and technology | biblio. citations | 20 mio records | 5900 journals | ISI Web of Science, DIALOG, ORBIT, DIMDI, DataStar, STN | CD-ROM, online | weekly | www.isinet.com http://isi2.isi-knowledge.com/portal.cgi |
| Ullmann's Encyclopedia of Industrial Chemistry | VCH | chemical technology | biblio. | 16 mio words | – | – | commercial; CD-ROM, online | 6th edition | www.mrw.interscience.wiley.com/ueic/ |

**Table 5-8.** (Cont.)

| Database | Producer | Content | Type | Size (records) | Source | Access (host) | Availability (Price) | Update | URL |
|---|---|---|---|---|---|---|---|---|---|
| Beilstein | Beilstein Information Systems, Inc. | organic chemistry. | numeric, structure | >8 mio subst., >5 mio reactions | 180 journals *Beilstein Handbook* | MDL Information Systems GmbH | commercial; online, Chemweb, Crossfire | quarterly | www.beilstein.com |
| DETHERM | Dechema e.V. FIZ Chemie Berlin GmbH | thermophysical properties | numeric, factual, biblio. | 442 000 records, 57 000 biblio. | journals, patent offices, proceedings, books | STN | in-house, online | twice a year | www.dechema.de |
| Gmelin | MDL Information Systems GmbH, Germany | inorganic and organometallic chemistry | struct., numeric | 1.4 mio | 110 journals, *Gmelin Handbook* | MDL | commercial; online | periodically | www.mdli.com |
| WebBook | National Institute of Standards and Technology | thermochem., spectra | numeric | 22 300 substances | – | – | online | – | http://webbook.nist.gov |

**Table 5-8.** (Cont.)

| Database | Producer | Content | Type | Size (records) | Source | Access (host) | Availability (Price) | Update | URL |
|---|---|---|---|---|---|---|---|---|---|
| SpecInfo | Chemical Concepts GmbH, Germany | spectral data | numeric, structure | 150 000 subst., 80 000 $^{13}$C-NMR, 850 $^{15}$N-NMR, 670 $^{17}$O-NMR, 1750 $^{19}$F-NMR, 2000 $^{31}$P-NMR, 17 000 IR, 65 000 MS | – | Chemical Concepts GmbH | online | periodically | www.chemical-concepts.com/products.htm |
| KnowItAll | Bio-Rad's Sadtler™ | software & database solutions for spectroscopy | numeric | IR, NMR, MS, NIR, and Raman data | – | Bio-Rad Laboratories, Inc. | commercial; CD-ROM | periodically | www.bio-rad.com |
| SDBS | National Institute of Advanced Industrial Science and Technology; Tsukuba, Ibaraki, Japan | spectral data of organic subst. | numeric | 30 300 compounds: ca. 20 500 MS, 13 700 $^1$H-NMR, 11 800 $^{13}$C-NMR, ca. 47 300 IR, ca. 3500 Raman, ca. 2000 ESR | experiments | National Institute of Advanced Industrial Science and Technology | free | irregular | http://www.aist.go.jp/RIODB/SDBS/menu-e.html |
| CSD | Cambridge Crystallographic Data Centre | organic, metalorganic crystal structures | numeric | 257 000 | experiments | Cambridge Crystallographic Data Centre | commercial; CD-ROM | periodically | www.ccdc.cam.ac.uk |

**Table 5-8.** (Cont.)

| Database | Producer | Content | Type | Size (records) | Source | Access (host) | Availability (Price) | Update | URL |
|---|---|---|---|---|---|---|---|---|---|
| ICSD | FIZ Karlsruhe, Germany; NIST, USA | crystal structure data of inorganic compounds | numeric, factual | 65 000 | journals | STN | online, CD-ROM | biannually | *www.fiz-informationsdienste.de* |
| PDB | Research Collaboratory for Structural Bioinformatics (RCSB) | macromolecular structure data on proteins, nucleic acids, protein–nucleic acid complexes, and viruses | numeric, biblio. | ~20 000 records | experiments | Research Collaboratory for Structural Bioinformatics | online, CD-ROM | periodically | *www.rcsb.org/pdb/* |
| BIOSIS | BIOSIS (USA) Biological Abstracts | biosciences/biomedical | biblio. | 13 mio | journals, patent offices, proceedings, books | STN | online | weekly | *www.biosis.org* |
| EMBL | European Bioinformatics Institute | nucleotide sequence database | biblio., substance, sequence | 20 mio nucleotide seq., 28 billion nucleotides | journals, author submissions | European Bioinformatics Institute | free | daily | *http://www.e-bi.ac.uk/embl/index.html* |

**Table 5-8.** (Cont.)

| Database | Producer | Content | Type | Size (records) | Source | Access (host) | Availability (Price) | Update | URL |
|---|---|---|---|---|---|---|---|---|---|
| GenBank | National Center for Biotechnology Information, USA | nucleic acid sequence | biblio., substance, sequence | 22 mio sequences 28 billion bases | journals, author submissions | STN | online | daily | *www.ncbi.nlm.nih.gov* |
| PIR | National Biomedical Research Foundation | protein sequence | biblio., substance, sequence | 280 000 protein sequences | journals, author submissions | National Biomedical Research Foundation | free | periodically | *http://pir.georgetown.edu* |
| SwissProt | EMBL and Swiss Institute of Bioinformatics | protein sequence | biblio., substance, sequence | 120 000 protein sequences, 44 mio amino acids | journals, author submissions | European Bioinformatics Institute | free | periodically | *http://www.ebi.ac.uk/swissprot/index.html* |
| REGISTRY | Chemical Abstracts Service (CAS), USA | chemical substances | structure | 22 mio struct., 25 mio sequences | 9000 journals, 37 patent offices, proceedings, books | STN | commercial; CD-ROM, online | daily | *www.cas.org* |

**Table 5-8.** (Cont.)

| Database | Producer | Content | Type | Size (records) | Source | Access (host) | Availability (Price) | Update | URL |
|---|---|---|---|---|---|---|---|---|---|
| CASREACT | Chemical Abstracts Service | chemical reactions | biblio., reaction, structure | >400 000 documents >6.6 mio reactions | journals, patents | STN | commercial; CD-ROM, online | weekly | www.cas.org |
| ChemIn-formRX | FIZ CHEMIE GmbH, Germany | chemical reactions | reaction, biblio., structure | 1.0 mio substances, 113 859 records, 689 029 single-step reactions, 377 491 multi-step reactions | 250 journals | FIZ | commercial; online | quarterly | www.mdli.com |
| ChemReact | InfoChem GmbH, Germany | chemical reactions | reaction, biblio., structure | 392 000 records | journals | STN | commercial; CD-ROM, online | irregularly | www.cas.org/ONLINE/DBSS/chemreactss.html |
| MARPAT | Chemical Abstracts Service (CAS), USA | Markush structures in patents | structure, Markush, biblio. | 180 000 records, 505 000 Markush struct. | patent offices | STN | commercial; CD-ROM, online | weekly | www.cas.org/ONLINE/DBSS/marpatss.html |

**Table 5-8.** (Cont.)

| Database | Producer | Content | Type | Size (records) | Source | Access (host) | Availability (Price) | Update | URL |
|---|---|---|---|---|---|---|---|---|---|
| INPADOC | European Patent Office, Vienna Branch Office, Austria | international patents | biblio. | 26 mio records, 35 mio citations, 59 mio legal status | patent offices | STN | commercial; online | weekly | www.european-patent-office.org/inpadoc/ |
| JAPIO | Japanese Patent Office, Japan | Japanese patent information | biblio. | 7.7 mio records, 5.1 mio images | patent abstracts, INPADOC | STN | commercial; CD-ROM, online | monthly | www.cas.org/ONLINE/DBSS/japioss.html |
| PATDPA | Deutsches Patent- und Markenamt, Germany | German patents | biblio. | 4.3 mio records, 500 000 technical drawings | patent documents | STN | commercial; CD-ROM, online | weekly | www.cas.org/ONLINE/DBSS/patdpass.html |
| WPINDEX | Derwent Information, Ltd., UK | international patents | biblio. | 13 mio | patent offices, patent documents | Thomson Inc. | online commercial | weekly | www.derwent.com/ |
| CORDIS | Community Research & Development Information Service | research | biblio. | variable | author submissions | Cordis | online | periodically | http://www.cordis.lu/en/home.html |

**Table 5-8.** (Cont.)

| Database | Producer | Content | Type | Size (records) | Source | Access (host) | Availability (Price) | Update | URL |
|---|---|---|---|---|---|---|---|---|---|
| CRIS/USDA | Current Research Information System; USDA/CSREES/ISTM | research projects in agriculture, food, nutrition, and forestry | biblio. | variable | author submissions | CRIS | online | annual | *http://cris.csrees.usda.gov/* |
| FEDRIP | National Technical Information Service (NTIS), USA | federal research projects | directory, biblio. | 220 000 | – | STN | online | monthly | *http://grc.ntis.gov/fedrip.htm* |
| NUMERIGUIDE | American Chemical Society, USA | property data | directory | 875 | – | STN | online | irregular | *www.stn-international.de/strndatabases/databases/numerigu.html* |
| UFORDAT | Umweltbundesamt, Germany | environmental projects | directory | 68 000 | – | FIZ Karlsruhe | online, CD-ROM | twice a year | *www.umweltbundesamt.de/index-e.htm* |

**Table 5-8.** (Cont.)

| Database | Producer | Content | Type | Size (records) | Source | Access (host) | Availability (Price) | Update | URL |
|---|---|---|---|---|---|---|---|---|---|
| MRCK | Merck & Co., Inc., USA | descriptions of chemicals, drugs, agricultural and natural products | substance, numeric | 10 000 | *Merck Index* (encyclopedia) | STN | | semiannually | *www.cas.org/ ONLINE/ DBSS/ mrckss.html* |
| TOCXENTER | Chemical Abstracts Service, USA. | toxicology | biblio. | 5.7 mio | journals, patents | STN | commercial; CD-ROM, online | weekly | *http://www.cas.org/ON-LINE/DBSS/ toxcen-terss.html* |

## Essentials

- The hierarical, network, relational, and object-oriented database models are the four fundamental ones.
- They are classified as bibliographic, factual, and structure databases.
- The Chemical Abstracts (CA) File is the main abstracting and indexing service for biochemistry, chemistry, and chemical engineering.
- Beilstein and Gmelin are the world's largest factual databases in chemistry.
- The Cambridge Structural Database (CSD) and the Inorganic Crystal Structure Database (ICSD) contain information obtained from X-ray structure analysis.
- Compounds are stored as connection tables (CT) in structure and reaction databases, e.g., Beilstein, Gmelin, CAS Registry, and CASREACT.
- INPADOC is the most comprehensive bibliographic database of scientific and technological patent documents.
- Scientific and chemical information is becoming available increasingly on the Internet.

## Selected Reading

- J.D. Ulmann, J. Widom, *A First Course in Database Systems*, Upper Saddle River, NJ, Academic Press, **1997**.
- T.M. Connolly, *Database Systems: A Practical Approach to Design, Implementation and Management*, Eds. T.M. Connolly ; C.E. Begg : A.D. Strachan, Wokingham, Addison–Wesley, **1996**.
- S.R. Heller, *Internet J. Chem.* **1998**, *1(32); www.ijc.com/articles/1998v1/32*
- G. Wiggins, *J. Chem. Inf. Comput. Sci.* **1998**, *38*, 956–965.

## Interesting Websites

- *www.stn-international.de*
- *http://info.cas.org*
- *http://www.ccdc.cam.ac.uk/prods/csd/csd.html*
- *http://pir.georgetown.edu/pirwww/dbinfo/resid.html*
- *http://www.thomson.com/scientific/scientific.jsp*
See tables 5.7, 5.8 and tools section 5.18

## Available Software

- STN, STN Easy, SciFinder: *http://www.cas.org/prod.html*
- SciFinder Scholar: *http://www.cas.org/SCIFINDER/SCHOLAR2002* (access only with license agreement)
- CrossFire Commander, AutoNom: *http://www.mimas.ac.uk/crossfire/download.html*

# References

[1] A. Silberschatz, H. F. Korth, S. Sudarshan, *Database Systems Concepts*, 3rd edition, New York, McGraw-Hill, **1997**.

[2] J. D. Ulmann, J. Widom, *A First Course in Database Systems*, Upper Saddle River, NJ, Academic Press, **1997**.

[3] G. Vossen, *Datenbankmodelle, Datenbanksprachen und Datenbankmanagement-Systeme*, 3. Auflage, R. Oldenburg Verlag, München, **1999**.

[4] Directory of Online Databases, Cuadra Associates, 2001 Wilshire Blvd., Suite 305, Santa Monica, CA 90403

[5] *www.stn-international.de/*

[6] C. J. Date, *An Introduction to Database Systems*, 6th edition, Addison–Wesley, New York, **1995**.

[7] T. M. Connolly, *Database Systems: A Practical Approach to Design, Implementation and Management*, Eds. T. M. Conolly ; C. E. Begg ; A. D. Strachan, Wokingham, Addison–Wesley, **1996**.

[8] CODASYL Data Description Language Committee. *Information Systems* **1987**, *3(4)*, 247–320.

[9] E. F. Codd, *Communications of the ACM*, **1970**, *13*, 377–387.

[10] G. Lausen, G. Vossen, *Models and Languages for Object-Oriented Databases*, Harlow, UK, Addison–Wesley, **1998**

[11] D. W. Embley, *Object Database Development – Concepts and Principles*, Reading, MA, Addison–Wesley, **1998**.

[12] A. Barth, "Online databases in chemistry" in *Encyclopedia of Computational Chemistry, Vol. 3*, P. von R. Schleyer, N. L. Allinger, T. Clark, J. Gasteiger, P. A. Kollman, H. F. Schaefer, P. R. Schreiner (Eds.), Wiley, Chichester, **1998**, 1968–1980.

[13] U. Boehme, S. Tesch. *The Chemical Education Journal (CEJ)*, **2002**, *5(2)*, *http://chem.sci.utsunomiya-u.ac.jp/ v5n2/wboehme/header.html*

[14] *http://www.galegroup.com*

[15] *www.cas.org/*

[16] *www.dialog.com*

[17] *www.questel.orbit.com*

[18] *www.nlm.nih.gov*

[19] *http://www.ncbi.nlm.nih.gov/PubMed/*

[20] Database description of Beilstein: *http://info.cas.org/ONLINE/DBSS/ beilsteinss.html* or *www.beilstein.com/ products/xfire/*

[21] Database description of Gmelin: *www.cas.org/ONLINE/DBSS/gmelinss.html* or *www.beilstein.com/products/ xfire/gmelin.shtml*

[22] More Information about this database is available at the MDL websites: *http:// www.mdli.com/*

[23] This section was written by Jürgen Vogt, Sektion für Spektren- und Strukturdokumentation, Universität Ulm, D-89069 Ulm, Germany.

[24] *http://www.cas.org/ONLINE/DBSS/ specinfoss.html*

[25] *http://www.ccdc.cam.ac.uk/prods/csd/ csd.html*

[26] Database description of ICSD: *http:// www.cas.org/ONLINE/DBSS/icsdss.html*

[27] *http://www.rcsb.org/pdb/holdings.html*

[28] *Nucleic Acids Res.* **1996**, *24*, Oxford Journals, Oxford, 1996 and A.D. Baxevanis, *Nucleic Acids Res.* **2003**, *31*, 1–12.

[29] C. Discala, X. Benigni, E. Barillot, G. Vaysseix, *Nucleic Acids Res.* **2000**, *28*, 8–9.

[30] *http://www.infobiogen.fr/services/dbcat*

[31] *www.biosis.org*

[32] *http://www.ncbi.nlm.nih.gov/Genbank/genbankstats.html*

[33] *http://www.ebi.ac.uk/embl/*

[34] S. Miyazaki, H. Sugawara, T. Gojobori, Y. Tateno, *Nucleic Acids Res.* **2003**, *31*, 13–16.

[35] *http://pir.georgetown.edu/pirwww/dbinfo/resid.html*

[36] *http://www.ebi.ac.uk/swissprot/index.html*

[37] V. V. Poroikov, D. A. Filimonov, W. D. Ihlenfeldt, T. A. Gloriozova, A. A. Lagunin, Y. V. Borodina, A. V. Stepanchikova, M. C. Nicklaus, *J. Chem. Inf. Comput. Sci.* **2003**, *43(1)*, 228–236.

[38] W. D. Ihlenfeldt, J. H. Voigt, B. Bienfait, F. Oellien, M. C Nicklaus, *J. Chem. Inf. Comput. Sci.* **2002**, *42(1)*, 46–57.

[39] J. H. Voigt, B. Bienfait, S. Wang, M. C. Nicklaus, *J. Chem. Inf. Comput. Sci.* **2001**, *41(3)*, 702–712.

[40] E. Zass, Databases on chemical reactions, in *Handbook of Chemoinformatics*, J. Gasteiger (Ed.), Wiley-VCH, Weinheim, **2003**, Chapter V, Section 8.

[41] *http://www.fiz-chemie.de/*

[42] A. Parlow, Ch. Weiske, J. Gasteiger, *J. Chem. Inf. Comput. Sci.* **1990**, *30*, 400–402

[43] J. T. Bohlen, ChemInform Electronic Journal and ChemInform RX Reaction Database, New Developments in *Software, Entwicklung in der Chemie*, Vol. 10, Gesellschaft Deutscher Chemiker, Frankfurt, **1996**, pp. 27–32; available online at *http://www2.chemie.uni-erlangen.de/external/cic/tagungen/workshop95/bohlen/index.html*

[44] This section was written by Oliver Sacher, Computer-Chemie-Centrum, Universität Erlangen-Nürnberg, Nägelsbachstraße 25, 91052 Erlangen, Germany.

[45] InfoChem Gesellschaft für chemische Information mbH, Landsberger Straße 408, D-81241 München, Germany; *http://www.infochem.de*

[46] *STN Patent Basics*, American Chemical Society, Washington DC, **2001**.

[47] C. Glander-Höbel, *Online Information Review*, **2001**, *25(4)*, 257–266.

[48] L. L. Wright, *Toxicology* **2001**, *157*, 89–110

[49] This section was written by Kristina Voigt, GSF – National Research Center for Environment and Health, Institute of Biomathematics and Biometry, Neuherberg, Germany.

[50] K. Voigt, Databases on environmental information, in *Handbook of Chemoinformatics*, J. Gasteiger (Ed.), Wiley-VCH, Weinheim, **2003**, Chapter V, Section 10.

[51] K. Voigt, J. Gasteiger, R. Brüggemann, *J. Chem. Inf. Comput. Sc.*, **2000**, *40(1)*, 44–49

[52] *http://chemfinder.cambridgesoft.com/*

[53] C. L. Russom, *Toxicology* **2002**, *173*, 75–88.

# 6

# Searching Chemical Structures

*Nikolay Kochev, Valentin Monev, and Ivan Bangov*

## Learning Objectives

- To become familiar with various methods and tools for full structure recognition and the search in structural datasets.
- To learn a more thorough approach to the solution of the substructure search problem.
- To become familiar with the basics of chemical structure similarity, similarity measures, and different approaches exploited within the similarity search process.

## 6.1
## Introduction

Large chemical databases, combinatorial libraries, and data warehouses have become indispensable tools in modern chemical research. Accordingly, structural information must be stored in these databases and searched in an appropriate manner.

We shall discuss here the methods that have been developed for enabling the computer to perceive both complete chemical structures and fragments of them, as well as their mutual similarity. This is very important in many fields of chemistry. The recognition of full structures is required routinely in everyday work with large databases.

Thus, the presence or absence of a given structure frequently needs to be checked and the retrieved structure (if any) has to be further processed.

The search for structural fragments (substructures) is very important in medicinal chemistry, QSAR, spectroscopy, and many other fields in the process of perception of *pharmacophore, chromophore,* or other *-phores*.

Similarity search appears as an extremely useful tool for computer-aided structure elucidation as well as for molecular design. Here the similarity property principle is involved. This may be stated as:

*"Structurally similar molecules are expected to exhibit similar physical properties or, similar biological activities."*

Any development of chemical structure perception methods must obey the following requirements:

- *Completeness and non-redundancy.* Does the strategy guarantee to find all and only those solutions which are in the data set?
- *Time complexity.* How long does it take to find the solution?
- *Memory complexity.* How much computer memory is needed to perform the searching?
- *Optimality.* Does the strategy find the highest-quality solution when there are several solutions?

Mathematical theory of labeled colored graphs is exclusively used to formalize the structure and substructure search problem. There is almost a one-to-one correspondence between the terms used in graph theory and the ones used in chemical structure theory. Formally a graph $G$ can be given by Eq. (1), where $V$ is the set of graph vertices and $E$ the set of edges.

$$G = (V, E) \tag{1}$$

Accordingly, a molecular structure can be represented by the molecular graph of Eq. (2).

$$S = (A, B) \tag{2}$$

Here $A = \{a_1, a_2, ..., a_n\}$ is the set of atoms $a_i$, and $B = \{b_{1,2}, b_{1,3}, ...b_{mn})$ is the set of chemical bonds, where the bond $b_{ij}$ is the bond between atoms $a_i$ and $a_j$. One can see that there is a one-to-one correspondence between a graph and a molecular structure representation. Figure 6-1 summarizes the various types of representation of a chemical graph.

## 6.2
## Full Structure Search

The problem of perception complete structures is related to the problem of their representation, for which the basic requirements are to represent as much as possible the functionality of the structure, to be unique, and to allow the restoration of the structure. Various approaches have been devised to this end. They comprise the use of molecular formulas, molecular weights, trade and/or trivial names, various line notations, registry numbers, constitutional diagrams (2D representations), atom coordinates (2D or 3D representations), topological indices, hash codes, and others (see Chapter 2).

Empirical molecular formulas and molecular weights usual identify a whole class of compounds (chemical isomers) rather than a single structure. Further-

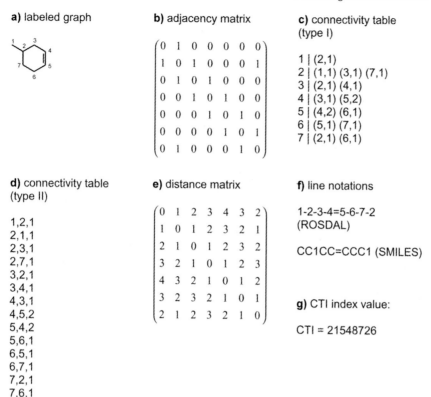

**a) labeled graph**

**b) adjacency matrix**

$$\begin{pmatrix} 0 & 1 & 0 & 0 & 0 & 0 & 0 \\ 1 & 0 & 1 & 0 & 0 & 0 & 1 \\ 0 & 1 & 0 & 1 & 0 & 0 & 0 \\ 0 & 0 & 1 & 0 & 1 & 0 & 0 \\ 0 & 0 & 0 & 1 & 0 & 1 & 0 \\ 0 & 0 & 0 & 0 & 1 & 0 & 1 \\ 0 & 1 & 0 & 0 & 0 & 1 & 0 \end{pmatrix}$$

**c) connectivity table (type I)**

1 | (2,1)
2 | (1,1) (3,1) (7,1)
3 | (2,1) (4,1)
4 | (3,1) (5,2)
5 | (4,2) (6,1)
6 | (5,1) (7,1)
7 | (2,1) (6,1)

**d) connectivity table (type II)**

1,2,1
2,1,1
2,3,1
2,7,1
3,2,1
3,4,1
4,3,1
4,5,2
5,4,2
5,6,1
6,5,1
6,7,1
7,2,1
7,6,1

**e) distance matrix**

$$\begin{pmatrix} 0 & 1 & 2 & 3 & 4 & 3 & 2 \\ 1 & 0 & 1 & 2 & 3 & 2 & 1 \\ 2 & 1 & 0 & 1 & 2 & 3 & 2 \\ 3 & 2 & 1 & 0 & 1 & 2 & 3 \\ 4 & 3 & 2 & 1 & 0 & 1 & 2 \\ 3 & 2 & 3 & 2 & 1 & 0 & 1 \\ 2 & 1 & 2 & 3 & 2 & 1 & 0 \end{pmatrix}$$

**f) line notations**

1-2-3-4=5-6-7-2
(ROSDAL)

CC1CC=CCC1 (SMILES)

**g) CTI index value:**

CTI = 21548726

**Figure 6-1.** Different forms of representation of a chemical graph: a) labeled (numbered) graph; b) adjacency matrix; c) connectivity table, type I; d) connectivity table, type II; f) line notations; g) structural index.

more, millions of structures might correspond to each molecular formula, i.e., they are highly degenerate. Hence, they are usually used as supplementary descriptors.

Trivial or trade names can be stored and searched as character strings. Their use is the simplest and most intuitive way of storing chemical information. However, being not subject to strict rules, their formation does not reflect accurately the molecular composition. Hence, the structure of the searched compound cannot be derived from them. Thus, a name such as "Flexricin" does not tell the user very much. Furthermore, many more than one trivial or trade name for a given compound usually exist.

The use of registry numbers is closed within a given company and related only to the database system for which they have been developed. Thus, in the 1960s the Chemical Abstracts Service (CAS) started identifying its compounds by giving them a unique registration number. A typical CAS Registry Number (RN) looks like this: 553-97-9. In a similar way, the Beilstein Registry Number (BRN) was introduced. Nowadays many chemical and pharmaceutical companies have their own registry numbers, though no correlation between these numbers exists. However, in many databases one can find either the CAS RN or the BRN, or both. The registry numbers provide a useful key to online searching, but have no correspondence to the structure of the search compound.

Plenty of line notations, such as Wiswesser (WLN) [1], Dyson–IUPAC, Hayward, Skolnik, Gremas, ROSDAL, SMILES [2], and many others have been proposed (see Section 2.3). Examples of the ROSDAL and SMILES notations are presented in Figure 6-1f; the SMILES notation is in widespread use among chemists. Consisting of character strings, these representations are compact and easy to use. Their creation is subject to different rules: a search procedure using them consists of a simple comparison of two strings. However, to be successful, these character strings must uniquely represent the corresponding chemical structures. This uniqueness is again related to the solution of the isomorphism problem [3] (see Section 2.5).

While the trivial and trade nomenclature in most cases has accidental character, the IUPAC Commission has worked out a series of rules [4] which allow the great majority of structures to be represented uniformly, though there still exists some ambiguity within this nomenclature. Thus, many structures can have more than one name. It is important that the rules of some dialects of the IUPAC systematic nomenclature are transformed into a program code. Thus, programs for generating the names from chemical structures, and vice versa (structures from names) have been created [5] (see Chapter II, Section 2 in the Handbook).

Thus, the conversion of chemical names into structures consists of parsing the name into longest text fragments. These fragments are submitted to lexical analysis and the derived lexical units are compared with a collection of predefined units in a dictionary.

Each predefined unit is related to its adjacency matrix representation. Finally, the derived units are assembled; thus, the structural information is derived from the compound name.

In order to derive unique names from structural information we have to generate unique (canonical) numbering for each structure, i.e., here again we have to solve the isomorphism problem. One of the most widely used programs for converting structural information into chemical names is AutoNom (from "Automatic Nomenclature") developed in the Nomenclature department of Beilstein [5]. The AutoNom algorithm is based on an analysis of the connection table and on selecting the smallest structural entities present in the structure, which are further transformed into functional groups. These groups are then transformed into a data tree by using the IUPAC nomenclature rules. The root of the tree is the parent structure, and the branches contain the corresponding chemical groups. It is apparent that a series of steps such as ring system perception, recognition of the functional groups, parent structure selection, and creation of the name tree must be completed in order for the IUPAC name to be generated (for more details see Chapter II, Section 2 in the Handbook).

A series of *topological indices* have been devised during the period since the 1950s. These are numbers derived from the connectivity of a given structure.

Hence, they take account of only the structure constitution (topology). One of the first and most frequently used topological indices is the Wiener index. It has the form of Eq. (3), where $D_{ij}$ are all the routes from atom $i$ to atom $j$.

$$W = \sum D_{ij} \tag{3}$$

Other quite frequently used indices are the Randic index and the information-topological indices such as the Bonchev index (see Chapter VIII, Section 1 in the Handbook). Up to now several hundred indices have been devised.

The advantage of using topological indices is that they occupy very little memory and the search process is extremely simple, a comparison between numbers. The main disadvantage is that most of them are not unique, or as is usually said they are *degenerate*, i.e., for distinct structures they produce the same values.

An index with extremely low degeneracy (i.e., where no degenerate structures have been observed yet) [6], the Charge-related Topological Index (*CTI*), has been proposed by one of the authors (I.B.) (Figure 6-1). It has a potential-like form (Eq. (4)).

$$CTI = \frac{\sum L_i L_j}{D_{ij}} \tag{4}$$

Here, $L_i$ and $L_j$ are local indices having the form shown in Eq. (5), where $Lo$ is a constant characterizing the $i$th atom (in some cases the atom valence can be used to this end), $N_H$ is the number of attached hydrogen atoms and $q_i$ is the charge density calculated by some fast method such as the Marsili–Gasteiger charge calculation method [7].

$$L_i = Lo - N_H + q_i \tag{5}$$

A second disadvantage in the use of topological indices is that whereas the process of transformation of connectivity into one number is straightforward, the reverse process of reconstruction of connectivity from the index is not possible.

Another useful measure for a structure lookup is the *hash code*. A hash code is a string generated according to rules which ensure that it represents the structure uniquely. Thus, the hash code of Ihlenfeldt and Gasteiger [8] for the structure of 2-methylbenzo-1,4-quinone is 6CAA42EOD61B68CC. The recognition power of such a code is dependent on its precision; thus the degeneracy (the failure in discriminating the structures) of a 32-bit hash code is much greater than in the case of a 64-bit hash code.

## 6.3
## Substructure Search

### 6.3.1
### Basic Ideas

The development of substructure search algorithms has attracted many efforts of software developers and chemists for several decades. Many information systems that manipulate large structure datasets need a fast and robust substructure search algorithm. The substructure search algorithm is usually the first step in the implementation of other important topological procedures for the analysis of chemical structures such as: identification of equivalent atoms, determination of maximal common substructure, ring detection, calculation of topological indices, etc. Accordingly, since 1960 many scientists have done intensive research in order to improve known substructure search software.

Substructure searching is the process of identifying parts of a given structure that are equivalent to a specified query substructure. In graph-theoretical terms substructure searching is the task of checking whether the query graph ($G_Q$) is isomorphic with a subgraph of another target graph ($G_T$). Sometimes the target graph is called a *reference* graph. In this chapter $G_Q$ denotes the query substructure and $G_T$ denotes the structure to be analyzed.

$G_Q$ is a substructure of $G_T$ (i.e., $G_Q$ is isomorphic with a subgraph of $G_T$) if and only if all the atoms of $G_Q$ can be mapped onto a subset of atoms of $G_T$ in such a way that the bonds of $G_Q$ map the corresponding bonds which connect the mapped atoms from $G_T$. Each mapping between $G_Q$ and $G_T$ can be considered as a function of the type M: $G_Q \rightarrow G_T$ and it can be presented by an array M = ($M_1$, $M_2$, ..., $M_n$) where $M_i$ is the number of target graph atoms, which is mapped onto the *i*th query graph atom. If $G_Q$ has $n$ atoms and $G_T$ has $m$ atoms the number of all mappings between $G_Q$ and $G_T$ is given by Eq. (6).

$$N_{maps} = \frac{m!}{(m - n)!} \tag{6}$$

Hence, searching for an isomorphism between graph $G_Q$ and a subgraph of $G_T$ is a Non-Polynomial (*NP*)-complete problem [9]. *NP*-completeness implies that, in the worst case the algorithm will have an exponential computational complexity, i.e., the computing time will be an exponential function of the input parameters (the number of atoms in the graphs). In some special cases, polynomial timing algorithms do exist, e.g., when the compared graphs do not contain cycles. However, in contemporary chemical databases cyclic structures occur very often, which means the software systems will have to face the problem of *NP*-completeness.

In the example of the pair of structures of Figure 6-2, $N_{maps}$ = 5! / (5−4)! = 120 mappings can be generated. Only two of them, mappings (2, 4, 3, 1) and (4, 2, 3, 1) represent possible isomorphisms between $G_Q$ and a substructure of $G_T$. Accord-

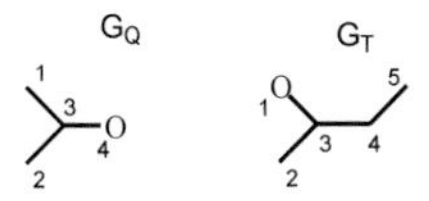

Mappings ($G_Q$ → $G_T$)

(1,2,3,4), (1,2,3,5),

(1,2,4,5), (2,1,3,4),

...

(5,4,3,1), (5,4,3,2)

**Figure 6-2.** Mappings between the query graph ($G_Q$) and the target graph ($G_T$). Notation such as (2, 1, 3, 4) means that atom 2 of the query subgraph $G_Q$ is mapped to atom 1 from the target graph $G_T$.

ingly the tuple (2, 4, 3, 1) corresponds to the following mapping 1 → 2, 2 → 4, 3 → 3, 4 → 1.

The simplest and least effective algorithm for substructure searching is to walk through all mappings (their number is $N_{maps}$) and to check for each candidate whether the atom types and bond types match. This is the so-called "brute force" approach. Since it has a factorial degree of computational complexity (Eq. (6)) it is practically inapplicable even for structures with ten atoms. For example, the number of mappings between a query structure with seven atoms and a target structure with ten atoms is 604 800.

To obtain an effective algorithm for substructure searching the factorial degree of the brute force algorithm has to be drastically decreased. In the next sections we discuss several approaches where combination leads to a much more effective and applicable approach for substructure searching. In the process of searching the isomorphism between $G_Q$ and a substructure of $G_T$, the partial mappings $G_Q$ → $G_T$ can be used as well. In these cases, not all atoms from $G_Q$ are mapped and, for those which are not, the array value $M_i$ is set to 0.

There are several basic strategies for the improvement of the performance of substructure search algorithms:

1. optimization of the hardware and software technologies used;
2. usage of various heuristics to improve the perception of the substructures isomorphic with the query graph or with the rejection of inappropriate target structure candidates as early as possible;
3. pre-processing of the most time-consuming operations that are independent of the query structure and storing them as an integral part of the database which can be used at search time.

It can be said that these three main strategies have been applied equally and very often in combination. Basically, the first approach implies the use of a faster computer or a parallel architecture. To some extent it sounds like a brute force approach but the exponential increase of the computer power observed since 1970 has made the hardware solution one of the most popular approaches. The Chemical Abstracts Service (CAS) [10] was among first to use the hardware solution by distributing the CAS database onto several machines.

The development of heuristic approaches applied to the second strategy has continued for a long period of time. Although the basic methods were invented a long time ago, they still find wide application and continue to be refined and improved. These approaches determine the most time-consuming part of the substructure searching algorithm, and therefore scientists have always had a great interest in the second strategy; even small improvements in these heuristics are considered worthy of investment of time and money. The effectiveness of algorithms is determined by the degree of implementation of such heuristics which decrease the exponential complexity of the substructure searching algorithm.

The pre-processing concepts have been a more recent development of substructure searching systems. These approaches have become popular since the mid-1980s, when the cost of the storage devices (hard disks and CD-ROMs) decreased.

The next sections deal exclusively with the second and the third approaches to the optimization of substructure search algorithms.

## 6.3.2
### Backtracking Algorithm

The basic approach to a fast search of an isomorphism among all mappings (mappings of the type $G_Q \rightarrow G_T$) is the so-called backtracking algorithm [10]. Essentially, the algorithm starts with an arbitrary query atom $Q_1$ which is mapped onto a target atom $T_1$ ($T_1$ and $Q_1$ are of the same type). Further, all neighbors ($Q_2$, $Q_3$,...) of $Q_1$ are tried to be mapped to some of the neighbors of $T_1$ ($T_2$, $T_3$, ...). If this step is successful the algorithm continues with the neighbors of $Q_2$, $Q_3$, ... etc. until all query atoms are mapped. During the mapping, equality of atom types and bond types is imposed. If at a given stage a query atom $Q$ cannot be mapped to any target atom, the algorithm backtracks to the last successfully mapped atom $Q'$ and tries to map it to a different partner from the target graph. If it is impossible to find an alternative mapping for $Q'$ a backtracking is performed again, and alternative mapping of atom $Q''$, which is the last successfully mapped atom before $Q'$, is attempted, and so on. The process stops when an isomorphism is found or when it is impossible to perform any more backtracking. The latter means that the algorithm has returned to the original atom $Q_1$ and did not find an alternative mapping for it. The backtracking algorithm searches an isomorphism $G_Q \rightarrow G_T$ in such a way that all checked mappings can be organized hierarchically as a tree (Figure 6-3). Very often this approach is named *atom-by-atom* searching, since at every step a new single atom is mapped. Figure 6-3 displays the search tree of mappings for the pair of structures from Figure 6-2, obtained by the application of the backtracking algorithm. It can be seen that in the worst case 16 mappings have to be checked for isomorphism, which is somewhat faster than the factorial approach (brute force algorithm), where 120 mappings are to be checked. In this example the algorithm starts with the first query atom. On the second level of the tree, one can see all possible mappings for atom 1 from $G_Q$; the mappings of the neighbor of atom 1 (atom 3) are visualized on the third level, etc.

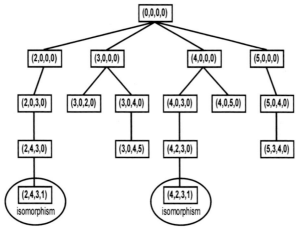

**Figure 6-3.** Search tree of mappings obtained by applying the backtracking algorithm for the pair of structures $G_Q$ and $Q_T$ (see the graphs in Figure 6-2). Array ($M_1$, $M_2$, $M_3$, $M_4$) denotes the mapping $1 \rightarrow M_1$, $2 \rightarrow M_2$, $3 \rightarrow M_3$, $4 \rightarrow M_4$.

In many real-world applications, the isomorphism would be found before all 16 mappings were checked. For the example from Figure 6-3, many algorithms would find the isomorphism at the fourth mapping following the leftmost path in the search tree (the bold line in Figure 6-4):

$$(0,0,0,0) - (2,0,0,0) - (2,0,3,0) - (2,4,3,0) - \text{isomorphism found } (2,4,3,1)$$

The latter is based on the presumption that the mapping atoms are chosen in increasing order according to their numbers. In this case no backtracking is done. Since the numbering in chemical graphs is random, there is no guarantee that the shortest way to the isomorphism will be walked. If the tree is searched through a path on the rightmost side, some backtracking has to be performed (Figure 6-4). This is an example of a sequence of mappings that is listed on the right-hand side when the tree is walked in depth from right to left.

Usually when an isomorphism exists between $G_Q$ and a subgraph of $G_T$, an entire tree scanning is not necessary and the isomorphism is found at an early stage. To prove that $G_Q$ is not a substructure of $G_T$ is a more time-consuming task, since it requires a traversing of all the mappings of the search tree. The backtracking approach is applied to many other tasks that require searching of a solution in a tree structure. This approach is typically implemented through the popular "depth-first" search [11] algorithm (Figure 6-4), where each node in the tree is expanded on the deepest level of the tree. Only when the search hits "dead end" (no isomorphism is found) does the search go back and expand the nodes at the shallower levels.

The depth-first search algorithm (the backtracking algorithm, respectively) has an exponential order of computational complexity $CC$ [11]: $CC = O(b^k)$. The ex-

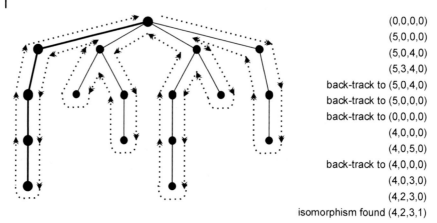

(0,0,0,0)
(5,0,0,0)
(5,0,4,0)
(5,3,4,0)
back-track to (5,0,4,0)
back-track to (5,0,0,0)
back-track to (0,0,0,0)
(4,0,0,0)
(4,0,5,0)
back-track to (4,0,0,0)
(4,0,3,0)
(4,2,3,0)
isomorphism found (4,2,3,1)

**Figure 6-4.** Backtracking approach realized as depth-first search algorithm. Dotted arrows trace the route used for traversing all mappings in the search tree. Each node in the tree corresponds to a mapping between $G_Q$ and $G_T$ (Figure 6-2).

ponential multiplier $b$ is the so-called branching factor, i.e., the number of branches which are generated by a single node in the search tree, and $k$ is the depth of the search tree. By applying these relationships to the case of substructure searching, an estimation of the computing complexity of the backtracking algorithm is obtained (Eq. (7)).

$$CC = mb^n \tag{7}$$

In the worst case, the backtracking algorithm will form a search tree of depth $n$, where $n$ is the number of atoms in the query graph. Also, in this case a separate sub-tree search process for each atom of the target graph will be initiated. That is why the linear multiplier $m$ is applied to Eq. (7).

In so far as the values of the branching factor vary, a mean value should be used for making an estimation of the computational complexity. The mean value $b$ is the most important one since it determines the speed of the algorithm. It depends on the specific heuristics applied in the algorithm and on the complexity of the query and the target graphs. For the example with the *GMA* (Generic Match Algorithm) [12] algorithm the branching factor was estimated by using statistical data from the literature ($b = 2$) and some specific data from the algorithm tests. These tests show that the backtracking occurs in more than 50% of the nodes. The complexity of *GMA* is estimated as $CC = m \times 2^{n/2} = m \times 1.141^n$. Another advantage of the depth-first search is that it requires the storage for only $b \times n$ nodes, which implies efficient memory management.

The backtracking algorithm is the core part of every software system that performs substructure searching. There are other approaches which have been applied both as alternatives to the backtracking algorithm or (most usually) in combination with it. Section 6.3.3 describes the approaches used for the optimization of the backtracking algorithm. They can be regarded as adjuncts to it and their main task is to decrease the mean value of the branching factor $b$.

### 6.3.3
### Optimization of the Backtracking Algorithm

The optimization of the backtracking algorithm usually consists of an application of several heuristics which reduce the number of candidate atoms for mapping from $G_Q$ to $G_T$. These heuristics are based on local properties of the atoms such as atom types, number of bonds, bond orders, and ring membership. According to these properties the atoms in $G_Q$ and $G_T$ are separated into different classes. This step is known in the literature as partitioning [13]. Table 6.1 illustrates the process of partitioning.

If any atom in $G_Q$ does not have a candidate from $G_T$ or the total number of candidates from its class is greater than the number of candidates from $G_T$, it is guaranteed that the isomorphism search process will fail; hence the algorithm can be stopped at this stage. In the structure example in Table 6-1, the partitioning step does not reject the possibility of isomorphism. The $G_Q$ atoms of class I match the $G_T$ atoms from classes I, II, and III. Since the $G_Q$ atom 3 has only one candidate for mapping, the remaining possible target candidates for atoms 1 and 2 from $G_Q$ are atoms 2, 5, and 4 from $G_T$. After the end of the partitioning process, the number of isomorphism checks needed is reduced to six (the number of all possible permutations without repetitions of two out of three). The partitioning process can be continued iteratively, taking into account the properties of the neighboring atoms in the graphs, after that of *their* neighbors, and so on. This approach is called relaxation [14]. Thus, in the example in Table 6-1 the target atom 5 of Figure 6-2 has been rejected in the process of relaxation as a possible candidate for a mapping to the query atoms 1 and 2. In this way only two mappings are possible: (2, 4, 3, 1) and (4, 2, 3, 1) (the ones that were obtained in the previous sections). The partitioning and relaxation algorithms run in polynomial time, but they are not sufficient to prove that a subgraph isomorphism exists, so the backtracking algorithm is needed again, at the final stage. However, their implementation at a preliminary stage has proven to be very effective in the reduction of the final complexity of the substructure searching algorithm.

**Table 6.1.** Application of partitioning approach for substructure search optimization. According to their local properties, the atoms of the graphs in Figure 6-2 are separated into several classes.

| Class description | Atoms from $G_Q$ | Atoms from $G_T$ |
| --- | --- | --- |
| C-atom with one single bond (class I) | 1, 2 | 2, 5 |
| C-atom with two single bonds (class II) | | 4 |
| C-atom with three single bonds (class III) | 3 | 3 |
| O-atom with one single bond (class IV) | 4 | 1 |

6.3.4
**Screening**

Although the optimized backtracking algorithm offers considerable improvement over the brute force approach, it still remains a heavy task to search a structural database with more than 50 000 compounds on conventional computers. Today, substructure searching is quite often performed on databases that contain a few hundred thousand or even a million structures. The third strategy (pre-processing the computationally expensive parts of the algorithm) for optimization of substructure searching allows this algorithm to be applied effectively on very large structural databases. This is done by a process termed "screening".

Screening systems normally use a predefined set of structural fragments called keys. For each key a preliminary (in a pre-process phase) substructure search is performed across the whole structural database. For each database compound a string of bits is constructed. Each bit of this string denotes the presence or absence of a key in the corresponding database compound. The $k$th bit in the bit-string is set to 1 if the $k$th key fragment is a substructure of the current database compound; otherwise the $k$th bit is set to 0. In the same way during the substructure searching a bit-string of the query structure is constructed. This step is usually very fast since a few hundred isomorphism checks are performed for a set of quite simple structural fragments. Further, the query bit-string is compared with each of the bit-strings from the database. The target compounds are screened as follows: each key which is present in the query structure must be present in the target structure (the corresponding bits are compared by using the fast bitwise operation AND, OR, and XOR). If at least one key that is present in the query graph is not present in the target graph, then this compound is pruned from a further processing. In this way a great many structures which are not likely to survive the isomorphism check are pruned early in the screening stage, thus escaping the much more time-consuming backtracking algorithm. This results in a much smaller set of structures being checked for structure isomorphism by means of the backtracking algorithm. For example, in a typical screening session more than 90 % of the database compounds which do not contain the query substructure are removed. Inasmuch as the time dependence of the screening procedure is linear, it decreases the time needed for substructure searching considerably (for example, this approach is 10 to 20 times faster than without the use of screening). Additionally, the screening procedure itself is very fast since it involves only a few isomorphism checks and is performed with the very fast bitwise operations.

The selection of keys is of prime importance in designing a screening system [15]. The basic principle is that one has to use middle-frequency keys. The most frequently occurring keys, e.g., the $-CH_2-$ fragments are not useful since they do not discriminate effectively between the different target structures. Nor are the most sparsely occurring structural keys of any use, since they will occur very seldom in the query structures so they will contribute nothing to the screening process.

## 6.4
## Similarity Search

### 6.4.1
### Similarity Basics

Similarity (fuzzy) searching is an alternative and complement [16] to exact search-ing. Similarity searching retrieves objects that are similar to a query, sorted in order of their decreasing similarity. High-ranked objects are likely to have similar proper-ties to the query and thus be of interest for property prediction. Pattern matching and signature analysis are names given to similarity searching that originate from other application areas.

In general, the similarity $S_{A,B}$ between two objects $A$ and $B$ is estimated by the number of matches or the overlap in the objects, with respect to one or more of their characteristics $\{X_{jA}\}$, $\{X_{jB}\}$, $j = 1, 2, ...n$. For identical objects, estimates of similarity $S_{A,B}$ take a *maximal* value. As a rule-of-thumb, in the mathematical ex-pressions for calculating $S_{A,B}$ (similarity measures), the numerator contains either a component-by-component multiplication $X_{jA}X_{jb}$, or conjunction $\cap$ (set theory), or the logical operator AND.

Accordingly, dissimilarity $D_{A,B}$ between two objects A and B is estimated by the number of mismatches or the difference between the objects, with respect to one or more of their characteristics $\{X_{jA}\}$, $\{X_{jB}\}$, $j = 1, 2, ...n$. For identical objects, the estimates of dissimilarity $D_{A,B}$ take a *minimal* value. Again, as a rule-of-thumb, in the mathematical expressions for calculating $D_{A,B}$ (dissimilarity measures), the nu-merator contains either a component-by-component subtraction $X_{jA}-X_{jb}$, or disjunction $\cup$ (set theory), or the logical operator XOR (exclusive OR).

"Similarity" is often used as a general term to encompass either similarity or dis-similarity or both (see Section 6.4.3, on similarity measures, below). The terms "proximity" and "distance" are used in statistical software packages, but have not gained wide acceptance in the chemical literature. Similarity and dissimilarity can in principle lead to different rankings.

Usually, the denominator, if present in a similarity measure, is just a normalizer; it is the numerator that is indicative of whether similarity or dissimilarity is being estimated, or both. The characteristics chosen for the description of the objects being compared are interchangeably called descriptors, properties, features, attri-butes, qualities, observations, measurements, calculations, etc. In the formulations above, the terms "matches" and "mismatches" refer to qualitative characteristics, e.g., binary ones (those which take one of two values: 1 (present) or 0 (absent)), while the terms "overlap" and "difference" refer to quantitative characteristics, e.g., those whose values can be arranged in order of magnitude along a one-dimen-sional axis.

## 6.4.2
### Similarity Measures

In order to compare two chemical (or any other) objects, e.g., two molecules, we need a measure. Plenty of similarity measures have been proposed; they are listed in Table 6-2. Generally speaking these measures can be divided into two cases: one of qualitative characteristics, and the other of quantitative characteristics. Here we consider these two cases.

Following Bradshaw [17], we can give the definition of a similarity measure as follows: Consider two objects A and B, $a$ is the number of features (characteristics) present in A and absent in B, $b$ is the number of features absent in A and present in B, $c$ is the number of features common to both objects, and $d$ is the number of features absent from both objects. Thus, $c$ and $d$ measure the present and the absent matches, respectively, i.e., similarity; while $a$ and $b$ measure the corresponding mismatches, i.e., dissimilarity. The total number of features is $n = a + b + c + d$.

The total number of bits set on A is $a + c$, and the total number of bits set on B is $b + c$. These totals form the basis of an alternative notation that uses $a$ instead of $a + c$, and $b$ instead of $b + c$ [16]. This notation, however, lumps together similarity and dissimilarity "components" – a disadvantage when interpreting a similarity measure.

Consequently, we can construct a similarity measure intuitively in the following way: all matches $c + d$ relative to all possibilities, i.e., matches plus mismatches $(c + d) + (a + b)$, yields $(c + d) / (a + b + c + d)$, which is called the simple matching coefficient [18], and equal weight is given to matches and mismatches. (Normalized similarity measures are called similarity indices or coefficients; see, e.g., Ref. [19].) When absence of a feature in both objects is deemed to convey no information, then $d$ should not occur in a similarity measure. Omitting $d$ from the above similarity measure, one obtains the Tanimoto (alias Jaccard) similarity measure (Eq. (8); see Ref. [16] and the citations therein):

$$T = c/a + b + c \tag{8}$$

For examples of different types of similarity measures, see Table 6-2. The Tanimoto similarity measure is monotonic with that of Dice (alias Sorensen, Czekanowski), which uses an arithmetic-mean normalizer, and gives double weight to the "present" matches. Russell/Rao (Table 6-2) add the matching absences to the normalizer in Tanimoto; the cosine similarity measure [19] (alias Ochiai) uses a "geometric mean" normalizer.

To construct dissimilarity measures, one uses mismatches: Here $a + b$ is the Hamming (Manhattan, taxi-cab, city-block) distance, and $\sqrt{(a + b)}$ is the Euclidean distance.

We should mention here that using just similarity or dissimilarity in a similarity measure might be misleading. Therefore, some composite measures using both similarity and dissimilarity have been developed. These are the Hamann and the Yule measures (Table 6-2). A simple product of (1 – Tanimoto) and squared Eucli-

**Table 6.2.** Different types of similarity measures derived from both qualitative and quantitative characteristics. $a$ is the number of features of an object A, $b$ is the number of features of object B, $c$ is the number of features common to A and B, $d$ is the number of features absent from A and B (see the text for more details).

| Type | Name(s) | For qualitative characteristics | For quantitative characteristics | |
|---|---|---|---|---|
| | | | Summation form | Integration form |
| Similarity measures | number of matches overlap | $c$ | $\sum_{j=1}^{n} X_{jA}X_{jB}$ | $S_{A,B} = \iint \Gamma_A^*(\mathbf{r},\mathbf{r}')\Omega(\mathbf{r},\mathbf{r}')\Gamma_B(\mathbf{r},\mathbf{r}')d\mathbf{r}d\mathbf{r}'$ |
| | simple matching coefficient | $(c+d)/(a+b+c+d)$ | | |
| | Tanimoto, Jaccard | $c/a+b+c$ | $\dfrac{\sum_{j=1}^{n} X_{jA}X_{jB}}{\left(\sum_{j=1}^{n} X_{jA}X_{jA} + \sum_{j=1}^{n} X_{jB}X_{jB} - \sum_{j=1}^{n} X_{jA}X_{jB}\right)}$ | |
| | Dice, Sorensen, Czekanowski, Hodgkin–Richards | $c/0.5[(a+c)+(b+c)]$ | $2\sum_{j=1}^{n} X_{jA}X_{jB} \Big/ \left(\sum_{j=1}^{n} X_{jA}X_{jA} + \sum_{j=1}^{n} X_{jB}X_{jB}\right)$ | $S_{A,B}/0.5(S_{A,A} + S_{B,B})$ |
| | cosine, Ochiai, Carbo | $c/\sqrt{(a+c)(b+c)}$ | $\sum_{j=1}^{n} X_{jA}X_{jB} \Big/ \sqrt{\sum_{j=1}^{n} X_{jA}X_{jA} \sum_{j=1}^{n} X_{jB}X_{jB}}$ | $S_{A,B}/\sqrt{S_{A,A}S_{B,B}}$ |
| | Russell/Rao | $c/(a+b+c+d)$ | | |
| | Rogers/Tanimoto | $(c+d)/(2a+2b+c+d)$ | | |
| | Baroni–Urbani/Buser | $\left(\sqrt{cd}+c\right)/\left(\sqrt{cd}+a+b+c\right)$ | | |
| | Kulczynski-2 | $\dfrac{1}{2}\left(\dfrac{c}{a+c}+\dfrac{c}{b+c}\right)$ | | |

**Table 6.2.** (cont.)

| Type | Name(s) | For qualitative characteristics | For quantitative characteristics | |
|---|---|---|---|---|
| | | | *Summation form* | *Integration form* |
| Dissimilarity measures | Hamming, Manhattan, taxi-cab, city-block distance | $(a+b)$ | | |
| | mean Hamming distance | $(a+b)/(a+b+c+d)$ | $\dfrac{1}{n}\sum_{j=1}^{n}\vert X_{jA}-X_{jB}\vert$ | |
| | Euclidean distance | $\sqrt{(a+b)}$ | $\sqrt{\sum_{j=1}^{n}(X_{jA}-X_{jB})^2}$ | |
| | squared mean, Euclidean distance | $(a+b)/(a+b+c+d)$ | $\dfrac{1}{n}\sum_{j=1}^{n}(X_{jA}-X_{jB})^2$ | |
| | power distance (user-defined $p$, $r$) | | $\left(\sum_{j=1}^{n}\vert X_{jA}-X_{jB}\vert^{p}\right)^{1/r}$ | |
| | Soergel distance | $(a+b)/(a+b+c)$ | $\sum_{j=1}^{n}\vert X_{jA}-X_{jB}\vert \Big/ \sum_{j=1}^{n}\max(X_{jA}X_{jB})$ | |
| | Chebychev distance | | $\max_{j=1}^{n}\vert X_{jA}-X_{jB}\vert$ | |
| | pattern difference | $ab/(a+b+c+d)^2$ | | |
| | variance | $(a+b)/4(a+b+c+d)$ | | |
| | size | $(a-b)^2/(a+b+c+d)^2$ $(a-b)/(a+b+c+d)$ | | |
| | shape | $-[(a-b)/(a+b+c+d)]^2$ | | |

$$D_{A,B} = \left(\iint\vert\Gamma_A(\mathbf{r},\mathbf{r}') - \Gamma_B(\mathbf{r},\mathbf{r}')\vert^2 d\mathbf{r}d\mathbf{r}'\right)^{1/2}$$

**Table 6.2.** (cont.)

| Type | Name(s) | For qualitative characteristics | For quantitative characteristics | |
|---|---|---|---|---|
| | | | Summation form | Integration form |
| Composite measures | Hamann | $(c + d - a - b)/(a + b + c + d)$ | | |
| | Yule | $(cd - ab)/(cd + ab)$ | | |
| | Pearson | $\dfrac{(cd - ab)}{\sqrt{(a + c)(b + c)(a + d)(b + d)}}$ | $\dfrac{\displaystyle\sum_{j=1}^{n}(x_{jA} - \bar{x}_{jA})(x_{jB} - \bar{x}_{jB})}{\sqrt{\displaystyle\sum_{j=1}^{n}(x_{jA} - \bar{x}_{jA})^2 \sum_{j=1}^{n}(x_{jB} - \bar{x}_{jB})^2}}$ | |
| | dispersion | $(cd - ab)/(a + b + c + d)^2$ | | |
| | McConnaughey | $(c^2 - ab)/(a + c)(b + c)$ | | |
| | Stiles | $\log_{10}\left(\dfrac{n(|cd - ab| - n/2)^2}{(a + c)(b + c)(a + d)(b + d)}\right)$ | | |
| | Dixon | $(1 - \text{Tanimoto}) \times$ (squared Euclidean distance) | | |
| | Grotch | $(a + b) - \mu c$ | | |
| Asymmetric measures ($a$ and $b$ are weighed unequally) | similarity measure | Tversky | $c/(\alpha a + \beta b + c)$ | |
| | composite measure | Tversky contrast model | $\theta c - \alpha a - \beta b$ | |
| | similarity measure | Simpson | $c/-[(a + c),(b + c)]$ | |

dean distance is used by Dixon (Table 6-2). The Grotch metric is $(a + b) - \mu c$, where $\mu$ weights the relative contribution of the similarity component.

Asymmetry in a similarity measure is the result of asymmetrical weighing of a dissimilarity component – multiplication is commutative by definition, difference is not. By weighing $a$ and $b$, one obtains asymmetric similarity measures, including the Tversky similarity measure $c / (\alpha a + \beta b + c)$, where $\alpha$ and $\beta$ are user-defined constants. The Tversky measure can be regarded as a generalization of the Tanimoto and Dice similarity measures; like them, it does not consider the absence matches $d$. A particular case is $c/(a + c)$, which measures the number of common features relative to all the features present in $A$, and gives zero weight to $b$.

We shall explore the quantitative measures further. They are presented in the third column of Table 6-2.

Overlap is usually expressed mathematically by a component-by-component multiplication $X_{jA}X_{jb}$ followed by summation (integration). Thus, the measures using quantitative characteristics could be reduced to those having qualitative characteristics in the following way: $\sum\limits_{j=1}^{n} X_{jA}X_{jB}$ to $c$, $\sum\limits_{j=1}^{n} X_{jA}X_{jA}$ to $(a + c)$, and $\sum\limits_{j=1}^{n} X_{jB}X_{jB}$ to $(b + c)$, which is easily verified by substituting 0 and 1 in these expressions. The number of absences $d$ does not figure in them, and there seems to be no way to estimate empty overlap within this mathematical apparatus. The summation forms of some of the similarity measures (the Tanimoto, Dice, and cosine coefficients) are presented in Table 6-2.

Replacing summation by integration, one obtains the integration forms of the above-described similarity measures (Table 6-2). Using different characteristics to describe the objects being compared, one obtains different similarity measures. A typical example is the Carbo [20] similarity measure which is given by Eq. (9). where $\rho_A(\mathbf{r})$ and $\rho_A(\mathbf{r}')$ are the electron density functions of quantum objects A and B, weighted by a positive definite operator $\Omega(\mathbf{r}, \mathbf{r}')$, chosen either as the Dirac function $\delta(\mathbf{r} - \mathbf{r}')$ or the Coulomb operator $|\mathbf{r} - \mathbf{r}'|^{-1}$, etc.

$$S_{A,B} = \iint \rho_A(\mathbf{r})\Omega(\mathbf{r},\mathbf{r}')\rho_B(\mathbf{r}')d\mathbf{r}d\mathbf{r}' \tag{9}$$

The resulting similarity measures are overlap-like $S_{A,B} = \int \rho_A(\mathbf{r}) \rho_B(\mathbf{r}) \, d\mathbf{r}$, Coulomb-like, etc. The Carbo similarity coefficient is obtained after geometric-mean normalization $S_{A,B}/\sqrt{S_{A,A}S_{B,B}}$ (cosine), while the Hodgkin–Richards similarity coefficient uses arithmetic-mean normalization $S_{A,B}/0.5\,(S_{A,A}+ S_{B,B})$ (Dice). The Cioslowski [18] similarity measure NOEL – Number of Overlapping ELectrons (Eq. (10)) – uses reduced first-order density matrices (one-matrices) rather than density functions to characterize A and B. No normalization is necessary, since NOEL has a direct interpretation, at the Hartree–Fock level of theory.

$$S_{A,B} = \iint \Gamma_A^*(\mathbf{r},\mathbf{r}')\Gamma_B(\mathbf{r},\mathbf{r}')d\mathbf{r}d\mathbf{r}' \tag{10}$$

The difference between two objects is usually expressed with component-by-component subtraction $X_{jA}-X_{jb}$, followed by summation (integration). The mathematical term for this is "distance in the $n$-dimensional descriptor space", and is frequently used as a synonym of dissimilarity. Distances are asymmetric unless special care is taken, e.g., by taking the absolute value of the difference, the square of the distance, an arithmetic mean, a geometric mean, etc. Distances are metric if they satisfy a number of mathematical conditions (of which symmetry is one). These were previously considered all-important for similarity measures, but now tend to be relaxed in favor of more pragmatic considerations such as ease of computation and general usefulness. Thus, in the limiting case of binary characteristics, $\sum_{j=1}^{n}|X_{jA} - X_{jB}|$ leads to $(a + b)$. The summation and integration forms of some dissimilarity measures are presented in Table 6-2.

Once we have the measures, we have to apply them to chemical objects. Objects of interest to a chemist include molecules, reactions, mixtures, spectra, patents, journal articles, atoms, functional groups, and complex chemical systems. Most frequently, the objects studied for similarity/dissimilarity are molecular structures.

In order to apply the similarity measures to the objects, the latter must be described by some *characteristics*.

Any set of characteristics can be used to describe the compared objects. Object characteristics can be roughly classified as global and local, with the latter providing sufficient local information for object alignment/superposition to be effected. Local similarity can only be estimated when local characteristics are used. Global characteristics are at the other extreme, providing overall descriptions of objects.

Examples of global characteristics are the atom pair (*ap*) and the topological torsion (*tt*). Atom pairs are defined as substructures of the form $AT_i - AT_j -$ (distance), where "(distance)" is the distance in bonds along the shortest path between an atom of type $AT_i$ and an atom of type $AT_j$. Atom types encode the species of the given atom, the number of non-hydrogen atoms attached to it, and the number of incident $\pi$-bonds. For instance, "n21o1005" is an atom pair of a nitrogen with two non-hydrogen neighbors and one $\pi$-bond, five bonds away from an oxygen with one neighbor and no $\pi$-bonds. Topological torsions are of the form $AT_i - AT_j - AT_k - AT_l$, where $i$, $j$, $k$, and $l$ are consecutively bonded distinct atoms and the atom types are as described above. All of the *ap*s and/or *tt*s in a molecule are counted to form a frequency vector.

Table 6-3 shows an example of a molecule parsed into atom pairs and topological torsions.

**Table 6.3.** Sample molecules acetone and isobutene described by atom pair (*ap*) descriptors. *ap*'s are defined as substructures of the form $AT_i - AT_j -$ distance, where (distance) is the distance in bonds along the shortest path between an atom of type $AT_i$ and an atom of type $AT_j$ (see text).

| | Unique ap | acetone | isobutene | Descriptor average |
|---|---|---|---|---|
| 1 | c10c1002 | 1 | 1 | 1 |
| 2 | c10o1102 | 2 | 0 | 1 |
| 3 | c10c1102 | 0 | 2 | 1 |
| 4 | c31c1001 | 2 | 2 | 2 |
| 5 | c31o1101 | 1 | 0 | 0.5 |
| 6 | c31c1101 | 0 | 1 | 0.5 |

### 6.4.3
### The Similarity Search Process

Similarity searching is the database implementation of the similarity concept. Some of the steps involved in similarity searching are overviewed next, in the context of chemoinformatics.

#### 6.4.3.1   Object Selection

The most common objects of interest to a chemist are molecules. Some sources of drug-like compounds are the MDL Drug Data Report (MDDR) a licensed database compiled from the patent literature containing about 115 000 compounds, as well as the database of the National Cancer Institute (NCI), containing about 250 000 compounds. Molecules in the MDDR are assigned a "therapeutic category" by the vendor. There are 647 therapeutic categories. MDDR-3D is also available. The MDDR is a commercial database, whereas the NCI database is freely available at *http://dtp.nci.nih.gov/docs/3D_database/structural_information/structural_data.html* and contains both structural information and biological data. The biological database is formed of three files, which contain data from different types of measurements – TGI, LC50, and GI50.

An example of a fragment-based search space is:

"…a large set of diverse fragments together with generic definitions of how the fragments can be combined to molecules … a chemistry space created by shredding the World Drug Index into small fragments. The space contains about 17 000 fragments which can be connected to each other via 12 different link types."[21]

This virtual search space can be searched using a feature tree descriptor.

Another approach employing the autocorrelation coefficients as descriptors was suggested by Gasteiger et al. [22]. They used the neural networks as a working tool for solving a similarity problem.

Reactions can be considered as composite systems containing reactant and product molecules, as well as reaction sites. The similarity of chemical structures is defined by generalized reaction types and by gross structural features. The similarity of reactions can be defined by physicochemical parameters of the atoms and bonds at the reaction site. These definitions provide criteria for searching reaction databases [23].

Mixtures containing up to several thousand distinct chemical entities are often synthesized and tested in mix-and-split combinatorial chemistry. The descriptor representation of a mixture may be approximated as the descriptor average of its individual component molecules, e.g., using atom-pair and topological torsion descriptors.

### 6.4.3.2 **Descriptor Selection and Encoding**

The atom pair, *ap*, and topological torsion, *tt*, descriptors are selected for illustrative purposes in the similarity searching context.

A limitation of the *ap* and *tt* descriptors is the specificity of the atom typing, e.g., benzoic acid and phenyltetrazole would not be perceived as very similar, even though carboxylates and tetrazoles are both anions at physiological pH.

A fuzzier atom type participating in these descriptors has been defined that is pharmacologically relevant – the physicochemical type at near-neutral pH [24], which is one of the following seven binding property classes: 1 = cation; 2 = anion; 3 = neutral hydrogen-bond donor; 4 = neutral H-bond acceptor; 5 = polar atom (atoms which are both donors and acceptors, e.g., hydroxy oxygen or either donor or acceptor via tautomerization, e.g., the nitrogens of imidazole); 6 = hydrophobe; 7 = other (nonpolar atoms in a polar environment or polar atoms that cannot accept or donate H-bonds). The physicochemical atom type, however, is too fuzzy for an atomic descriptor for the purpose of identifying common substructures (symmetric local similarity based on atoms and topological distances as descriptors). In this case, an atom type has been defined as a string containing the chemical element (all halogens equivalent to the "element" Hal), the number of incident $\pi$-bonds, and the physicochemical type.

Two other atomic properties have been used in the definition of atom type, thereby increasing its fuzziness relative to that in the *ap* and *tt* descriptors – atomic log *P* contribution (yielding hydrophobic pairs, *hps*, and torsions, *hts*) and partial atomic charges (charge pairs, *cps*, and charge torsions, *cts*).

Increasing the fuzziness of object description reduces the number of descriptors used and broadens the scope of a similarity search. At the same time, increasing fuzziness may reduce the discriminatory power of descriptors to unacceptable levels. Therefore it is desirable to be able to control the degree of fuzziness of descriptors.

### 6.4.3.3 **Similarity Measure Selection**

In general, different similarity measures yield different rankings, except when they are monotonic. Improved results are obtained by using data fusion methods to combine the rankings resulting from different coefficients.

Empirically, the Dice coefficient has worked better than cosine similarity in retrieving actives and is the standard choice for use with the *ap* and *tt* descriptors.

Asymmetric similarity measures allow fuzzy super- and substructure searching. A substructure search is defined as looking for structures containing the given query and a superstructure search is defined as looking for structures embedded in the given query. In both cases asymmetric local similarity is estimated.

### 6.4.3.4 **Query Object Specification**

The user either enters, or copies, a query object at search time, using the graphical user interface.

### 6.4.3.5 **Similarity Scores**

To evaluate the performance of the descriptors one needs a database of compounds for which the biological activities are known, e.g. .either the MDDR or the NCI databases. Queries are selected that are typical of a drug-like molecule and from therapeutic categories that

1. contain a large enough number of actives (e.g., $>50$) for reasonable statistics,
2. have several chemical classes present in them, and
3. are fairly specific

so that most of the molecules probably work by the same mechanism.

The atom pair (*ap*) and topological torsion (*tt*) descriptors and their fuzzy binding property analogs *bp* and *bt* are again selected for illustrative purposes [24, 25].

The connection table of the query object (similarity probe) is processed to obtain the set of atom pairs, and then the database file is scanned to evaluate the similarity between the query and each of the database structures. The maximum number of structures that the program will select is specified, as well as the minimum similarity score that a database compound must show to be selected. Within these limits, the program will select from the database the structures that are most similar (with the highest similarity value) to the query and will create an output file of compound numbers and similarity values, sorted by decreasing similarity, for the selected compounds.

### 6.4.3.6 **Application Areas**

In chemical similarity searching, browsing of ranked similar objects may be used for evaluation of the uniqueness of proposed or newly synthesized compounds, finding starting materials or intermediates in synthesis design, or handling of chemical reactions and mixtures – finding the right chemicals for one's needs, even if one does not know exactly what one is looking for.

"Direct" property prediction is a standard technique in drug discovery. "Reverse" property prediction can be exemplified with chromatography application databases that contain separations, including method details and assigned chemical structures for each chromatogram. Retrieving compounds present in the database that are similar to the query allows the retrieval of suitable separation conditions for use with the query (method selection).

Automated, miniaturized, and parallelized synthesis and testing (combinatorial chemistry/high-throughput screening) are accelerating the development of a complex of methods for data mining and computer screening (virtual screening) of object libraries. Clusters of objects are recognized (cluster analysis) on the basis of the estimation of the distances in the descriptor space (dissimilarities). In the case of object selection, classes that are as diverse as possible are selected so that all the different types of properties (e.g., bioactivities) within a larger collection are sampled using as few objects as possible (diversity analysis). Key chemical features and the spatial relationships among them that are considered to be responsible for a desired biological activity may be identified (pharmacophore recognition) using local similarity, e.g., via common substructures in sets of active molecules; pharmacophore searching in 3D databases (see below) may be carried out using a pharmacophore as the query. Shape similarity of ligands to a receptor site (ligand docking) may be used for finding structures that fit into proteins.

### 6.5
### Three-Dimensional Structure Search Methods

Chemists know that 2D representation of molecular moieties gives a very rough picture of their real-world structure. While for some practical applications this representation is sufficient, for most modern investigations in all areas of molecular design, 3D structure representation and 3D structure search are highly mandatory [26]. However, the presence of one more degree of freedom (Z-coordinate) and the free rotations around single bonds (conformational flexibility; see Section 2.9 of this Textbook and Chapter II, Section 7.2 in the Handbook) is conducive to an immense complexity related to these phenomena. Nevertheless, extensive work has been done since the early 1990s in this direction.

The first task was the creation of large 3D chemical structure databases. By devising so-called fast Automatic 3D model builder, software such as the CORINA [27, 28] and CONCORD [29, 30] programs resulted in a boom in 3D database development (see Section 2.9 in this book and Chapter II, Section 7.1 in the Handbook). A subsequent step was the development of fast

3D search approaches. In general this step followed the existing 2D search methods, such as atomly-atom-mapping, maximal common substructure, 2D keys, fragments, etc., by substituting them with their 3D counterparts and extending them with some new descriptors such as interatomic distances, atom-pair descriptors and pharmacophore triplets [31]. Further, conformational flexibility has been considered, although in a constrained form. Thus, in ROTATE (see Section 2.9 in this book and Chapter II, Section 7.2 in the Handbook), information from the Cambridge Structural Database (CSD) is used as constraints for the conformers generated.

Full structure search can be developed by using similar approaches to those employed in the case of 2D structure search. Thus, some topological indices can be modified in such a way that they include geometrical information. For example, the global index given by Eq. (4) can be modified to Eq. (11), where $R_{ij}$ are real interatomic distances.

$$CTI = \frac{\sum L_i L_j}{R_{ij}} \tag{11}$$

The RDF code discussed in Chapter VIII, Sections 2 and 3 of the Handbook) also can be used to this end.

3D similarity search methods are quite well developed. Thus, methods which attempt to find overlapping parts (atoms and functional groups) of the molecular moieties studied were reported first [31]. As discussed above for the case of 2D searching, these methods are of combinatorial complexity. To reduce this complexity some field-based methods have been introduced. In this case, the overlap of the fields of two structures is considered as a similarity measure.

3D substructure search is usually known as pharmacophore searching in QSAR. Generally speaking, there are two major approaches to it: topological and chemical function queries. These two techniques are based on a slightly different philosophy and usually provide different results [31].

In all of the 3D search methods the conformational flexibility creates considerable difficulties. Large databases of multiple conformations for each structure have been developed which make the solution of this problem possible.

**Essentials:**

- A substructure search algorithm is usually the first step in the implementation of other important topological procedures for the analysis of chemical structures such as identification of equivalent atoms, determination of maximal common substructure, ring detection, calculation of topological indices, etc.
- The search for structural fragments (substructures) is very important in medicinal chemistry, QSAR, spectroscopy and many other fields in the process of pharmacophore, chromophore or other -phore perceptions.
- The similarity property principle states: "structurally similar molecules are expected to exhibit similar physical properties or biological activities."

- Similarity search appears as an extremely useful tool for computer-aided structure elucidation as well as molecular design.
- 3D substructure search is usually known as pharmacophore searching in QSAR. In all of the 3D search methods the conformational flexibility creates considerable difficulties.

## Selected Reading

- J.L. Wisniewski, Nomenclature: automatic generation and conversion, in *Encyclopedia of Computational Chemistry*, eds. P. von R. Schleyer, N.L. Allinger, T. Clark, J. Gasteiger, P.A. Kollman, H.F. Schaefer III, P.R. Schreiner, Wiley, Chichester, UK, **1998**, pp. 1881–1894.
- M.F. Lynch, *Chemical Information Systems*, eds. J.E. Ash and E. Hyde, Ellis Horwood, **1985**, pp. 88–93.
- P. Willett, Structure similarity measures for database searching, in *Encyclopedia of Computational Chemistry*, eds. P. von R. Schleyer, N.L. Allinger, T. Clark, J. Gasteiger, P.A. Kollman, H.F. Schaefer III, P.R. Schreiner, Wiley, Chichester, UK, **1998**, pp. 2748–2756.
- P. Willet, *Three-Dimensional Chemical Structure Handling*, Wiley, New York, **1991**.

## Interesting Websites

- *www2.chemi.uni-erlangen.de*
- *http://dtp.nci.nih.gov/docs/3D_database/structural_information/structural_data.html*
- *http://cactus.nci.nih.gov/ncidb2/download.html*
- *www.daylight.com*
- *www.queryplus.com*
- *www.isinet.com/isi/hot/essays/chemicalliterature/15.html*
- *http://citeseer.nj.nec.com/santini99similarity.html*
- *www.mol-net.de*
- *www.tripos.com*

## Available Software

DayCart™ is a software cartridge which offers a range of operation on an Oracle database, such as complete structure, similarity, and substructure search. The software can be obtained from Daylight Chemical Information Systems, Inc. (Mission Viejo CA); URL: *www.daylight.com*

CACTVS is a chemical information system which provides 2D and 3D complete structure, substructure, and similarity search on plain files of structures. The software can be obtained from Molecular Networks GmbH (Erlangen, Germany); URL: *www.mol-net.de*

C@ROL is a web-based warehouse with possibilities for full structure, substructure, and similarity searching. The software can be obtained from Molecular Networks GmbH (Erlangen, Germany); URL: *www.mol-net.de*

CORINA and CONCORD are 3D-builder programs which can be obtained from Molecular Networks GmbH (Erlangen, Germany) URL; URL: *www.mol-net.de* – and from Tripos, Inc., St. Louis, MO, USA; URL: *http://www.tripos.com*, respectively.

# References

[1] W.J. Wiswesser, *A Line-Formula Chemical Notation*, Thomas Crowell, New York, **1954**.

[2] D. Weininger, *J. Chem. Inf. Comput. Sci.* **1988**, *28*, 31–36.

[3] R.C. Read, D.G. Corneil, *J. Graph Theory* **1977**, *1*, 339–363.

[4] International Union of Pure and Applied Chemistry, Commission on the Nomenclature of Organic Chemistry, *Nomenclature of Organic Chemistry, Sections A, B, C, D, E, F, and H*, **1979**, Pergamon, Oxford.

[5] J.L. Wisniewski, Nomenclature: automatic generation and conversion, in *Encyclopedia of Computational Chemistry*, eds. P. von R. Schleyer, N.L. Allinger, T. Clark, J. Gasteiger, P.A. Kollman, H.F. Schaefer III, P.R. Schreiner, Wiley, Chichester, UK, **1998**, 1881–1894.

[6] P.A. Demirev, A. Dyulgerov, I.P. Bangov *J .Math. Chem.* **1991**, *8*, 367–382.

[7] J. Gasteiger, M. Marsili, *Tetrahedron* **1980**, *36*, 3219–3228.

[8] W.D. Ihlenfeldt, J. Gasteiger, *J. Comput. Chem.* **1994**, *15*, 793–813.

[9] M. Garey, D. Johnson, *Computers and Intractability; A Guide to the Theory of NP-Completeness*; W.H. Freeman, New York, **1979**.

[10] L.C. Ray, R.A. Kirsch, *Science* **1957**, *126*, 814–819.

[11] S.J. Russel, P. Norvig (eds.), *Artificial Intelligence, a Modern Approach*, Prentice-Hall, New Jersey, **1996**, Chapter 3.

[12] J. Xu, *J. Chem. Inf. Comput. Sci.* **1996**, *36*, 25–34.

[13] T.K. Ming, S.J. Tauber, *J. Chem. Doc.* **1971**, *11*, 47–51.

[14] A. von Scholley, *J. Chem. Inf. Comput. Sci.* **1984**, *24*, 235–241.

[15] M.F. Lynch, *Chemical Information Systems*, J.E. Ash, E. Hyde eds., Ellis Horwood, Chichester, **1985**, 88–93.

[16] P. Willett, J.M. Barnard, G.M. Downs, *J. Chem. Inf. Comput. Sci.* **1998**, *38*, 983-996.

[17] J. Bradshaw, *www.daylight.com/meetings/emug01/Bradshaw/Similarity/YAMS.html*.

[18] P.H.A. Sneath, R.R. Sokal, *Numerical taxonomy*, W.H. Freeman and Co., San Francisco, **1973**.

[19] J. Cioslowski, in *Encyclopedia of Computational Chemistry*, P. von R. Schleyer, N.L. Allinger, T. Clark, J. Gasteiger, P.A. Kollman, H.F. Schaefer III, P. R. Schreiner (eds.), Wiley, Chichester, UK, **1998**, 892-905.

[20] R. Carbo, M. Arnau, L. Leyda, *Int. J. Quant. Chem.* **1980**, *17*, 1185-1189.

[21] M. Rarey, M. Stahl, *J. Comp.-Aid. Mol. Des.* **2001**, *15*, 497-520.

[22] J. Sadowski, M. Wagener, J. Gasteiger, *Angew. Chem. Int. Ed. Engl.* **1996**, *34*, 2674-2677.

[23] J. Gasteiger, W.-D. Ihlenfeldt, R. Fick, J.R. Rose, *J. Chem. Inf. Comput. Sci.*, **1992**, *32*, 700-717.

[24] S.K. Kearsley, S. Sallamack, E.M. Fluder, J.D. Andose, R.T. Mosley, R.P. Sheridan, *J. Chem. Inf. Comput. Sci.* **1996**, *36*, 118-127.

[25] R.D. Hull, E.M. Fluder, S.B. Singh, R.B. Nachbar, S.K. Kearsley, R.P. Sheridan, *J. Med. Chem.* **2001**, *44*, 1185-1191.

[26] P. Willett, "Three-Dimensional Chemical Structure Handling", Wiley, New York, **1991**.

[27] J. Sadowski, J. Gasteiger, *Chemical Reviews* **1993**, *7*, 2567-2581.

[28] CORINA Version 2.6 can be accessed at *http://www2.chemie.uni-erlangen.de/ services/* and is available from Molecular Networks GmbH, Erlangen, Germany *(http://www.mol-net.de)*.

[29] R. S. Pearlman, *Chem. Des. Auto. News* **1987**, *2*, 6.

[30] CONCORD is available from Tripos, Inc., St. Louis, MO, USA *(http:// www.tripos.com)*.

[31] O. Güner, D.R. Henry, "*3D Structure Searching*" in *Encyclopedia of Computational Chemistry*, P. von R. Schleyer, N.L. Allinger, T. Clark, J. Gasteiger, P.A. Kollman, H.F. Schaefer III, P.R. Schreiner, eds., Wiley, Chichester, UK, **1998**, 2988-3003.

# 7
# Calculation of Physical and Chemical Data

## Learning Objectives

- To be able to calculate molecular properties by additivity schemes based on con-
tributions by structural subunits
- To become familiar with the estimation of thermochemical data
- To understand the estimation of average drug–receptor binding energies
- To become familar with the algorithm for charge calculation by partial equalization
of orbital electronegativity (PEOE) and by a modified Hückel Molecular Orbital
method
- To appreciate residual electronegativity as a measure of the inductive effect
- To follow a simple scheme for calculating the polarizability effect
- To know how linear equations can be used for calculation of enthalpies of gas-
phase reactions
- To understand the basic concepts of force field calculations
- To see the contributions to the molecular mechanics potential energy function and
their mathematical representation
- To get an overview of the currently available software for molecular mechanics cal-
culations with their strengths, weaknesses, and application areas
- To understand the importance of investigating the dynamical behavior of mole-
cules
- To have an overview of the algorithms and basic concepts used to perform molec-
ular dynamics simulations
- To consider exemplary state-of-the-art applications of MD simulations
- To become familiar with the different quantum mechanical methods
- To know which properties can be derived from quantum mechanical methods
- To ponder on the future of quantum mechanics in chemoinformatics

**7.1**
**Empirical Approaches to the Calculation of Properties**

*J. Gasteiger*

**7.1.1**
**Introduction**

Molecular structures consist of atoms that are held together by bonds. The forces between molecules are typically one or two orders of magnitude lower than the forces that hold atoms together in molecules. The picture of an atom in a molecule is intuitively quite appealing and it raises the question of whether atoms in their bound state in a molecule bring with them properties that can still be discerned. Or, conversely, can we break up a molecular property into contributions of its constituent atoms and calculate molecular properties by adding these contributions from its atoms? It will be shown that this can be done with satisfactory accuracy for only a few molecular properties. However, the idea of calculating molecular properties by contributions from its atoms can be extended to considering contributions from structural units larger than atoms, such as those of bonds or of groups.

In fact, there is a hierarchy in calculating molecular properties by additivity of atomic, bond, or group properties, as was pointed out some time ago by Benson [1, 2]. The larger the substructures that have to be considered, the larger the number of increments that can be derived and the higher the accuracy in the values obtained for a molecular property.

A basic assumption in such additivity schemes is that the interactions between the atoms of a molecule are of a rather short-range nature. This fact can be expressed in a more precise manner: The law of additivity can be expressed in a chemical equation [1]. Let us consider the atoms (or groups) X and Y attached to a common skeleton, S, and also the redistribution of these atoms on that skeleton as expressed by Eq. (1).

$$X–S–X \ + \ Y–S–Y \quad \rightleftharpoons \quad 2 \ X–S–Y \tag{1}$$

The law of additivity then says that the sum of the properties of the molecules on the right-hand side is the same as the sum of the properties on the left-hand side of Eq. (1).

When additivity of atomic properties is valid then the skeleton S disappears and Eq. (1) can be rewritten as Eq. (2).

$$X–X \ + \ Y–Y \quad \rightleftharpoons \quad 2 \ X–Y \tag{2}$$

The sum of the properties of the diatomic species $X_2$ and $Y_2$ is the same as twice the property of XY. This is the zero-order approximation to additivity rules.

If $S$ is a single atom or a group of atoms with the bonds attached to the same atom (such as a $CH_2$ group), then we have the additivity of bond properties, the first-order approximation, as given by Eq. (3).

$$X-CH_2-X + Y-CH_2-Y \rightleftarrows 2\ X-CH_2-Y \tag{3}$$

When group additivity is valid, $S$ consists of a group with the bonds of X and Y attached to two adjacent atoms as in Eq (4).

$$X-CH_2-CH_2-X + Y-CH_2-CH_2-Y \rightleftarrows 2\ X-CH_2-CH_2-Y \tag{4}$$

This is the second-order approximation to additivity rules.

Equations (2)–(4) clearly illustrate the increase in distance in the interactions between atoms X and Y in going from the additivity of atomic, to bond, and further to group contributions.

## 7.1.2
### Additivity of Atomic Contributions

Clearly, there is one molecular property that can be exactly calculated from the contributions of its constituent atoms: the molecular weight, or, more correctly, the molecular mass, which is exactly the sum of the masses of its constituent atoms.

However, there are other molecular properties, $P_m$, such as molar volume, molar refraction [3], diamagnetic susceptibility [4], and parachor [5], that can be obtained to sufficient accuracy from contributions, $p_i$, of its $N$ atoms (Eq. (5)).

$$P_m = \sum_{i=1}^{N} p_i \tag{5}$$

Investigations to find such additive constituent properties of molecules go back to the 1920s and 1930s with work by Fajans [6] and others. In the 1940s and 1950s the focus had shifted to the estimation of thermodynamic properties of molecules such as heat of formation, $\Delta H_f^\circ$, entropy $S^\circ$, and heat capacity, $C_p^\circ$.

As Benson [1] pointed out, the additivity of atomic contributions is an insufficient approximation for estimating enthalpy, $\Delta H_f^\circ$, for it would lead to unacceptably large errors. However, the error in the estimation of the molar heat capacity $\Delta C_p^\circ$, by Eq. (5) is not greater than $\pm 14.0$ J/mol K and is usually closer to $\pm 6.0$ J/mol K. In a similar fashion the error $\Delta S^\circ$ for Eq. (5) is rarely higher than 20.0 J/mol K and is usually in the region of 8.0 J/mol K.

For any molecule, additivity of atomic properties requires as many variables as there are different atom types contained in the molecule. For example, for acetic acid, $C_2H_4O_2$, three different atomic increments are needed, one each for a carbon, a hydrogen, and an oxygen atom.

### 7.1.2.1 **Hybridization States**

The next step towards increasing the accuracy in estimating molecular properties is to use different contributions for atoms in different hybridization states. This simple extension is sufficient to reproduce mean molecular polarizabilities to within 1–3 % of the experimental value. The estimation of mean molecular polarizabilities from atomic refractions has a long history, dating back to around 1911 [7]. Miller and Savchik were the first to propose a method that considered atom hybridization in which each atom is characterized by its state of atomic hybridization [8]. They derived a formula for calculating these contributions on the basis of a theoretical interpretation of variational perturbation results and on the basis of molecular orbital theory.

Kang and Jhon [9] showed that mean molecular polarizabilities, $\bar{\alpha}$, can be estimated from atomic hybrid polarizabilities, $\alpha_i$, by a simple additivity scheme summing over all $N$ atoms (Eq. (6)).

$$\bar{\alpha} = \sum_{i=1}^{N} \alpha_i \tag{6}$$

Miller [10] later revised these atomic contributions, $\alpha_i$, somewhat, on the basis of new experimental data.

In this scheme, a carbon atom can attain four different values, $\alpha_i$, one for an $sp^3$ carbon atom, two for an $sp^2$ carbon atom depending on whether this carbon atom is attached to at least one hydrogen atom or whether it is attached to three other $sp^2$ carbon atoms, and one for an sp carbon atom.

Table 7-1 lists some comparisons between experimental mean molecular polarizabilities and those estimated by Eq. (6). In this scheme, the estimation of mean molecular polarizability for acetic acid needs five values, values for $sp^3$-C, for $sp^2$-C, for $sp^3$-O, for $sp^2$-O, and for a hydrogen atom.

**Table 7.1.** Experimental mean molecular polarizabilities and values calculated by Eq. (6).

| Molecule | $\bar{a}$ [$\text{Å}^3$] | |
|---|---|---|
| | *exp.* | *calc.* |
| $H_2$ | 0.79 | 0.77 |
| $CH_4$ | 2.60 | 2.61 |
| $n\text{-}C_5H_{12}$ | 9.95 | 9.95 |
| $neo\text{-}C_5H_{12}$ | 10.20 | 9.95 |
| $cyclo\text{-}C_6H_{12}$ | 10.99 | 11.01 |
| $CH_2{=}CH_2$ | 4.26 | 4.25 |
| $C_6H_6$ | 10.39 | 10.43 |
| $CH_3F$ | 2.62 | 2.52 |
| $CF_4$ | 2.92 | 2.25 |
| $CCl_4$ | 10.47 | 10.32 |
| $NH_3$ | 2.26 | 2.13 |
| Aniline | 11.58–12.12 | 11.91 |
| Acetic acid | 5.05–5.15 | 5.17 |
| Pyridine | 9.14–9.47 | 9.72 |

7.1.3
**Additivity of Bond Contributions**

The next higher order of approximation, the first-order approximation, is obtained by estimating molecular properties by the additivity of bond contributions. In the following, we will concentrate on thermochemical properties only.

Values for $C_p°$ are now usually within $\pm 4.0$ J/mol K and rarely as poor as $\pm 8.0$ J/mol·K. The $S°$ values are usually well within $\pm 3.0$ J/mol K and rarely deviate by more than 6.0 J/mol K. The $\Delta H_f°$ values are usually within $\pm 10.0$ kJ/mol and seldom deviate by more than $\pm 20.0$ kJ/mol. In many cases, the values for the first member of a series of compounds such as $CH_4$, $H_2O$, or $NH_3$ deviate quite strongly.

Furthermore, such a scheme cannot distinguish between the values for isomeric hydrocarbons because these compounds have the same number and type of bonds. To summarize, such a scheme seems to be sufficient for the estimation of $C_p°$ and $S°$, but does not give sufficiently accurate values for the heat of formation, $\Delta H_f°$.

7.1.4
**Additivity of Group Contributions**

The second-order approximation for the estimation of molecular properties consists of summing the contributions of groups. A group consists of a central atom and its directly bonded neighboring atoms. A molecule is fragmented into all such monocentric groups, a value is taken for each of these groups, and these contributions are summed to obtain the molecular property. The values for these group contributions can be obtained from a multilinear regression analysis of the properties of a series of molecules.

Figure 7-1 shows the groups that are obtained for alkanes, and the corresponding notation of these groups as introduced by Benson [1]. Table 7-2 contains the group contributions to important thermochemical properties of alkanes. Results obtained with these increments and more extensive tables can be obtained from Refs. [1] and [2].

With group additivity, the estimated values for $C_p°$ and $S°$ now agree on the average to within $\pm 1.2$ J/mol K, which is well within the experimental error. The agreement for $\Delta H_f°$ is now much improved against the results from bond additivity and is on the average within $\pm 1.7$ kJ/mol of the experimental value, with exceptional cases going as high as $\pm 12.0$ kJ/mol.

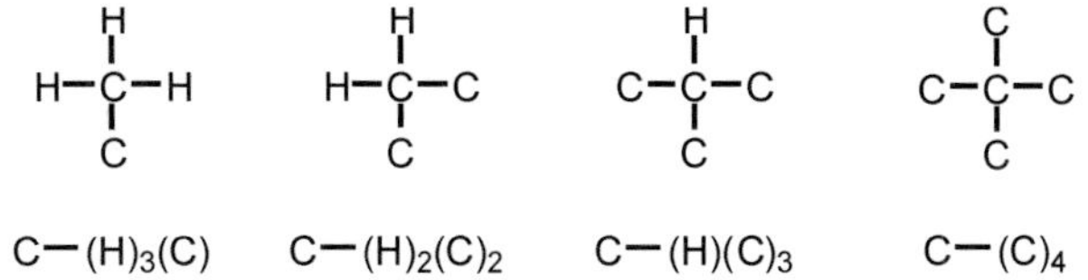

$$C-(H)_3(C) \qquad C-(H)_2(C)_2 \qquad C-(H)(C)_3 \qquad C-(C)_4$$

**Figure 7-1.** The four groups contained in alkanes, and their linear code as introduced by Benson [1].

**Table 7.2.** Group contributions to $C_p°$, $S°$, and $\Delta H_f°$ for ideal gases at 25 °C, 1 atm, for alkanes.

| Group | Contribution to $C_p$ [J/mol K] | $S$ [J/mol K] | $\Delta H_f°$ [kJ/mol] | $\Delta H_a$ [kJ/mol] |
|-------|--------------------------------|---------------|------------------------|------------------------|
| C-(H)$_3$(C) | 25.95 | 127.30 | −42.19 | 1412.31 |
| C-(H)$_2$(C)$_2$ | 22.81 | 39.43 | −20.72 | 1172.08 |
| C-(H) (C)$_3$ | 18.71 | −50.53 | −6.20 | 936.20 |
| C-(C)$_4$ | 18.21 | −146.93 | 8.16 | 704.42 |

Table 7-2 also contains, besides the values for heats of formation, $\Delta H_f°$, values for the heats of atomization, $\Delta H_a$, as these are easier to interpret.

Heats of atomization can be obtained from heats of formation by addition of values for converting elements in their standard state into gaseous atoms. For carbon, the standard state is defined as graphite; thus the sublimation enthalpy of graphite (715.4 kJ/mol) has to be added for each carbon atom in a molecule to obtain heats of atomization from heats of formation. The standard state of hydrogen is gaseous hydrogen in its diatomic form. Thus, half of the bond dissociation energy of a hydrogen molecule (218.1 kJ/mol) has to be added for each hydrogen atom in a molecule. Heats of atomization can be interpreted better because they refer to the enthalpy required to transform a molecule into its constituent atoms in the gas phase.

In order to develop a quantitative interpretation of the effects contributing to heats of atomization, we will introduce other schemes that have been advocated for estimating heats of formation and heats of atomization. We will discuss two schemes and illustrate them with the example of alkanes. Laidler [11] modified a bond additivity scheme by using different bond contributions for C–H bonds, depending on whether hydrogen is bonded to a primary $(E(C–H)_p)$, secondary $(E(C–H)_s)$, or tertiary $(E(C–H)_t)$ carbon atom. Thus, in effect, Laidler also used four different kinds of structure elements to estimate heats of formation of alkanes, in agreement with the four different groups used by Benson.

Another scheme for estimating thermochemical data, introduced by Allen [12], accumulated the deviations from simple bond additivity in the carbon skeleton. To achieve this, he introduced, over and beyond a contribution from a C–C and a C–H bond, a contribution G(CCC) every time a consecutive arrangement of three carbon atoms was met, and a contribution D(CCC) whenever three carbon atoms were bonded to a central carbon atom. Table 7-3 shows the substructures, the symbols, and the contributions to the heats of formation and to the heats of atomization.

Inspection of the values for the structure elements and their contribution to the heats of formation again allows interpretation: The B-terms correspond to the energies to break these bonds, and a sequence of three carbon atoms introduces stability into an alkane whereas the arrangement of three carbon atoms around a central carbon atom leads to the destabilization of an alkane.

**Table 7.3.** The Allen scheme: substructures, notations, and contributions to heats of formation and heats of atomization (values in kJ/mol).

| | C–C | C–H | C–C–C | C–C–C<br>$\|$<br>C |
|---|---|---|---|---|
| | B(CC) | B(CH) | G(CCC) | D(CCC) |
| $\Delta H_f^\circ$ | 18.80 | −17.33 | −4.6 | 0.25 |
| $\Delta H_a$ | 338.90 | 414.29 | 4.6 | −0.25 |

All three schemes, the Benson, the Laidler, and the Allen scheme, use four structure contributions for the estimation of thermochemical data of alkanes. As might be guessed, they are numerically equivalent; all three schemes provide the same accuracy. This is shown below by Eqs. (7)–(10) for the interconversion of the various contributions.

$$C-(C)(H)_3 = \tfrac{1}{2}E(C-C) + 3E(C-H)_p$$
$$= \tfrac{1}{2}B(C-C) + 3B(C-H) \tag{7}$$

$$C-(C)_2(H)_2 = E(C-C) + 2E(C-H)_s$$
$$= B(C-C) + 2B(C-H) + G(CCC) \tag{8}$$

$$C-(C)_3(H) = \tfrac{3}{2}E(C-C) + E(C-H)_t$$
$$= \tfrac{3}{2}B(C-C) + B(C-H) + 3G(CCC) + D(CCC) \tag{9}$$

$$C-(C)_4 = 2E(C-C) = 2B(C-C) + 6G(CCC) + 4D(CCC) \tag{10}$$

Any one of these additivity schemes can be used for the estimation of a variety of thermochemical molecular data, most prominently for heats of formation, with high accuracy [13]. A variety of compilations of thermochemical data are available [14–16]. A computer program based on Allen's scheme has been developed [17, 18] and is included in the PETRA package of programs [19].

The use of group contribution methods for the estimation of properties of pure gases and liquids [20, 21] and of phase equilibria [22] also has a long history in chemical engineering.

## 7.1.5
## Effects of Rings

The entire discussion on atom, bond, and group additivity schemes has tacitly assumed that there are no rings in organic structures. Rings can, however, exert drastic deviations from atom, bond, or group additivity for physical or chemical data in general, and thermochemical data in particular. In the following we will again limit our discussion on heats of formation and heats of atomization.

Rings can either stabilize or destabilize molecules beyond what is to be expected from a simple additivity scheme. Stabilization comes from aromatic ring systems

| 115.6 | 111.9 | 115.2 | 73.6 |

**Figure 7-2.** Strain energies [kJ/mol] of three-membered ring systems.

such as those in benzene, naphthalene derivatives, or heterocyclic systems, e.g., furan, pyrrole, thiophene, or pyridine compounds.

Destabilization is observed in small rings such as three- and four-membered ring systems, due to bond angle strain. Interestingly, there is not much difference in ring strain of small rings between carbocyclic and heterocyclic structures containing carbon, oxygen, or nitrogen atoms, as shown in Figure 7-2; only rings containing sulfur atoms show less strain. Additional strain results from the introduction of multiple bonds into small rings. Medium-sized rings such as nine- and ten-membered ring systems are destabilized due to eclipsed conformations. Combinations of rings, particularly if small rings are involved, can introduce additional strain.

All of these effects can be accounted for by extensions of an additivity scheme, when special increments are attributed to monocyclic structures and the combination of two ring systems having one, two, or three atoms in common [23]. Combination of a table containing values for these ring fragments with an algorithm for the determination of the smallest set of smallest rings (SSSR) [24] (see also Section 2.5.1) allows such a procedure to be performed automatically.

### 7.1.6
### Drug–Receptor Binding Energies

Until now, we have concentrated on the calculation of fundamental thermochemical values of molecules in the gas phase. With the next example, we aim to show that simple additivity schemes can be quite useful in a completely different area, in drug design for the study of ligand–receptor interactions. Andrews et al. [25] have analyzed binding constants and the derived binding energies for the binding of a series of 200 drug and enzyme inhibitors to their respective protein receptors. They reasoned that the individual functional groups in a molecule contributed an intrinsic value to the binding of a drug to its receptor. They came up with a simple additivity model based on contributions from atoms and groups in a molecule to estimate its average binding energy to a receptor. Thus, drugs that match their receptors exceptionally well have a measured binding energy that substantially exceeds the calculated average value – examples are biotin and estradiol. Conversely, if the observed binding energy is much less than the calculated average binding energy, then the drug matches the receptor less well than the average – examples are methotrexate and thyroxine.

**Table 7.4.** Intrinsic binding energies [kJ/mol].

| Atom/group | Value |
|---|---|
| $C_{sp}^3$ | 3.3 |
| $C_{sp}^2$ | 2.9 |
| N | 5.0 |
| $N^+$ | 48.1 |
| O, S | 4.6 |
| Halogen | 5.4 |
| OH | 10.5 |
| $COO^-$ | 34.3 |
| C=O | 14.2 |
| $OPO_3^-$ | 41.8 |

In order to analyze the binding energies, allowance had first to be made for entropy effects. A value of 58.2 kJ/mol was estimated to be attributed to the overall loss of rotational and translational entropy and a value of 2.9 kJ/mol for each degree of conformational freedom. The remaining binding energies were submitted to a multi-linear regression analysis considering those atoms and groups contained in Table 7-4. The values of the intrinsic binding energies are also contained in Table 7-4.

Figure 7-3 shows a comparison of observed binding energies with the calculated average binding energies for a few drugs.

The remarkable achievement of this work is that such a simple additivity scheme can reproduce a quantitative value for such a complicated process as the binding of a ligand to its protein receptor, which involves a series of events such as desolvation of the ligand and of the protein from water molecules, conformational changes in the ligand and the protein, and alignment of functional groups in the ligand with binding sites in the protein.

Clearly, the assertion that the deviations of the observed binding energies from the calculated average binding energies reflect whether a ligand is particularly well or poorly fitted for binding is just a working hypothesis. Nevertheless, it has to be observed that even 20 years after publication of this method, average binding energies calculated by the Andrews scheme are still used for screening potential drug candidates.

### 7.1.7
### Attenuation Models

Until now, we have discussed the use of additivity schemes to estimate global properties of a molecule such as its mean molecular polarizability, its heat of formation, or its average binding energy to a protein receptor.

Many phenomena ask for local, site-specific properties of a molecule such as the partial charge on a specific atom in a molecule or the hydrogen bond donor ability of a certain OH group. It would be highly desirable to have methods as simple as an additivity model to estimate such site-specific molecular properties.

normal binders

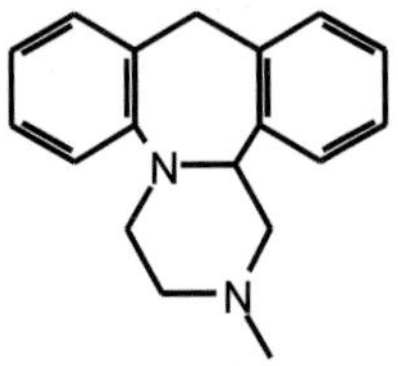

mianserin
52.0
*49.1*

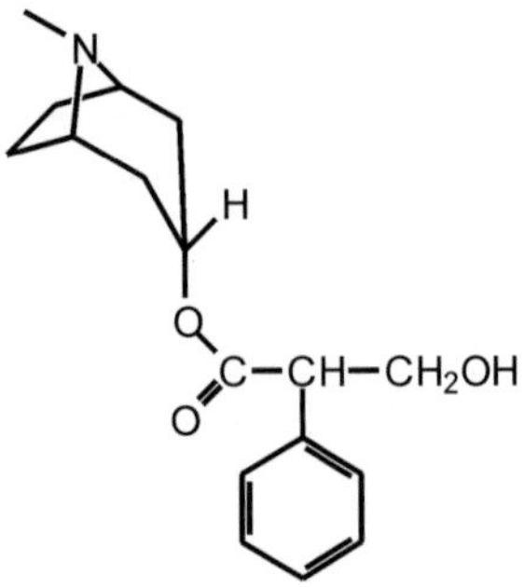

atropine
53.2
*52.0*

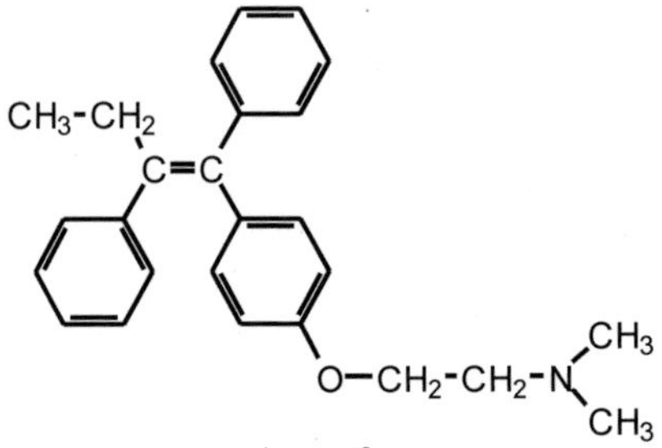

lamosifen
51.1
*49.5*

exceptionally strong binders

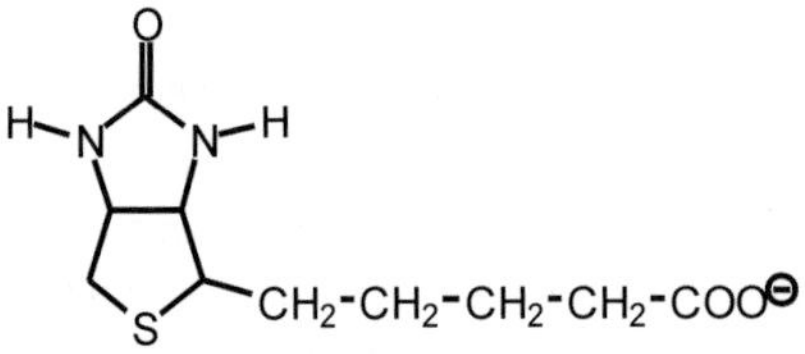

biotin
85.2
*16.6*

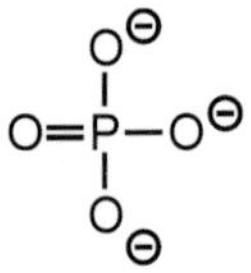

phosphate
32.0
*−16.6*

exceptionally weak binder

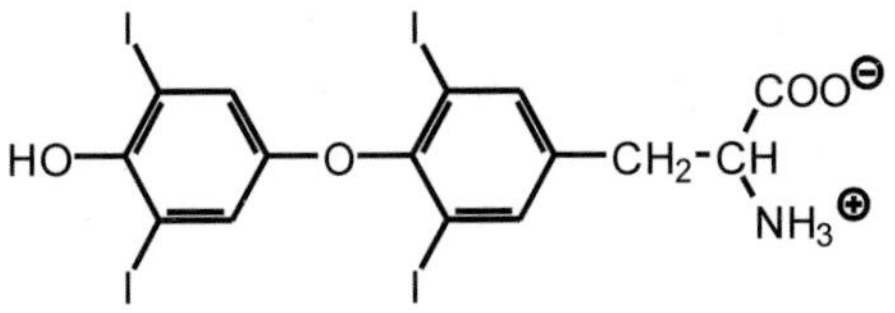

thyroxine
42.8
*81.9*

**Figure 7-3** Binding energies [kJ/mol] of some drugs (observed and calculated *average* values).

Clearly, simple additivity of atom properties will no longer suffice, as the contribution of an atom will diminish the further it is away from the atom whose property has to be estimated. In the following, we present two methods of accounting for the influence of one atom on another, attenuated over the distance between the two atoms.

### 7.1.7.1 Calculation of Charge Distribution

The electron distribution in a molecule is one of the most important factors influencing its physical, chemical, or biological properties. Detailed information on the electron distribution of a molecule can be obtained from quantum mechanical calculations of various degrees of sophistication (see Section 7.4 and Chapter 7, Sections 2 and 4 in the Handbook). However, chemists have always liked the idea of dissecting the electron distribution of a molecule and assigning the pieces thus obtained to the individual atoms. Thus, the picture of a molecule consisting of atoms carrying partial charges has emerged and was used to explain many phenomena. Intuitively appealing as this picture is, there is no unequivocal criterion to decide how to assign the electron distribution to the various atoms. The most widely used method is a Mulliken population analysis [26]. However, the values obtained with this method are heavily dependent on the level of quantum mechanical approach being taken and the basis set chosen. Several other methods have been proposed but none has yet found general acceptance. Thus, it has to be concluded that no unanimously accepted method for the definition of partial atomic charges is available, making it difficult, when using any other method for the calculation of charges, to select widely accepted data for comparison. In the end, any method for the calculation of partial atomic charges can only be judged by how valuable the resulting charges are in reproducing experimental data.

We set out in 1975 to develop a rapid empirical method for the calculation of the charge distribution in organic molecules. We were interested in calculating partial charges for the molecules encountered in a synthesis design study [27]. As we had to deal with hundreds of molecules of sizable complexity in such a synthesis design study, we needed a fast method. At that time computers were much slower than they are now, and the method we developed and which is presented below fulfilled the intended purpose. Now, more than 25 years later, with the enormous increase in computing power one might think that this method should have outlived its merits. However, with the advent of combinatorial chemistry and high-throughput screening (see Chapter 10, Section 4.2 in the Handbook), people are studying huge datasets of molecules, particularly in virtual screening procedures, with $10^6$–$10^9$ molecules. Thus, a rapid method for the calculation of the charge distribution in molecules is still of interest and the PEOE method presented below is applied routinely to datasets with millions of structures.

### 7.1.7.1.1 The PEOE Method

At the outset of the Partial Equalization of Orbital Electronegativities (PEOE) method [28] is the electronegativity concept in the form of Eq. (11) presented by Mulliken, who put it on a sound theoretical basis [29].

$$\chi_v = \frac{1}{2}(I_v + E_v) \tag{11}$$

In this equation, the electronegativity of an atom is related to its ionization potential, $I$, and its electron affinity, $E$. Mulliken already pointed out that in this definition the ionization potential, $I_v$, and the electron affinity, $E_v$, of valence states have to be used. This idea was further elaborated by Hinze et al. [30, 31], who introduced the concept of orbital electronegativity.

The electronegativity of an atom further depends on the charge in this orbital and also on the charge in the other orbitals of this atom. For this dependence of orbital electronegativity on the total charge, Q, irrespective of whether part of it resides in the orbital considered or in the other orbitals of this atom, we selected a polynomial of degree two (Eq. (12)).

$$\chi_{iv} = a_{iv} + b_{iv}Q_i + c_{iv}Q_i^2 \tag{12}$$

Values for these coefficients, $a$, $b$, $c$, of Eq. (12) can be obtained from the ionization potentials and electron affinities of the neutral, the cationic, and the anionic states of an orbital.

Table 7-5 gives these coefficients for the various valence states of a variety of atoms.

**Table 7.5.** Parameters for the dependence of orbital electronegativity on charge (Eq. (12)).

| Atom | Valence state[a] | $a_v$ | $b_v$ | $c_v$ |
|------|------------------|-------|-------|-------|
| H    |                  | 7.17  | 6.24  | −0.56 |
| C    | tetetete         | 7.98  | 9.18  | 1.88  |
|      | trtrtr$\pi$      | 8.79  | 9.32  | 1.51  |
|      | didi$\pi\pi$     | 10.39 | 9.45  | 0.73  |
| N    | te$^2$tetete     | 11.54 | 10.82 | 1.36  |
|      | tr$^2$trtr$\pi$  | 12.87 | 11.15 | 0.85  |
|      | di$^2$di$\pi\pi$ | 15.68 | 11.7  | −0.27 |
| O    | te$^2$te$^2$tete[b] | 14.18 | 12.92 | 1.39 |
|      | tr$^2$trtr$\pi$  | 17.07 | 13.79 | 0.47  |
| F    | s$^2$p$^2$p$^2$p[c] | 14.66 | 13.85 | 2.31 |
| Cl   | s$^2$p$^2$p$^2$p[d] | 11.00 | 9.69  | 1.35 |
| Br   | s$^2$p$^2$p$^2$p[e] | 10.08 | 8.47  | 1.16 |
| J    | s$^2$p$^2$p$^2$p[f] | 9.90  | 7.96  | 0.96 |
| S    | te$^2$te$^2$tete | 10.14 | 9.13  | 1.38  |

[a] te = tetrahedral (sp$^3$); tr = trigonal (sp$^2$); di = diagonal (sp);
[b] the hybridization was adjusted to a bond angle of 106° (methanol);
[c] s-character of 13 %;
[d] s-character of 17 %;
[e] s-character of 17 %;
[f] s-character of 23 %.

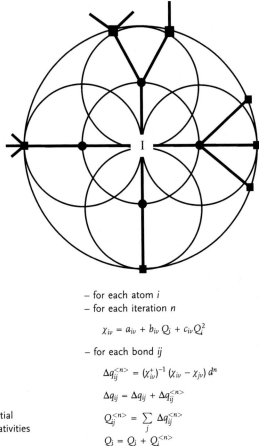

– for each atom $i$
– for each iteration $n$

$$\chi_{iv} = a_{iv} + b_{iv}\, Q_i + c_{iv} Q_i^2$$

– for each bond $ij$

$$\Delta q_{ij}^{<n>} = (\chi_{iv}^+)^{-1} (\chi_{iv} - \chi_{jv})\, d^n$$

$$\Delta q_{ij} = \Delta q_{ij} + \Delta q_{ij}^{<n>}$$

$$Q_{ij}^{<n>} = \sum_{j} \Delta q_{ij}^{<n>}$$

$$Q_i = Q_i + Q_i^{<n>}$$

**Figure 7-4.** The procedure for Partial Equalization of Orbital Electronegativities (PEOE).

Upon bond formation, the electronegativities of the orbitals involved tend to equalize by shifting electrons from the orbital with lower electronegativity to the one with higher electronegativity. On theoretical grounds equalization of orbital electonegativities has to be assumed [32]. However, their electronegativities will change when atoms are put into a molecular environment. To this effect, Partial Equalization of Orbital Electronegativities (PEOE) was invoked [28] as outlined in Figure 7-4.

The attenuation factor, $d^n$, with $n$ being the iteration number and the damping factor, $d$, having a value of $\frac{1}{2}$, was chosen bearing in mind that the inductive effect is considered to diminish by a factor of 2–3 with each intervening bond. The electronegativity of the positive state, $\chi_{iv}^+$ , is needed to scale electronegativity to charge values.

The PEOE method basically consists of only three loops, one running over all the atoms of a molecule, a second loop through each iteration, and a third loop running over all the directly bonded neighbors of an atom. The cycle is usually stopped after six iterations as the charge shifts have then become quite minute. This explains the high speed in calculating the charge distribution in organic molecules. In each iteration only the directly bonded neighbors of an atom are considered. However, as the first loop ensures that this is done for each atom, the consequence is that in the second iteration the influence of atoms two bonds away is considered implicitly, and so on (see Figure 7-4).

It has already been said that the merits of a method for charge calculation can be assessed mainly by its usefulness in modeling experimental data. Charges from the PEOE procedure have been correlated with C1s-ESCA shifts [28], dipole moments [33], and $^{13}$C NMR shifts [34], to name but a few.

The PEOE procedure has been incorporated into practically all molecular modeling packages, e.g., SYBYL of Tripos and Catalyst of Accelrys, because of its high speed and the quality of the charge values obtained.

Variants of this method have been developed taking account of the 3D structure of molecules. However, we have not felt it necessary to follow this path; for the simplicity of the method the values obtained are remarkably good.

#### 7.1.7.1.2 Residual Electronegativity

The PEOE method leads to only partial equalization of orbital electronegativities. Thus, each atom of a molecule retains, on the basis of Eq. (12), a residual electronegativity that measures its potential to attract further electrons. It has been shown that the values of residual electronegativities can be taken as a quantitative measure of the inductive effect [35].

#### 7.1.7.1.3 $\pi$-Charge Distribution

The underlying principle of the PEOE method is that the electronic polarization within the $\sigma$-bond skeleton as measured by the inductive effect is attenuated with each intervening $\sigma$-bond. The electronic polarization within $\pi$-bond systems as measured by the resonance or mesomeric effect, on the other hand, extends across an entire $\pi$-system without any attenuation. The simple model of an electron in a box expresses this fact. Thus, in calculating the charge distribution in conjugated $\pi$-systems an approach different from the PEOE method has to be taken.

Our first approach took resort in simple resonance theory [36, 37]. For each conjugated $\pi$-system all resonance structures were generated, such as those shown in Figure 7-5.

The resonance structures indicate how the free electron pair on nitrogen is shifted across the $\pi$-system generating charges at quite specific positions. In order to estimate the charge distribution of the entire $\pi$-system, the individual resonance structures such as the ones shown in Figure 7-5 had to be weighted according to the importance of their contribution to the overall picture. To this effect, the electronegativity concept was again invoked, this time using the values of the $\pi$-orbital electronegativities.

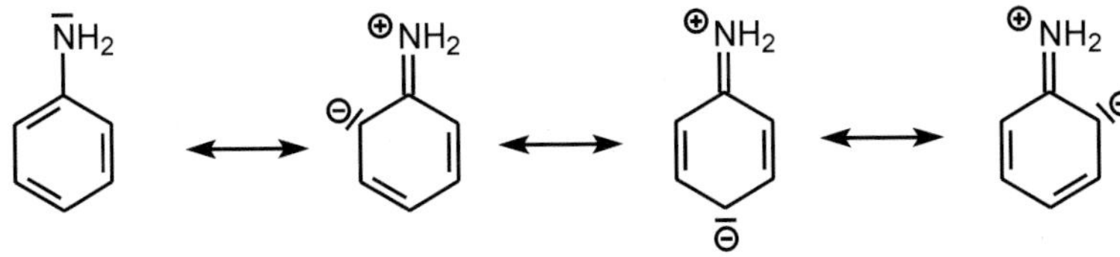

**Figure 7-5.** Resonance structures of aniline.

In spite of the success of this method it was later felt that the calculation of the charge distribution in conjugated $\pi$-systems should be put on a less empirical basis. To achieve this, a modified Hückel Molecular Orbital (HMO) approach (Section 7.4) was developed. Again, the charge distribution in the $\sigma$-skeleton is first calculated by the PEOE method.

Then the Hückel matrix for the conjugated $\pi$-system is constructed. The $\alpha$-values of the Hückel matrix of each atom $i$ of the conjugated system are adjusted to the $\sigma$-charge distribution by Eq. (13).

$$\alpha_i = \alpha_{o,i} + 0.5 \, q_{\sigma,i} \tag{13}$$

The initial values, $\alpha_{o,i}$, are derived by correlations with dipole moments of a series of conjugated systems. The exchange integrals $\beta_{ij}$ are taken from Abraham and Hudson [38] and are considered as being independent of charge. The $\pi$-charges are then calculated from the orbital coefficients, $c_{ij}$, of the HMO theory according to Eq. (14).

$$q_{\pi,i} = n_i - \sum_j n_i c_{ij}^2 \tag{14}$$

The quality of the $\pi$-charge values thus obtained has been demonstrated by the calculation of dipole moments of a series of 80 conjugated systems [39].

### 7.1.7.2 Polarizability Effect

In Section 7.1.2 a method for the calculation of mean molecular polarizability was presented. Mean molecular polarizability can be calculated from additive contributions of the atoms in their various hybridization states in a molecule (see Eq. (6)). Mean molecular polarizability, $\bar{\alpha}$, expresses the magnitude of the dipole moment, $\mu$, induced into a molecule under the influence of an external field, $E$ (Eq. (15))

$$\mu = \bar{\alpha} \cdot E \tag{15}$$

In many chemical applications, however, it would be more interesting to know how polarizability can stabilize a charge introduced *into* a molecule. Thus, rather than the global molecular property, mean molecular polarizability, a local, site-specific value for polarizability is needed.

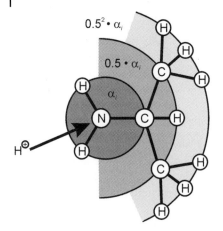

$0.5^2 \cdot \alpha_i$

$0.5 \cdot \alpha_i$

$\alpha_i$

**Figure 7-6.** Graphical representation of the procedure for calculating the effective polarizability on the nitrogen atom of 2-aminopropane.

To obtain such a property, again an attenuation model was developed [40]. The atom hybridization state polarizability values [10] were taken for estimating the stabilization of a charge introduced into a molecule but they were taken with a magnitude reflecting the distance, in number of bonds, of an atom from the site where the charge is located. Figure 7-6 indicates this with the generation of a positive charge on a nitrogen atom of an alkylamine upon protonation.

An effective polarizability, reflecting the stabilization of a positive charge on the protonated nitrogen atom through polarizability, is calculated by Eq. (16).

$$\alpha_{eff,i} = \sum_{j=1}^{N} d^f \cdot \alpha_j \tag{16}$$

$$f = 0 \quad \text{if} \quad i = j$$

$$f = n_{ij} - 1 \quad \text{if} \quad i \neq j$$

The contribution of an atom $j$ to the polarizability effect is attenuated by the number of bonds, $n_{ij}$, between this atom and the site of protonation, $i$.

The method for calculating effective polarizabilities was developed primarily to obtain values that reflect the stabilizing effect of polarizability on introduction of a charge into a molecule. That this goal was reached was proven by a variety of correlations of data on chemical reactivity in the gas phase with effective polarizability values. We have intentionally chosen reactions in the gas phase as these show the predominant effect of polarizability, uncorrupted by solvent effects.

Thus, the values calculated for effective polarizability at the nitrogen atom for a series of 49 amines carrying only alkyl groups was correlated directly with their proton affinities, a reaction that introduces a positive charge on the nitrogen atom by protonation (Figure 7-7) [40].

Once amines that also carry heteroatoms were included in the study, a dataset of 80 proton affinities was obtained. For those alkyl amines the inductive effect as quantified by residual electronegativity had also to be taken into account. A simple

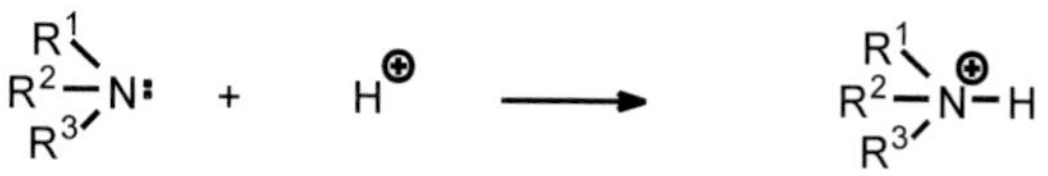

alkyl amines only (49 cpds): PA (kJ/mol) = 205.6 + 2.82 $\alpha_{\text{eff, N}}$
alkyl amines + heteroatom substituted alkyl amines (80 cpds):
PA (kJ/mol) = 1435.5 + 12.5 $\alpha_{\text{eff, N}}$ − 116.3 $\chi_{\text{r, N}}$

**Figure 7-7.** Equations for the calculation of proton affinities (PA) of simple alkyl amines and of heteroatom-substituted alkyl amines.

two-variable equation, comprising effective polarizability $\alpha_{\text{eff}}$ and residual electronegativity, $\chi$, of the nitrogen atom could correlate all the proton affinities of the amines available at that time [40].

Additional gas-phase reactivity data, such as gas-phase acidities of alcohols [41], proton affinities of alcohols and ethers [41], and proton affinities of carbonyl compounds [42] could equally well be described by similar equations.

# References

[1] S. W. Benson, J. H. Buss, *J. Chem. Phys.* **1958**, *29*, 546–572.

[2] S. W. Benson, *Thermochemical Kinetics*, 2nd edition, Wiley, New York, **1976**.

[3] K. G. Denbigh, *Trans. Faraday Soc.* **1940**, *36*, 936–948.

[4] S. S. Bhatnagar, K. M. Mathur, *Physical Principles and Applications of Magneto-chemistry*, Macmillan, London, **1935**.

[5] S. Sugden, *J. Chem. Soc.* **1924**, *125*, 32.

[6] K. Fajans, *Ber. Deut. Chem. Ges.* **1920**, *53B*, 643–665; *Z. Phys. Chem.* **1921**, *99*, 395–415; *Ber. Deut. Chem. Ges.* **1922**, *55B*, 2826–2838.

[7] F. Eisenlohr, *Z. Phys. Chem. (Leipzig)* **1911**, *75*, 585–607.

[8] K. Miller, J. A. Savchik, *J. Am. Chem. Soc.* **1979**, *101*, 7206–7213.

[9] Y. K. Kang, M. S. Jhon, *Theor. Chim. Acta* **1982**, *61*, 41–48.

[10] K. Miller, *J. Am. Chem. Soc.* **1990**, *112*, 8533–8542.

[11] K. J. Laidler, *Can. J. Chem.* **1956**, *34*, 626–648.

[12] T. L. Allen, *J. Chem. Phys.* **1959**, *31*, 1039–1049.

[13] J. Gasteiger, P. Jacob, U. Strauss, *Tetrahedron* **1979**, *35*, 139–146.

[14] J. D. Cox, G. Pilcher, *Thermochemistry of Organic and Organometallic Compounds*, Academic Press, London, **1970**.

[15] J. B. Pedley, R. D. Naylor, S. P. Kirby, *Thermochemical Data of Organic Compounds*, 2nd edition, Chapman and Hall, London, **1986**.

[16] *http://www.nist.gov/srd/thermo.htm*

[17] J. Gasteiger, *Comput. Chem. (Oxford, UK)* **1978**, *2*, 85–88.

[18] J. Gasteiger, *Tetrahedron* **1979**, *35*, 1419–1426.

[19] *http://www2.chemie.uni-erlangen.de/software/petra/index.html http://www.mol-net.de*

[20] R. C. Reid, J. M. Prausnitz, B. E. Polling, *The Properties of Gases and Liquids*, 4th edition, McGraw-Hill, New York, **1988**.

[21] W. Cordes, J. Rarey, *Fluid Phase Equil.* **2002**, *201(2)*, 397–421.

[22] a) R. Wittig, J. Lohmann, J. Gmehling, *Ind. Eng. Chem. Res.* **2003**, *42*, 183–188; b) J. Gmehling, R. Wittig, J. Lohmann, R. Joh, *Ind. Eng. Chem. Res.* **2002**, *41*, 1678–1688; c) S. Horstmann, K. Fischer, J. Gmehling, *Fluid Phase Equil.* **2000**, *167*, 173–186.

[23] J. Gasteiger, O. Dammer, *Tetrahedron* **1978**, *34*, 2939–2945.

[24] J. Gasteiger, C. Jochum, *J. Chem. Inf. Comput. Sci.* **1979**, *19*, 43–48.

[25] P. R. Andrews, D. J. Craik, J. L. Martin, *J. Med. Chem.* **1984**, *27*, 1648–1657.

[26] R. S. Mulliken, *J. Chem. Phys.* **1955**, *23*, 1833–1840.

[27] J. Gasteiger, C. Jochum, *Topics Curr. Chem.* **1978**, *74*, 93–126.

[28] J. Gasteiger, M. Marsili, *Tetrahedron* **1980**, *36*, 3219–3228.

[29] R. S. Mulliken, *J. Chem. Phys.* **1934**, *2*, 782–793.

[30] J. Hinze, H. H. Jaffé, *J. Am. Chem. Soc.* **1962**, *84*, 540–546.

[31] B. Bergmann, J. Hinze, *Angew. Chem.* **1996**, *108*, 162–176; *Angew. Chem. Int. Ed. Engl.* **1996**, *35*, 150–163.

[32] W. J. Mortier, K. van Genechten, J. Gasteiger, *J. Am. Chem. Soc.* **1985**, *107*, 829–835.

[33] J. Gasteiger, M. D. Guillen, *J. Chem. Research (S)* **1983**, 304–305, *J. Chem. Research (M)* **1983**, 2611–2624.

[34] J. Gasteiger, I. Suryanarayana, *Magn. Reson. Chem.* **1985**, *23*, 156–157.

[35] M. G. Hutchings, J. Gasteiger, *Tetrahedron Lett.* **1983**, *24*, 2541–2544.

[36] M. Marsili, J. Gasteiger, *Croat. Chem. Acta* **1980**, *53*, 601–614.

[37] J. Gasteiger, H. Saller, *Angew. Chem.* 1985, 97, 699–701; Angew. Chem. *Int. Ed. Engl.* **1985**, *24*, 687–689.

[38] R. J. Abraham, B. Hudson, *J. Comput. Chem.* 1985, 6, 173–181.

[39] M. Fato, T. Kleinöder, University of Erlangen-Nürnberg, unpublished results.

[40] J. Gasteiger, M. G. Hutchings, *J. Chem. Soc. Perkin 2* **1984**, 559–564.

[41] J. Gasteiger, M. G. Hutchings, *J. Am. Chem. Soc.* **1984**, 106, 6489–6495.

[42] M. G. Hutchings, J. Gasteiger, *J. Chem. Soc. Perkin 2* **1986**, 447–454.

**7.2**
**Molecular Mechanics**

*H. Lanig*

7.2.1
**Introduction**

The title of this contribution – Molecular Mechanics – could also be Force Field Methods. Nowadays the two terms essentially mean the same. We can go a into little more detail and define molecular mechanics as the calculation of the static properties of a molecule or a group of molecules, such as structure, energy, or electrostatics. If we are interested in dynamic properties like the time evolution of a molecular system, resulting in a trajectory of snapshots, we have to use molecular dynamics. Finally, if we need to know thermodynamic properties like enthalpies, or include entropy or free energy, an alternative to sampling the conformational space by molecular dynamics is to apply Monte Carlo simulations. The latter method does not concern time evolution at all, but is generally considered to generate statistically meaningful thermodynamic ensembles much more effectively.

Why "force field"? In many situations it is necessary to know about the forces between atoms. This is the case for molecular dynamics, but also for many molecular mechanics applications. According to Eq. (17), the forces $F$ are calculated as the negative derivative of the potential energy $E$ with respect to the coordinates $r_i$:

$$F = -\frac{\partial E}{\partial r_i} \tag{17}$$

The force acting on an atom is therefore the negative energy gradient. Molecular mechanics methods are often used because they are fast and can be applied to a large number of molecules with many atoms. The reason for this advantage is the so-called Born–Oppenheimer approximation, which separates the movement of the electrons from the much slower movement of the nuclei and makes it possible to describe the energy of a molecule or molecular system as a function of the atomic coordinates. We will see this in more detail in Section 7.2.3. This simplification does not necessarily mean that the results of a molecular mechanics calculation are less reliable than, for example, high-level quantum mechanical data. Force field parameters are often adjusted to correctly reproduce experiments, which normally represent thermodynamic averages and not a particular geometry or conformation of the molecule under consideration. A second important point to mention is the transferability of the parameter set, which is generated using a limited set of experimental data and small test molecules. Nevertheless, these parameters can be applied to a much greater number of problems and larger molecular systems.

## 7.2.2
## No Force Field Calculation Without Atom Types

A minimum input for a quantum mechanical calculation consists of the geometry of the molecule under investigation (e.g., in internal or Cartesian coordinates), the atomic number of each nucleus, and the overall charge and spin state. Information about the distribution of the electrons, in terms of electron density or a wavefunction, or the better-interpretable partial atomic charges is calculated on the basis of molecular geometry. In the context of force field methodology, input of the total charge and the spin of a molecule is not mandatory, because these types of calculations do not deal with electrons. To represent electrostatics, not even partial atomic charges are necessary if, e.g.. bond dipoles are used. In contrast to quantum mechanics, molecular mechanics needs more information than the atomic number alone. In fact, each atom has to be described in a more detailed way.

The concept of atom types allows a differentiation in terms of hybridization state, local environment, or specific conditions such as strain in small-ring systems. Allinger and co-workers, the developers of the MM2, MM3, and MM4 force fields for "small molecules" (see Section 7.2.4.1.1), defined within the MM3 parameterization more than 15 different atom types for carbon alone. These are, e.g., $sp^3$ alkane, $sp^2$ alkene, $sp^2$ carbonyl, $sp^2$ cyclopropene, sp alkyne and so on, all necessary to make MM3 applicable (this means to obtain reasonable results) to a diverse set of molecules. Mindful readers will immediately see the difficulty behind this approach: the more atom types are defined, the more parameters for the contributions to the potential energy function (bonds, angles, dihedrals, etc.; see Section 7.2.3) have to be developed. More general force fields would therefore assign, e.g., only one generic $sp^2$ carbon atom type, sacrificing accuracy for general applicability. Another trend is that force fields for specific classes of molecules use more atom types than parameterizations for general use. AMBER, which has been designed for biomolecules such as proteins and DNA (see Section 7.2.4.2.1), provides templates for the amino acid histidine in three different protonation states with different atom type assignments.

## 7.2.3
## The Functional Form of Common Force Fields

A force field does not consist only of a mathematical expression that describes the energy of a molecule with respect to the atomic coordinates. The second integral part is the parameter set itself. Two different force fields may share the same functional form, but use a completely different parameterization. On the other hand, different functional forms may lead to almost the same results, depending on the parameters. This comparison shows that force fields are empirical; there is no "correct" form. Because some functional forms give better results than others, most of the implementations within the various available software packages (academic and commercial) are very similar.

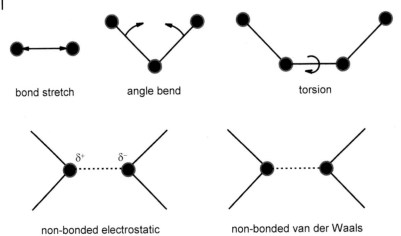

Figure 7-8. Bonded (upper row) and non-bonded (lower row) contributions to a typical molecular mechanics force field potential energy function. The latter two types of interactions can also occur within the same molecule.

The mathematical formulation of a typical molecular mechanics force field, which is also called the potential energy function (PEF), is shown in Eq. (18). Do not worry yet about the necessary mathematical expressions – they will be explained in detail in the following sections:

$$V_{PEF} = \sum V_{bonded} + \sum V_{non-bonded}$$

$$V_{PEF} = \sum V_{bonds} + \sum V_{angles} + \sum V_{torsions} + \sum V_{electrostatic} + \sum V_{vander Waals} \qquad (18)$$

The PEF is a sum of many individual contributions. It can be divided into bonded (bonds, angles, and torsions) and non-bonded (electrostatic and van der Waals) contributions V, responsible for intramolecular and, in the case of more than one molecule, also intermolecular interactions. Figure 7-8 shows schematically these types of interactions between atoms, which are included in almost all force field implementations.

The following sections give an overview of the functional form of the PEF and a short explanation of the various contributions to the total force field energy of a molecule or molecular system.

### 7.2.3.1 Bond Stretching

Figure 7-9 shows two mathematical functions used to describe the potential energy curve for the variation of the distance between two bound atoms within a molecule.

To calculate the bonded interaction of two atoms, a Morse function is often used. It has the form described by Eq. (19).

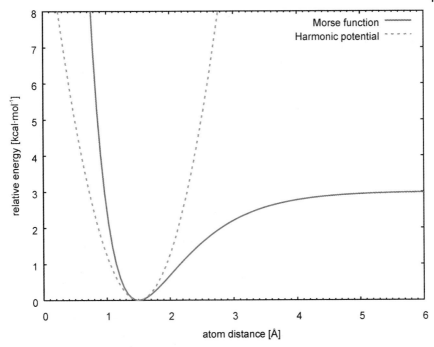

**Figure 7-9.** Variation of the potential energy of the bonded interaction of two atoms with the distance between them. The solid line comes close to the experimental situation by using a Morse function; the broken line represents the approximation by a harmonic potential.

$$V(l) = D_e\{1 - \exp[-a(l - l_0)]\}^2$$

$$a = \omega\sqrt{\left(\frac{\mu}{2D_e}\right)} \qquad (19)$$

$$\omega = \sqrt{\frac{k}{\mu}}$$

For each pair of interacting atoms ($\mu$ is their reduced mass), three parameters are needed: $D_e$ (depth of the potential energy minimum), $k$ (force constant of the particular bond), and $l_0$ (reference bond length). The Morse function will correctly allow the bond to dissociate, but has the disadvantage that it is computationally very expensive. Moreover, force fields are normally not parameterized to handle bond dissociation. To circumvent these disadvantages, the Morse function is replaced by a simple harmonic potential, which describes bond stretching by Hooke's law (Eq. (20)).

$$V(l) = \frac{k}{2}(l - l_0)^2 \qquad (20)$$

In this case, only two parameters ($k$ and $l_0$) per atom pair are needed, and the computation of a quadratic function is less expensive. Therefore, this type of expression is used especially by biomolecular force fields (AMBER, CHARMM, GROMOS) dealing with large molecules like proteins, lipids, or DNA.

Compared with the Morse potential, Hooke's law performs reasonably well in the equilibrium area near $l_0$, where the shape of the Morse function is more or less quadratic (see Figure 7-9 in the minimum-energy region). To improve the performance of the harmonic potential for non-equilibrium bond lengths also, higher-order terms can be added to the potential according to Eq. (21).

$$V(l) = \frac{k}{2}(l - l_0)^2 \left[1 - k'(l - l_0) - k''(l - l_0)^2 - ...\right] \tag{21}$$

These anharmonicity corrections are used by force fields like MM3/MM4, MMFF, and CFF, which intend to reproduce bond length for small organic molecules with high precision. It should be noted that adding only a cubic term to the bond-stretching function (as in MM2) might result in an "explosion" of the molecule if the starting geometry is far from equilibrium (remember that the cubic function tends to minus infinity with increasing bond length). To get around this problem MM3 also adds a quartic term with a positive force constant. An alternative way realized in certain software packages or implementations is that the geometry optimization is started with a harmonic potential and the higher-order terms are added later when the bond lengths are already close to the equilibrium values and the dissociation of a bond is therefore impossible.

### 7.2.3.2 Angle Bending

As for bond stretching, the simplest description of the energy necessary for a bond angle to deviate from the reference value is a harmonic potential following Hooke's law, as shown in Eq. (22).

$$V(\theta) = \frac{k}{2}(\theta - \theta_0)^2 \tag{22}$$

For every type of angle including three atoms, two parameters (force constant $k$ and reference value $\theta_0$) are needed. Also, as in the bond deformation case, higher-order contributions such as that given by Eq. (23) are necessary to increase accuracy or to account for larger deformations, which no longer follow a simple harmonic potential.

$$V(\theta) = \frac{k}{2}(\theta - \theta_0)^2 \left[1 - k'(\theta - \theta_0) - k''(\theta - \theta_0)^2 - ...\right] \tag{23}$$

Force fields like MM3, MM4, CFF, or MMFF therefore use cubic and/or quartic or even higher contributions, up to the sixth power. Special attention has to be paid in the case of reference angles approaching 180°, e.g., for molecules with linear fragments such as acetylene compounds. In this circumstance, replacing Eq. (23) by a cosine function like $E_\theta = k_\theta(1 + \cos\theta)$ avoids differentiation problems when calculating gradients and second-order derivatives.

### 7.2.3.3 Torsional Terms

To account for barriers of rotation about chemical bonds, i.e., the energetics of twisting the 1,4-atoms attached to the bonds formed by the atoms 2–3, a three-term torsion energy function like that in Eq. (24) is used, in the given form or slightly modified, in almost every force field.

$$V(\omega) = \sum_{n=1}^{N} \frac{V_n}{2}[1 + \cos(n\omega - \gamma)] \qquad (24)$$

$V_n$ is often called the barrier of rotation. This is intuitive but misleading, because the exact energetic barrier of a particular rotation is the sum of all $V_n$ components and other non-bonding interactions with the atoms under consideration. The multiplicity $n$ gives the number of minima of the function during a 360° rotation of the dihedral angle $\omega$. The phase $\gamma$ defines the exact position of the minima.

The origin of a torsional barrier can be studied best in simple cases like ethane. Here, rotation about the central carbon–carbon bond results in three staggered and three eclipsed stationary points on the potential energy surface, at least when symmetry considerations are not taken into account. Quantum mechanically, the barrier of rotation is explained by anti-bonding interactions between the hydrogens attached to different carbon atoms. These interactions are small when the conformation of ethane is staggered, and reach a maximum value when the molecule approaches an eclipsed geometry.

It is noteworthy that it is not obligatory to use a torsional potential within a PEF. Depending on the parameterization, it is also possible to represent the torsional barrier by non-bonding interactions between the atoms separated by three bonds. In fact, torsional potentials and non-bonding 1,4-interactions are in a close relationship. This is one reason why force fields like AMBER downscale the 1,4-non-bonded Coulomb and van der Waals interactions.

Figure 7-10 shows that even a simple energy function like the one given in Eq. (24) allows the definition of quite complex rotation potentials. To calculate torsional barriers and rotation frequencies accurately, especially for unsymmetrical substituted molecules, force fields like MM4 use terms up to the sixth order ($N = 6$). These higher-order terms are generally small and only important when one is interested in high-quality reproduction of experimental data for organic and inorganic molecules.

### 7.2.3.4 Out-of-Plane Bending

To ensure that the arrangement of four atoms in a trigonal planar environment (e.g., a $sp^2$–hybridized carbon atom) remains essentially planar, a quadratic term like $V(\theta) = (k/2)\theta^2$ is used to achieve the desired geometry. By calculating the angle $\theta$ between a bond from the central atom and the plane defined by the central

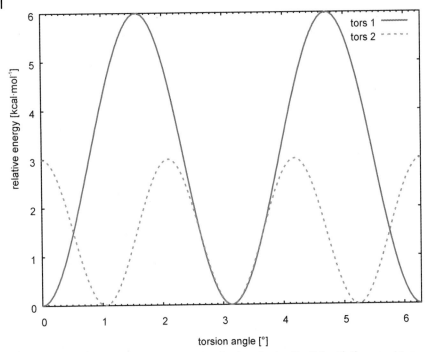

**Figure 7-10.** Two examples of torsional potentials plotted using Eq. (24) with the parameters $n = 2$, $V_n = 6.0$, $\gamma = \pi$ (solid line), and $n = 3$, $V_n = 3.0$, $\gamma = 0$ (broken line). All other $V_n$ contributions are zero.

atom and the remaining two atoms, the term out-of-plane bend becomes the obvious one (see Figure 7-11). The use of a harmonic potential ensures that the additional energy contribution is zero when the angle is zero. Alternatively, the height $h$ of the considered atom above the defined plane may also serve as a variable for the potential.

Another way is to define an improper torsion angle $\omega$ (for atoms 1–2–3–4 in Figure 7-11) in combination with a potential like $V(\omega) = k(1-\cos 2\omega)$, which has its minima at $\omega = 0$ and $\pi$. This of course implies the risk that, if the starting geometry is far from reality, the optimization will perhaps lead to the wrong local minimum.

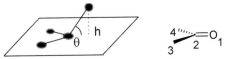

**Figure 7-11.** Modeling out-of-plane bending contributions by defining a plane using three atoms of a molecule and calculating the height $h$ or the angle $\theta$ corresponding to the fourth atom.

### 7.2.3.5 Electrostatic Interactions

The interaction between two charges $q_i$ and $q_j$ separated by the distance $r_{ij}$ in a medium with a dielectric constant $\varepsilon$ is given by Coulomb's law, which sums the energetic contributions over all pairs $ij$ of point charges within a molecule (Eq. (25)).

$$V(q) = \sum_{i<j}^{N} \frac{1}{4\pi\varepsilon_0\varepsilon} \frac{q_i q_j}{r_{ij}} \tag{25}$$

$N$ is the number of point charges within the molecule and $\varepsilon_0$ is the dielectric permittivity of the vacuum. This form is used especially in force fields like AMBER and CHARMM for proteins. As already mentioned, Coulombic 1,4-non-bonded interactions interfere with 1,4-torsional potentials and are therefore scaled (e.g., by 1:1.2 in AMBER). Please be aware that Coulombic interactions, unlike the bonded contributions to the PEF presented above, are not limited to a single molecule. If the system under consideration contains more than one molecule (like a peptide in a box of water), non-bonded interactions have to be calculated between the molecules, too. This principle also holds for the non-bonded van der Waals interactions, which are discussed in Section 7.2.3.6.

An alternative method to the point charge model presented above is to apply the so-called multipole expansion, which is based on electric moments or multipoles. These are charges, dipoles, quadrupoles, octopoles, etc. Based on this principle, MM2, MM3, or MM4 avoid the very time-consuming Coulomb double-sum of interacting charges by a dipole–dipole formulation of the electrostatic energy. Here, dipoles oriented along the polar bonds in a neutral molecule are used to calculate the interaction energy. Despite the fact that many charge–charge interactions are substituted by only a few dipole–dipole interactions (which also saves computational time), the latter approach compares well with the Coulomb form. However, if a molecule contains charged groups, these non-zero monopoles must be included in the electrostatics calculation via additional charge–charge and charge–dipole interaction terms.

In contrast to the point charge model, which needs atom-centered charges from an external source (because of the geometry dependence of the charge distribution they cannot be parameterized and are often pre-calculated by quantum mechanics), the relatively few different bond dipoles are parameterized. An elegant way to calculate charges is by the use of so-called bond increments $\delta_{ij}$ (Eq. (26)), which are defined as the charge contribution of each atom $j$ bound to atom $i$.

$$q_i = q_i^0 + \sum_i \delta_{ij} \tag{26}$$

This model (e.g., implemented in MMFF) allows the charges on each atom $i$ to be modified by the electronegativity of the bonded atoms $j$. The additional charge $q^0$ is normally zero, but it ensures that the correct total charge of a molecule or fragment is maintained.

### 7.2.3.6 Van der Waals Interactions

The rare gas atoms reveal through their deviation from ideal gas behavior that electrostatics alone cannot account for all non-bonded interactions, because all multipole moments are zero. Therefore, no dipole–dipole or dipole–induced dipole interactions are possible. Van der Waals first described the forces that give rise to such deviations from the expected behavior. This type of interaction between two atoms can be formulated by a Lennard-Jones [12–6] function (Eq. (27)).

$$V(r_{ij}) = \varepsilon_{ij} \left[ \left( \frac{R_{ij}^*}{r_{ij}} \right)^{12} - 2 \left( \frac{R_{ij}^*}{r_{ij}} \right)^{6} \right] \qquad (27)$$

The energy function $V(r_{ij})$ passes a minimum at $R_{ij}^*$, which is the sum of the van der Waals radii of the atoms $i$ and $j$; $\varepsilon_{ij}$ is the potential well depth of the atom pair; and $r_{ij}$ is the distance between the interacting atoms. Figure 7-12 shows a plot of Eq. (27) for $\varepsilon_{ij} = 2.0$ kcal $\cdot$ mol$^{-1}$ and $R_{ij}^* = 1.5$ Å. Using $\partial V(r)/\partial r = 0$ at the distance $r = R_{ij}^*$, it can be shown that the collision diameter $\sigma$, where the interaction energy is zero, amounts to $R_{ij}^*/2^{1/6}$.

Qualitatively, the first term of Eq. (27) represents the electron exchange repulsion as a result of the Pauli principle, and the second long-range term accounts for the attractive dispersion interaction. The [12–6] formulation is only qualitatively

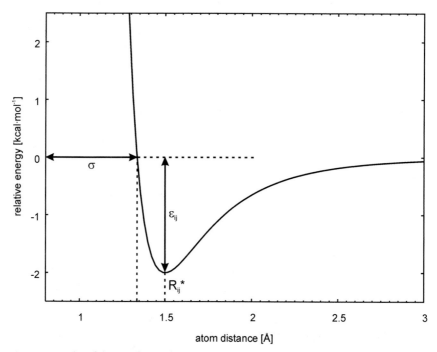

**Figure 7-12.** Plot of the van der Waals interaction energy according to the Lennard-Jones potential given in Eq. (27) ($\varepsilon_{ij} = 2.0$ kcal $\cdot$ mol$^{-1}$, $R_{ij}^* = 1.5$ Å). The calculated collision diameter $\sigma$ is 1.34 Å.

correct, because the repulsive part is too large compared with experimental results. The reason why it is used anyway, especially in force fields like AMBER and CHARMM for proteins, is the fact that the power of 12 can be calculated quickly from the power of 6. Saving time is sometimes more important than physical correctness. In fact, possible problems arising from this "wrong" functional form are corrected by the concerted parameterization of all terms in the potential energy function. As in the case of the non-bonding electrostatic interactions, 1,4 van der Waals contributions are scaled in AMBER by a factor of $\frac{1}{2}$. Interactions between 1,2- and 1,3-atom pairs are completely neglected, because these contributions are considered implicitly in the bond and angle terms.

If computing time does not play the major role that it did in the early 1980s, the [12−6] Lennard-Jones potential is substituted by a variety of alternatives meant to represent the "real" situation much better. MM3 and MM4 use a so-called Buckingham potential (Eq. (28)), where the repulsive part is substituted by an exponential function:

$$V(r) = \varepsilon_{ij} \left[ \frac{6}{\alpha - 6} e^{-\alpha\left(\frac{r}{R^*} - 1\right)} - \frac{\alpha}{\alpha - 6} \left(\frac{R^*}{r}\right)^6 \right] \tag{28}$$

Additionally to $\varepsilon_{ij}$ and $R^*_{ij}$, a third adjustable parameter $\alpha$ was introduced. For $\alpha$-values between 14 and 15, a form very similar to the Lennard-Jones [12−6] potential can be obtained. The Buckingham type of potential has the disadvantage that it becomes attractive for very short interatomic distances. A Morse potential may also be used to model van der Waals interactions in a PEF, assuming that an adapted parameter set is available.

The fact that for non-bonding interactions every atom interacts with every other atom in the molecule or molecular system (leaving cutoffs and other simplifications out) results in the necessity for a large number of parameters. For $N$ different atom types within a molecule, $N(N − 1)/2$ parameters for all non-identical atom type combinations are necessary. To circumvent the difficult and time-consuming process of parameter generation, it is common practice to apply mixing rules. Starting from the well-known collision diameter $\sigma$ and the well depth $\varepsilon$ for the pure species A and B, parameters for mixed interactions can be calculated according to Lorentz–Berthelot (Eq. (29)).

$$\sigma_{AB} = \frac{1}{2}(\sigma_{AA} + \sigma_{BB})$$

$$\varepsilon_{AB} = \sqrt{(\varepsilon_{AA}\varepsilon_{BB})} \tag{29}$$

As Eq. (29) shows, $\sigma_{AB}$ corresponds to the arithmetic mean and $\varepsilon_{AB}$ to the geometric mean of the homodimeric parameters AA and BB.

All of the contributions to the energy function presented above assume that pairwise interactions are sufficient to describe the situation within a molecule or molecular system. Whether or not multi-centered interactions are negligible is controversial. On the other hand, failure or success of a force field with its functional form and corresponding parameter set is not a matter of mathematics or exhaustive analytical considerations. It is simply the ability of the PEF to reproduce experimental data that makes it useful. It is only a model to describe the real world!

### 7.2.3.7 Cross-Terms

The interaction between two or more internal coordinates of a molecule can be reflected by the presence of cross-terms within the PEF of a force field. Let us take the water molecule as a simple but representative example: decreasing the bond angle between the two hydrogen atoms results in a stretching of the corresponding C–H bonds. Figure 7-13 depicts several such interactions, including, e.g., bond–bond, bond–angle, bond–torsion, and angle–torsion combinations.

Intensive use of cross-terms is important in force fields designed to predict vibrational spectra, whereas for the calculation of molecular structure only a limited set of cross-terms was found to be necessary. For the above-mentioned example, the coupling of bond-stretching ($l_1$ and $l_2$) and angle-bending ($\theta$) within a water molecule (see Figure 7-13, top left) can be calculated according to Eq. (30).

$$V(l_1, l_2, \theta) = \frac{k_{l_1, l_2, \theta}}{2} [(l_1 - l_{1,0}) + (l_2 - l_{2,0})](\theta - \theta_0) \tag{30}$$

In the so-called Urey–Bradley force fields, angle bending is not achieved by an explicit potential, but rather by using a 1,3-non-bonded interaction between the atoms 1 and 3 forming that angle. This is modeled by a harmonic potential describing the distance between these non-covalently bound atoms (Eq. (31)).

$$V(r_{1,3}) = \frac{k_{r_{1,3}}}{2} (r_{1,3} - r_{1,3}^0)^2 \tag{31}$$

Extensive use of cross-terms is made by Hagler's quantum mechanically derived force field CFF, because one of the intentions of this type of PEF is to obtain ac-

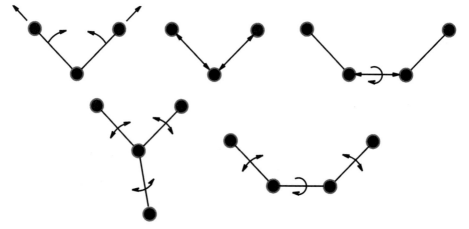

**Figure 7-13.** Cross-terms combining internal vibrational modes such as bond stretch, angle bend, and bond torsion within a molecule.

curate vibrational spectra and a high predictive value for non-equilibrium compounds such as geometrically distorted and strained systems and transition states.

The mathematical form of the PEF is in almost every case a compromise between speed and accuracy. As computer power continually increases, ideally following Moore's Law, and the cost/performance ratio is getting better, one might think that there is no longer a need to sacrifice accuracy to save computational time. This is not really true, because in direct proportion to the CPU speed is the rise in the scientists' interest in calculating larger and larger molecules (in fact, their interest always rises faster than the CPU speed).

## 7.2.4
## Available Force Fields

In this section there is an attempt to give a compact overview of currently available force fields; however, it is by no means complete, as in this context it is not possible to mention every feature or functionality (minimization, dynamics, polarization, periodic boundaries, free energies, treatment of long-range electrostatics, water models, NMR restraints, QM/MM capabilities, etc.) implemented in the various packages. For further information, readers should consult the references or web sites cited. Mindful readers may recognize that the well-known, mainly commercial modeling packages, which all include implementations of one or more force fields, have largely been omitted. The reason is that the focus of this compilation is on the different parameterizations, and not on their implementation.

There are several excellent publications in the literature which compare force fields, their application areas, and their pros and cons [1–5]. Available force field parameters are published in a comprehensive and very extensive form, e.g., within the *Reviews in Computational Chemistry* series [6, 7].

### 7.2.4.1   Force Fields for Small Molecules
<<URLs following headings 7.2.4.x.y in Sections 7.2.4.1 and 7.2.4.2 should be in the heading style>>

#### 7.2.4.1.1   MM2/MM3/MM4
*http://europa.chem.uga.edu/allinger/mm2mm3.html*

The parameterizations provided by Norman Allinger's group are probably the best-known implementations of the molecular mechanics concept and are widely used as a synonym for force field calculations in general. MM2 was developed by Allinger and co-workers in 1977 and is designed primarily for hydrocarbons [8]. It was extended by many functional groups to cover almost all kinds of small organic molecules [9]. One has to differentiate between an update of the parameters only (modified or new functional groups) and changes in the functional form (e.g., treatment of conjugated systems or atoms with more than four bonds) of the potential energy function. The latest version of this program, MM2(91), was just a parameter update; the last change in the functional form of

MM2 was, according the web site of the authors, released as MM2(87). The various MM2 flavors are superseded by MM3, with significant improvements in the functional form [10]. It was also extended to handle amides, polypeptides, and proteins [11]. The last release of this series was MM3(96). Further improvements followed by starting the MM4 series, which focuses on hydrocarbons [12], on the description of hyperconjugative effects on carbon–carbon bond lengths [13], and on conjugated hydrocarbons [14] with special emphasis on vibrational frequencies [15]. For applications of MM2 and MM3 in inorganic systems, readers are referred to the literature [16–19].

Variations of these releases are implemented in almost every commercial or academic software package, which cannot be listed in this context. A comprehensive comparison of several force fields focusing the calculation of conformational energies of organic molecules has been published by Pettersson and Liljefors [1].

### 7.2.4.1.2 TINKER

*http://dasher.wustl.edu/tinker*

TINKER is a modular molecular modeling package designed for molecular mechanics, dynamics, and several other energy-based structural manipulation calculations. The software in its actual version is distributed with several parameter sets (AMBER94/96, CHARMM27, MM2(1991), MM3(2000), OPLS-AA, and OPLS-UA), which makes it an ideal tool for their comparison on the investigation system. In addition, the authors are actively developing their own protein force field (also called TINKER) which is based on polarizable atomic multipole electrostatics [20]. In the actual release, only the TINKER parameters for the polarizable water model are included. According to the authors, this model is "equal or better to the best available water models for many bulk and cluster properties" [21]. Worth mentioning is the large number of modules provided for tasks like potential surface scanning, global optimization, solvent treatment, normal mode analysis, and many more.

### 7.2.4.1.3 UFF

*http://franklin.chm.colostate.edu/mmac/uff.html*

The Universal Force Field, UFF, is one of the so-called "whole periodic table" force fields. It was developed by A. Rappé, W. Goddard III, and others. It is a set of simple functional forms and parameters used to model the structure, movement, and interaction of molecules containing any combination of elements in the periodic table. The parameters are defined empirically or by combining atomic parameters based on certain rules. Force constants and geometry parameters depend on hybridization considerations rather than individual values for every combination of atoms in a bond, angle, or dihedral. The equilibrium bond lengths were derived from a combination of atomic radii. The parameters [22, 23], including metal ions [24], were published in several papers.

The authors emphasize on their web pages that UFF is not designed to be used in conjunction with partial atomic charges, as it is the default option in several software packages. A second point is that UFF is often used to model biological systems. In fact, it was never designed for peptides, proteins, or nucleic acids. The published force field, unlike eventually modified or extended implementations, also lacks hydrogen bonding terms.

### 7.2.4.1.4 MOMEC

*http://www.uni-heidelberg.de/institute/fak12/AC/comba/*

MOMEC is a molecular mechanics program for strain energy minimization of inorganic and coordination compounds. It has been developed by Peter Comba (University of Heidelberg, Germany), and Trevor Hambley (University of Sydney, Australia) especially for transition metal and rare earth compounds. Special emphasis is placed on the modeling of coordination geometries by including intraligand repulsion, electrostatic interactions, ligand field-based electronic functions, and the possibility of combining various potentials. Additional 1,3-non-bonded interaction terms can be added, and various bonded or non-bonded contributions to the energy function may be selectively activated or turned off. A module for the refinement of Jahn–Teller-distorted hexacoordinated compounds is provided.

The modeling of inorganic compounds in general is gaining more and more interest [25–28]. The authors of MOMEC addressed this in a monograph describing how molecular modeling techniques can be applied to metal complexes and how the results can be interpreted [29]. The current force field parameter set is available on the author's web site.

### 7.2.4.1.5 COSMOS

*http://www.cosmos-software.de/*

Many problems in force field investigations arise from the calculation of Coulomb interactions with fixed charges, thereby neglecting possible mutual polarization. With that obvious drawback in mind, Ulrich Sternberg developed the COSMOS (Computer Simulation of Molecular Structures) force field [30], which extends a classical molecular mechanics force field by semi-empirical charge calculation based on bond polarization theory [31, 32]. This approach has the advantage that the atomic charges depend on the three-dimensional structure of the molecule. Parts of the functional form of COSMOS were taken from the PIMM force field of Lindner et al., which combines self-consistent field theory for $\pi$-orbitals ($\pi$-SCF) with molecular mechanics [33, 34].

The authors of COSMOS performed their parameterization with the intention of minimizing the number of necessary parameters, as in the case of UFF [24]. As in most of the other force field implementations presented in this chapter, Sternberg introduced no specific hydrogen bonding term to the PEF, because this type of interaction is essentially caused by electrostatics. COSMOS was recently parameterized and validated to handle zinc complexes with fluctuating atomic charges, so it should be useful for research groups working on zinc-containing enzyme systems [35]. Other (tested) areas of application are interaction energies and hydrogen bond geometries for pairs of small molecules, and DNA/RNA base pairs. The software can be obtained from the authors upon request [36].

### 7.2.4.2  **Force Fields for Biomolecules**

#### 7.2.4.2.1  **AMBER**
*http://amber.scripps.edu*

One of the most popular force fields for modeling proteins and nucleic acids is AMBER (Assisted Model Building with Energy Refinement), developed under the leadership of Peter Kollman at UCSF in collaboration with other groups in industry and academics. The 2002 version is AMBER 7. It comes with a variety of parameterizations. The default parameter set for versions 5 and 6 was the Cornell et al. (1994) force field [37], named parm94, which is still widely used. Compared with the original Weiner et al. (1984 and 1986) force fields, parm91X [38, 39], explicit hydrogen bonding terms were omitted and no united-atom counterpart of the parameterization was provided. Minor modifications and improvements led to parm96 [40] and parm98 [41]. By modifying the charge-fitting procedure and attaching polarizable dipoles to the atoms, the authors created a polarizable variant of parm99 [42]. The problem of missing parameters for small organic molecules (such as inhibitors, substrates, or ligands) was addressed by providing a so-called "general amber force field" (gaff). In combination with auxiliary programs such as *antechamber*, the generation of the necessary input files for non-standard molecules is facilitated.

Modifications of the various versions of the AMBER parameter sets are implemented in numerous commercial and academic software packages. They are often referred to as AMBER*. In every case, the user should read the documentation provided critically, and check the implementation by comparing the results of the implementation with "original" data.

#### 7.2.4.2.2  **CHARMM**
*http://yuri.harvard.edu*

CHARMM, which stands for Chemistry at HARvard Macromolecular Mechanics, is designed for macromolecular simulations including energy minimization, molecular dynamics simulations, and normal mode calculations. The development of the code was started and is coordinated by Martin Karplus (Cambridge, MA) and continues throughout the world with contributing developers in over 20 universities, research institutes, and companies. Originally described by Brooks et al. in 1983 [43] and Nilsson in 1986 [44], the energy function used an extended atom model, which treated all hydrogen atoms as part of the corresponding heavy atom [43]. Polar hydrogen atoms were included to better represent hydrogen bonds only by Coulomb and van der Waals interactions (CHARMM19 parameter set for proteins) [45]. Although the increase in computer power allowed the development of several all-atom parameter sets for proteins [46], lipids [47], and nucleic acids [48–50] (the version from July 2002 is named c30a1), the united-atom implementation is still used for complex and time-consuming applications such as folding simulations or free energy calculations on model proteins [51]. Please note that CHARMM (with the last M a capital letter) refers to the academic version. The commercial, somewhat modified, counterpart is named CHARMm and is available within several Accelrys products, which were formerly distributed by Biosym–MSI as part of the Quanta or Insight II packages [52].

### 7.2.4.2.3 **GROMOS**
*http://www.igc.ethz.ch/gromos/*

GROMOS (Groningen Molecular Simulation) is a molecular dynamics computer simulation package for the study of biomolecular systems with the purpose of energy minimization and the simulation of molecules in solution or solid state by molecular dynamics, stochastic dynamics, or path-integral methods. The development has been performed from the early 1980s onward by Wilfred van Gunsteren and Herman Berendsen in Groningen (The Netherlands), and from 1990 onward in Zürich (Switzerland). The ongoing improvements in the software and the force field have not been published in every detail, but there are two major versions that should be mentioned here: GROMOS87 [53] and GROMOS96 [54], which contains a complete rewrite of the molecular dynamics, stochastic dynamics, and energy minimization part of the package. The latest versions are the general 43A1 force field for application in aqueous or apolar solution, and the 43B1 force field for the simulation of isolated molecules in the gas phase. Apart from the GROMOS96 manual [54], the force field parameters have not been published in comprehensive form. Schuler et al. reported the re-parameterization of the 96-force field regarding aliphatic carbons, to simulate the behavior of aliphatic hydrocarbons in the condensed phase necessary to apply GROMOS to lipid systems [55]. Despite the fact that computational resources continue to increase, the united-atom version of the GROMOS96 force field is still important and improved, e.g., for calculating alkanes in the liquid phase [56].

### 7.2.4.2.4 **OPLS**
*http://zarbi.chem.yale.edu*

The OPLS (Optimized Potentials for Liquid Simulations) force fields calculate the potential energy of a molecular system using harmonic terms for bond stretching and angle bending, a Fourier series for each torsional angle, and Coulomb and Lennard-Jones interactions between atoms separated by three or more bonds [57]. Originally, the intention of the developers was to reproduce properties of pure fluids (e.g., water and liquid organics), and the dynamic behavior of organic molecules dissolved in a macroscopic environment. Later, the parameterizations have been expanded to treat proteins also.

The original potential function, OPLS-UA, used a partially united-atom (UA) model with implicit hydrogen atoms attached to aliphatic carbons [58]. The optimization of the very sensitive non-bonding parameters was done by Monte Carlo simulations of pure liquids and led to the development of the water potential functions TIP3P and TIP4P, which are still two of the most often used water models [59]. To apply the OPLS-UA force field to peptides and proteins, the well-established non-bonding parameters were merged with the bond-stretching, angle-bending, and torsional parts of the AMBER united-atom force field developed by Weiner et al. [38]. The combination obtained was called the OPLS/AMBER force field [60].

The fact that the additional sites in all-atom models allow better charge distributions and torsional energetics resulted in the development of an all-atom model (OPLS-AA) for hydrocarbons, which reduced the average error of the free energies of hydration of alkanes considerably [61]. The extension of this parameter set by organic functional groups finally added all the organic components necessary to simulate proteins [62]. Again, non-bonded parameters obtained by Monte Carlo simulations were combined with the bond-stretching and angle-bending terms from the AMBER all-atom force field [39]. Because of their strong coupling, the non-bonded and torsional contributions to the PEF were simultaneously parameterized.

### 7.2.4.2.5  ECEPP

*http://www.tc.cornell.edu/reports/NIH/resource/CompBiologyTools/eceppak/*

The Empirical Conformational Energy Program for Peptides, ECEPP [63, 64], is one of the first empirical interatomic potentials whose derivation is based both on gas-phase and X-ray crystal data [65]. It was developed in 1975 and updated in 1983 and 1992. The actual distribution (dated May, 2000) can be downloaded without charge for academic use.

The intention of the software is to compute the energetically most favored conformations of a polypeptide or protein. Because conformational energy calculations are computationally very expensive, the authors chose the unusual way (compared with other commonly used force fields like AMBER, CHARMM, GROMOS, or OPLS) of not optimizing amino acid bond lengths and angles. In fact, the bond lengths and angles are fixed to averages computed over observed values from X-ray and neutron diffraction studies on oligopeptide crystals. According to the authors, the use of rigid geometry for the amino acid residues reduces the number of variables and simplifies the calculations to a great extent while still providing a reasonably accurate representation of the properties of the polypeptide chain [65].

### 7.2.4.2.6  CVFF/CFF

*http://struktur.kemi.dtu.dk/cff/cffhome.html*

The consistent force field (CFF) has been designed for the calculation of structures, energies, and vibrational frequencies of both small organic molecules and large biomolecular systems, including peptides, proteins, nucleic acids, carbohydrates, and lipids, in vacuo and in the condensed phase [66]. It differs from empirical force fields in that its force constants are derived from a generalizable quantum mechanical procedure. The authors call this a "quantum mechanical force field" (QMFF), because "observables" generated ab initio describe the energy hypersurface of a family of molecules [67]. Scaling of the QMFF force constants gives the final values for CFF, because it is well known that force constants derived from Hartree–Fock theory tend to be too large, and bond lengths tend systematically to be too small.

Special emphasis was placed on the calculation of spectroscopic properties and properties of distorted molecules. The potential energy function of CFF is dominated by anharmonic contributions up to quartic terms and cross-terms representing coupling interactions, especially in combination with torsional motions. Additionally, large amounts of quantum mechanically derived data for strained molecules were used during the parameterization process [67].

### 7.2.4.2.7 **MMFF**

Inadequate availability of experimental data can considerably inhibit the development of improved energy functions for more accurate simulations of energetic, structural, and spectroscopic properties. This has led to the development of class II force fields such as CFF and the Merck Molecular Force Field (MMFF), which are both based primarily on quantum mechanical calculations of the energy surface. The purpose of MMFF, which has been developed by Thomas Halgren at Merck and Co., is to be able to handle all functional groups of interest in pharmaceutical design.

The current version, MMFF94 [68–70], has been implemented in various programs such as the academic version of CHARMM (giving CHARMM extensive small-molecule functionality), but also in several commercial software packages like CHARMm via QUANTA or Cerius2, MacroModel, Spartan, or Sybyl. The MMFF94 parameters are available as supplementary material for ref. [68] via the URL *ftp://ftp.wiley.com/public/journals/jcc/suppmat/17/490/*. A validation suite for MMFF94 has been posted to the Computational Chemistry List Web archive maintained by the Ohio Supercomputer Center [71].

# References

[1] I. Pettersson, T. Liljefors, Molecular mechanics calculated conformational energies of organic molecules: a comparison of force fields, in *Reviews in Computational Chemistry, Vol. 9*, K. B. Lipkowitz, D. B. Boyd (Eds.), VCH Publishers, New York, **1996**, pp. 167–189.

[2] D. J. Price, C. L. Brooks III, *J. Comput. Chem.*, **2002**, *23*, 1045–1057.

[3] A. D. MacKerell, Jr., Protein force fields, in *The Encyclopedia of Computational Chemistry, Vol. 3*, P. v. R. Schleyer, N. L. Allinger, T. Clark, J. Gasteiger, P. A. Kollman, H. F. Schaefer III, P. R. Schreiner (Eds.); John Wiley & Sons, Chichester, **1998**, pp. 2191–2200.

[4] K. Gundertofte, J. Palm, I. Pettersson, A. Stamvik, *J. Comput. Chem.* **1991**, *12*, 200–208.

[5] K. Gundertofte, T. Liljefors, P.-O. Norrby, I. Pettersson, *J. Comput. Chem.* **1996**, *17*, 429–449.

[6] M. Jalaie, K. B. Lipkowitz, Published force field parameters for molecular mechanics, molecular dynamics, and Monte Carlo simulations, in *Reviews in Computational Chemistry, Vol. 14*, K. B. Lipkowitz, D. B. Boyd (Eds.), Wiley-VCH, New York, **2000**, pp. 441–486.

[7] E. Osawa, K. B. Lipkowitz, Published force field parameters, in *Reviews in Computational Chemistry, Vol. 6*, K. B. Lipkowitz, D. B. Boyd (Eds.), VCH, New York, **1995**, pp. 355–381.

[8] N. L. Allinger, *J. Am. Chem. Soc.* **1977**, *99*, 8127–8134.

[9] U. Burkert, N. L. Allinger, *Molecular Mechanics*, ACS Monograph No. 177, American Chemical Society, Washington, DC, **1982**, **1986**.

[10] N. L. Allinger, Y. H. Yuh, J.-H. Lii, *J. Am. Chem. Soc.* **1989**, *111*, 8551–8565.

[11] J.-H. Lii, N. L. Allinger, *J. Comput. Chem.* **1991**, *12*, 186–199.

[12] N. L. Allinger, K. Chen, J.-H. Lii, *J. Comput. Chem.* **1996**, *17*, 642–668.

[13] N. L. Allinger, K. Chen, J. A. Katzenellenbogen, S. R. Wilson, G. M. Anstead, *J. Comput. Chem.* **1996**, *17*, 747–755.

[14] N. Nevins, J.-H. Lii, N. L. Allinger, *J. Comput. Chem.* **1996**, *17*, 695–729.

[15] N. Nevins, N. L. Allinger, *J. Comput. Chem.* **1996**, *17*, 730–746.

[16] B. P. Hay, *Coord. Chem. Rev.* **1993**, *126*, 177–236 and references therein.

[17] P. C. Yates, A. K. Marsden, *Computers & Chemistry* **1994**, *18*, 89–94.

[18] A. Albinati, F. Lianza, H. Berger, P. S. Pregosin, H. Ruegger, R. W. Kunz, *Inorg. Chem.* **1993**, *32*, 478–486.

[19] B. P. Hay, J. R. Rustad, *J. Am. Chem. Soc.* **1994**, *116*, 6313.

[20] M. J. Dudek, J. W. Ponder, *J. Comput. Chem.* **1995**, *16*, 791–816.

[21] Y. Kong, J. W. Ponder, *J. Chem. Phys.* **1997**, *107*, 481–492.

[22] A. K. Rappé, C. J. Casewit, K. S. Colwell, W. A. Goddard III, W. M. Skiff, *J. Am. Chem. Soc.* **1992**, *114*, 10024–10035.

[23] C. J. Casewit, K. S. Colwell, A. K. Rappé, *J. Am. Chem. Soc.* **1992**, *114*, 10035–10046, 10046–10053.

[24] A. K. Rappé, K. S. Colwell, C. J. Casewit, *Inorg. Chem.* **1993**, *32*, 3438–3450.

[25] J. C. A. Boeyens, P. Comba, *Coord. Chem. Rev.* **2001**, *212*, 3–10.

[26] P. Comba, *Coord. Chem. Rev.* **2000**, 200–202, 217–245.

[27] P. Comba, *Coord. Chem. Rev.* **1999**, *182*, 343–371.

[28] P. Comba, Molecular mechanics modeling of transition metal compounds, in *NATO Science Series, Series E: Applied Sciences, Vol. 360*, Implications of Molecular and Materials Structure for New Technologies, **1999**, pp. 71–86.

[29] P. Comba, T. Hambley, *Molecular Modeling of Inorganic Compounds*, 2nd ed., Wiley-VCH, Weinheim, **2000**.

[30] M. Möllhoff, U. Sternberg, *J. Mol. Model.* **2001**, *7*, 90–102.

[31] U. Sternberg, *Mol. Phys.* **1988**, *63*, 249–267.

[32] U. Sternberg, F.-T. Koch, M. Möllhoff, *J. Comput. Chem.* **1994**, *15*, 524–531.

[33] H. J. Lindner, *Tetrahedron* **1974**, *30*, 1127–1132.

[34] M. Kroeker, H. J. Lindner, *J. Mol. Model.* **1996**, *2*, 376–378.

[35] U. Sternberg, F.-T. Koch, M. Bräuer, M. Kunert, E. Anders, *J. Mol. Model.* **2001**, *7*, 54–64.

[36] Further information can be obtained by email from sternberg@ioq.physik.uni–jena.de.

[37] W. D. Cornell, P. Cieplak, C. I. Bayly, I. R. Gould, K. M. Merz, Jr., D. M. Ferguson, D.C. Spellmeyer, T. Fox, J. W. Caldwell, P. A. Kollman, *J. Am. Chem. Soc.* **1995**, *117*, 5179–5197.

[38] S. J. Weiner, P. A. Kollman, D. A. Case, U. C. Singh, C. Ghio, G. Alagona, S. Profeta, Jr., P. J. Weiner, *J. Am. Chem. Soc.* **1984**, *106*, 765–784.

[39] S. J. Weiner, P. A. Kollman, D. T. Nguyen, D. A. Case, *J. Comput. Chem.* **1986**, *7*, 230–252.

[40] P. A. Kollman, R. Dixon, W. Cornell, T. Fox, C. Chipot, A. Pohorille, The development/application of a 'minimalist' organic/biochemical molecular mechanic force field using a combination of ab-initio calculations and experimental data, in *Computer Simulation of Biomolecular Systems, Vol. 3*, A. Wilkinson, P. Weiner, W. F. van Gunsteren (Eds.), Elsevier, Amsterdam, **1997**, pp. 83–96.

[41] T. E. Cheatham III, P. Cieplak, P. A. Kollman, *J. Biomol. Struct. Dyn.* **1999**, *16*, 845–862.

[42] P. Cieplak, J. Caldwell, P. Kollman, *J. Comput. Chem.* **2001**, *22*, 1048–1057.

[43] B. R. Brooks, R. E. Bruccoleri, B. D. Olafson, D. J. States, S. Swaminathan, M. Karplus, *J. Comp. Chem.* **1983**, *4*, 187–217.

[44] L. Nilsson, M. Karplus, *J. Comput. Chem.* **1986**, *7*, 591–616.

[45] E. Neria, M. Karplus, *J. Chem. Phys.* **1996**, *105*, 10812–10818.

[46] A. D. MacKerell, Jr., D. Bashford, M. Bellott, R. L. Dunbrack, Jr., J. D. Evanseck, M. J. Field, S. Fischer, J. Gao, H. Guo, S. Ha, D. Joseph-McCarthy, L. Kuchnir, K. Kuczera, F. T. K. Lau, C. Mattos, S. Michnick, T. Ngo, D. T. Nguyen, B. Prodhom, W. E. Reiher III, B. Roux, M. Schlenkrich, J. C. Smith, R. Stote, J. Straub, M. Watanabe, J. Wiorkiewicz-Kuczera, D. Yin, M. Karplus, *J. Phys. Chem. B*, **1998**, *102*, 3586–3616.

[47] S. E. Feller, A. D. MacKerell, Jr., *J. Phys. Chem. B* **2000**, *104*, 7510–7515.

[48] A. D. MacKerell, Jr., N. Banavali, N. Foloppe, *Biopoly (Nucleic Acid Sci.)* **2001**, *56*, 257–265.

[49] N. Foloppe, A. D. MacKerell, Jr., *J. Comput. Chem.* **2000**, *21*, 86–104.

[50] A. D. MacKerell, Jr., D. Alexander, N. K. Banavali, *J. Comput. Chem.* **2000**, *21*, 105–120.

[51] E. M. Boczko, C. L. Brooks III, *Science*, **1995**, *269*, 393–396.

[52] F. A. Momany, R. Rone, *J. Comput. Chem.* **1992**, *13*, 888–900.

[53] W. F. van Gunsteren, H. J. C. Berendsen, *Groningen Molecular Simulation (GROMOS) Library Manual*, Biomos, Nijenborgh 16, Groningen, The Netherlands, **1987**.

[54] W. F. van Gunsteren, S. R. Billeter, A. A. Eising, P. H. Hünenberger, P. Krüger, A. E. Mark, W. R. P. Scott, I. G. Tironi, *Biomolecular Simulation: The Gromos 96 Manual and User Guide*, Vdf Hochschulverlag AG, ETH Zürich, **1996**.

[55] L. D. Schuler, X. Daura, W. F. van Gunsteren, *J. Comput. Chem.* **2001**, *22*, 1205–1218.

[56] X. Daura, A. E. Mark, W. F. van Gunsteren, *J. Comput. Chem.* **1998**, *19*, 535–547.

[57] W. L. Jorgensen, OPLS force fields, in *The Encyclopedia of Computational Chemistry, Vol. 3*, P. v. R. Schleyer, N. L. Allinger, T. Clark, J. Gasteiger, P. A. Kollman, H. F. Schaefer III, P. R. Schreiner (Eds.); John Wiley & Sons, Chichester, **1998**, pp. 1986–1989.

[58] W. L. Jorgensen, J. D. Madura, *Mol. Phys.* **1985**, *56*, 1381–1392.

[59] W. L. Jorgensen, J. Chandrasekhar, J. D. Madura, R. W. Impey, M. L. Klein, *J. Chem. Phys.* **1983**, *79*, 926–935.

[60] W. L. Jorgensen, J. Tirado-Rives, *J. Am. Chem. Soc.* **1988**, *110*, 1657–1666.

[61] G. Kaminski, E. M. Duffy, T. Matsui, W. L. Jorgensen, *J. Phys. Chem.* **1994**, *98*, 13077–13082.

[62] W. L. Jorgensen, D. S. Maxwell, J. Tirado-Rives, *J. Am. Chem. Soc.* **1996**, *118*, 11225–11236.

[63] G. Nemethy, H. A. Scheraga, *J. Phys. Chem.* **1983**, *87*, 1883–1891.

[64] G. Nemethy, K. D. Gibson, K. A. Palmer, C. N. Yoon, G. Paterlini, A. Zagari, S. Rumsey, H. A. Scheraga, *J. Phys. Chem.* **1992**, *96*, 6472–6484.

[65] D. R. Ripoll, H. A. Scheraga, ECEPP: Empirical Conformational Energy Program for Peptides, in *The Encyclopedia of Computational Chemistry, Vol. 2*, P. v. R. Schleyer, N. L. Allinger, T. Clark, J. Gasteiger, P. A. Kollman, H. F. Schaefer III, P. R. Schreiner (Eds.); John Wiley, Chichester, **1998**, pp. 813–815.

[66] J. R. Maple, Force fields: CFF, in *The Encyclopedia of Computational Chemistry, Vol. 2*, P. v. R. Schleyer, N. L. Allinger, T. Clark, J. Gasteiger, P. A. Kollman, H. F. Schaefer III, P. R. Schreiner (Eds.); John Wiley, Chichester, **1998**, pp. 1025–1028.

[67] J. R. Maple, M.-J. Hwang, K. J. Jalkanen, T. P. Stockfisch, A. T. Hagler, *J. Comput. Chem.* **1998**, *19*, 430–458.

[68] T. A. Halgren, *J. Comput. Chem.* **1996**, *17*, 490–519.

[69] T. A. Halgren, *J. Comput. Chem.* **1996**, *17*, 520–552, 553–586, 616–641.

[70] T. A. Halgren, R. B. Nachbar, *J. Comput. Chem.* **1996**, *17*, 587–615.

[71] *http://www.ccl.net/cca/data/*

**7.3**
**Molecular Dynamics**

*Harald Lanig*

7.3.1
**Introduction**

To investigate the behavior of molecules, the energy is a very useful property. As we have seen in Section 7.2 on molecular mechanics, it can be calculated by using a relatively simple mechanical model, which approximates atoms as masses and bonds as connecting springs applying forces on the atoms (so this is called a force field). The simplicity of this method ensures the rapid calculation of an energy corresponding to a certain geometry of a molecule, also denoted as a conformation. This so-called single-point calculation gives us information about the static properties of molecules, which, besides the molecular mechanics, can include, for example, derivatives of the potential energy and molecular orbital energies, including the coefficients for the ground and electronically excited states. The molecular structure on which a single-point calculation is based normally reflects a stationary point on the potential energy surface of the molecule (e.g., a minimum or a transition state). If not, one way to remove excess strain is to perform a geometry optimization. Unfortunately, this type of calculation only leads to the next local minimum on the energy hypersurface; it is not possible to overcome potential energy barriers. If one is interested in other local minima and/or the global minimum of a molecule, one possible type of conformational analysis is molecular dynamics (MD) simulations.

A molecular dynamics simulation samples the phase space of a molecule (defined by the position of the atoms and their velocities) by integrating Newton's equations of motion. Because MD accounts for thermal motion, the molecules simulated may possess enough thermal energy to overcome potential barriers, which makes the technique suitable in principle for conformational analysis of especially large molecules. In the case of small molecules, other techniques such as systematic, random, Genetic Algorithm-based, or Monte Carlo searches may be better suited for effectively sampling conformational space.

7.3.2
**The Continuous Movement of Molecules**

Even at 0 K, molecules do not stand still. Quantum mechanically, this unexpected behavior can be explained by the existence of a so-called zero-point energy. Therefore, simplifying a molecule by thinking of it as a collection of balls and springs which mediate the forces acting between the atoms is not totally unrealistic, because one can easily imagine how such a mechanical model wobbles around, once "activated" by an initial force. Consequently, the movement of each atom influences the motion of every other atom within the molecule, resulting in a complex mathematical problem of coupled differential equations. This can be solved only by numerical integration and not analytically.

By looking more closely at how molecules move, we find that bonds between two atoms can vibrate, angles between three atoms can bend, and torsions between four atoms can twist. These types of elementary motions can be combined for groups of atoms, leading to the motion of substituents (e.g., the rotation of a methyl group) or even whole domains (e.g., in proteins). If we want to simulate how these motions occur, our protocol must allow the sampling of the fastest possible movement of an atom within the system under consideration. These are the vibrations of bonds involving a hydrogen atom (e.g., a C–H bond in a methyl group), taking between 10 and 100 fs. Therefore, the integration steps when the equations of motion are being solved numerically must be at least one order of magnitude smaller than the fastest motion, i.e., about 1 fs. Otherwise, one would run into problems concerning the numerical stability of the algorithms used. Considering the fact that the rotation around a single bond needs about 100 ps (of course, the number depends strongly on the rotational barrier modulated not only by the atoms involved, but also on the environment), the simulation of this elementary process needs about $10^5$ integration steps. The time necessary for one step depends mainly on the size of the molecule, because the energy of the whole system has to be recalculated for the actual geometry. If one is interested in complex systems like proteins, the calculation of the energy for a specific geometry (a single point) may increase up to 1 s. Additionally, complex motions in proteins occur on a much larger time scale. The folding of some proteins from the denatured state to the active conformation may last about 1 s (or $10^{15}$ integration steps). Taking these facts into account it is easy to understand that the simulation of such an event is not possible with the computer power and algorithms currently available.

### 7.3.3
### Methods

The development of methods of describing the motion of atoms in molecules accurately started by using Newton' s equations of motion. It quickly became clear that the calculation of the non-bonding interactions is not only the most time-consuming step during the energy evaluation, but the long-range electrostatic contributions are also very error-prone, especially when cutoffs are applied. Therefore much effort has been expended on improving the efficient calculation of the long-range forces. During recent years, first-principle or *ab-initio* molecular dynamics seems to have been the major target of software development [1]. There are several excellent books and reviews available covering different aspects of molecular dynamics simulations, development of methods, and applications [2–8].

## 7.3.3.1 Algorithms

To describe mathematically a molecule in the gas phase or surrounded by a macro-scopic solvent environment (in terms of a supermolecule approach), we first need an empirical potential energy function $U_i$. For the treatment of proteins or DNA, well-known examples are AMBER, CHARMM, GROMOS, or OPLS (see Section 7.2). A typical potential energy function used for proteins comprises bonding terms representing bond lengths $l$, angles $\theta$, and torsions $\omega$, as well as non-bonding terms such as electrostatic and van der Waals contributions (Eq. (32)):

$$
U_i = \sum_{bonds} \frac{k_i}{2}(l_i - l_{0,i})^2 + \sum_{angles} \frac{k_i}{2}(\Theta_i - \Theta_{0,i})^2
$$

$$
+ \sum_{torsions} \frac{V_n}{2}(1 + \cos(n\omega - \gamma))
$$

$$
+ \sum_{i=1}^{N}\sum_{j=i+1}^{N}\left(4\varepsilon_{ij}\left[\left(\frac{\sigma_{ij}}{r_{ij}}\right)^{12} - \left(\frac{\sigma_{ij}}{r_{ij}}\right)^{6}\right] + \frac{q_i q_j}{4\pi\varepsilon_0 r_{ij}}\right) \tag{32}
$$

The parameters necessary for the calculation of a force field energy are the force constants $k$ for bonds and angles, as well as their corresponding reference bond lengths $l_0$ and angles $\theta_0$. The energy contribution of a torsion is described via the "barrier height" $V$, the torsion angle $\omega$, the multiplicity $n$, and the phase shift $\gamma$. The non-bonding van der Waals interactions are characterized by the atom-pair collision parameters $\sigma$, the atom–atom distance $r$, and the corresponding potential well depth $\varepsilon$. For the calculation of the non-bonded electrostatic interactions, the charges $q$ on the interacting atoms as well as the distance $r$ between them are necessary. This issue is described in detail in Section 7.2 on molecular mechanics.

The starting point of an MD simulation is an initial set of coordinates, which may originate from X-ray crystallographic or NMR investigations. To remove bad contacts and initial strain (usually van der Waals overlap of non-bonded atoms), which may disturb the subsequent MD simulation, the structure is normally geometry-optimized using the same potential energy function. After assigning velocities $v$, which typically represent a low-temperature Maxwell distribution, the simulation is started by calculating the acceleration $a_i$ for every atom $i$ according to Newton's law $F_i = m_i\,a_i$, written as in Eq. (33).

$$
-\frac{\partial U(x_i)}{\partial x_i} = F_i = m_i a_i = m_i \frac{\partial^2 x_i}{\partial t^2} \tag{33}
$$

In this case, $F_i$ is the force acting on the Cartesian coordinate $x_i$ for $i = 1, ..., 3N$ for the $N$ atoms in the molecule or molecular system, $m_i$ is the atomic mass of atom $i$, $U$ is the potential energy function given in Eq. (32), and $t$ is the time. The total energy $E$ of the system is the sum of all kinetic ($\frac{1}{2}\,mv^2$) and potential energy $U(x)$ contributions (Eq. (34)).

$$
E = \frac{1}{2}m\left(\frac{\partial x}{\partial t}\right)^2 + U(x) \tag{34}
$$

If we wish to know the position $x_i(t+\Delta t)$ of an atom $i$ at the time $t + \Delta t$, it can be calculated based on the known position $x_i(t)$ at the time $t$, e.g., by the Leapfrog algorithm [9] (which is a modification of the well-known Verlet [10] algorithm) given in Eq. (35).

$$x_i(t + \Delta t) = x_i(t) + v_i(t + \Delta t/2)\Delta t$$

$$v_i(t + \Delta t/2) = v_i(t - \Delta t/2) + \frac{\partial^2 x_i(t)}{\partial t^2}\Delta t \qquad (35)$$

By defining the atomic displacement and velocities at a distinct time (which is not necessarily the same), these equations can be integrated. Various other integration schemes, which cannot be discussed in this context, are available in the literature and implemented in different simulation software packages [11]. Numerically solving differential equations is of course error-prone. Fortunately, the aim of a dynamics simulation is not to predict exactly the atomic positions after a certain time, but to give statistically meaningful results. Therefore, different simulations with different starting geometry will always give deviating trajectories, but one hopes to obtain comparable statistics.

The temperature $T$ of a system is related to the mean kinetic energy of all atoms $N$ via Eq. (36), where $k_B$ is the Boltzmann constant and $<v_i^2>$ the average of the squared velocities of atom $i$.

$$\frac{1}{2}\sum m_i\langle v_i^2\rangle = \frac{3}{2}Nk_BT \qquad (36)$$

### 7.3.3.2 Ways to Speed up the Calculations

The computer time required for a molecular dynamics simulation grows with the square of the number of atoms in the system, because of the non-bonded interactions defined in the potential energy function (Eq. (32)). They absolutely dominate the time necessary for performing a single energy evaluation and therefore the whole simulation. The easiest way to speed up calculations is to reduce the number of non-bonded interactions by the introduction of so-called cutoffs. They can be applied to the van der Waals and electrostatic interactions by simply defining a maximum distance at which two atoms are allowed to interact through space. If the distance is greater than this, the atom pair is not considered when calculating the non-bonded interactions. Several cutoff schemes have been introduced, from a simple sphere to switched or shifted cutoffs, which all aim to reduce the distortions in the transition region that are possibly destabilizing the simulation.

A second idea to save computational time addresses the fact that hydrogen atoms, when involved in a chemical bond, show the fastest motions in a molecule. If they have to be reproduced by the simulation, the necessary integration time step $\Delta t$ has to be at least 1 fs or even less. This is a problem especially for calculations including explicit solvent molecules, because in the case of water they do not only increase the number of non-bonded interactions, they also increase the number of fast-moving hydrogen atoms. This particular situation is taken into account

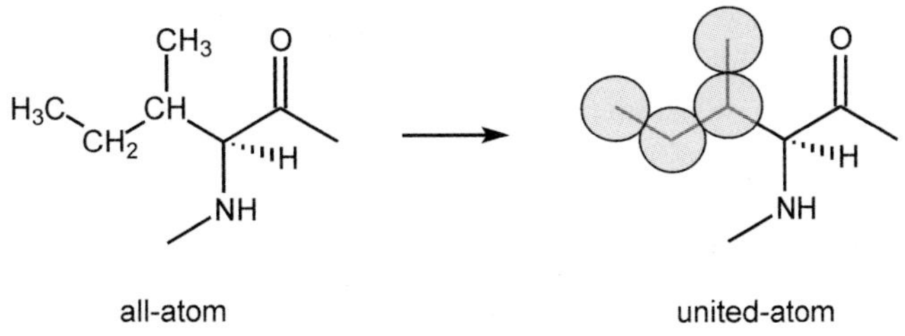

CH3 O

H3C
CH2 CH

'''H

NH

all-atom → united-atom

O

'''H

NH

**Figure 7-14.** All-atom and united-atom representation of the amino acid isoleucine. In this example, 13 atoms, which are able to form explicit non-bonding interactions, are reduced to only four pseudo-atoms.

by the development of water models like TIP3P, which are intended to reproduce the experimentally known bulk water properties with very limited computational effort [12]. On the other hand, it is often unnecessary to include hydrogen motions in the simulation, because they do "not contribute" to the much slower motions of the rest of the protein. A commonly used method to constrain the motion of hydrogen atoms is called SHAKE, which allows the integration step to be increased to 1–2 fs [13]. In principle, constraining bonds can go even further by acting on all covalent bonds, or fixing the bond angles as well, leaving only the dihedral angles free. Of course, these dramatic restrictions on the potential energy function may lead to serious sampling errors, especially in long-term simulations.

Another way to speed up calculations should be mentioned only briefly in this context – the definition of so-called united-atom force field parameterizations. Here, reducing the number of explicitly considered interactions is also achieved by the fact that hydrogen atoms bound to carbon (as in the case of aliphatic side chains of amino acids) do not have any "mechanistic function". Their "structural function" is mainly steric in nature, which offers an elegant method of simplification: decreasing the number of atoms by contracting the hydrogens of an explicitly defined methyl or methylene group to pseudo-carbon (united) atoms with increased van der Waals radii and modified charges. In the case of isoleucine, going from an all-atom to an united-atom representation reduces the number of non-bonded interactions considerably (Figure 7-14).

It is important to notice that the united-atom simplification cannot be applied to functional hydrogens which are involved in the formation of a hydrogen bond or a salt bridge. This would destroy interactions important for the structural integrity of the protein. Removing the hydrogen at the $\alpha$-carbon of the peptide backbone is also dangerous, because it prevents racemization of the amino acid.

### 7.3.3.3 Solvent Effects

Although there are examples of enzymes which maintain their catalytic activity even when crystallized, they normally work in their natural (i.e., aqueous) environment. This is the reason why the majority of the simulations are carried out applying a technique that accounts for solvent effects. But what is the effect of a solvent?

On one hand, there are the dielectric properties, which are especially important for polar solvents like water. Bulk properties can, on the other hand, only be modeled by using a supermolecule approach with explicitly defined solvent molecules.

One of the reasons why water is such an outstanding solvent is its high dielectric constant, which effectively reduces charge–charge interactions by electrostatic shielding. To account for this effect in a relatively simple manner, the first modification introduced into the Coulomb term, which describes electrostatic interactions, was a re-scaling of the dielectric permittivity of free space $\varepsilon_0$ by a factor $D$. This damps the long-range electrostatic interactions according to the relationship $\varepsilon = D\varepsilon_0$. Using the macroscopic value for water ($D = 78.0$), $\varepsilon$ then amounts to $78.0\varepsilon_0$. An alternative way is to introduce a distance dependence into the electrostatic interactions by defining an effective dielectric constant $\varepsilon = D\varepsilon_0 r_{ij}$, which modifies to the Coulomb term of Eq. (32) according to Eq. (37).

$$E_{coul} = \sum_{i=1}^{N} \sum_{j=i+1}^{N} \left( \frac{q_i q_j}{4\pi D\varepsilon_0 r_{ij}^2} \right) \tag{37}$$

By using an effective, distance-dependent dielectric constant, the ability of bulk water to reduce electrostatic interactions can be mimicked without the presence of explicit solvent molecules. One disadvantage of all vacuum simulations, corrected for shielding effects or not, is the fact that they cannot account for the ability of water molecules to form hydrogen bonds with charged and polar surface residues of a protein. As a result, adjacent polar side chains interact with each other and not with the solvent, thus introducing additional errors.

It is often the case that the solvent acts as a bulk medium, which affects the solute mainly by its dielectric properties. Therefore, as in the case of electrostatic shielding presented above, explicitly defined solvent molecules do not have to be present. In fact, the bulk can be considered as "perturbing the molecule in the gas phase", leading to so-called continuum solvent models [14, 15]. To represent the electrostatic contribution to the free energy of solvation, the generalized Born (GB) method is widely used. Within the GB equation, $\Delta G_{el}$ equals the difference between $G_{el}$ and the vacuum Coulomb energy (Eq. (38)):

$$\Delta G_{el} = -\left(1 - \frac{1}{\varepsilon}\right) \sum_{i=1}^{N} \sum_{j=i+1}^{N} \frac{q_i q_j}{r_{ij}} - \frac{1}{2}\left(1 - \frac{1}{\varepsilon}\right) \sum_{i=1}^{N} \frac{q_i^2}{a_i} \tag{38}$$

The total electrostatic free energy $G_{el}$ of a system is given by the sum of the Coulomb energy and the Born free energy of solvation (Eq. (39)):

$$G_{el} = \sum_{i=1}^{N} \sum_{j=i+1}^{N} \frac{q_i q_j}{r_{ij}} - \frac{1}{2}\left(1 - \frac{1}{\varepsilon}\right) \sum_{i=1}^{N} \frac{q_i^2}{a_i} \tag{39}$$

Within Eqs. (38) and (39), $q_i$ and $a_i$ are the charge and the radius of the $i$th of $N$ particles, respectively. The dielectric permittivity of the system is described by $\varepsilon$.

The GB equation is suitable for the description of solvent effects in molecular mechanics and dynamics [16], as well as in quantum mechanical calculations [17, 18]. An excellent review of implicit solvation models, with more than 900 references, is given by Cramer and Truhlar [19].

An approach between a continuum model and an explicit solvation model is the Langevin dipole method, introduced by Warshel and Levitt [20, 21]. In the region beyond the solvent-accessible surface of the molecule, a grid of rotatable point dipoles is defined; these represent the dipoles of the solvent molecules, which are not explicitly present. By using the Langevin equation (not shown), the size and direction of each dipole $\mu_i$ can be determined, considering the fact that the electric field $E_i$ at each dipole has contributions from the solute and all other dipoles present in the system. The free energy of Langevin dipoles can be calculated according to Eq. (40)

$$\Delta G_{sol} = -\frac{1}{2}\sum_i \mu_i \cdot E_i^0 \tag{40}$$

Note that $\mathbf{E}_i^0$ represents the field caused by the solute charges only.

Another way of calculating the electrostatic component of solvation uses the Poisson–Boltzmann equations [22, 23]. This formalism, which is also frequently applied to biological macromolecules, treats the solvent as a high-dielectric continuum, whereas the solute is considered as an array of point charges in a constant, low-dielectric medium. Changes of the potential $\phi$ within a medium with the dielectric constant $\varepsilon$ can be related to the charge density $\rho$ according to the Poisson equation (Eq. (41)).

$$\nabla^2\phi(\mathbf{r}) = -\frac{4\pi\rho(\mathbf{r})}{\varepsilon} \tag{41}$$

Note that the mathematical symbol $\nabla$ stands for the second derivative of a function (in this case with respect to the Cartesian coordinates $\partial^2/\partial x^2 + \partial^2/\partial y^2 + \partial^2/\partial z^2$); therefore the relationship stated in Eq. (41) is a second-order differential equation. Only for a constant dielectric Eq.(41) can be reduced to Coulomb's law. In the more interesting case where the dielectric is not constant within the volume considered, the Poisson equation is modified according to Eq. (42).

$$\nabla\varepsilon(\mathbf{r})\nabla\phi(\mathbf{r}) = -4\pi\rho(\mathbf{r}) \tag{42}$$

If there are ions in the solution, they will try to change their location according to the electrostatic potential in the system. Their distribution can be described according to Boltzmann. Including these effects and applying some mathematics leads to the final linearized Poisson–Boltzmann equation (Eq. (43)).

$$\nabla\varepsilon(\mathbf{r})\nabla\phi(\mathbf{r}) - \kappa'\phi(\mathbf{r}) = -4\pi\rho(\mathbf{r}) \tag{43}$$

In the non-linear differential equation Eq. (43), $\kappa'$ is related to the inverse Debye–Hückel length. The method briefly outlined above is implemented, e.g., in the program DelPhi of Honig et al, which uses a finite difference approach to solve the Poisson–Boltzmann equation [24].

The explicit definition of water molecules seems to be the best way to represent the bulk properties of the solvent correctly. If only a thin layer of explicitly defined solvent molecules is used (due to limited computational resources), difficulties may rise to reproduce the bulk behavior of water, especially near the border with the vacuum. Even with the definition of a full solvent environment the results depend on the model used for this purpose. In the relative simple case of TIP3P and SPC, which are widely and successfully used, the atoms of the water molecule have fixed charges and fixed relative orientation. Even without internal motions and the charge polarization ability, TIP3P reproduces the bulk properties of water quite well. For a further discussion of other available solvent models, readers are referred to Chapter VII, Section 1.3.2 of the Handbook. Unfortunately, the more sophisticated the water models are (to reproduce the physical properties and thermodynamics of this outstanding solvent correctly), the more impractical they are for being used within molecular dynamics simulations.

### 7.3.3.4 Periodic Boundary Conditions

The problems already mentioned at the solvent/vacuum boundary, which always exists regardless of the size of the box of water molecules, led to the definition of so-called periodic boundaries. They can be compared with the unit cell definition of a crystalline system. The unit cell also forms an "endless system without boundaries" when repeated in the three directions of space. Unfortunately, when simulating liquids the situation is not as simple as for a regular crystal, because molecules can diffuse and are in principle able to leave the unit cell.

For biomolecular systems, the protein under consideration is placed in the center of a box of explicitly defined water molecules. This box is normally regular (all angles are equal to 90°, but other geometries, e.g., a truncated octahedron, can be used also) and surrounded by its periodic images. Therefore, the box of real water molecules no longer has a border with the vacuum, which reduces related artifacts and additionally improves the bulk properties of the simulated solvent. To conserve the number of atoms (i.e., the total mass) in the system, a water molecule leaving the real box must be added again. One very important condition for periodic boundary calculation is that an atom of the real molecule must not interact with another real atom and its image at the same time (minimum image convention). For that reason, spherical cutoffs for the non-bonding interactions should be defined which have to be smaller than half the smallest box dimension.

### 7.3.4
### Constant Energy, Temperature, or Pressure?

Molecular dynamics simulations can produce trajectories (a time series of structural snapshots) which correspond to different statistical ensembles. In the simplest case, when the number of particles $N$ (atoms in the system), the volume $V$, and the energy $E$ are conserved, we call this a microcanonical ensemble (abbreviated NVE, because these three components remain constant). The temperature of the system is calculated rather than specified, and the constant energy is able to flow between kinetic and potential energy. This is also called free dynamics.

What does constant energy mean? Consider a ball rolling on a surface with hilltops and valleys. Rolling down the hill, the ball gets faster and faster, increasing its kinetic energy. At the same time, the potential energy decreases. Climbing up the next hill, the ball slows down, losing kinetic and gaining potential energy again. In each situation, the ball remains on the same hypersurface and the sum of the potential and kinetic energies is constant. Because the ball gets very slow when reaching the hilltop (in an ideal case it stops and reverses direction), the time during which it has much potential energy is long compared with the time when it has much kinetic energy (in this case the ball is very fast). If we transfer this three-dimensional picture to a molecule with a multidimensional hypersurface, it is more than unlikely that a molecule stays a longer time with all its energy concentrated as strain (to be exact, potential or steric energy) rather than kinetic energy. From this point of view, constant-energy dynamics may not be the best way to describe the behavior of molecules.

Instead of constant-energy dynamics, an alternative approach is to hold the temperature constant. As already mentioned, the temperature of a system corresponds to its kinetic energy (see Eq. (36)), whereas the total energy is the sum of the kinetic and potential energy components. In terms of statistical mechanics, constant temperature MD produces a canonical ensemble (NVT). Practically, if the system cools down (i.e., the potential energy is rising due to internal strain), energy has to be added to keep the temperature (i.e., the kinetic energy) constant. Alternatively, if the molecule is in a low-energy conformation, excess kinetic energy has to be removed to reduce the temperature again. To control temperature, Berendsen introduced a method whereby the system is weakly coupled to an external heat bath, which acts as a source of thermal energy [25]. Here, the rate of the temperature change is proportional to the temperature difference between bath ($T_0$) and system (Eq. (44)).

$$\frac{dT(t)}{dt} = \frac{1}{\tau_T}(T_0 - T(t)) \tag{44}$$

By applying Eq. (45), the velocities $v$ are scaled by the factor $\lambda$ to come closer to the desired temperature $T_0$; $\lambda$ is defined in Eq. (45).

$$\lambda = \left[1 + \frac{\Delta t}{\tau_T}\left(\frac{T_0}{T(t)} - 1\right)\right]^{1/2} \tag{45}$$

$T(t)$ corresponds to the actual temperature at the time $t$, $\Delta t$ is the integration time step, and the relaxation time $\tau_T$ represents the strength of the coupling (smaller values mean stronger coupling to the bath). If the coupling is too strong ($\tau_T$ smaller than 0.1 ps), the resulting ensemble is isokinetic rather than canonical, and time-dependent properties of ensemble averages cannot be calculated from this trajectory. Nevertheless, the advantage of this approach is that the system temperature is able to fluctuate around the target temperature.

However, it is common practice to sample an isothermal isobaric ensemble (NPT, constant pressure and constant temperature), which normally reflects standard laboratory conditions well. Similarly to temperature control, the system is coupled to an external bath with the desired target pressure $P_0$. By rescaling the dimensions of the periodic box and the atomic coordinates by the factor $\mu$ at each integration step $\Delta t$ according to Eq. (46), the volume of the box and the forces of the solvent molecules acting on the box walls are adjusted.

$$\mu = \left[ 1 - \frac{\Delta t}{\tau_P} (P_0 - P(t)) \right]^{1/3} \tag{46}$$

Also in this case, $\tau_P$ corresponds to a relaxation time which determines the coupling of the modulated variable to the external bath. The pressure scaling can be applied isotropically, which means that the factor is the same in all three spatial directions. More realistic is an anisotropic pressure scaling, because the box dimensions also change independently during the course of the simulation.

### 7.3.5
### Long-Range Forces

The potential energy function given in Eq. (32) contains bonded and non-bonded terms to evaluate the energy of a molecule or molecular system. The non-bonded Coulomb and van der Waals contributions need special attention, because every atom interacts with every other atom in the system. Mathematically expressed by a double sum, it is obvious that the major part of the computing time is consumed by these sums. To speed up simulations, these interactions were simply truncated at a fixed distance, typically 8–10 Å. From another look at the potential energy function in Eq. (32), it becomes obvious that the truncation after such a short distance is only a minor problem in the case of the van der Waals interactions. The Lennard-Jones potential uses the distance $r_{ij}$ between two interacting atoms $i$ and $j$ in the power of $-6$ and $-12$, so the function becomes very small at the cutoff distance. The problem is much more serious for the electrostatic part of the non-bonded interaction, because Coulomb's law uses the reciprocal of the distance $r_{ij}$, resulting in long-range electrostatic interactions. The unscreened interaction of two full charges on opposite sides of a protein is still significant. At 100 Å, the electrostatic interaction is greater than 3 kcal mol$^{-1}$ [26].

Ways to circumvent the above-mentioned problems have been to simply increase the cutoff distance to larger values, to use more than one cutoff value with different update frequencies, or to define more sophisticated cutoff schemes. In the last case, a truncation of the non-bonded interactions was replaced by shifting the interaction energies to zero or by additionally applying a switched sigmoidal function. The common intention of these procedures is to limit the distortions at the border introduced by neglecting the long-range Coulomb interactions.

In periodic boundary conditions, one possible way to avoid truncation of electrostatic interaction is to apply the so-called Particle Mesh Ewald (PME) method, which follows the Ewald summation method of calculating the electrostatic energy for a number of charges [27]. It was first devised by Ewald in 1921 to study the energetics of ionic crystals [28]. PME has been widely used for highly polar or charged systems. York and Darden applied the PME method already in 1994 to simulate a crystal of the bovine pancreatic trypsin inhibitor (BPTI) by molecular dynamics [29]. They compared the PME method with equivalent simulations based on a 9 Å residue-based cutoff and found that for PME the averaged RMS deviations of the non-hydrogen atoms from the X-ray structure were considerably smaller than in the non-PME case. Also, the atomic fluctuations calculated from the PME dynamics simulation were in close agreement with those derived from the crystallographic temperature factors. In the case of DNA, which is highly charged, the application of PME electrostatics leads to more stable dynamics trajectories with geometries closer to experimental data [30]. A theoretical and numerical comparison of various particle mesh routines has been published by Deserno and Holm [31].

## 7.3.6
### Application of Molecular Dynamics Techniques

How can we apply molecular dynamics simulations practically? This section gives a brief outline of a typical MD scenario. Imagine that you are interested in the response of a protein to changes in the amino acid sequence, i.e., to point mutations. In this case, it is appropriate to divide the analysis into a static and a dynamic part. What we need first is a reference system, because it is advisable to base the interpretation of the calculated data on changes compared with other simulations. By taking this relative point of view, one hopes that possible errors introduced due to the assumptions and simplifications within the potential energy function may cancel out. All kinds of simulations, analyses, etc., should always be carried out for the reference and the model systems, applying the same simulation protocols.

A typical molecular dynamics simulation comprises an equilibration and a production phase. The former is necessary, as the name implies, to ensure that the system is in equilibrium before data acquisition starts. It is useful to check the time evolution of several simulation parameters such as temperature (which is directly connected to the kinetic energy), potential energy, total energy, density (when periodic boundary conditions with constant pressure are applied), and their root-mean-square deviations. Having these and other variables constant at the end of the equilibration phase is the prerequisite for the statistically meaningful sampling of data in the following production phase.

Figure 7-15 shows the time evolution of the temperature, total energy, and potential energy for a 300 ps simulation of the tetracycline repressor dimer in its induced (i.e., ligand-bound) form. Starting from the X-ray structure of the monomer in a complex with one molecule of tetracycline and a magnesium ion (protein database

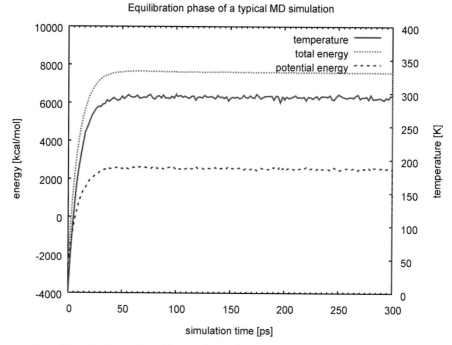

**Figure 7-15.** Heating and equilibration phase of a typical MD simulation. In the ideal case, the temperature should fluctuate around the desired value (here 298 K), and the potential energy should remain constant. Remember that the total energy is the sum of potential and kinetic energy, the latter being directly coupled to the temperature of the system.

entry 2TRT), the dimer was produced first according to the crystallographic symmetry transformations provided in the file. After careful initial geometry optimization to remove bad contacts and internal strain, the system was heated to 298 K by weak coupling (2 ps) to an external heat bath. All calculations were performed with AMBER, applying the parm94 parameter set of Cornell et al. [32].

After 75 ps, the simulation has already reached its final temperature, which remains constant due to the coupling to the heat bath. The total energy and the potential energy show the same principle behavior. It is important to look at both kinetic and potential energy contributions to the total energy of the system, because, especially when simulating a constant-energy ensemble, a redistribution of the energy between these two forms occurs. Although the total energy remains constant, a steady decrease in the potential energy may point to slow conformational changes within the protein.

Assuming that an equilibrium is now well established, the simulation may be restarted (not newly started) to begin with the sampling of structural and thermodynamic data. In our model case, data acquisition was performed for 3 ns (trajectory data plot not shown). For the production phase, also, the time evolution of the variables mentioned above should be monitored to detect stability problems or con-

**Figure 7-16.** Superimposition of the X-ray structure of the tetracycline repressor class D dimer (dark, protein database entry 2TRT) with the calculated geometrical average of a 3 ns MD simulation (light trace). Only the protein backbone $C_\alpha$ trace is shown. The secondary structure elements and the tertiary structure are almost perfectly reproduced and maintained throughout the whole production phase of the calculation.

formational changes that may occur during the course of the calculation. A way often used to obtain an impression of the simulated protein structure is to calculate an average protein geometry over all the extracted snapshots. This can be done by a few lines of Perl programming, or by using a special tool which is normally provided with every simulation package (in the case of AMBER, the trajectory analysis tool *carnal* can be used for this purpose). An overlay of the average protein geometry of our MD simulation with the corresponding X-ray structure is shown in Figure 7-16. For simplification, only the protein backbones are shown, with no additional information.

The use of an averaged dynamics structure for comparison with the X-ray geometry has the advantage that we obtain an impression on how the protein is most likely to look over the simulation time. In contrast, comparing selected snapshots extracted from the trajectory may reflect conformationally extreme situations caused by local fluctuations or distortions, which are not at all representative. In our test case (Figure 7-16), we can see clearly that the simulation reproduces the geometry of the X-ray starting structure very well, confirming the applicability of the protocol used. As already stated above, the average MD geometry can now serve as a geometrical reference for subsequent simulations with modifications, e.g., in the amino acid sequence or inducer structure. A closer look at Figure 7-16 reveals another interesting possibility of geometrically analyzing the two structures: by calculating the RMS differences of the protein backbone for every residue pair, one should in principle be able to localize regions of low and high geometrical deviation. This information gives important hints about areas of different stability within the protein, which leads us to the final step our analysis.

An average structure calculated from an MD trajectory shows us a representative geometry around which a protein is fluctuating during the course of the simulation. This structure may differ from the crystallographic starting point, for example, but these deviations do not say anything about the intensity of the fluctuations and the mobility of the involved atoms or residues. Figure 7-17 shows a residue-

based fluctuation analysis of our MD production trajectory. For each $i$ of the $N$ snapshots, the RMS deviation of every atom $j$ of a residue from a reference structure (e.g., the X-ray or the average geometry shown in Figure 7-16) is calculated according to Eq. (47) and summed.

$$RMS_j^{fl} = \sqrt{\frac{1}{N} \sum_{i=1}^{N} \left(r_{i,j} - r_{i,av}\right)^2} \qquad (47)$$

Plotting the MD fluctuations helps to identify sequence regions of high and low mobility. In contrast to the average structures, fluctuations provide a dynamic picture of the simulation. Why is this important? Consider a region of large deviation between the X-ray and the average structure, e.g., the large loops within the two DNA binding heads of the tetracycline repressor (at the bottom in Figure 7-16). These deviations indicate that, compared with the crystal structure, the simulation predicts a different average geometry of these loops over time. However, nothing is said about the dynamics of these loops. They can simply have another orientation,

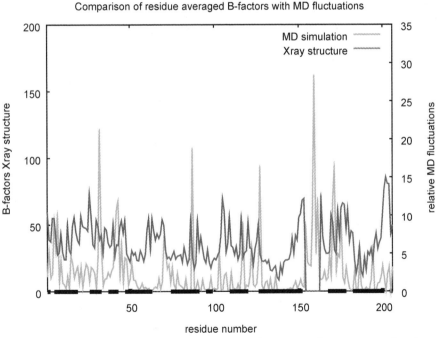

**Figure 7-17.** Analysis of the dynamic characteristics of an MD simulation in terms of residue-based fluctuations (light grey), which can be compared with the crystallographic temperature factors (dark grey). For better guidance through the protein, the secondary structure elements of our model system (ten helices) are marked on the abscissa of the plot. Please note that the area between the amino acids 156 and 164 is structurally unresolved (no B-factors available). The corresponding region within the simulation shows the highest fluctuations, therefore giving a possible explanation for the problems during the structure refinement.

but stay almost immobile. This is not very likely of course, but the example shows the difference between a static and a dynamic point of view.

The relative molecular dynamics fluctuations shown in Figure 7-17 can be compared with the crystallographic $B$-factors, which are also called temperature factors. The latter name, especially, indicates the information content of these factors: they show how well defined within the X-ray structure the position of an atom is. Atoms with high "temperature" have an increased mobility. In principle, this is the same information as is provided by the molecular dynamics fluctuations. Using Eq. (48), the RMS fluctuation of an atom $j$ can be converted into a $B$-factor:

$$B_j = \frac{8}{3}\pi^2(RMS_j^{fl})^2 \tag{48}$$

The comparison of both data sources qualitatively shows a similar picture. Regions of high mobility are located especially between the secondary structure elements, which are marked on the abscissa of the plot in Figure 7-17. Please remember that the fluctuations plotted in this example also include the amino acid side chains, not only the protein backbone. This is the reason why the side chains of large and flexible amino acids like lysine or arginine can increase the fluctuations dramatically, although the corresponding backbone remains almost immobile. In these cases, it is useful to analyze the fluctuations of the protein backbone and side chains individually.

# References

[1]  R. Car, M. Parrinello, *Phys. Rev. Lett.* **1985**, *55*, 2471–2474.

[2]  P. Deuflhard, J. Hermans, B. Leimkühler, A. E. Mark, S. Reich, R. D. Skeel (Eds), *Computational Molecular Dynamics: Challenges, Methods, Ideas (Lecture Notes in Computational Science and Engineering, Vol. 4)*, Springer, Berlin, **1999**.

[3]  D. Frenkel, B. Smits, *Understanding Molecular Simulation*, Academic Press, San Diego, **1996**.

[4]  J. M. Haile, *Molecular Dynamics Simulation*, John Wiley, Chichester, **1997**.

[5]  D. C. Rapaport, *The Art of Molecular Dynamics Simulation*, Cambridge University Press, **1995**.

[6]  D. W. Heermann, *Computer Simulation Methods in Theoretical Physics*, Springer, Berlin, **1986**.

[7]  W. van Gunsteren, H. Berendsen, *Angew. Chem., Int. Ed. Engl.* **1990**, *29*, 992–1023.

[8]  P. Kollman, *Acc. Chem. Res.* **1996**, *29*, 461–469.

[9]  R. W. Hockney, *Methods Comput. Phys.* **1970**, *9*, 136–211.

[10]  L. Verlet, *Phys. Rev.* **1967**, *159*, 98–103.

[11]  M. P. Allen, D. J. Tildesley, *Computer Simulation of Liquids*, Clarendon Press, Oxford, **1987**.

[12]  W. L. Jorgensen, J. Chandrasekhar, J. D. Madura, R. W. Impey, M. L. Klein, *J. Chem. Phys.* **1983**, *79*, 926–935.

[13]  W. van Gunsteren, H. Berendsen, *Mol. Phys.* **1977**, *34*, 1311–1327.

[14]  M. Orozco, F. J. Luque, F. Javier, *Chem. Rev.* **2000**, *100*, 4187–4225.

[15]  P. E. Smith, M. B. Pettitt, *J. Phys. Chem.* **1994**, *98*, 9700–9711.

[16]  W. C. Still, A. Tempczyk, R. C. Hawley, T. Hendrickson, *J. Am. Chem. Soc.* **1990**, *112*, 6127–6129.

[17]  P. Winget, J. D. Thompson, J. D. Xidos, C. J. Cramer, D. G. Truhlar, *J. Phys. Chem. A*, **2002**, *106*, 10707–10717.

[18]  J. Li, T. Zhu, C. J. Cramer, D. G. Truhlar, *J. Phys. Chem A*, **2000**, *104*, 2178–2182, and references cited therein.

[19]  C. J. Cramer, D. G. Truhlar, *Chem. Rev.* **1999**, *99*, 2161–2200.

[20]  A. Warshel, M. Levitt, *J. Mol. Biol.* **1976**, *103*, 227–249.

[21]  A. Warshel, *Computer Modeling of Chemical Reactions in Enzymes and Solutions*, John Wiley, New York, **1991**.

[22]  B. Honig, A. Nicholls, *Science* **1995**, *268*, 1144–1149.

[23]  M. K. Gilson, K. Sharp, B. Honig, *J. Comput. Chem.* **1987**, *9*, 327–335.

[24]  W. Rocchia, E. Alexov, B. Honig, *J. Phys. Chem. B*, **2001**, *105*, 6507–6514.

[25]  H. J. C. Berendsen, J. P. M. Postma, W. F. van Gunsteren, A. Di Nola, J. R. Haak, *J. Chem. Phys.* **1984**, *81*, 3684–3690.

[26]  L. Pedersen, T. Darden, Molecular dynamics; techniques and applications to proteins, in *The Encyclopedia of Computational Chemistry, Vol. 3*, P. v. R. Schleyer, N. L. Allinger, T. Clark, J. Gasteiger, P. A. Kollman, H. F. Schaefer III, P. R. Schreiner (Eds.), John Wiley, Chichester, **1998**, pp. 1650–1659.

[27]  T. A. Darden, D. York, L. Pedersen, *J. Chem. Phys.* **1993**, *98*, 10089–10092.

[28]  P. Ewald, *Annalen der Physik*, **1921**, *64*, 253–287.

[29] D. M. York, A. Wlodawer, L. G. Pedersen, T. A. Darden, *Proc. Nat. Acad. Sci. USA,* **1994**, *91*, 8715–8718.

[30] T. E. Cheatham III, J. L. Miller, T. Fox, T. A. Darden, P. A. Kollman, *J. Am. Chem. Soc.* **1995**, *117*, 4193–4194.

[31] M. Deserno, C. Holm, *J. Chem. Phys,* **1998**, *109*, 7678–7693, 7694–7701.

[32] W. D. Cornell, P. Cieplak, C. I. Bayly, I. R. Gould, K. M. Merz, Jr., D. M. Ferguson, D. C. Spellmeyer, T. Fox, J. W. Caldwell, P. A. Kollman, *J. Am. Chem. Soc.* **1995**, *117*, 5179–5197.

## 7.4
## Quantum Mechanics

*Tim Clark*

At first sight, quantum mechanical calculations, which are generally thought of as being very computation-intensive, are not widely applicable to chemoinformatics. However, recent progress in computer soft- and hardware has made it possible to use semi-empirical molecular orbital (MO) techniques for tens or hundreds of thousands of molecules [1, 2], and the more expensive density functional theory (DFT) or *ab-initio* methods for tens to hundreds of compounds [3]. The speed and accuracy of modern programs make it possible to optimize molecular structures quantum mechanically starting with structures generated automatically using programs such as CORINA [4]. Quantum mechanical techniques are usually used to obtain accurate molecular properties, such as electrostatic potentials or polarizabilities, that are only available with much lower resolution from classical mechanical techniques (Chapter VII, Section 1 in the Handbook) or those (like ionization potentials, electron affinities, etc.) that can be obtained only quantum mechanically. This section is designed to introduce the quantum mechanical techniques used for such studies in a qualitative way that provides a sound basic understanding without resort to more complicated theoretical descriptions, which are available elsewhere. The computational chemistry textbooks by Leach [5], Cramer [6], or Jensen [7] and the theoretical one by Szabo and Ostlund [8] provide more detail.

The theoretical methods used commonly can be divided into three main categories, semi-empirical MO theory, DFT and *ab-initio* MO theory. Although it is no longer applied often, Hückel molecular orbital (HMO) theory will be employed to introduce some of the principles used by the more modern techniques.

### 7.4.1
### Hückel Molecular Orbital Theory

HMO theory is named after its developer, Erich Hückel (1896–1980), who published his theory in 1930 [9] partly in order to explain the unusual stability of benzene and other aromatic compounds. Given that digital computers had not yet been invented and that all Hückel's calculations had to be done by hand, HMO theory necessarily includes many approximations. The first is that only the $\pi$-molecular orbitals of the molecule are considered. This implies that the entire molecular structure is planar (because then a plane of symmetry separates the $\pi$-orbitals, which are antisymmetric with respect to this plane, from all others). It also means that only one atomic orbital must be considered for each atom in the $\pi$-system (the p-orbital that is antisymmetric with respect to the plane of the molecule) and none at all for atoms (such as hydrogen) that are not involved in the $\pi$-system. Hückel then used the technique known as *linear combination of atomic orbitals* (LCAO) to build these atomic orbitals up into molecular orbitals. This is illustrated in Figure 7-18 for ethylene.

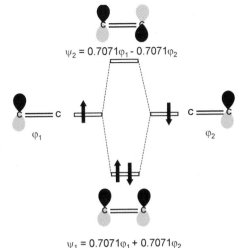

$\psi_2 = 0.7071\varphi_1 - 0.7071\varphi_2$

$\varphi_1$ $\varphi_2$

**Figure 7-18.** Schematic representation of the LCAO scheme in a $\pi$-only calculation for ethylene. The AOs $\varphi_1$ and $\varphi_2$ are combined to give the bonding MO $\psi_1$ and its antibonding equivalent $\psi_2$.
The outlined boxes show energy levels and the black arrows (indicating spin-up or -down) the electrons.

$\psi_1 = 0.7071\varphi_1 + 0.7071\varphi_2$

Within HMO theory, ethylene consists of only two orbitals, one p-orbital for each carbon perpendicular to the molecular plane. These two atomic orbitals, which are usually called the *basis set*, must be combined in order to form molecular orbitals. This is done by considering the energies of the atomic orbitals (in this case equal) and their *overlap*, which determines how they interact with each other. Thus, the two atomic orbitals (AOs) can be combined with the same phase to give the stabilized, bonding MO, $\Psi_1$. Reversing the phase of one of the AOs gives the antibonding, destabilized MO, $\Psi_2$. As the $\pi$-system of ethylene has two electrons, these can both occupy $\Psi_1$, leaving $\Psi_2$ unoccupied. Unoccupied MOs are usually denoted as *vacant* or *virtual* orbitals. The net energy balance of combining the two carbon atoms (represented in HMO theory only by the two p-orbitals) to give the simple ethylene $\pi$-system is twice the stabilization energy of $\Psi_1$ relative to the energy of the starting AOs. This stabilization energy corresponds to the $\pi$-bond energy.

However, even such a simple "$\pi$-only" theory is intractable without computers, so that Hückel was forced to make further approximations. Firstly, the energies of the contributing AOs in an all-carbon $\pi$-system are all set to the same value, denoted $\alpha$. In a real $\pi$-system, these energies would often vary slightly because of the effects of the other ($\sigma$-) orbitals, which are neglected in HMO theory. The second approximation is that the overlap between carbons that are bonded to each other is also constant (and is given a value denoted as $\beta$), but is zero between all carbons that are not bonded to each other. This is clearly a drastic approximation because it takes neither variations in bond lengths nor non-bonded overlap into account. The final major approximation of HMO theory is that it is a so-called "one-electron theory", which does not mean that only one electron is considered, but rather that the energy of each individual electron is calculated without allowing it to interact with the others (i.e., as if it were alone in the molecule). In reality, electrons inter-

**1**    **Figure 7-19.** Numbering scheme for 1,3-butadiene, **1**.

act with each other in a very complex manner. They are negatively charged, so that they repel each other, but they are also free to move in a correlated fashion that helps to minimize their electrostatic repulsion. Put simply, electrons avoid each other as much as possible. Thus, the calculation of the electron–electron repulsion is one of the most arduous tasks in MO theory. Hückel solved this problem by ignoring it.

Once these approximations have been made, HMO theory becomes very simple. Using 1,3-butadiene, **1**, as an example, we can work through an HMO calculation in order to outline the process involved. Firstly, we assign numbers to the carbon atoms, as shown in Figure 7-19.

We can now assign the four carbon p-orbitals, one to each carbon. For simplicity, we will label them $\psi_{1-4}$ with the subscript corresponding to the number of the carbon atom to which the AO belongs. We will use the symbol $\psi$ to denote AOs and $\Psi$ for MOs. We can now write the *Hückel matrix* as a square matrix involving the AOs $\psi_{1-4}$, as shown in Figure 7-20.

The diagonal elements of the Hückel matrix represent the energies of the contributing AOs, which in this case are all $\alpha$. Each of the bonds (in this case $\psi_1-\psi_2$, $\psi_2-\psi_3$ and $\psi_3-\psi_4$) is assigned the overlap energy $\beta$ and all other elements of the matrix are set to zero. This matrix is then diagonalized to give the eigenvalues, which are the energies of the MOs, and the eigenvectors, which are the coefficients of the individual AOs in the LCAO representation of the MOs. The eigenvalues and eigenvectors of the Hückel matrix for 1,3-butadiene are also shown in Figure 7-20. The lowest MO, $\Psi_1$, for instance, has an energy of $(\alpha - 1.618\beta)$ and is described by Eq. (49).

$$\Psi_1 = 0.3717\psi_1 + 0.6015\psi_2 + 0.6015\psi_3 + 0.3717\psi_4 \tag{49}$$

The energies and forms of the four MOs are shown in Figure 7-21.

Note that the sums of the squares of the coefficients in a given MO must equal 1 (e.g., $0.3717^2 + 0.6015^2 + 0.3717^2 + 0.6015^2 = 1.0$ for $\Psi_1$) because each of the AOs represents a probability distribution of finding the electron at a given point in space. The total probability of finding an electron in all space for an MO must be unity, exactly as for its constituent AOs. We now can see that the LCAO approximation is only one of many possibilities to describe the electron density ($\equiv$ probability) for MOs. We do not have to express the electron density as a linear combination of the electron densities of AOs centered at the atoms. We could also

|        | $\varphi_1$ | $\varphi_2$ | $\varphi_3$ | $\varphi_4$ |
|--------|------|------|------|------|
| $\varphi_1$ | $\alpha$ | $\beta$ | 0 | 0 |
| $\varphi_2$ | $\beta$ | $\alpha$ | $\beta$ | 0 |
| $\varphi_3$ | 0 | $\beta$ | $\alpha$ | $\beta$ |
| $\varphi_3$ | 0 | 0 | $\beta$ | $\alpha$ |

## Hückel matrix

|  | $\Psi_1$ | $\Psi_2$ | $\Psi_3$ | $\Psi_4$ |
|---|---|---|---|---|
| eigenvalue | $\alpha-1.618\beta$ | $\alpha-0.618\beta$ | $\alpha+0.618\beta$ | $\alpha+1.618\beta$ |
| $\varphi_1$ | 0.3717 | 0.6015 | 0.6015 | 0.3717 |
| $\varphi_2$ | 0.6015 | 0.3717 | -0.3717 | -0.6015 |
| $\varphi_3$ | 0.6015 | -0.3717 | -0.3717 | 0.6015 |
| $\varphi_3$ | 0.3717 | -0.6015 | 0.6015 | -0.3717 |

## eigenvalues and eigenvectors

**Figure 7-20.** The Hückel matrix (above) and the eigenvalues and eigenvectors for 1,3-butadiene.

describe it by dividing the space in and around the molecule into tiny elements and giving each of these an electron density value, by using linear combinations of plane waves in space or by linear combinations of spherical Gaussian functions distributed throughout the space in and around the molecule. The LCAO approximation is, however, computationally very convenient and is the way that we often think about MOs in chemistry. The $\pi$-MOs of butadiene shown in Figure 7-21, for instance, are instantly recognizable as such because they are portrayed as LCAO combinations of AOs. The "real" electron density plots shown in Figure 7-22 are often less familiar than the "LCAO picture".

As mentioned above, HMO theory is not used much any more except to illustrate the principles involved in MO theory. However, a variation of HMO theory, extended Hückel theory (EHT), was introduced by Roald Hoffmann in 1963 [10]. EHT is a one-electron theory just like HMO theory. It is, however, three-dimensional. The AOs used now correspond to a *minimal basis set* (the minimum number of AOs necessary to accommodate the electrons of the neutral atom and retain spherical symmetry) for the valence shell of the element. This means, for instance, for carbon a 2s-, and three 2p-orbitals ($2p_x$, $2p_y$, $2p_z$). Because EHT deals with three-dimensional structures, we need better approximations for the Hückel matrix than

**Eigenvalue**

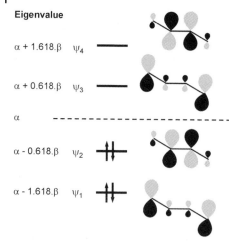

$\alpha + 1.618.\beta$   $\psi_4$  ——

$\alpha + 0.618.\beta$   $\psi_3$  ——

$\alpha$   — — — — — — — — —

$\alpha - 0.618.\beta$   $\psi_2$

$\alpha - 1.618.\beta$   $\psi_1$

**Figure 7-21.** The MOs and energy levels given by HMO theory for 1,3-butadiene. The occupation of the orbitals is shown for the neutral molecule.

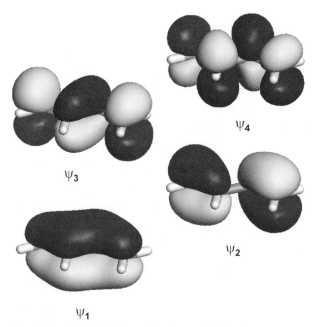

$\psi_4$

$\psi_3$

$\psi_2$

$\psi_1$

**Figure 7-22.** Isodensity plots of the four $\pi$-orbitals of 1,3-butadiene. The nomenclature corresponds to that in Figures 7-21 and 7-22 /correct?/.

those used in HMO. Rather than assigning each diagonal term the same $\alpha$-value, they are assigned the so-called *valence shell ionization energy* (VSIE) for the appropriate AOs. The VSIEs are essentially ionization potentials from the AO in question. In EHT, the overlap integrals are actually calculated (because they require little computer time) and used to evaluate the off-diagonal elements of the Hückel matrix, which is diagonalized as before to obtain the eigenvalues and eigenvectors. Until recently, EHT was used very extensively to treat and understand the bonding in transition metal complexes. However, the advance of DFT for these systems (see below) has caused a decline in the importance of EHT, which is now mostly used for many of the model studies on polymers, solids, etc., and for developing the first versions of new calculational methods based on MO theory.

## 7.4.2
### Semi-empirical Molecular Orbital Theory

The term "semi-empirical" is used to describe variations of molecular orbital theory in which significant simplifying approximations are used (similar to, but not as extreme as, those used in HMO theory). However, the effects of these approximations are compensated by parameterizing the variables used within the theory to make the results as close as possible to experimental values. The first semi-empirical methods arose out of the necessity to use severe approximations in order to be able to do the calculations at all. They were developed early in the history of digital computers, when speed, memory, and disk space were all very limited. The first such theory, now called Pople–Pariser–Parr (PPP) theory [11], was a $\pi$-only theory much like HMO, but considered electron–electron repulsion. The first significant three-dimensional semi-empirical MO theory was *complete neglect of differential overlap* (CNDO), introduced by Pople and Segal in 1965 [12]. By 1965, digital computers were available for research, so that fewer severe approximations were used than in HMO or PPP. However, the electron–electron repulsion still represented a problem because of its complexity. The approximation used to treat electron–electron interactions in CNDO and many subsequent approaches is known variously as the *mean field approximation, self-consistent field* (SCF) theory and *Hartree–Fock* (HF) theory. Of these names, the mean field approximation is probably the most descriptive, but the term SCF has become the most common. Because the problem of calculating the electron–electron interaction energy in a many-electron system cannot be solved exactly, we must use approximations. SCF theory treats each electron as if it interacts with the mean (over time) field of all the other electrons in the molecule. This means that the remaining electrons in the molecule are "short-sighted" (i.e., they do not react to the instantaneous position of the electron being considered). Thus, calculating the energy of each electron individually becomes essentially a one-electron problem to which we "only" have to add the field caused by the remaining electrons. This approximation neglects the fact that the movements of the electrons are correlated to reduce their mutual repulsion (i.e., each electron reacts to the instantaneous positions of all the others). Thus, SCF theory makes the computational task manageable at the cost of overestimating the electron–electron repulsion energy.

In 1965, however, the computational resources needed for the full SCF approach were not yet available. Practical MO theories therefore still needed approximations. The main problem is the calculation and storage of the four-center integrals, denoted $(\mu\nu \mid \lambda\sigma)$, needed to calculate the electron–electron interactions within the

SCF approximation. The indices $\mu$, $v$, $\lambda$, and $\sigma$ denote four atomic orbital centers, so that the number of such orbitals that needs to be calculated increases proportionally ("scales with") $N^4$, where $N$ is the number of AOs. This was an intractable task in 1965, so Pople, Santry, and Segal introduced the approximation that only integrals in which $\mu = v$ and $\lambda = \sigma$ (i.e., $(\mu\mu \mid \lambda\lambda)$) would be considered and that, furthermore, all AOs would be treated the same way (as if they were s-orbitals), so that Eq. (50) applies, where $\mu$ is centered on atom $A$, $\lambda$ on atom $B$, and $\gamma_{AB}$ depends only on the identities of $A$ and $B$, and can thus be treated as a parameter.

$$(\mu\mu|\lambda\lambda) = \gamma_{AB} \tag{50}$$

One early approximation, due to Pariser and Parr [12], was to treat the one-center term $\gamma_{AA}$ as the difference between the ionization potential $IP_A$ and the electron affinity $EA_A$ of $A$ (Eq. (51)).

$$\gamma_{AA} = IP_A - EA_A \tag{51}$$

Two-center terms were then given by Eq. (52).

$$\gamma_{AB} = \frac{\gamma_{AA} + \gamma_{BB}}{2 + r_{AB}(\gamma_{AA} + \gamma_{BB})} \tag{52}$$

This gives $\gamma_{AB} = (\gamma_{AA} + \gamma_{BB})/2$ at an interatomic distance, $r_{AB}$, of zero and $\gamma_{AB} \approx 1/r_{AB}$ at longer interatomic distances. These expressions (Eqs. (50)–(52)) are given to indicate the simplicity of the CNDO technique, which was used to calculate electronic properties such as dipole moments or excitation energies, usually using experimental geometries. There are many variations of the expressions given in Eqs. (51) and (52), but they are of comparable simplicity. Similarly, simplified expressions were also used for the one-electron integrals.

However, the CNDO method showed systematic weaknesses that were directly attributable to the approximations outlined above, so that it was superseded by the *intermediate neglect of diatomic differential overlap* (INDO) method, introduced by Pople, Beveridge, and Dobosh in 1967 [13]. The approximation outlined in Eq. (50) proved to be too severe and was replaced by individual values for the possible different types of interaction between two AOs. These individual values, often designated $G_{ss}$, $G_{sp}$, $G_{pp}$ and $G_p{}^2$ in the literature, can be adjusted to give better agreement with experiment than was possible for CNDO. However, in INDO the two-center terms remain of the same type as those given in Eqs. (51) and (52) (again, there are many variations). This approximation leads to systematic weaknesses, for instance in treating interactions between lone pairs.

In order to overcome these weaknesses, Pople and co-workers reverted to a more complete approach that they first proposed in 1965 [14], *neglect of diatomic differential overlap* (NDDO). In NDDO, all four-center integrals $(\mu v \mid \lambda\sigma)$ are considered in which $\mu$ and $v$ are on one center, as are $\lambda$ and $\sigma$ (but not necessarily on the same one as $\mu$ and $v$). Furthermore, integrals for which the two atomic centers are different are treated in an analogous way to the one-center integrals in INDO, resulting in a much improved description of lone pair–lone pair interactions than for the earlier methods. NDDO forms the basis for almost all modern semi-empirical methods, which, with a few exceptions were developed by M. J. S. Dewar and his school.

The first semi-empirical techniques developed by Dewar and his group were designated MINDO/1–3 and were based on INDO. Many of the integral approximations inherent in the original INDO were replaced and the methods were parameterized to reproduce a wider range of experimental data, notably energies and geometries. The MINDO methods are, however, now largely obsolete and will not be discussed further.

The seminal method for most modern semi-empirical MO techniques is MNDO, which was published by Dewar and Thiel in 1977 [15]. MNDO is an NDDO method in which Dewar and Thiel introduced a new multipole-based formalism for calculating the two-electron integrals. It was parameterized to reproduce experimental heats of formation, geometries, dipole moments, and ionization potentials. It proved to be very superior to the MINDO methods for most calculated quantities. However, MNDO has one weakness that severely limits its usefulness; it does not reproduce hydrogen bonding. This weakness was fixed pragmatically in MNDO/H by Burstein and Isaev, [16] who simply modified the core–core repulsion potential with additional Gaussian functions in order to obtain hydrogen bonds. This "fix" was adopted by the Dewar group for their next method, AM1 [17], which is otherwise identical to MNDO. AM1, in turn, was found to have significant weaknesses for nitro and hypervalent compounds. These weaknesses were addressed by Stewart in a new parameterization, named PM3 [18], which is otherwise identical to AM1. However, MNDO, MNDO/H, AM1, and PM3 are quantum mechanically essentially identical. Their differences are restricted to classical "correcting" potentials between atoms and to which parameters are treated as variables in the parameterization procedure.

The first quantum mechanical improvement to MNDO was made by Thiel and Voityuk [19] when they introduced the formalism for adding d-orbitals to the basis set in MNDO/d. This formalism has since been used to add d-orbitals to PM3 to give PM3-tm and to PM3 and AM1 to give PM3(d) and AM1(d), respectively (all three are available commercially but have not been published at the time of writing). Voityuk and Rösch have published parameters for molybdenum for AM1(d) [20] and AM1 has been extended to use d-orbitals for Si, P, S and Cl in AM1* [21]. Although PM3, for instance, was parameterized with special emphasis on hypervalent compounds but with only an s,p-basis set, methods such as MNDO/d or AM1*, that use d-orbitals for the elements Si–Cl are generally more reliable.

7.4.3
### *Ab-Initio* Molecular Orbital Theory

Whereas it is generally sufficient (at least for the published methods) to specify the semi-empirical MO technique used in order to define the exact method used for the calculations, *ab-initio* theory offers far more variations, so that the exact "level" of the calculation must be specified. The starting point of most *ab-initio* jobs is an SCF calculation analogous to those discussed above for semi-empirical MO calculations. In *ab-initio* theory, however, all necessary integrals are calculated correctly, so that the calculations are very much (by a factor of about 1000) more time-consuming than their semi-empirical counterparts.

The major factor determining the level of an *ab-initio* SCF calculation is the quality of the basis set. It should be clear from the above discussion that the MOs obtained within the LCAO approximation can only be linear combinations of the AOs used (i.e., of the basis set). If, for instance, our AOs are very compact, we will not be able to describe very diffuse MOs from them using linear combinations. Therefore, the nature and number of the AOs comprising the basis set ("basis functions") affects the quality of the LCAO-SCF electron density. Semi-empirical techniques usually use minimal basis sets, as described above (an exception are the methods with d-orbitals for Si–Cl). Furthermore, the AOs used by semi-empirical techniques are usually *Slater-type orbitals* (STOs). This form for the AOs, first proposed by Slater [22], gives a good description of the electron density for many cases. However, it suffers the practical difficulty that integrals involving STOs are very difficult and time-consuming to calculate. Most *ab-initio* techniques do not, therefore, use STOs directly, but rather attempt to mimic them by linear combinations of Gaussian functions. These have the advantage that the integrals can be calculated very quickly and efficiently (because the product of two Gaussians is a third), but one of their important disadvantages for molecular calculations is that they do not describe the electron density far from a nucleus as well as STOs do.

The simplest *Gaussian-type orbital* (GTO) basis sets are minimal basis sets in which the STO electron density is mimicked by a fixed linear combination of several (usually three to six) Gaussian functions. A typical, and once very popular, minimal basis set is known as STO-3G (*Slater-type orbitals approximated by 3 Gaussian functions*) [23]. The STO-3G basis set was determined by fitting three Gaussian functions so that they reproduced the electron density of the corresponding STO as closely as possible. STO-3G, which is almost never used any more, was the standard basis set for geometry optimizations in the early 1970s when *ab-initio* methods were beginning to become practicable. Minimal basis sets suffer, however, from a major drawback. Because there is only one basis function of each type, they cannot be used to produce either more compact or more diffuse MOs than the constituent AOs. However, $\sigma$-bonding orbitals, for instance, are very much more compact than lone pairs. This is because the former are spatially associated with two positively charged nuclei and the latter with only one. Quite generally, very stable MOs are much more compact than weakly bound ones. Minimal basis sets, however, cannot reproduce this variation in the spatial extent of MOs, and can therefore give very poor results.

The solution to this problem is to use more than one basis function of each type; some of them compact and others diffuse. Linear combinations of basis functions of the same type can then produce MOs with spatial extents between the limits set by the most compact and the most diffuse basis functions. Such basis sets are known as *double* $\zeta$ ($\zeta$ is the usual symbol for the exponent of the basis function, which determines its spatial extent) if all orbitals are split into two components, or *split valence* if only the valence orbitals are split. A typical early split valence basis set was known as 6-31G [24]. This nomenclature means that the core (non-valence) orbitals are represented by six Gaussian functions and the valence AOs by two sets of three (compact) and one (more diffuse) Gaussian functions. Split valence basis sets generally give much better results than minimal ones, but at a cost. Remember that the number of two-electron integrals is proportional to $N^4$, where $N$ is the number of basis functions. Whereas STO-3G has only five basis functions for carbon, 6-31G has nine, resulting in more than a tenfold increase in the size of the calculation.

Although split valence basis sets give far better results than minimal ones, they still have systematic weaknesses, such as a poor description of three-membered rings. This results from their inability to polarize the electron density to one side of an atom. Consider the $\pi$-bond shown in Figure 7-23.

Two factors affect the stability of this orbital. The first is the stabilizing influence of the positively charged nuclei at the center of the AOs. This factor requires that the center of the AO be as close as possible to the nucleus. The other factor is the stabilizing overlap between the two constituent AOs, which requires that they approach each other as closely as possible. The best compromise is probably to shift the center of each AO slightly away from its own nucleus towards the other atom, as shown in Figure 7-23a. However, these slightly shifted positions are only correct for this particular MO. Others may require a slight shift in the opposite direction. The solution is therefore to add basis functions that allow a shift of the electron density without moving the centers of the basis functions away from their nuclei. Such basis functions must be of lower symmetry than the ones they polarize. Thus,

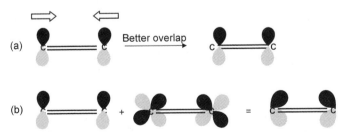

**Figure 7-23.** The polarizing effect of d-orbitals. a) The energy of the $\pi$-MO can be lowered slightly by shifting the centers of the AOs slightly towards the center of the bond. b) The lobes can be directed towards each other by mixing some d-character into the p-orbitals. The effect is exaggerated for clarity.

polarization of s-orbitals requires p-functions; p- needs d-, and so on. Figure 7-23b shows the effect of polarizing d-orbitals on the $\pi$-MO. In an older notation that is still used, 6-31G* denotes the 6-31G basis set with added d-orbitals for non-hydrogen atoms [25]. Now 6-31G* is more commonly called 6-31G(d) and the same basis with p-functions added to hydrogen (earlier known as 6-31G**) is 6–31G(d,p). Note that the fact that d-orbitals improve the bonding description of certain types of molecule does not mean "d-orbitals are involved in bonding". The question as to which AOs are involved in chemical bonds has no meaning outside the LCAO approximation and therefore it has no physical significance.

Basis sets can be extended indefinitely. The highest MOs in anions and weakly bound lone pairs, for instance, are very diffuse; maybe more so than the most diffuse basis functions in a split valence basis set. In this case, extra diffuse functions must be added to give a *diffuse augmented basis set*. An early example of such a basis set is 6-31+G* [26]. Basis sets may also be split more than once and have many sets of polarization functions.

The size of the basis set is, however, only one criterion for judging the level of an *ab-initio* calculation. The situation is best illustrated by what has become known as a "Pople diagram" [27], as shown in Figure 7-24.

The different levels of *ab-initio* theory are represented on two axes. The vertical one indicates the size of the basis set, which we have already discussed. However, the diagram shows that we can never reach the correct result (top right-hand

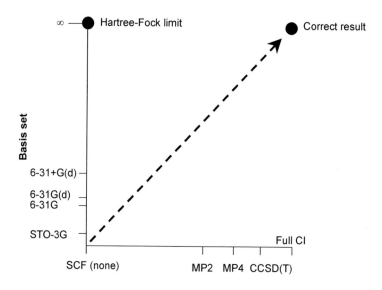

**Correlation treatment**

**Figure 7-24.** The "Pople diagram". The vertical axis gives the size of the basis set and the horizontal axis the correlation treatment. The basis sets and methods given are chosen from the examples discussed in the text. Their positions on the axes (but not the order) are arbitrary.

corner) simply by extending the basis set in SCF calculations. This would only take us to the top left-hand corner, which is marked "Hartree–Fock Limit". To understand this HF limit, we should consider the case of a molecule for which we know the correct total energy. If we were to perform an STO-3G calculation using the SCF method (HF/STO-3G), we would obtain energy very much less negative than the target value. Extending the basis set, for instance to HF/6-31G, would give us a more negative energy, which would, however, still not be negative enough. Adding d-orbitals would improve the energy by a small amount, but the improvement in the energy with every improvement in the basis set would become smaller and smaller until we converged to a limit (strictly speaking, with an infinitely large basis set), after which no further improvement in the energy can be obtained by basis set improvements. This is all very understandable, but the important point is that when we reach this energy limit, our calculated energy is still not as negative as the correct result. This is the HF limit and results from the fact that we use the SCF approximation, which overestimates electron–electron repulsion. The missing energy, usually called the *correlation energy*, is the result of the mean field approximation outlined above, which means that SCF theory neglects the fact that electrons do their best to avoid each other. Most *ab-initio* techniques that try to compensate for this error do so by applying corrections on the basis of the SCF wavefunction. These methods are generally called *post-SCF*. There are too many different methods for considering electron correlation to describe them in detail here, so we will limit our discussion to the general principles of the three most common techniques.

*Configuration interaction* (CI) solves the problem of electron correlation by considering more than a single occupation scheme for the MOs and by mixing the microstates obtained by permuting the electron occupancies over the available MOs. In its simplest form, a CI calculation consists of a preliminary SCF calculation, which gives the MOs that are used unchanged throughout the rest of the calculation. Microstates are then constructed by moving electrons from occupied orbitals to vacant ones according to preset schemes. The CI matrix is then calculated, in which the diagonal elements represent the energies of the microstates and the off-diagonal ones their interactions. This matrix is diagonalized in order to obtain the energies of the different states (ground and excited states) of the molecule as linear combinations of the microstates. Once again, the energies are given by the eigenvalues and the coefficients for the linear combinations by the eigenvectors. This procedure results in a stabilization of the ground state, but also gives energies and wavefunctions for excited states. The problem is that if we were to consider every possible arrangement of all the electrons in all the MOs (a full CI), the calculations would become far too large even for moderate-sized molecules with a fairly large basis set (because there are so many virtual orbitals). Thus, two types of restriction are usually used; only a limited number of MOs around the HOMO–LUMO gap are included in the CI, and only certain types of rearrangement (excitation) of the electrons are used. The most economical form is that in which only microstates in which one electron is promoted from the ground state to a virtual orbital (single excitations) are used. This is abbreviated as CIS and has traditionally been used for calculating spectra. Adding all double excitations (in which two electrons are promoted) gives CISD, and so on. Many more specific selection schemes for CI calculations have also been proposed. CI calculations are used very extensively with semi-empirical MO methods to calculate spectra and the properties of excited states, but they are not used as commonly with *ab-initio* techniques. One reason for this is that limited CI calculations are not *size-consistent*. This means that if, for instance, we perform a CISD calculation including the three highest occupied and three lowest unoccupied orbitals for benzene, we consider all the $\pi$-MOs, whereas for the benzene dimer the same calculation would neglect half of them. This leads to a better treatment of the monomer than the dimer.

A more practical way of considering electron correlation is to use perturbation theory to apply a correction to the SCF energy. Such an approach was first proposed by Møller and Plesset [28] for atoms and was extended by Pople et al. [29] to molecules. Because it is a perturbational treatment, Møller–Plesset (MP) theory can be applied considering the perturbation series to include different numbers of terms (i.e., to different orders). Second-order MP theory (MP2) is often used for geometry optimizations and fourth-order (MP4) for refining calculated energies. The reason, for instance, that MP3 theory is used less often is that the MP series tends to oscillate, so that only using only the even-numbered orders gives results that are more consistent. MP techniques are size-consistent and computationally efficient, so that their use is very common.

A further group of related techniques for considering electron correlation is formed by the *coupled cluster* (CC) methods and quadratic CI. These techniques represent the corrected wavefunction $\psi$ as the result of applying a so-called cluster operator to the HF wavefunction. The cluster operator can be built up from a series of operators that consider excitations of 1, 2, 3, ..., $n$ electrons, where $n$ is the total number of electrons in the molecule. Thus, CC techniques can be truncated like MP methods, but are more accurate. However, they are also computationally more expensive. Coupled cluster calculations using single and double excitations (CCSD) are common, but very often an additional perturbational term to take some triple excitations into account is used to give CCSD(T). CCSD(T) calculations (or the closely related QCISD(T) technique) represent about the best that is currently possible using an HF wavefunction as the starting point (reference wavefunction). More sophisticated techniques use more than one reference wavefunction, but these will not be discussed here.

Let us now return to the Pople diagram (Figure 7-24). The level at which electron correlation is taken into account is shown on the horizontal axis. Only calculations that use an adequately large basis set and consider electron correlation properly approach the correct result. In practice, computational resources are always limited, so that it is best to use methods around the diagonal of the Pople diagram. Very large basis set Hartree–Fock calculations (i.e., without any consideration of electron correlation) or sophisticated correlation corrections with a small basis set do not usually give good results. However, one important feature of *ab-initio* techniques can be seen clearly from the Pople diagram; *ab-initio* calculations can be improved systematically by improving the basis set, the correlation treatment, or both. Thus, in cases where no experimental data are available for comparison (and where we would have no idea of the reliability of semi-empirical calculations), the reliability of *ab-initio* calculations can be tested by checking then at the next higher level of calculation. If the results do not change much, they are probably reliable. If they do, the next higher level must also be tried until the results become consistent. This advantage is unique to *ab-initio* calculations, which, however, have other disadvantages. They rapidly become very expensive computationally as the size of the molecule or the level of calculation increases and only the most sophisticated *ab-initio* techniques give good results for transition metals. Density functional theory (DFT) does not suffer from these two disadvantages, but it cannot be improved systematically. Nevertheless, since 1990 DFT has become probably the most important technique for calculating molecular structures and properties.

### 7.4.4
### Density Functional Theory

The MO techniques discussed all calculate molecular orbitals in order to derive the energy and properties of molecules. This is not strictly necessary. The first Hohenberg–Kohn theorem [30] states that all the electronic properties of a molecule can be derived from its electron density. Thus, we need to solve only two problems; how do we find the electron density and how do we derive the properties we need from it? The solution to the first question lies in Hohenberg and Kohn's second theorem, which states that all densities other than the correct one will give a worse energy. This is a so-called variational theorem that allows us to look for the density that corresponds to the lowest energy possible. Now we have a real "Catch 22" because we do not know how to calculate the energy from the density. Kohn and Sham [31] proposed a solution to this problem in 1965. They suggested calculating the kinetic energy of the non-interacting electron density that corresponds to the real one exactly, and treating the correction from this energy to that of the real, interacting system approximately. The correction to the non-interacting kinetic energy is known as the *exchange correlation* (XC) energy and is calculated as a function of the electron density. As the electron density itself is a function, the XC energy is a function of a function, which is known as a functional; hence the name "density functional theory". The basic principles are described more fully in the monograph by Koch and Holthausen [32]. We now have the basis for an excellent theory, especially because parts of the XC energy are those that are most expensive to calculate in the normal SCF procedure. The problem, however, is that we do not know the functional(s) that translate the electron density into the XC energy. There are now many alternative functionals available, but there is no way to say that functional A is better than functional B. Thus, the major advantage of *ab-initio* theory, the ability to improve it systematically, is lost in DFT. There are, however, three basic types of functional.

The *local density approximation* (LDA) is the oldest and simplest of the functional types still in use. It is based on the idea of a uniform electron gas, a homogeneous arrangement of electrons moving against a positive background charge distribution that makes the total system neutral. This construct is abstract and not very realistic, but we do know the exact form of the exchange part of the XC functional for it and have accurate results to simulate for the correlation part. Importantly, the XC energy depends only on the electron density itself at a given position and so is easy to calculate. LDA calculations are thus very fast and often give good geometries. They tend, however, to give systematic errors in the energy and generally make bonds too strong. LDA calculations are therefore used less often for molecular applications than more sophisticated functionals.

The *generalized gradient approximation* (GGA) gives better results. GGA functionals are usually divided into exchange and correlation functionals, which are often derived separately and may be combined in different ways. The most important practical feature of GGA functionals is that they depend not only on the value of the electron density itself, but also on its derivative (gradient) with respect to the position in space. The inclusion of the first derivative of the density allows GGA functionals to treat the inhomogeneities in the electron density better than LDA functionals. Koch and Holthausen [32] give an up-to-date list of GGA exchange and correlation functionals.

The final class of density functional methods considered here, the *hybrid functionals*, does not use a third type of functional. Hybrid functionals are simply a combination of a GGA correlation functional with an exchange contribution that comes partly from an exchange functional and partly from HF theory, where the exchange energy is calculated exactly. The relative proportions of the HF exchange energy and those of the two GGA functionals vary between hybrid methods and are usually parameterized to fit a set of experimental data. Hybrid methods are generally the most accurate but suffer the disadvantage that calculating the HF exchange energy requires the four-center integrals. Hybrid DFT calculations are thus more expensive computationally than GGA.

DFT calculations offer a good compromise between speed and accuracy. They are well suited for problem molecules such as transition metal complexes. This feature has revolutionized computational inorganic chemistry. DFT often underestimates activation energies and many functionals reproduce hydrogen bonds poorly. Weak van der Waals' interactions (dispersion) are not reproduced by DFT; a weakness that is shared with current semi-empirical MO techniques.

### 7.4.5
### Properties from Quantum Mechanical Calculations

All the techniques described above can be used to calculate molecular structures and energies. Which other properties are important for chemoinformatics? Most applications have used semi-empirical theory to calculate properties or descriptors, but *ab-initio* and DFT are equally applicable. In the following, we describe some typical properties and descriptors that have been used in quantitative structure–activity (QSAR) and structure–property (QSPR) relationships.

### 7.4.5.1 **Net Atomic Charges**

Net atomic charges have often been used as descriptors, especially for properties related to individual atoms. Although atomic charges are an established part of our understanding of chemistry, neither are they physically measurable nor can they be defined uniquely. In short, net atomic charges do not exist. The result of this situation is that there are very many different methods of calculating them. The problem is that the molecular electron density must be assigned to individual atoms but that the borders between atoms are not defined. There are three different ways of dividing the electrons and many different variations of each way.

*Population analysis* methods of assigning charges rely on the LCAO approximation and express the numbers of electrons assigned to an atom as the sum of the populations of the AOs centered at its nucleus. The simplest of these methods is the Coulson analysis usually used in semi-empirical MO theory. This analysis assumes that the orbitals are orthogonal, which leads to the very simple expression for the electronic population $P_i$ of atom $i$ that is given by Eq. (53), where $N_{occ}$ is the number of occupied MOs, $i_{first}$ and $i_{last}$ are the first and last atomic orbitals centered on atom $i$, respectively, $n_j$ is the occupancy number of molecular orbital $j$, and $c_{i,k}$ is the coefficient of AO $k$ in MO $j$.

$$P_i = \sum_{j=1}^{Nocc} \sum_{k=i_{first}}^{i_{last}} n_j c_{j,k}^2 \tag{53}$$

The net atomic charge is simply the sum of the electronic population of the atom and its nuclear charge. The Mulliken population analysis used in most *ab-initio* and DFT programs is similar, but corrects for the fact that the AOs are not orthogonal. Both schemes give instinctively reasonable results as long as the basis set is small but results may vary widely if large basis sets are used. This is because the most diffuse functions in a large basis set may describe regions of space close to nuclei other than the one on which the AO is centered. In this case, electron density close to the second atom will be assigned to the first and thus lead to charges that change strongly when the basis set is expanded. Thus, Coulson or Mulliken charges are often very useful for semi-empirical MO techniques, but less so for *ab-initio* or DFT. Weinhold's *natural population analysis* (NPA) [33] is often used to overcome this problem. It is a population analysis of the same type as the others, but is based on natural atomic orbitals (NAOs); this makes it less dependent on the extent of the basis set. NAOs are obtained by diagonalizing the one-atom blocks of the density matrix, and help to iron out the effects of very extensive basis sets. They are processed further within the framework of the NPA to natural hybrid orbitals (NHOs) and to natural bond orbitals (NBOs). These, in turn are used to analyze the bonding pattern in the molecule to give results that are relatively independent of the size of the basis set.

A second group of methods for assigning net atomic charges from the results of quantum mechanical calculations relies on dividing up the space around the atoms within a molecule and integrating the electron density within the space assigned to each molecule. All that is needed for these methods is a way of defining the borders between atoms. An early example of this approach was reported by Streitwieser et al. [34], who projected the electron density of a molecule onto a plane and then drew dividing lines between the atoms. This technique is, however, limited to small molecules. The *atoms-in-molecules* (AIM) approach of Bader [35], in which the borders between atoms are defined by features of the Laplacian of the electron density, is now better known. AIM charges, however, have found little use in chemoinformatics, partly because their calculation is time-consuming.

The final group of methods used to calculate net atomic charges does not derive them from the electron density, but rather from the electrostatic potential around the molecule. These *molecular–electrostatic–potential* (MEP) *derived charges* are calculated by least-squares fitting of a set of net atomic charges so that they reproduce the calculated MEPs at a grid of points around the molecule as closely as possible. The CHELP [36] and RESP [37] techniques are well known for *ab-initio* and DFT calculations and MNDO-ESP [38] or VESPA [39] charges can be derived from semi-empirical calculations. Because MEP-derived charges are designed to reproduce the electrostatic properties of molecules as well as possible, they are inherently attractive for describing physical properties. However, in practice the simple Coulson or Mulliken charges have been used more frequently. MEP-derived charges, however, do occur in many QSPR models as the sums of all the MEP-derived charges on atoms of a given element in the molecule.

### 7.4.5.2 Dipole and Higher Multipole Moments

Molecular dipole moments are often used as descriptors in QPSR models. They are calculated reliably by most quantum mechanical techniques, not least because they are part of the parameterization data for semi-empirical MO techniques. Higher multipole moments are especially easily available from semi-empirical calculations using the *natural atomic orbital–point charge* (NAO-PC) technique [40], but can also be calculated reliably using *ab-initio* or DFT methods. They have been used for some QSPR models.

### 7.4.5.3 Polarizabilities

The molecular electronic polarizability is one of the most important descriptors used in QSPR models. Paradoxically, although it is an electronic property, it is often easier to calculate the polarizability by an additive method (see Section 7.1) than quantum mechanically. *Ab-initio* and DFT methods need very large basis sets before they give accurate polarizabilities. Accurate molecular polarizabilities are available from semi-empirical MO calculations very easily using a modified version of a simple variational technique proposed by Rivail and co-workers [41]. The molecular electronic polarizability correlates quite strongly with the molecular volume, although there are many cases where both descriptors are useful in QSPR models.

#### 7.4.5.4 Orbital Energies

The eigenvalues (energies) of the highest occupied (HOMO) and lowest unoccupied molecular orbitals (LUMO) have often been used directly as descriptors. These values, which are most often derived from semi-empirical MO calculations, are equivalent to the ionization potential and the electron affinity within Koopmans' theorem, so that these quantities also appear in models as descriptors. Some common descriptors, such as covalent acidity and basicity [42], are differences between calculated orbital energies and constants that represent the orbital energies of interacting partners. A further variation that is useful when specific interactions develop at the atomic center of interest is to use the energy of the relevant localized MO, rather than HOMO or LUMO energies [43].

#### 7.4.5.5 Surface Descriptors

The MEP at the molecular surface has been used for many QSAR and QSPR applications. Quantum mechanically calculated MEPs are more detailed and accurate at the important areas of the surface than those derived from net atomic charges and are therefore usually preferable [1]. However, any of the techniques based on MEPs calculated from net atomic charges can be used for full quantum mechanical calculations, and vice versa. The best-known descriptors based on the statistics of the MEP at the molecular surface are those introduced by Murray and Politzer [44]. These were originally formulated for DFT calculations using an isodensity surface. They have also been used very extensively with semi-empirical MO techniques and solvent-accessible surfaces [1, 2]. The *charged polar surface area* (CPSA) descriptors proposed by Stanton and Jurs [45] are also based on charges derived from semi-empirical MO calculations.

#### 7.4.5.6 Local Ionization Potential

In the spirit of Koopmans' theorem, the local ionization potential, $IP_L$, at a point in space near a molecule is defined [46] as in Eq. (54), where HOMO is the highest occupied MO, $p_i$ is the electron density due to MO $i$ at the point being considered, and $\varepsilon_I$ is the eigenvalue of MO $i$.

$$IP_L = \frac{\sum\limits_{i=1}^{HOMO} -p_i\varepsilon_i}{\sum\limits_{i=1}^{HOMO} p_i} \tag{54}$$

This quantity is found to be related to the local polarization energy and is complementary to the MEP at the same point in space, making it a potentially very useful descriptor. Reported studies on local ionization potentials have been based on HF *ab-initio* calculations. However, they could equally well use semi-empirical methods, especially because these are parameterized to give accurate Koopmans' theorem ionization potentials.

## 7.4.6
### Quantum Mechanical Techniques for Very Large Molecules

The quantum mechanical techniques discussed so far are typically applied to moderate-sized molecules (up to about 100 atoms for *ab-initio* or DFT and up to 500 for semi-empirical MO techniques). However, what about very large systems, such as enzymes or DNA, for which we need to treat tens of thousand of atoms? There are two possible solutions to this problem, depending on the application.

### 7.4.6.1  Linear Scaling Methods

The problem with most quantum mechanical methods is that they scale badly. This means that, for instance, a calculation for twice as large a molecule does not require twice as much computer time and resources (this would be linear scaling), but rather $2^n$ times as much, where $n$ varies between about 3 for DFT calculations to 4 for Hartree–Fock and very large numbers for *ab-initio* techniques with explicit treatment of electron correlation. Thus, the size of the molecules that we can treat with conventional methods is limited. Linear scaling methods have been developed for *ab-initio*, DFT and semi-empirical methods, but only the latter are currently able to treat complete enzymes. There are two different approaches available.

"Divide and conquer" (D&C) methods, such as those proposed by Yang et al. [47] or Merz et al. [48], solve the scaling problem by dividing the molecule into smaller parts, performing individual calculations on the parts, and then putting the molecule back together again. In order to be able to do this, some overlapping parts ("buffer regions") must be calculated twice, but D&C nevertheless allows the calculation of very large systems. The scaling advantage can be illustrated: assuming a realistic scaling behavior for a semi-empirical calculation, then a calculation twice as large requires $2^3$ (= 8) times the computer resources. If this larger calculation is simply divided into two of the original size, it would need only twice the computer resources. There is, however, an overhead involved in D&C techniques that makes them competitive with conventional calculations only for molecules of a given, relatively large, size.

The second technique, proposed by Stewart [49], relies on using localized molecular orbitals. The details of this technique go beyond the scope of this chapter (they can be found in the "Handbook", Chapter VII, Section 2). The *localized molecular orbital* (LMO) technique relies on the fact that, in principle, the molecular orbitals of a system can be represented as localized orbitals representing individual bonds, lone pairs, etc. If these LMOs are used throughout the calculation from the beginning, the number of interactions that must be considered between individual orbitals is reduced considerably because LMOs only interact with other LMOs in their immediate vicinity. Once again, the LMO technique only becomes competitive with conventional calculations for relatively large molecules. Nevertheless, geometry optimizations on small proteins and single-point calculations (i.e., without geometry optimization) on large enzymes are possible.

### 7.4.6.2 Hybrid QM/MM Calculations

For many applications, especially studies on enzyme reaction mechanisms, we do not need to treat the entire system quantum mechanically. It is often sufficient to treat the center of interest (e.g., the active site and the reacting molecules) quantum mechanically. The rest of the molecule can be treated using classical molecular mechanics (MM; see Section 7.2). The quantum mechanical technique can be *ab-initio*, DFT or semi-empirical. Many such techniques have been proposed and have been reviewed and classified by Thiel and co-workers [50] Two effects of the MM environment must be incorporated into the quantum mechanical system. The first is the simple mechanical effect of the MM part of the calculation. This can be simulated by adding additional classical potentials analogous to those used in the MM part, between the QM and the MM systems. The second effect is the electrostatic polarization of the QM part by the MM environment. This is also a relatively simple perturbation that usually uses net atomic charges within the MM part of the calculation. The most significant problem to be solved is how to couple the QM system to the MM environment. There are many schemes to achieve this, ranging from so-called "link atoms" (fictitious hydrogens used to saturated the QM and MM systems) [51, 52] pseudopotentials and pseudo-atoms [53, 54], and fixed hybrid orbitals at the boundaries [55] to the ONIOM superimposition technique [56], which is used predominantly for large conventional molecules rather than very large biological systems.

### 7.4.7
### The Future of Quantum Mechanical Methods in Chemoinformatics

The principal advantage of quantum mechanical methods for calculating molecular energies, structures, and properties is their accuracy and the detailed information that they can provide. The major disadvantage is the time required to do the calculations. Unfortunately, if we want more accurate results, we must also use more computer time. This fast becomes the limiting factor for large datasets. Nevertheless, it is realistic to be able treat hundreds of thousands of compounds with semi-empirical MO theory or hundreds with *ab-initio* or DFT. In this respect, loosely coupled clusters of cheap PC-based computers have made applications routine that were impossible only a few years ago. The semi-empirical optimization of 53 000 compounds required a multi-million dollar supercomputer with 128 processors in 1997 [56], but only five years later it was routinely being done on a $20 000 cluster of PCs. Thus, the question is not whether we can afford to use quantum mechanical methods but rather whether they provide us with new information that we cannot obtain more simply. In many cases, the quality of the experimental data available may not justify using high-quality quantum mechanically derived descriptors. However, the extra information and detail provided by quantum mechanics is becoming important for accurate work in which specific interactions play a dominant role. One of the first applications in which this is likely to be the case are docking studies, in which polarization of the ligand plays an important role. Typically, quantum mechanical techniques will be used for detailed studies at relatively late stages of investigations in the near future. Another important recent development is the creation of a Quantum Bioinformatics Database [57]. It will eventually contain data calculated using the D&C technique within semi-empirical MO theory for all of the protein structures contained in the Protein Data Bank. Again, such an application would have been unthinkable only a few years ago. Our current situation is that we are learning to treat both very large numbers of drug-size compounds or complete databases of very large molecules with quantum mechanical techniques. Now, the semi-empirical methods are taking the lead, but soft- and hardware development will make more and more applications accessible to *ab-initio* and DFT and extend the capabilities of semi-empirical methods to even larger systems.

## References

[1] T. Clark, Quantum chemoinformatics: an oxymoron? (Part 1), in *Chemical Data Analysis in the Large; The Challenge of the Automation Age*, M. Hicks (Ed.), Logos Verlag, Berlin, 2002. Also available as *Proceedings of the Beilstein Workshop 2000: Chemical Data Analysis in the Large*, May 22–26, **2000**, Bozen, Italy, pp. 88–99 (*http://www.beilstein-institut.de/englisch/1200/veran/index.php3? bild=events*)

[2] T. Clark, Quantum chemoinformatics: an oxymoron? (Part 2), in *Rational Approaches to Drug Design*, H.-D. Höltje, W. Sippl (Eds.), Prous Science, Barcelona, **2001**, pp. 29–40.

[3] See, for instance, S. Hannongbua, K. Nivesanond, L. Lawtrakul, P. Pungpo, P. Wolschann, *J. Chem. Inf. Comput. Sci.*, **2001**, *41*, 848–855.

[4] CORINA (*http://www2.chemie.uni-erlangen.de/services/3d.html*)

[5] A. J. Leach, *Molecular Modelling*, Longman, Harlow, UK, **1996**.

[6] C. J. Cramer, *Essentials of Computational Chemistry*, Wiley, New York, **2002**.

[7] F. Jensen, *Introduction to Computational Chemistry*, Wiley, Chichester, **1999**.

[8] A. Szabo, N. S. Ostlund, *Modern Quantum Chemistry*, Macmillan, New York, **1982**.

[9] E. Hückel, *Z. für Physik* **1930**, *60*, 423–456.

[10] R. Hoffmann, *J. Chem Phys.* **1963**, *39*, 1397–1412.

[11] R. Pariser, R. G. Parr, *J. Chem. Phys.* **1953**, *21*, 466–471, 767–776; J. A. Pople, *Trans. Faraday Soc.* **1953**, *49*, 1375–1385.

[12] J. A. Pople, G. A. Segal, *J. Chem. Phys.* **1965**, *43*, S136–S149.

[13] J. A. Pople, D. L. Beveridge, P. A. Dobosh, *J. Chem. Phys.* **1967**, *47*, 2026–2033.

[14] J. A. Pople, D. P. Santry, G. A. Segal, *J. Chem. Phys.* **1965**, *43*, S129–S135.

[15] M. J. S. Dewar, W. Thiel, *J. Am. Chem. Soc.*, **1977**, *99*, 4899–4907, 4907–4917; MNDO: W. Thiel, *Encyclopedia of Computational Chemistry, Vol. 3*, P. v. R. Schleyer, N. L. Allinger, T. Clark, J. Gasteiger, P. A. Kollman, H. F. Schaefer III, P. R. Schreiner (Eds.), Wiley, Chichester, **1998**, p. 1599.

[16] K. Y. Burstein, A. N. Isaev, *Theor. Chim. Acta* **1984**, *64*, 397–401.

[17] M. J. S. Dewar, E. G. Zoebisch, E. F. Healy, J. J. P. Stewart, *J. Am. Chem. Soc.* **1985**, *107*, 3902–3909; AM1: A. J. Holder, *Encyclopedia of Computational Chemistry, Vol. 1*, P. v. R. Schleyer, N. L. Allinger, T. Clark, J. Gasteiger, P. A. Kollman, H. F. Schaefer III, P. R. Schreiner (Eds.), Wiley, Chichester, **1998**, pp. 8–11.

[18] J. J. P. Stewart, *J. Comput. Chem.*, *10*, **1989**, 209–220; 221–264; PM3: J. J. P. Stewart, *Encyclopedia of Computational Chemistry, Vol. 3*, P. v. R. Schleyer, N. L. Allinger, T. Clark, J. Gasteiger, P. A. Kollman, H. F. Schaefer III, P. R. Schreiner (Eds), Wiley, Chichester, **1998**, pp. 2080–2086.

[19] W. Thiel, A. A. Voityuk, *Theoret. Chim. Acta* **1992**, *81*, 391–404; **1996**, *93*, 315–315; *Int. J. Quant. Chem.* **1994**, *44*, 807; *J. Mol. Struct.* **1994**, *313*, 141–154; *J. Phys. Chem.* **1996**, *100*, 616–626; MNDO/d: W. Thiel, *Encyclopedia of Computational Chemistry, Vol. 3*, P. v. R. Schleyer, N. L. Allinger, T. Clark, J. Gasteiger, P. A. Kollman, H. F. Schaefer III, P. R. Schreiner (Eds.), Wiley, Chichester, **1998** pp. 1604–1606.

[20] A. A. Voityuk, N. Rösch, *J. Phys. Chem. A*, **2000**, *104*, 4089–4094.

[21] P. Winget, C. Selçuki, A. Horn, B. Martin, T. Clark, *J. Mol. Model.*, **2003**, *9*, in press.

[22] J. C. Slater, *Phys. Rev.* **1930**, *36*, 57–64.

[23] W. J. Hehre, R. F. Stewart, J. A. Pople, *J. Chem. Phys.* **1969**, *51*, 2657–2664.

[24] W. J. Hehre, R. Ditchfield, J. A. Pople, *J. Chem. Phys.* **1972**, *56*, 2257–2261.

[25] P. C. Hariharan, J. A. Pople, *Chem. Phys. Lett.* **1972**, *16*, 217–219.

[26] T. Clark, J. Chandrasekhar, G. W. Spitznagel, P. v. R. Schleyer, *J. Comput. Chem.* **1983**, *4*, 294–301.

[27] J. A. Pople, *J. Chem. Phys.* **1965**, *13*, S229–S230.

[28] C. Møller, M. S. Plesset, *Phys. Rev.* **1934**, *46*, 618–622.

[29] J. A. Pople, J. S. Binkley, R. Seeger, *Int. J. Quantum Chem. Symp.* **1976**, *10*, 1–19.

[30] P. Hohenberg, W. Kohn, *Phys. Rev. B*, **1964**, *136*, 864–873.

[31] W. Kohn, L. J. Sham, *Phys. Rev. A*, **1965**, *137*, 1697–1705.

[32] W. Koch, M. C. Holthausen, *A Chemist's Guide to Density Functional Theory*, 2nd edition, Wiley-VCH, Weinheim, **2001**.

[33] A. E. Reed, L. A. Curtiss, F. Weinhold, *Chem. Rev.* **1988**, *88*, 899–926.

[34] A. Streitwieser, Jr.; J. B. Collins, J. M. McKelvey, D. Grier, J. Sender, A. G. Toczko, *PNAS* **1979**, *76*, 2499–2502.

[35] R. F. W. Bader, *Atoms in Molecules: A Quantum Theory*, Oxford University Press, Oxford, **1994**.

[36] M. M. Francl, C. Carey, L. E. Chirlian, D. M. Gange, *J. Comput. Chem.* **1996**, *17*, 367–383.

[37] W. D. Cornell, P. Cieplak, C. I. Bayly, P. A. Kollman, *J. Am. Chem. Soc.* **1993**, *115*, 9620–9631.

[38] B. H. Besler, K. M. Merz, Jr., P. A. Kollman, *J. Comput. Chem.* **1990**, *11*, 431–439.

[39] B. Beck, T. Clark, R. C. Glen, *J. Mol. Model.* **1995**, *1*, 176–187.

[40] G. Rauhut, T. Clark, *J. Comput. Chem.* **1993**, *14*, 503–509; B. Beck, G. Rauhut, T. Clark, *J. Comput. Chem.* **1994**, *15*, 1064–1073.

[41] J.-L. Rivail, A. Cartier, *Chem. Phys. Lett.* **1979**, *61*, 469–72; G. Schürer, P. Gedeck, M. Gottschalk, T. Clark, *Int. J. Quant. Chem.* **1999**, *75*, 17–31.

[42] C. J. Cramer, G. R. Famini, A. H. Lowrey, *Acc. Chem. Res.* **1993**, *26*, 599–605.

[43] M. Hennemann, T. Clark, *J. Mol. Model.* **2002**, *8*, 95–101.

[44] J. S. Murray, P. Lane, T. Brinck, M. E. Grice, P. Politzer, *J. Phys. Chem.* **1993**, *97*, 9369–9373.

[45] D. T. Stanton, P. C. Jurs, *Anal. Chem.* **1990**, *62*, 2323–2329.

[46] J. S. Murray, Z. Peralta-Inga, P. Politzer, Peter, *Int. J. Quant. Chem.* **2001**, *83*, 245–254.

[47] P. Politzer, J. S. Murray, M. C. Concha, *Int. J. Quant. Chem.* **2002**, *88*, 19–27.

[48] W. Pan, T.-S. Lee, W. Yang, *J. Comput. Chem.* **1998**, *19*, 1101–1109.

[49] A. Van Der Vaart, V. Gogonea, S. L. Dixon, K. M. Merz, Jr., *J. Comput. Chem.* **2000**, *21*, 1494–1504.

[50] D. Bakowies, W. Thiel, *J. Phys. Chem.* **1996**, *100*, 10580–10594.

[51] J. J. P. Stewart, *Int. J. Quant. Chem.* **1996**, *58*, 133–146.

[52] J. Aaqvist, A. Warshel, *Chem. Rev.* **1993**, *93*, 2523–2544.

[53] M. J. Field, P. A. Bash, M. Karplus, *J. Comput. Chem.* **1990**, *11*, 700–733.

[54] Y. Zhang, Yingkai, T.-S. Lee, W. Yang, *J. Chem. Phys.* **1999**, *110*, 46–54.

[55] I. Antes, W. Thiel, *J. Phys. Chem. A* **1999**, *103*, 9290–9295.

[56] M. Svensson, S. Humbel, R. D. J. Froese, T. Matsubara, S. Sieber, K. Morokuma, *J. Phys. Chem.* **1996**, *100*, 19357–19363.

[57] B. Beck, A. Horn, J. E. Carpenter, T. Clark, *J. Chem. Inf. Comput. Sci.* **1998**, *38*, 1214–1217.

[58] K. M. Merz, Jr., *http://qbiodb.chem.psu.edu/*

**Essentials**

- Additivity schemes allow the calculation of important molecular properties.
- Additivity schemes for estimating molecular properties play an important role in chemical engineering.
- The accuracy of an additivity scheme can be increased by going from atomic contributions through bond contributions to group contributions.
- Heats of formation can be estimated with reasonable accuracy by additivity of group increments and corrections for ring effects.
- Average binding energies from an additivity scheme can be used to recognize weak and strong binding of drugs to the corresponding receptor.
- The PEOE method allows a rapid calculation of the charge distribution in $\sigma$-bonded systems.
- The PEOE method in conjunction with a modified Hückel Molecular Orbital (HMO) method allows charge calculation in conjugated $\pi$-systems.
- Residual electronegativity values obtained by the PEOE method are useful quantitative measures of the inductive effect.
- The polarizability effect can be calculated by a simple attenuation model.
- Fundamental enthalpies of gas-phase reactions such as proton affinities or gas-phase acidities can be correlated with the values of the inductive and the polarizability effect.
- Molecular mechanics calculations are a very useful tool for the spatial and energetic description of small molecules as well as macroscopic systems like proteins or DNA.
- Current trends in ongoing development are in areas such as the treatment of polarization or applications to transition metal systems.
- In combination with quantum mechanical methods, a QM/MM methodology allows the description of reaction mechanisms including whole enzymes.
- Molecules should never be treated as static systems which are not able to undergo conformational changes.
- Molecular dynamics simulations provide information about the motion of molecules, which facilitates the interpretation of experimental results and allows the statistically meaningful sampling of (thermodynamic) data.
- The development of efficient algorithms and the sophisticated description of long-range electrostatic effects allow calculations on systems with 100 000 atoms and more, which address biochemical problems like membrane-bound protein complexes or the action of "molecular machines".

**Selected Reading**

- S.W. Benson, *Thermochemical Kinetics*, 2nd edition, Wiley, New York, **1976**.
- J.D. Cox, G. Pilcher, *Thermochemistry of Organic and Organometallic Compounds*, Academic Press, London, **1970**.

- R.C. Reid, J.M. Prausnitz, B.E. Polling, *The Properties of Gases and Liquids*, 4th edition, McGraw-Hill, New York, **1988**.
- K. Miller, *J. Am. Chem. Soc.* **1990**, *112*, 8533–8542.
- P.R. Andrews, D.J. Craik, J.L. Martin, *J. Med. Chem.* **1984**, *27*, 1648–1657.
- J. Gasteiger, M. Marsili, *Tetrahedron*, **1980**, *36*, 3219–3228.
- J. Gasteiger, M.G. Hutchings, *J. Chem. Soc. Perkin 2* **1984**, 559–564.
- A.R. Leach, *Molecular Modelling – Principles and Applications*, 2nd edition, Pearson Education, Harlow, UK, **2001**.
- C.J. Cramer, *Essentials of Computational Chemistry – Theories and Models*, Wiley, Chichester, **2002**.
- J.M. Haile, *Molecular Dynamics Simulation*, Wiley, Chichester, **1997**

**Interesting Web Sites**

- *http://www.nist.gov/srd/thermo.htm*
- *http://www2.chemie.uni-erlangen.de/software/petra/index.html*
- *http://www.mol-net.de*
- NIH Center for Molecular Modeling: *http://cmm.info.nih.gov/*
- Molecular mechanics across chemistry: *http://franklin.chm.colostate.edu/mmac/*
- Chemistry and biology applications on Linux: *http://SAL.KachinaTech.COM/Z/2/*
- Software used in molecular modeling and molecular dynamics: *http://www.ahpcc. unm.edu/~aroberts/main/mol__mod__software.htm*
- A molecular dynamics primer with examples in Fortran90: *http://www.fisica. uniud.it/~ercolessi/md/*

**Available Software**

- Tinker

A molecular modeling and simulation package with various implemented force field parameterizations. Free of charge for academic use. Available for different platforms.

  *http://dasher.wustl.edu/tinker*

- gOpenMol

Visualization and analysis of structure and dynamics simulation results. Free of charge for academic use. Available for different platforms. Imports TINKER results and accepts various file formats.

  *http://www.csc.fi/gopenmol/*

- DYNAMO

A Fortran90 library for the simulation of molecular systems using molecular mechanics (MM) and hybrid quantum mechanics/molecular mechanics (QM)/ MM) potential energy functions.
 *http://www.ibs.fr/ext/labos/LDM/projet6/*

- MOSCITO

Molecular dynamics simulation package with various force field implementations, special support for AMBER. Parallel version and X11 trajectory viewer available.
 *http://ganter.chemie.uni-dortmund.de/MOSCITO/*

# 8
# Calculation of Structure Descriptors

*Lothar Terfloth*

## Learning Objectives

- To understand what structure descriptors are.
- To know what QSAR and QSPR are, and the steps in QSAR/QSPR.
- To find out how to distinguish between the different kinds of molecular descriptors.
- To understand the recommendations for structure descriptors in order to be able to apply them in QSAR or drug design in conjunction with statistical methods or machine learning techniques.
- To become familiar with the properties of these descriptors.
- To know which are the frequently used descriptors.

## 8.1
## Introduction

A challenging task in material science as well as in pharmaceutical research is to custom tailor a compound's properties. George S. Hammond stated that "the most fundamental and lasting objective of synthesis is not production of new compounds, but production of properties" (Norris Award Lecture, 1968). The molecular structure of an organic or inorganic compound determines its properties. Nevertheless, methods for the direct prediction of a compound's properties based on its molecular structure are usually not available (Figure 8-1). Therefore, the establishment of Quantitative Structure–Property Relationships (QSPRs) and Quantitative Structure–Activity Relationships (QSARs) uses an indirect approach in order to tackle this problem. In the first step, numerical descriptors encoding information about the molecular structure are calculated for a set of compounds. Secondly, statistical and artificial neural network models are used to predict the property or activity of interest based on these descriptors or a suitable subset.

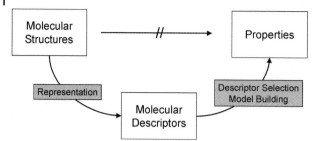

**Figure 8-1.** The general QSPR/QSAR problem.

This method of establishing a relationship between a molecular structure and its properties is inductive. It depends on a set of compounds with known properties or activities which is used for model building.

In general, a QSPR/QSAR study starts from a structure database. The molecular structure of each compound is entered and stored, providing information about – at least – the molecule's topology (suitable formats are discussed in Sections 2.4 and 2.9). If molecular descriptors are derived from the compound's 3D structure, both experimental and calculated geometries are used. Calculated geometries are submitted to a conformational analysis in order to restrict the study to low-energy conformations. Based on the structure database, a variety of descriptors can be calculated. Optional descriptor subsets are selected. Statistical methods like multi-linear regression analysis, or artificial neural networks such as backpropagation neural networks, are applied to build models. These models relate the descriptors with the property or activity of interest. Finally, the models are validated with an external data set which has not been used for the construction of the model. The steps of a typical QSPR/QSAR study are summarized as:

- structure entry (or start from an existing structure database),
- descriptor calculation,
- descriptor selection,
- model building,
- model validation.

Iteration of the steps, descriptor selection, model building, and model validation in combination with an optimization algorithm allows one to select a descriptor subset having maximum predictivity.

The method of building predictive models in QSPR/QSAR can also be applied to the modeling of materials without a unique, clearly defined structure. Instead of the connection table, physicochemical data as well as spectra reflecting the compound's structure can be used as molecular descriptors for model building.

In this chapter the focus is on structure descriptors. After a definition of this term their properties are described and an overview is given of some frequently used structure descriptors.

### 8.1.1
### Definition of the Term "Structure Descriptor"

A structure descriptor is a mathematical representation of a molecule resulting from a procedure transforming the structural information encoded within a symbolic representation of a molecule. This mathematical representation has to be invariant to the molecule's size and number of atoms, to allow model building with statistical methods and artificial neural networks.

The information content of a structure descriptor depends on two major factors: a) the molecular representation of the compound; b) the algorithm which is used for the calculation of the descriptor.

### 8.1.2
### Classification of Structure Descriptors

Structure descriptors can be distinguished by the data type (Table 8-1) of the descriptor and the molecular representation of the compound (Table 8-2).

## 8.2
## Structure Keys and 1D Fingerprints

Structural keys describe the chemical composition and structural motifs of molecules represented as a Boolean array. If a certain structural feature is present in a molecule or a substructure, a particular bit is set to 1 (true), otherwise to 0 (false). A bit in this array may encode a particular functional group (such as a carboxylic acid or an amidelinkage), a structural element (e.g., a substituted cyclohexane), or at least $n$ occurrences of a particular element (e.g., a carbon atom). Alternatively, the structural key can be defined as an array of integers where the elements of this array contain the frequency of a specific feature in the molecule.

In order to perform a database search the structural key of the query molecule or substructure is compared with the stored structural keys of the database entries. This implies that each array element in the structural key has to be defined initi-

**Table 8-1.** Classification of molecular descriptors by descriptor's data type.

| Data type | Example of a descriptor with this type of data |
|---|---|
| Boolean | compound has at least one aromatic ring |
| Integer number | number of heteroatoms |
| Real number | molecular weight |
| Vector | dipole moment |
| Tensor (3 x 3 matrix) | electric polarizability |
| Scalar field | electrostatic potential |
| Vector field | gradient of the electrostatic potential, i.e., force |

**Table 8-2.** Classification of descriptors by the dimensionality of their molecular representation.

| Molecular representation | Descriptor | Example(s) |
|---|---|---|
| 0D | atom counts, bond counts, molecular weight, sum of atomic properties | molecular weight, average molecular weight number of: atoms, hydrogen atoms, carbon atoms, heteroatoms, non-hydrogen atoms, bonds, multiple bonds, double bonds, triple bonds, aromatic bonds, rotatable bonds, rings, 3-membered rings, 4-membered rings, 5-membered rings, 6-membered rings, 7-membered rings; sum of atomic van der Waals volumes |
| 1D | fragment counts | number of: primary C (sp$^3$), secondary C (sp$^3$), tertiary C (sp$^3$), quaternary C (sp$^3$), secondary C (sp$^3$) in a ring, tertiary C (sp$^3$) in a ring, quaternary C (sp$^3$) in a ring, unsubstituted aromatic C, substituted C, primary C (sp$^2$, =CH$_2$), secondary C (sp$^2$, =CHR), tertiary C (sp$^2$, =CR$_2$), allene groups (=C=), terminal C (sp), internal C (sp), isocyanates, thiocyanates, isothiocyanates, amides (aliphatic/aromatic; primary, secondary, tertiary), amines (aliphatic/aromatic; primary, secondary, tertiary), ammonium groups, N in diazo groups, carbamates, N in hydrazines, nitriles, imines, enamines, hydroxylamines, oximes, nitroso groups, nitro groups, imides, hydroxyl groups, phenols, alcohols (aliphatic/aromatic, primary, secondary, tertiary), ethers, carboxylic acids, esters, thiols, thioketones, thioesters, sulfoxides, sulfones, sulfates, disulfides, sulfonic acids, sulfonamides, number of H-bond donor atoms, number of H-bond acceptor atoms unsaturation index, hydrophilic factor, molar refractivity (Ghose Crippen), fragment-based polar surface area |
| 2D | topological descriptors | Zagreb index, Wiener index, Balaban *J* index, connectivity indices chi ($\chi$), kappa ($\kappa$) shape indices, molecular walk counts, BCUT descriptors, 2D autocorrelation vector |
| 3D | geometrical descriptors | molecular eccentricity, radius of gyration, *E*-state topological parameter, 3D Wiener index, 3D Balaban index, 3D MoRSE descriptor, radial distribution function (RDF code), WHIM descriptors, GETAWAY descriptors, 3D autocorrelation vector |
| 3D – surface properties | | mean molecular electrostatic potential, hydrophobicity potential, hydrogen-bonding potential |
| 3D – grid properties | | Comparative Molecular Field Analysis (CoMFA) |
| 4D | | 3D coordinates + sampling of conformations |

ally. Therefore, this key is inflexible and can become extremely long. The choice and number of patterns included in the key affect the search speed across the database. Long keys slow down searching whereas short keys may screen out many structures that are of no interest. On the other hand, such a structural key can be optimized for the compounds present in the database to be investigated.

Therefore fingerprints were developed to overcome the shortcomings of structural keys. A fingerprint is a Boolean array, but in contrast to a structural key the meaning of any particular bit is not predefined. Initially all bits of a fingerprint with a fixed size are set to 0. In the second step, a list of patterns is generated for each atom (zero-bond path), for each pair of adjacent atoms and the bond connecting them (one-bond path), and for each group of atoms joined by longer pathways (usually paths of up to seven or eight bonds). Each pattern of the compound is assigned a unique set of bits (typically four or five bits per pattern) along the fingerprint by a hashcoding algorithm. The set of bits thus obtained is added to the fingerprint with a logical OR. Assuming a pattern is a substructure of a molecule, each bit in the pattern's fingerprint will be set in the molecule's fingerprint.

Using fingerprints a database can be screened in the same way as with structural keys by simple Boolean operations. Compared with structural keys, fingerprints have a higher information density without loosing specificity. Later the concept of folding a fingerprint was developed, whereby the fingerprint is split into two equal halves and then the two halves are connected by a logical OR.

In MDL's structure database systems (ISIS and MACCS), 166 search keys and 960 extended search keys are available [1, 2]. Several types of molecular fingerprints exist, depending on the substructures used to generate the fingerprints. MDL fingerprints represent the substructures defined by the MACCS search keys. Daylight fingerprints can be calculated with the Daylight Fingerprint Toolkit [3].

Following the "similar structure – similar property principle", high-ranked structures in a similarity search are likely to have similar physicochemical and biological properties to those of the target structure. Accordingly, similarity searches play a pivotal role in database searches related to drug design. Some frequently used distance and similarity measures are illustrated in Section 8.2.1.

## 8.2.1
### Distance and Similarity Measures

The similarity between compounds is estimated in terms of a distance measure $d_{st}$ between two different objects $s$ and $t$. The objects $s$ and $t$ are described by the vectors $x_s = (x_{s1}, x_{s2}, ..., x_{sm})$ and $x_t = (x_{t1}, x_{t2}, ..., x_{tm})$ where $m$ denotes the number of real variables and $x_{sj}$ and $x_{tj}$ are each the $j$th element of the corresponding vector. For calculation of the distance and similarity of two compounds, the variables $x_j$ should have a comparable magnitude. Otherwise scaling or normalization of the variables has to be performed. Two of the most prominent distance measures are given by Eqs. (1) and (2):

- Euclidean distance:

$$d_{st} = \sqrt{\sum_{j=1}^{m} (x_{sj} - x_{tj})^2}$$  (1)

- Manhattan distance:

$$d_{st} = \sum_{j=1}^{m} |x_{sj} - x_{tj}|$$  (2)

A similarity measure $s_{st}$ ($0 \le s_{st} \le 1$; $s_{ss} = 0$; $s_{st} = s_{ts}$) can be calculated from the distance measure $d_{st}$ by the functions in Eq. (3), where $d_{st}$ is the distance measure for the objects $s$ and $t$ and $d_{max}$ is the maximum distance between a pair of objects from the data set.

$$s_{st} = \frac{1}{1 + d_{st}} \quad \text{or} \quad s_{st} = 1 - \frac{d_{st}}{d_{max}}$$  (3)

As the scalar product of two vectors is related to the cosine of the angle included by these vectors by Eq. (4), a frequently used similarity measure is the cosine coefficient (Eq. (5)).

$$\mathbf{x}_s \cdot \mathbf{x}_t = |\mathbf{x}_s||\mathbf{x}_t|\cos(\angle \mathbf{x}_s, \mathbf{x}_t)$$  (4)

$$\text{cosine coefficient: } s_{st} = \cos(\angle \mathbf{x}_s, \mathbf{x}_t) = \frac{\mathbf{x}_s \cdot \mathbf{x}_t}{|\mathbf{x}_s||\mathbf{x}_t|} = \frac{\sum_{j=1}^{m} x_{sj} \cdot x_{tj}}{\sqrt{\sum_{j=1}^{m} x_{sj}^2 \cdot \sum_{j=1}^{m} x_{tj}^2}}$$  (5)

The calculation of a distance measure for two objects $s$ and $t$ represented by binary descriptors $\mathbf{x}_s$ and $\mathbf{x}_t$ with $m$ binary values is based on the frequencies of common and different components. For this purpose we define the frequencies $a$, $b$, $c$, and $d$ as follows:

$a$: number of components with $x_{sj} = x_{tj} = 1$
$b$: number of components with $x_{sj} = 1$ and $x_{tj} = 0$
$c$: number of components with $x_{sj} = 0$ and $x_{tj} = 1$
$d$: number of components with $x_{sj} = 0$ and $x_{tj} = 0$

The frequencies $a$ and $d$ reflect the similarity between two object $s$ and $t$, whereas $b$ and $c$ provide information about their dissimilarity (Eqs. (6)–(8)):

- Hamming distance: $d_{st} = b + c$  (6)

- Tanimoto coefficient: $d_{st} = \dfrac{a}{a + b + c}$  (7)

- cosine coefficient: $s_{st} = \dfrac{a}{\sqrt{(a + b) \cdot (a + c)}}$  (8)

If the binary descriptors for the objects $s$ and $t$ are substructure keys the Hamming distance (Eq. (6)) gives the number of different substructures in $s$ and $t$ (components that are 1 in either $s$ or $t$ but not in both). On the other hand, the Tanimoto coefficient (Eq. (7)) is a measure of the number of substructures that $s$ and $t$ have in common (i.e., the frequency $a$) relative to the total number of substructures they could share (given by the number of components that are 1 in either $s$ or $t$).

## 8.3
## Topological Descriptors

A huge variety of topological descriptors are frequently applied in modeling physical, chemical, or biological properties of organic compounds. Topological descriptors represent the constitution of these compounds and can be computed from their molecular graph (see Section 2.4). The structural diagram of molecules can be considered as a mathematical graph. Each atom is represented by a vertex (or node) in the graph. Accordingly, the bonds are described by the edges. As an example, the graphical representation of a molecule of 2-methylbutane (**1**) is shown in Figure 8-2: it consists of five nodes, four edges, and the adjacency relationships implied in the structure.

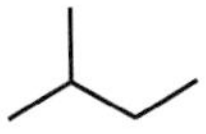

(1)

**Figure 8-2.** Edge graph of the carbon skeleton of 2-methylbutane (**1**).

In general, a graph $G$ is represented by a set of vertices $V(G)$ which are interconnected by a set of edges $E(G)$ [4, 5]. An edge connecting two vertices gives the relationship between these two objects. Considering organic compounds as graphs, application of the theorems of graph theory makes it possible to generate so-called graph invariants. In chemoinformatics these graph invariants are called topological descriptors.

Because of the fundamental role of graph theory for the understanding of topological descriptors, some terms from graph theory are defined below. Then a selection of topological descriptors are discussed.

### 8.3.1
### Some Fundamentals of Graph Theory

A simple graph can be thought of as a triple $G = (V,E,I)$, where $V$ and $E$ are disjoint finite sets and $I$ is an incidence relation such that each element of $E$ is incident with exactly two distinct elements of $V$ and no two elements of $E$ are incident with the same pair of elements of $V$. One gets general graphs, hypergraphs, infinite graphs, directed graphs, oriented graphs, etc., by variation of these requirements. We call $V$ the vertex set and $E$ the edge set of $G$. An edge connecting two vertices $v_i$ and $v_j$ is denoted as $e_{i,j}$.

The number, $n$, of elements in the vertex set $V(G) = \{v_1, v_2, ..., v_n\}$ gives the number of vertices $n$ in the graph $G$. Analogously the number of elements in the edge set $E(G)$, $m$, gives the number of edges $m$ in the graph $G$.

Two vertices $v_i$ and $v_j$ of a graph $G$ are adjacent if they are incident with a common edge $e_{i,j}$. Two distinct edges of a graph $G$ are called adjacent if they have at least one vertex in common.

The degree, $d(v)$, of a vertex $v$ is the number of edges with which it is incident. The set of neighbors, $N(v)$, of a vertex $v$ is the set of vertices which are adjacent to $v$. The degree of a vertex is also the cardinality of its neighbor set.

An alternating sequence of vertices and edges is called a walk. Each edge in a walk has to be incident with the vertices immediately preceding and succeeding it in the sequence. The length of a walk is given by the number of edges in the sequence defining the walk.

A path is a walk with no repeated vertices and its length is equal to the number of edges in the path. Assuming $u$ and $v$ to be vertices, the distance from $u$ to $v$, written $d(u,v)$, is the minimum length of any path from $u$ to $v$.

### 8.3.2
### The Adjacency Matrix

Given a graph $G$ with $n$ vertices, the adjacency matrix $A$ is a square $n \times n$ symmetric matrix (Eq. (9)).

$$
A = \begin{bmatrix}
a_{1,1} & a_{1,2} & a_{1,3} & \cdots & a_{1,n} \\
a_{2,1} & a_{2,2} & & & \vdots \\
a_{3,1} & & \ddots & & \vdots \\
\vdots & & & \ddots & \vdots \\
a_{n,1} & \cdots & \cdots & \cdots & a_{n,n}
\end{bmatrix}
\quad \text{with} \quad a_{i,j} = \begin{cases} 1 \; \forall \, i \neq j \wedge e_{i,j} \in E(G) \\ 0 \; \forall \, i = j \vee e_{i,j} \notin E(G) \end{cases}
\tag{9}
$$

The adjacency matrix of 2,2-dimethylbutane (**2**) (Figure 8-3) is given by Eq. (10).

$$
A = \begin{bmatrix}
0 & 1 & 0 & 0 & 0 & 0 \\
1 & 0 & 1 & 0 & 1 & 1 \\
0 & 1 & 0 & 1 & 0 & 0 \\
0 & 0 & 1 & 0 & 0 & 0 \\
0 & 1 & 0 & 0 & 0 & 0 \\
0 & 1 & 0 & 0 & 0 & 0
\end{bmatrix}
\tag{10}
$$

 **(2)**

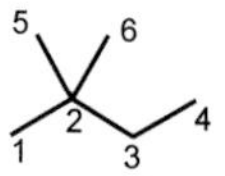

**Figure 8-3.** Hydrogen-depleted graph of 2,2-dimethylbutane (**2**).

8.3.3
## The Laplacian Matrix

Before we start to calculate the Laplacian matrix we define the diagonal matrix *DEG* of a graph *G*. The non-diagonal elements are equal to zero. The matrix element in row *i* and column *i* is equal to the degree of vertex $v_i$.

Then the Laplacian matrix *L* of a simple graph *G* can be calculated from the diagonal matrix *DEG* and the adjacency matrix *A* following Eq. (11).

$$L(G) = DEG(G) - A(G) \tag{11}$$

with

$$l_{i,j} = \begin{cases} deg(v_i) & \forall\, i = j \\ -1 & \forall\, i \neq j \,\wedge\, e_{i,j} \in E(G) \\ 0 & \forall\, i \neq j \,\wedge\, e_{i,j} \notin E(G) \end{cases}$$

As both the diagonal matrix *DEG* and the adjacency matrix *A* are symmetric it follows that the Laplacian matrix (Eq. (12)) is also a symmetric one.

$$\text{Laplacian matrix of 2: } L = \begin{bmatrix} 1 & -1 & 0 & 0 & 0 & 0 \\ -1 & 4 & -1 & 0 & -1 & -1 \\ 0 & -1 & 2 & -1 & 0 & 0 \\ 0 & 0 & -1 & 1 & 0 & 0 \\ 0 & -1 & 0 & 0 & -1 & 0 \\ 0 & -1 & 0 & 0 & 0 & -1 \end{bmatrix} \tag{12}$$

8.3.4
## The Distance Matrix

The distance matrix *D* of a graph *G* with *n* vertices is a square *n* x *n* symmetric matrix as represented by Eq. (13), where $d_{i,j}$ is the distance between the vertices $v_i$ and $v_j$ in the graph (i.e., the number of edges on the shortest path).

$$D = \begin{bmatrix} d_{1,1} & d_{1,2} & d_{1,3} & \cdots & d_{1,n} \\ d_{2,1} & d_{2,2} & & & \vdots \\ d_{3,1} & & \ddots & & \vdots \\ \vdots & & & \ddots & \vdots \\ d_{n,1} & \cdots & \cdots & \cdots & d_{n,n} \end{bmatrix} \text{ with } d_{i,j} = \begin{cases} d_{i,j} & \forall\, i \neq j \\ 0 & \forall\, i = j \end{cases} \tag{13}$$

Hence Eq. (14) gives the distance matrix of **2**.

$$D = \begin{bmatrix} 0 & 1 & 2 & 3 & 2 & 2 \\ 1 & 0 & 1 & 2 & 1 & 1 \\ 2 & 1 & 0 & 1 & 2 & 2 \\ 3 & 2 & 1 & 0 & 3 & 3 \\ 2 & 1 & 2 & 3 & 0 & 2 \\ 2 & 1 & 2 & 3 & 2 & 0 \end{bmatrix} \tag{14}$$

In graph theory, the conversion of the adjacency matrix into the distance matrix is known as the "all pairs shortest path problem".

### 8.3.5
### The Wiener Index

For historical reasons the Wiener index, $W$, is introduced in this section. It was defined in 1947 and is still a starting point for the invention of new topological indices.

The Wiener index was originally defined only for acyclic graphs and was initially called the path number [6]. "The path number, $W$, is defined as the sum of the distances between any two carbon atoms in the molecule in terms of carbon–carbon bonds". Hosoya extended the Wiener index and defined it as the half-sum of the off-diagonal elements of a distance matrix $D$ in the hydrogen-depleted molecular graph of Eq. (15), where $d_{i,j}$ is an element of the distance matrix $D$ and gives the shortest path between atoms $i$ and $j$.

$$W(G) = \frac{1}{2} \sum_{i=1}^{N} \sum_{\substack{j=1 \\ j \neq i}}^{N} d_{i,j} \tag{15}$$

Because of the symmetry of the distance matrix, the Wiener index can be expressed as Eq. (16).

$$W(G) = \sum_{i=1}^{N} \sum_{j>i}^{N} d_{i,j} \tag{16}$$

With Eq. (16) the Wiener index of compound **2** can be calculated from the distance matrix as shown in Eq. (17)

$$
\begin{aligned}
W &= 1 + 2 + 3 + 2 + 2 & (i &= 1; 1 < j \leq 6) \\
&\quad + 1 + 2 + 1 + 1 & (i &= 2; 2 < j \leq 6) \\
&\qquad + 1 + 2 + 2 & (i &= 3; 3 < j \leq 6) \\
&\qquad\quad + 3 + 3 & (i &= 4; 4 < j \leq 6) \\
&\qquad\qquad + 2 & (i &= 5; 5 < j \leq 6) \\
&= 28
\end{aligned} \tag{17}
$$

### 8.3.6
### The Randic Connectivity Index

The Randic connectivity index, $X$, is also called the connectivity index or branching index, and is defined by Eq. (18) [7], where $b$ runs over the bonds $i$–$j$ of the molecule, and $\delta_i$ and $\delta_j$ are the vertex degrees of the atoms incident with the considered bond.

$$X_R = {}^1X = \sum_b (\delta_i\delta_j)_b^{-1/2} \tag{18}$$

### 8.3.7
### Topological Autocorrelation Vectors

In order to transform the information from the structural diagram into a representation with a fixed number of components, an autocorrelation function can be used [8]. In Eq. (19) $a(d)$ is the component of the autocorrelation vector for the topological distance $d$. The number of atoms in the molecule is given by $N$.

$$a(d) = \sum_{j=i}^{N} \sum_{i=1}^{N} \delta(d_{ij} - d)p_j p_i \qquad \delta = \begin{cases} 1 \ \forall \ d_{ij} = d \\ 0 \ \forall \ d_{ij} = d \end{cases} \tag{19}$$

We denote the topological distance between atoms $i$ and $j$ (i.e., the number of bonds for the shortest path in the structure diagram) $d_{ij}$, and the properties for atoms $i$ and $j$ are referred to as $p_i$ and $p_j$, respectively. The value of the autocorrelation function $a(d)$ for a certain topological distance $d$ results from summation over all products of a property $p$ of atoms $i$ and $j$ having the required distance $d$.

A range of physicochemical properties such as partial atomic charges [9] or measures of the polarizability [10] can be calculated, for example with the program package PETRA [11]. The topological autocorrelation vector is invariant with respect to translation, rotation, and the conformer of the molecule considered. An alignment of molecules is not necessary for the calculation of their autocorrelation vectors.

### 8.3.8
### Feature Trees

The concept of feature trees as molecular descriptors was introduced by Rarey and Dixon [12]. A similarity value for two molecules can be calculated, based on molecular profiles and a rough mapping. In this section only the basic concepts are described. More detailed information is available in Ref. [12].

A molecule is represented by a tree which Rarey and Dixon called a feature tree, within which the nodes are fragments of the molecule. The atoms belonging to one node are connected in the molecular graph. A node consists at least of one atom.

Edges in the feature tree connect two nodes which have atoms in common or which have atoms connected in the molecular graph. Rings are collapsed into single nodes.

It is possible to represent molecules with feature trees at various levels of resolution. The maximum simplification of a molecule is its representation as a feature tree with a single node. On the other hand, each acyclic atom forms a node at the highest level. Due to the hierarchical nature of feature trees, all levels of resolution can be derived from the highest level. A subtree is replaced by a single node which represents the union of the atom sets of the nodes belonging to this subtree.

Two different kinds of features can be assigned to a node: steric or chemical. The steric features are the number of atoms in the fragment and an estimate for the van der Waals volume. For atoms belonging to several fragments only the corresponding fractional amount is taken into account. The chemical features are stored in an array and denote that the fragment has the ability to form interactions. The atom type profile considers the number of carbon or nitrogen atoms in different hybridization states as well as the number of oxygen, phosphorus, sulfur, fluorine, chlorine, bromine, iodine, or other non-hydrogen atoms. The FlexX interaction profile comprises hydrogen donors, hydrogen acceptors, aromatic ring centers, aromatic ring atoms, and a hydrophobic interaction.

For a pair of feature values a similarity value within the range from 0 (dissimilar) to 1 (identical) is calculated. For the comparison of two feature trees, the trees have to be matched against each other. The similarity value of the feature trees results from a weighted average of the similarity values of all matches within the two feature trees to be compared.

### 8.3.9
### Further Topological Descriptors

Some further topological descriptors are the Kier–Hall connectivity indices [13] and the electrotopological state index (or *E*-state index) [14]. A comprehensive overview of topological molecular descriptors is given by Todeschini and Consonni [15].

### 8.4
### 3D Descriptors

### 8.4.1
### 3D Structure Generation

Physical, chemical, and biological properties are related to the 3D structure of a molecule. In essence, the experimental sources of 3D structure information are X-ray crystallography, electron diffraction, or NMR spectroscopy. For compounds without experimental data on their 3D structure, automatic methods for the conversion of the connectivity information into a 3D model are required (see Section 2.9 of this Textbook and Part 2, Chapter 7.1 of the Handbook) [16].

Two of the widely used programs for the generation of 3D structures are CONCORD and CORINA. CONCORD was developed by Pearlman and co-workers [17, 18] and is distributed by TRIPOS [19]. The 3D-structure generator CORINA originates from Gasteiger's research group [20–23] and is available from Molecular Networks [24].

These programs generate one low-energy conformation for each molecule.

### 8.4.2
### 3D Autocorrelation

In the calculation of a 3D autocorrelation vector the spatial distance is used as given by Eq. (20).

$$a(d_l, d_u) = \sum_{j=1}^{N} \sum_{i=1}^{N} \delta(d_{ij}, d_l, d_u) p_j p_i \qquad \delta(d_{ij}, d_l, d_u) = \begin{cases} 1 \ \forall \ d_l < d_{ij} \le d_u \\ 0 \ \forall \ d_{ij} \le d_l \ \vee \ d_{ij} > d_u \end{cases} \qquad (20)$$

Here, the component of the autocorrelation vector $a$ for the distance interval between the boundaries $d_l$ (lower) and $d_u$ (upper) is the sum of the products of property $p$ for atoms $i$ and $j$, respectively, having a Euclidian distance $d$ within this interval.

In contrast to the topological autocorrelation vector, it is possible to distinguish between different conformations of a molecule using 3D autocorrelation vectors.

The calculation of autocorrelation vectors of surface properties [25] is similar (Eq. (21), with the distance $d(x_i x_j)$ between two points $x_i$ and $x_j$ on the molecular surface within the interval between $d_l$ and $d_u$; a certain property $p$, e.g., the electrostatic potential (ESP) at a point $x$ on the molecular surface; and the number of distance intervals $l$).

$$a(d_l, d_u) = \frac{1}{l} \sum_{j=1}^{N} \sum_{i=1}^{N} \delta(d(x_i, x_j), d_l, d_u) p(x_j) p(x_i) \qquad \text{and}$$

$$\delta(d(x_i, x_j), d_l, d_u) = \begin{cases} 1 \ \forall \ d_l < d(x_i, x_j) \le d_u \\ 0 \ \forall \ d(x_i, x_j) \le d_l \ \vee \ d(x_i, x_j) > d_u \end{cases} \qquad (21)$$

The component of the autocorrelation vector for a certain distance interval between the boundaries $d_l$ and $d_u$ is the sum of the products of the property $p(x_i)$ at a point $x_i$ on the molecular surface with the same property $p(x_j)$ at a point $x_j$ within a certain distance $d(x_i, x_j)$ normalized by the number of distance intervals $l$. All pairs of points on the surface are considered only once.

### 8.4.2.1  Example: Xylene Isomers
The 3D autocorrelation vector of the three xylene isomers in Figure 8-4 differ only with respect to the component relating to the two methyl groups. For o-xylene it is

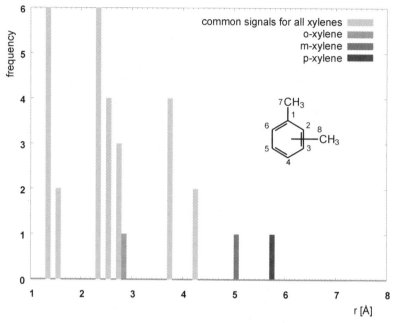

**Figure 8-4.** Comparison of 3D autocorrelation vectors of *o*-, *m*-, and *p*-xylene (without hydrogen atoms); Atomic property *p* = 1.

**Table 8.3.** Frequency of interatomic distances in *o*-xylene.

| Distance [Å] | Frequency | Atom pairs |
|---|---|---|
| 1.38 | 6 | 1-2, 2-3, 3-4, 4-5, 5-6, 1-6 |
| 1.51 | 2 | 1-7, 2-8 |
| 2.39 | 6 | 1-3, 2-4, 3-5, 4-6, 1-5, 2-6 |
| 2.50 | 4 | 1-8, 2-7, 3-8, 6-7 |
| 2.76 | 3 | 1-4, 2-5, 3-6 |
| 2.89 | 1 | 7-8 |
| 3.78 | 4 | 3-7, 4-8, 5-7, 6-8 |
| 4.27 | 2 | 4-7, 5-8 |

observed at a distance of about 2.9 Å and in *m*-xylene it occurs at about 5.1 Å (Table 8-3). The methyl groups have a maximum distance of about 5.8 Å in *p*-xylene.

When the resolution of the autocorrelation vector is decreased, some signals, e.g., those for the methyl groups in *m*- and *p*-xylene, may collapse. In such a case, one cannot distinguish between these two isomers.

8.4.3

### 3D Molecule Representation of Structures Based on Electron Diffraction Code (3D MoRSE Code)

The 3D MoRSE code is closely related to the molecular transform. The molecular transform is a generalized scattering function. It can be used to predict the intensity of the scattered radiation *i* for a known molecular structure in X-ray and electron diffraction experiments. The general molecular transform is given by Eq. (22), where $i(s)$ is the intensity of the scattered radiation caused by a collection of N atoms located at points $r_i$.

$$i(s) = \sum_{i=1}^{N} f_i \cdot e^{2\pi \cdot i \cdot r_i \cdot s} \tag{22}$$

The form factor $f_i$ takes the directional dependence of scattering from a spherical body of finite size into account. The reciprocal distance s depends on the scattering angle and the wavelength $\lambda$ as given by Eq. (23).

$$s = 4\pi \sin(\vartheta/2)/\lambda \tag{23}$$

This theoretical foundation of electron diffraction as given by Wierl [26] and modification of this equation partly following suggestions made by Soltzberg and Wilkins [27] gives Eq. (24), with the atomic property p for the atoms i and j, a reciprocal distance s, and the distance $r_{ij}$ between the atoms i and j.

$$i(s) = \sum_{i=2}^{N} \sum_{j=1}^{N} p_i p_j \frac{\sin sr_{ij}}{sr_{ij}} \tag{24}$$

In order to obtain descriptors with uniform length the intensity distribution $i(s)$ is calculated for a fixed number of discrete values of the reciprocal distance s.

This structure encoding method has been applied both for the classification of a data set comprising 31 corticosteroids, for which affinity data were available in the literature, binding to the corticosteroid-binding globulin (CBG) receptor, and for the simulation of infrared spectra [28, 29].

8.4.4

### Radial Distribution Function Code

Steinhauer and Gasteiger [30] developed a new 3D descriptor based on the idea of radial distribution functions (RDFs), which is well known in physics and physicochemistry in general and in X-ray diffraction in particular [31]. The radial distribution function code (RDF code) is closely related to the 3D-MoRSE code. The RDF code is calculated by Eq. (25), where *f* is a scaling factor, N is the number of atoms in the molecule, $p_i$ and $p_j$ are properties of the atoms i and j, B is a smoothing parameter, and $r_{ij}$ is the distance between the atoms i and j; *g(r)* is usually calculated at a number of discrete points within defined intervals [32, 33].

① removal of hydrogen atoms

② structure scan

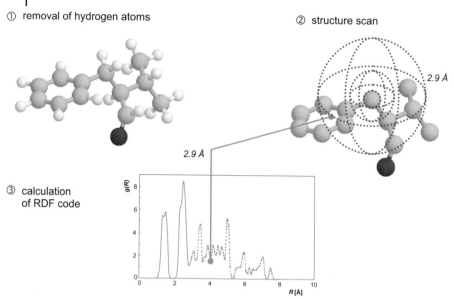

2.9 Å

③ calculation
of RDF code

2.9 Å

**Figure 8-5.** Procedure of encoding a structure with an RDF code.

$$g(r) = f \sum_{i=1}^{N-1} \sum_{j=i+1}^{N} p_i\, p_j\, e^{-B(r - r_{ij})^2} \tag{25}$$

The process of encoding is illustrated in Figure 8-5.

A slightly simplified interpretation of the radial distribution function for an ensemble of atoms is a kind of probability distribution of the individual interatomic distances. The exponential term additionally contains the smoothing parameter $B$, which can be regarded as a temperature factor that defines the movement of the atoms. Inclusion of characteristic atomic properties $p_i$ and $p_j$, into the calculation of the RDF code makes it possible to adapt the information represented to the requirements in different tasks. Choosing $p_i = p_j = 1$ is the simplest case, which solely considers the structure of a molecule. For the study of a set of ligands binding to a receptor it may be advantageous to use properties describing the atomic partial charges or their capability to act as an electron donor or acceptor.

The length or dimension of the RDF code is independent of the number of atoms and the size of a molecule, unambiguous regarding the three-dimensional arrangement of the atoms, and invariant against translation and rotation of the entire molecule.

Both regioisomers and different conformations can be distinguished with the RDF code. (Figures 8-6 and 8-7).

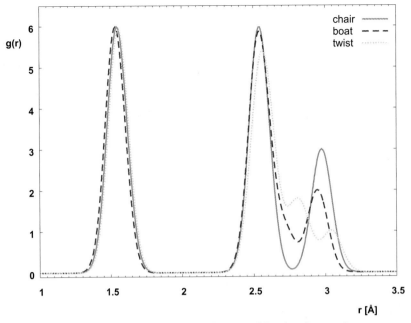

**Figure 8-6.** Comparison of the radial distribution function of the chair, boat, and twist conformations of cyclohexane (hydrogen atoms are not considered).

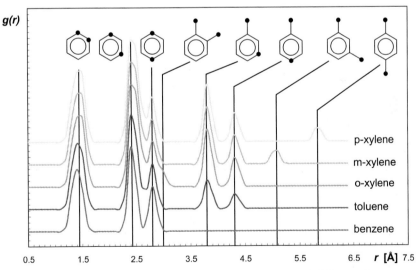

**Figure 8-7.** Comparison of the RDF code for aromatic compounds with different substitution patterns (hydrogen atoms are not considered).

## 8.5
## Chirality Descriptors [34]

### 8.5.1
### Quantitative Descriptions of Chirality

The Cahn–Ingold–Prelog (CIP) rules stand as the official way to specify chirality of molecular structures [35, 36] (see also Section 2.8), but can we measure the chirality of a chiral molecule? Can one say that one structure is *more chiral* than another? These questions are associated in a chemist's mind with some of the experimentally observed properties of chiral compounds. For example, the racemic mixture of one pair of specific enantiomers may be more clearly separated in a given chiral chromatographic system than the racemic mixture of another compound. Or, the difference in pharmacological properties for a particular pair of enantiomers may be greater than for another pair. Or, one chiral compound may rotate the plane of polarized light more than another. Several theoretical quantitative measures of chirality have been developed and have been reviewed elsewhere [37–40].

In other approaches, chirality descriptors were developed with the intention not of *measuring* chirality but of *describing* chirality in a way that correlations could be established with observable properties. These descriptors have different values for opposite enantiomers, in order that chirality-dependent properties can be predicted from them. They are usually multidimensional.

### 8.5.2
### Continuous Chirality Measure (CCM)

One example of a quantitative measure of molecular chirality is the continuous chirality measure (CCM) [39, 40]. It was developed in the broader context of continuous symmetry measures. A chiral object can be defined as an object that lacks improper elements of symmetry (mirror plane, center of inversion, or improper rotation axes). The farther it is from a situation in which it would have an improper element of symmetry, the higher its continuous chirality measure.

Continuous chirality measure is then defined as follows: given a configuration of points $\{P_i\}_{i=1}^{n}$, its chirality content is determined by finding the nearest configuration of points $\{\hat{P}_i\}_{i=1}^{n}$ which has an improper element of symmetry, and by calculating the distance between the two sets using Eq. (26).

$$S'(G) = \frac{1}{n} \sum_{i=1}^{n} \| P_1 - \hat{P}_i \|^2 \tag{26}$$

Here, $G$ is a given symmetry group, $P_i$ are the points of the original configuration, $\hat{P}_i$ are the corresponding points in the nearest $G$-symmetric configuration, and $n$ is the total number of the configuration points. For convenience, the expanded scale is $S = 100S'$.

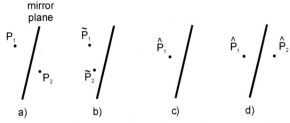

**Figure 8-8.** a) A non-mirror-symmetric pair of points; b) folding the pair of points shown in (a) results in a non-coincident cluster of two points; c) the noncoincident cluster is averaged; d) the averaged point in c) is unfolded to a mirror-symmetric pair.

In order to calculate $S$, one has to find the nearest configuration of $\hat{P}_i$ points that is $G$-symmetric. In the majority of cases, $S$ is the distance of a chiral object from a reflection mirror. The *folding/unfolding procedure* was developed for finding that configuration.

The $S(G_{achiral})$ thus obtained is the minimal chirality measure of the given configuration, on a continuous scale of $0 \le S \le 100$.

This procedure is illustrated with an example consisting of a much simplified situation (Figure 8-8). Consider a 2D space, with only two points ($P_1$ and $P_2$), and a predetermined mirror axis. In order to find the nearest mirror-symmetric configuration ($\hat{P}_1$ and $\hat{P}_2$), the identity operation is performed on $P_1$, and the mirror reflection is performed on $P_2$, yielding $\tilde{P}_1$ and $\tilde{P}_2$ respectively. The two new points, $\tilde{P}_1$ and, $\tilde{P}_2$ are then averaged. The resulting point, $\hat{P}_1$, is mirror-reflected, resulting in $\hat{P}_2$, and obtaining the mirror-symmetric configuration.

Application of the CCM to small sets ($n \le 6$) of enzyme inhibitors revealed correlations between the inhibitory activity and the chirality measure of the inhibitors, calculated by Eq. (26) for the entire structure or for the substructure that interacts with the enzyme (pharmacophore) [41]. This was done for arylammonium inhibitors of trypsin, $D_2$-dopamine receptor inhibitors, and organophosphate inhibitors of trypsin, acetylcholine esterase, and butyrylcholine esterase. Because the CCM values are equal for opposite enantiomers, the method had to be applied separately to the two families of enantiomers ($R$- and $S$-enantiomers).

### 8.5.3
### Chirality Codes

Aires-de-Sousa and Gasteiger proposed chirality codes that represent molecular chirality by spectrum-like, fixed-length codes including information about geometry, properties of the atoms, and chiral arrangement [42, 43]. The chirality codes do not rely on the $R/S$ classification to distinguish between enantiomers, but on the ranking of atoms according to physicochemical properties. Instead of a single chirality sense (or signal) for the whole molecule, different chirality signals are computed for different combinations of four atoms, and incorporated into the spectrum-like representation. Enantiomeric structures give rise to chirality codes with symmetrical values.

Neural networks were trained on the basis of these codes to predict chirality-dependent properties in enantioselective reactions [42] and in chiral chromatography [43]. A detailed description of the chirality codes is given in the Tutorial in Section 8.6.

## 8.6
## Tutorial: Conformation-Independent and Conformation-Dependent Chirality Codes [34]

### 8.6.1
### Introduction

Chirality codes are used to represent molecular chirality by a fixed number of descriptors. These descriptors can then be correlated with molecular properties by way of statistical methods or artificial neural networks, for example. The importance of using descriptors that take different values for opposite enantiomers resides in the fact that observable properties are often different for opposite enantiomers.

In most common chiral molecules, chirality arises from chiral tetravalent atoms. A conformation-independent chirality code (CICC) was developed that encodes the molecular chirality originating from a chiral tetravalent atom [42]. For more generality, a conformation-dependent chirality code (CDCC) is used [43]. CDCC treats a molecule as a rigid set of points (atoms) linked by bonds, and it accounts for chirality generated by chirality centers, chirality axes, or chirality planes.

### 8.6.2
### Conformation-Independent Chirality Code (CICC)

The calculation of the chirality code starts from a molecular structure, requires some preparatory calculations, and follows several steps that are described in detail below.

#### 8.6.2.1  Preparatory Calculations
The chirality code of a molecule is based on atomic properties and on the 3D structure. Examples of atomic properties are partial atomic charges and polarizabilities, which are easily accessible by fast empirical methods contained in the PETRA package. Other atomic properties, calculated by other methods, can in principle be used. It is convenient, however, if the chosen atomic property discriminates as much as possible between non-equivalent atoms. 3D molecular structures are easily generated by the CORINA software package (see Section 2.13), but other sources of 3D structures can be used as well.

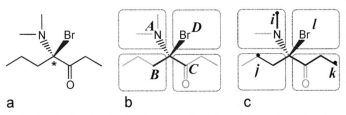

a               b             c

**Figure 8-9.** a) Example of a chiral molecule; (b) The atoms A, B, C, and D are those directly bonded to the chiral center. The neighborhood of atom A is the set of atoms whose distance (in number of bonds) to A is less than their distance to B, C, and D; b) example of a combination of four atoms (*i*, *j*, *k*, and *l*), each at a different ligand (A, B, C, or D) of the chiral center.

### 8.6.2.2. Neighborhoods of Atoms Bonded to the Chiral Center

The neighborhoods of the atoms directly bonded to the chiral center must be defined. The neighborhood of an atom A, directly bonded to the chiral center, is defined as the set of atoms whose distance (in number of bonds) to A is less than their distance to any of the other three atoms bonded to the chiral center (Figure 8-9). In cyclic structures different neighborhoods can overlap.

### 8.6.2.3 Enumeration of Combinations

Once the neighborhoods are defined, combinations of four atoms are enumerated. In a given combination, each atom (of the four atoms A, B, C, and D that are directly bonded to the chiral center) must belong to a different neighborhood – Figure 8-9c. Depending on what is considered to be relevant for the specific application, all combinations are enumerated or some are neglected. For example, it can be decided to neglect combinations containing hydrogen atoms, or combinations containing two atoms that are more than six bonds away from each other.

### 8.6.2.4 Characterization of Combinations

For each combination of atoms *i*, *j*, *k*, and *l*, each of them belonging to a different neighborhood of the four atoms A, B, C, and D, a value of $e_{ijkl}$ is defined through Eq. (27), where $a_i$ is a property of atom *i*, and $r_{ij}$ is a distance between atoms *i* and *j*.

$$e_{ijkl} = \frac{a_i a_j}{r_{ij}} + \frac{a_i a_k}{r_{ik}} + \frac{a_i a_l}{r_{il}} + \frac{a_j a_k}{r_{jk}} + \frac{a_j a_l}{r_{jl}} + \frac{a_k a_l}{r_{kl}} \tag{27}$$

In order to consider the 3D structure but make the chirality code independent of a specific conformer, $r_{ij}$ is taken as the sum of the bond lengths between atoms *i* and *j* on the path with a minimum number of bond counts.

Furthermore, a chirality signal, $s_{ijkl}$, is defined that can attain values of +1 or –1. For the computation of $s_{ijkl}$, atoms *i*, *j*, *k*, and *l* are ranked according to decreasing atomic property $a_i$ (when the property of two atoms is the same, the properties of the neighbors A, B, C, or D are used for ranking).

Then, the 3D coordinates of A are used for atom $i$, those of B for $j$, those of C for $k$, and those of D for $l$. The first three atoms (in the order established by the ranking) define a plane; if they are ordered clockwise and the fourth atom is behind the plane, the chirality signal, $s_{ijkl}$, obtains a value of $+1$; for the opposite geometric arrangement, $s_{ijkl}$ obtains a value of $-1$.

The value of $e_{ijkl}$ embodies the conformation-independent 3D arrangement of the atoms of the ligands of a chirality center in distance space and thus cannot distinguish between enantiomers. This distinction is introduced by the descriptor $s_{ijkl}$.

### 8.6.2.5 Generation of the Code

The two values, $e$ and $s$, calculated for all the combinations of four atoms (each one sampled from a different ligand of a chiral center) are then combined to generate a *conformation-independent chirality code*, $f_{CICC}$, using Eq. (28).

$$f_{CICC}(u) = \sum_i \sum_j \sum_k \sum_l s_{ijkl} \exp\left[-b\left(u - e_{ijkl}\right)^2\right] \tag{28}$$

$f_{CICC}(u)$ is calculated at a number of discrete points with defined intervals to obtain the same number of descriptors, irrespective of the size of the molecule. The actual range of $u$ used in an application is chosen according to the range of atomic properties related to the range of observed interatomic distances for the given molecules.

The number of discrete points of $f_{CICC}(u)$ determines the resolution of the chirality code; $b$ is a smoothing factor which in practice controls the width of the peaks obtained by a graphical representation of $f_{CICC}(u)$ versus $u$. An example of a chirality code for the enantiomers of **3** is shown in Figure 8-10.

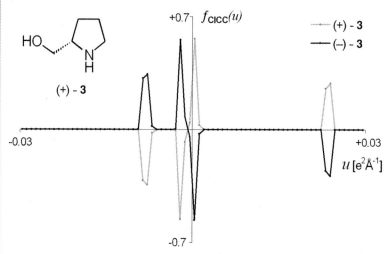

**Figure 8-10.** Graphical representation of $f_{CICC}(u)$ versus $u$ for $(+)$-**3** and $(-)$-**3** sampled at 75 evenly distributed points between $-0.03$ e$^2$ Å$^{-1}$ and $+0.03$ e$^2$ Å$^{-1}$. Hydrogen atoms not bonded to chiral carbon atoms were not considered.

Applications of the CICC to the prediction of enantiomeric preference in enantioselective reactions are described in refs. [37] and [42].

### 8.6.3
### Conformation-Dependent Chirality Code (CDCC)

#### 8.6.3.1 Overview
The conformation-dependent chirality code constitutes a more general description
of molecular chirality, which is formally comparable with the CICC [43]. The main
difference is that chiral carbon atoms are now not explicitly considered, and combinations of *any* four atoms are now used, independently of the existence or non-
existence of chiral centers, and of their belonging or not belonging to ligands of
chiral centers.

#### 8.6.3.2 Enumeration of combinations
As for the CICC, restrictions can be imposed onto the combinations that are considered. Again, it can be decided to neglect combinations containing a specific type
of atoms (e.g. hydrogen atoms) or pairs of atoms farther away from each other than
a specified threshold.

#### 8.6.3.3 Ranking of the Four Atoms in a Combination
Atoms $A$, $B$, $C$, and $D$ are ranked according to decreasing atomic property (and
renamed, according to ranking, in the order $i$, $j$, $k$, and $l$). When the atomic property of two atoms is the same, they are ranked according to a set of rules based on
geometric arguments and atomic properties. If a set of four atoms is an achiral set
(i.e., there is an element of symmetry making the mirror image of the set super-
imposable on it) then it is not further considered. The property, $a_i$, of an atom
should have values that allow one to distinguish between non-equivalent atoms.
For that purpose, partial atomic charges and polarizabilities calculated by
PETRA have been used because this software rapidly assigns highly selective values to the atoms of large molecules and sizable data sets. Furthermore, it was
decided to rank atoms using atomic physicochemical properties since these
have a greater influence on the physical, chemical, or biological behavior of molecules than other, conventionally used, values (such as atomic numbers in the
CIP rules).

#### 8.6.3.4 Characterization of Combinations
Each combination of four atoms ($A$, $B$, $C$, and $D$) is characterized by two parameters, $e$ and $c$. As for the CICC, $e$ is a parameter that depends on atomic properties and on distances, and is calculated by Eq. (27), with $r_{ij}$ again being the sum of
bond lengths between atoms on the path with the minimum number of bond
counts. However $c$ is now a geometric parameter (dependent on the conformation)

that takes real values, and it takes opposite values for the corresponding set of four atoms in opposite enantiomers.

For each combination of atoms $i$, $j$, $k$, and $l$, $c$ is defined by Eq. (29), where $x_j$, $y_j$, and $z_j$ are the coordinates of atom $j$ in Cartesian space defined in such a way that atom $i$ is at position (0, 0, 0), atom $j$ lies on the positive side of the $x$-axis, and atom $k$ lies on the $xy$-plane and has a positive $y$-coordinate. On the right-hand side of Eq. (29), the numerator represents the volume of a rectangular prism with edges $x_j$, $|y_k|$, and $|z_l|$, while the denominator is proportional to the surface of the same solid. If $x_j$, $y_k$, or $z_l$ has a very small absolute value, the set of four atoms is deviating only slightly from an achiral situation. This is reflected in $c$, which would then take a small absolute value; the value of $c$ is conformation-dependent because it is a function of the 3D atomic coordinates.

$$c_{ijkl} = \frac{x_j y_k z_l}{x_j y_k + x_j |z_l| + y_k |z_l|} \tag{29}$$

### 8.6.3.5 Generation of the Code

The two values, $e$ and $c$, calculated for all combinations of four atoms, are then combined to generate a *conformation-dependent chirality code*, $f_{CDCC}$, using Eq. (30), where $n$ is the number of atoms in each molecule, and $c$ introduces the conformation dependence:

$$f_{CDCC}(u) = \sum_i^n \sum_j^{n-1} \sum_k^{n-2} \sum_l^{n-3} c_{ijkl} \exp\left[ -b(u - e_{ijkl})^2 \right] \tag{30}$$

$f_{CDCC}(u)$ is calculated at a number of discrete values of $u$, with defined intervals to obtain the same number of descriptors, irrespective of the size of the molecule.

The number of discrete values of $f_{CDCC}(u)$ determines the resolution of the chirality code. Again, $b$ is a smoothing factor. An example with the conformation-dependent chirality codes for the enantiomers of 4 in two different conformations is shown in Figure 8-11.

### 8.6.3.6 Example of an Application

Twenty-eight chiral compounds were separated from their enantiomers by HPLC on a teicoplanin chiral stationary phase. Figure 8-12 shows some of the structures contained in the data set. This is a very complex stationary phase and modeling of the possible interactions with the analytes is impracticable. In such a situation, learning from known examples seemed more appropriate, and the chirality code looked quite appealing for representing such data.

Therefore the 28 analytes and their enantiomers were encoded by the conformation-dependent chirality code (CDCC) and submitted to a Kohonen neural network (Figure 8-13). They were divided into a test set of six compounds that were chosen to cover a variety of skeletons and were not used for the training. That left a training set containing the remaining 50 compounds.

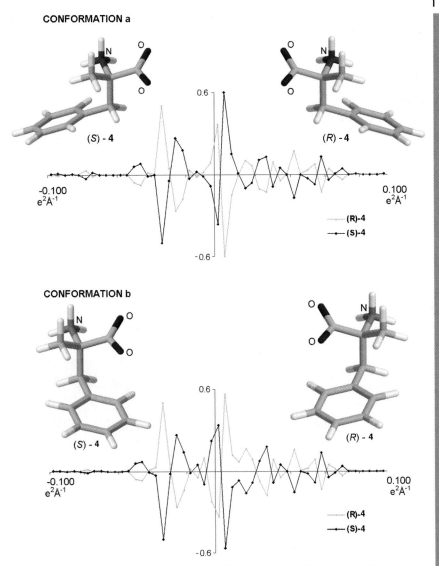

**CONFORMATION a**

(S) - 4

(R) - 4

0.6

-0.100
e²Å⁻¹

0.100
e²Å⁻¹

-0.6

----- (R)-4
——•—— (S)-4

**CONFORMATION b**

(S) - 4

(R) - 4

0.6

-0.100
e²Å⁻¹

0.100
e²Å⁻¹

-0.6

----- (R)-4
——•—— (S)-4

**Figure 8-11.** 3D structure and representation of $f_{CDCC}(u)$ versus $u$ for (R)-**4** and (S)-**4** at two different conformations (**a** and **b**) sampled at 50 evenly separated values between −0.100 e² Å⁻¹ and +0.100 e² Å⁻¹. Partial atomic charge was used as the atomic property.

The resulting trained Kohonen NN surface, shaded according to the mapping of the training set, is shown in Figure 8-14. The map resulting from the training shows that the first eluted enantiomers were mapped into neurons (black shading) in a characteristic region clearly different from that of the neurons excited by the last eluted enantiomers (gray). The objects of the test set were also mapped into it and are identified by F (<u>F</u>irst eluted) and L (<u>L</u>ast eluted). It can be seen that all the

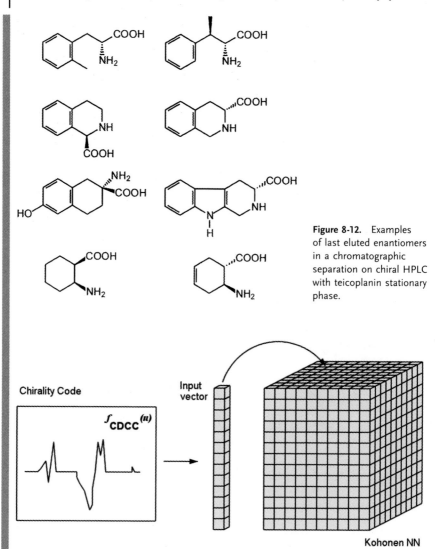

**Figure 8-12.** Examples of last eluted enantiomers in a chromatographic separation on chiral HPLC with teicoplanin stationary phase.

Kohonen NN

**Figure 8-13.** Training of a Kohonen neural network with a chirality code. The number of weights in a neuron is the same as the number of elements in the chirality code vector. When a chirality code is presented to the network, the neuron with the most similar weights to the chirality code is excited (this is the *winning* or *central* neuron) (see Section 9.5.3).

test objects were correctly classified (an object is classified according to the most frequent class of the winning and adjacent neurons). In this case, partial atomic charges were used as the atomic property for the computation of $e$, $f_{CDCC}(u)$ was sampled at 41 evenly distributed values of $u$ between $-0.060$ $e^2$ $Å^{-1}$ and $+0.060$ $e^2$ $Å^{-1}$. In Eq. (30), combinations of four atoms with interatomic distances greater than eight bonds were neglected. The resulting 41-dimensional vectors

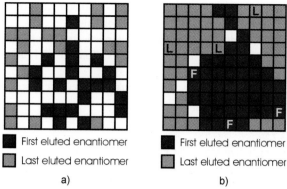

First eluted enantiomer
Last eluted enantiomer

a)

First eluted enantiomer
Last eluted enantiomer

b)

**Figure 8-14.** a) Mapping of chiral amino acids encoded by CDCC into a 10 × 10 Kohonen NN with a toroidal surface. After training, the 50 molecules from the training set were mapped and the neurons were shaded accordingly (some neurons were empty and some were excited by more than one molecule). b) The empty neurons of a) were shaded in by counting the number of neighbor neurons belonging to each class and choosing the winning class accordingly. Test structures of types F (First eluted) and L (Last eluted) were presented to the trained network in order to be classified. All test examples were correctly classified.

were normalized by their vector sum. These optimized parameters were found after screening different possibilities and evaluating their ability to cluster the two classes of objects *from the training set*.

## 8.7
## Further Descriptors

Before the comparative molecular field analysis (CoMFA), BCUT descriptors, 4D-QSAR, and HYBOT descriptors are discussed in more detail, some further descriptors are listed briefly.

- The representation of molecules by molecular surface properties was introduced in Section 2.10. Different properties such as the electrostatic potential, hydrogen bonding potential, or hydrophobicity potential can be mapped to this surface and serve for shape analysis [44] or the calculation of surface autocorrelation vectors (refer to Section 8.4.2).
- Quantum chemical descriptors such as atomic charges, HOMO and LUMO energies, HOMO and LUMO orbital energy differences, atom–atom polarizabilities, super-delocalizabilities, molecular polarizabilities, dipole moments, and energies such as the heat of formation, ionization potential, electron affinity, and energy of protonation are applicable in QSAR/QSPR studies. A review is given by Karelson et al. [45].
- EVA descriptors were proposed by Ferguson et al. [46, 47]. The EVA descriptor (EigenVAlue) extracts structural information from infrared spectra. The eigenva-

lues obtained from the normal coordinate matrix correspond to the fundamental vibrational frequencies of the molecule.

- WHIM descriptors (Weighted Holistic Invariant Molecular descriptors) are discussed in detail in the Handbook, Chapter XIII, Section 2 [48, 49].

A comprehensive overview of spatial molecular descriptors is given by Todeschini and Consonni [15].

### 8.7.1
### Comparative Molecular Field Analysis (CoMFA)

Besides the aforementioned descriptors, grid-based methods are frequently used in the field of QSAR (quantitative structure–activity relationships) [50]. A molecule is placed in a box and for an orthogonal grid of points the interaction energy values between this molecule and another small molecule, such as water, are calculated. The grid map thus obtained characterizes the molecular shape, charge distribution, and hydrophobicity.

After an alignment of a set of molecules known to bind to the same receptor a comparative molecular field analysis (CoMFA) makes it possible to determine and visualize molecular interaction regions involved in ligand–receptor binding [51]. Further on, statistical methods such as partial least squares regression (PLS) are applied to search for a correlation between CoMFA descriptors and biological activity. The CoMFA descriptors have been one of the most widely used set of descriptors. However, their apex has been reached.

### 8.7.2
### BCUT Descriptors

The BCUT descriptors (Burden–CAS–University of Texas eigenvalues) [52], are commonly used, are eigenvalue-based, and include 3D information also.

The BCUT descriptors have been proposed for the investigation of the similarity or diversity of large databases. They are an extension of the Burden approach and consider three classes of matrices whose diagonal elements correspond to values related to the atomic charge, atomic polarizability-related values, and atomic hydrogen bonding abilities. In addition, the off-diagonal terms include functions of interatomic distance, bond orders, etc. The Burden matrix $b$ represents the hydrogen-depleted graph of a compound. The diagonal elements $b_{i,i}$ are the atomic number $Z_i$ of the atoms. The off-diagonal elements $b_{i,j}$ represent the bond order (0.1, 0.2, 0.3, or 0.15 for a single, double, triple, or aromatic bond respectively). The value for terminal bonds is augmented by 0.01. All off-diagonal elements for combinations $i$ and $j$ which cannot be related to a bond in the compound are set to the value 0.001. The ordered sequence of the $n$ smallest eigenvalues of the $b$ matrix can be used as a structure descriptor.

## 8.7.3
## 4D-QSAR

Hopfinger et al. [53, 54] have constructed 3D-QSAR models with the 4D-QSAR analysis formalism. This formalism allows both conformational flexibility and freedom of alignment by ensemble averaging, i.e., the fourth dimension is the "dimension" of ensemble sampling. The 4D-QSAR analysis can be seen as the evolution of Molecular Shape Analysis [55, 56].

In 4D-QSAR, a grid is used to determine the regions in 3D space responsible for binding. Nevertheless, neither a probe nor interaction energy is used.

Partitioning of the compounds according to seven different interaction pharmacophore elements (IPEs) into specific atom/region types follows as the second step. Conformational ensemble profiles (CEPs) are constructed using molecular dynamics simulations (MDSs). Grid cell occupancy descriptors (GCODs) are calculated based on the selected IPEs for each conformation stored in the CEP. Each conformation of a specific compound is aligned in the reference grid. The frequency (how often it can be found) of a specific IPE in a particular grid cell is recorded. The grid occupancy data are reduced by a Partial Least Squares (PLS) regression analysis. Finally, a model for the remaining GCODs and the biological activity data is established.

## 8.7.4
## HYBOT Descriptors

Interactions between hydrogen-bond donor and acceptor groups in different molecules play a pivotal role in many chemical and biological problems. Hydrogen bonds can be studied with quantum chemical calculations and empirical methods. *Ab initio* calculations allow the calculation of the energy and geometric structures of small molecules with high accuracy. Nevertheless, this methodology is not yet applicable to large molecules. Raevsky et al. developed an approach called HYBOT (HYdrogen BOnd Thermodynamics) in order to describe hydrogen bonding quantitatively with an empirical correlation model [57–61]. The thermodynamics of H-bond complex formation is considered. The HYBOT descriptors are based on a database with over 13 500 thermodynamic measurements of hydrogen bonding systems. The relative proton acceptor and proton donor strengths of compounds are estimated using a common H-bond scale. The free energy $\Delta G$ and the enthalpy $\Delta H$ of the H-bond complex formation is given by Eq. (31), where $\Delta S$ is the entropy of complexation and $T$ the temperature in Kelvin.

$$\Delta G = \Delta H - T \cdot \Delta S \tag{31}$$

$\Delta G$ and $\Delta H$ can be expressed as a multiplicative function of hydrogen bonding in different polar and nonpolar solvents by means of enthalpy acceptor factors $E_a$, enthalpy donor factors $E_d$, free energy acceptor factors $C_a$, and free energy donor factors $C_d$ (Eqs. (32) and (33), where $k_1$, $k_2$, and $k_3$ [kcal/mol] are regression coefficients).

$$\Delta H = k_1 \cdot E_a \cdot E_d \tag{32}$$

$$\Delta G = k_2 \cdot C_a \cdot C_b + k_3 \tag{33}$$

Based on known experimental $\Delta G$ values, $C_a$ and $C_b$ were estimated. The system phenol–HMPA in $CCl_4$ was chosen as the reference standard; the H-bond free energy factors $C_b$ and $C_a$ were set at $-2.50$ (phenol) and $4.00$ (HMPA) respectively.

The HYBOT descriptors were successfully applied to the prediction of the partition coefficient log $P$ (*n*-octanol/water) for small organic compounds with one acceptor group from their calculated polarizabilities and the free energy acceptor factor $C_a$ as well as properties like solubility log $S$, the permeability of drugs (Caco-2, human skin), and for the modeling of biological activities.

## 8.8
### Descriptors that are not Structure-Based

The theme of this book has a strong emphasis on the structure of chemical compounds, its representation, and its correlation with properties. However, there are compounds whose structure is either unknown or ill-defined. This is true, for example, for many polymers, particularly those that have been prepared from several components, or for many technical materials, such as glues, washing powder, etc.

One way of representing such compounds is to characterize them by the components from which they were made, the relative ratios that were utilized, and the reaction conditions that were employed to make them.

Another, probably more broadly applicable, technique is to represent a chemical compound by some of its properties. Figure 8-15 is an extension of Figure 8-1 and shows that when no structure descriptors can be derived because the structure is not known, then a compound can be represented by a second property (2) or, better, a series of properties, in order to predict the property 1 of interest.

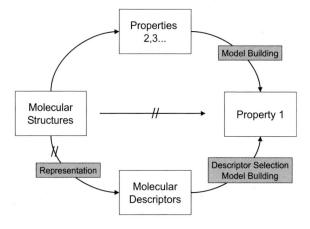

**Figure 8-15.** Extension of the QSAR method by descriptors not based on structure.

Properties that come foremost to mind to represent a compound are physical ones, because most of them can be measured easily and with high accuracy. Clearly, the more properties are used to characterize a compound, the better a model can be established for the prediction of the property of interest. Furthermore, one should select such properties which one knows or assumes to have a strong influence on the property that one wants to predict.

A particularly good selection of physical properties may be spectra, because they are known to depend strongly on the chemical structure. In fact, different types of spectra carry different kinds of structural information. $^{13}C$ NMR spectra characterize individual carbon atoms in their molecular environment. They therefore correspond quite closely to fragment-based descriptors, as underlined by the success of approaches to predict $^{13}C$ NMR spectra by fragment codes (see Section 10.2.3).

Infrared spectra are strongly dependent on the 3D structure of a compound, as reflected by the success of attempts to simulate infrared spectra from 3D structure representations (see Section 10.2.5). Infrared spectra should therefore be taken as representations of a chemical compound for modeling properties that are suspected to be dependent on the 3D structure of a compound. $^{1}H$ NMR spectra also depend on the chemical environment of the atoms (in this case, hydrogen atoms), in a molecule, corresponding to a fragment code, but also carry some 3D structure information.

Compounds need not to be represented by only one spectrum. Several different types of spectra can be used simultaneously to represent a chemical compound, thus taking advantage of the different information they contain.

The advantages of spectra for structure representation are their high information content and their easy, accurate, and reproducible measurement. On top of that, spectrometers provide this spectral information already in electronic form and therefore directly amenable to further processing.

Other important physical properties are melting points, boiling points, viscosity, molar refractivity, and data from differential scanning calorimetry (DSC). Furthermore, properties like glass temperature, average chain length, and distribution of the molecular weight as well as dispersity which are available from gel permeation chromatography (GPC) may be used to characterize polymers.

The representation of a chemical compound need not restrict itself to such properties.

Chemical compounds can also be represented by chemical data, or even by biological data. In fact, Briem et al. have used the results of a battery of biological assays to represent a compound for the modeling of other biological data [62–64].

## 8.9
### Properties of Structure Descriptors

Table 8-4 summarizes the most important invariance properties of different structure descriptors.

Topological descriptors and 3D descriptors calculated in "distance space", such as 3D autocorrelation, surface autocorrelation, and radial distribution function

**Table 8-4.** Invariance properties of molecular descriptors.

| Descriptor | Molecular representation | Mathematical representation | Invariance properties[a] |
|---|---|---|---|
| Molecular weight | 0D | scalar | ncd |
| Atom-type counts | 0D | scalar | ncd |
| Fragment counts | 1D | scalar | ncd |
| Topological information indices | 2D | scalar | ncd |
| Molecular profiles | 2D | vector | ncd |
| Substituent constants | 3D | scalar | ncd/lcd |
| WHIM descriptors | 3D | vector | hcd |
| 3D-MoRSE descriptors | 3D | vector | ncd/icd |
| Surface/volume descriptors | 3D | scalar | hcd/icd |
| Quantum-chemical descriptors | 3D | scalar | icd/hcd |
| Interaction energy values | 4D | lattice | hcd |

[a] ncd: no conformational dependency; lcd: low conformational dependency; icd: intermediate conformational dependency; hcd: high conformational dependency.

code, are translationally and rotationally invariant. Nevertheless, all descriptors calculated in distance space cannot be applied to distinguish between enantiomers.

For the calculation of grid-based descriptors it is necessary to align the ligands in the first step. If the molecules have the same skeleton, the alignment step is not critical. Otherwise, the alignment may be biased by the user.

**Essentials**

- A structure descriptor is a mathematical representation of a molecule resulting from a procedure transforming the structural information encoded within a symbolic representation of a molecule.
- The abbreviation QSAR stands for quantitative structure–activity relationships. QSPR means quantitative structure–property relationships. As the properties of an organic compound usually cannot be predicted directly from its molecular structure, an indirect approach is used to overcome this problem. In the first step numerical descriptors encoding information about the molecular structure are calculated for a set of compounds. Secondly, statistical methods and artificial neural network models are used to predict the property or activity of interest, based on these descriptors or a suitable subset. A typical QSAR/QSPR study comprises the following steps: structure entry (or start from an existing structure database), descriptor calculation, descriptor selection, model building, model validation.
- Molecules can be represented by structure descriptors in a hierarchical manner with respect to a) the descriptor data type, and b) the molecular representation of the compound.

- The QSPR/QSAR methodology can also be applied to materials and mixtures where no structural information is available. Instead of descriptors derived from the compound's structure, various physicochemical properties, including spectra, can be used. In particular, spectra are valuable in this context as they reflect the structure in a sensitive way.

### Selected Reading

- R. Todeschini, V. Consonni, Handbook of Molecular Descriptors, in *Methods and Principles in Medicinal Chemistry, Vol. 11*, R. Mannhold, H. Kubinyi, H. Timmerman (eds.), Wiley-VCH, Weinheim, **2000**.
- Chapter VIII in *Handbook of Chemoinformatics*, J. Gasteiger (ed.), Wiley-VCH, Weinheim, **2003**.

### Available Software

ADAPT (QSAR program including descriptor generation, variable selection, and modeling); P.C. Jurs, PennState University, University Park, PA 16802, USA: *http://research.chem.psu.edu/pcjgroup/ADAPT.html*

ASP (quantitative similarity calculation based on molecular shape, electrostatic potentials, and lipophilic information; within the TSAR3D package); Accelrys Inc.: *http://www.accelrys.com/products/tsar/asp.html*

AUTOCORR; Molecular Networks GmbH, Nägelsbachstrasse 25, D-91052 Erlangen, Germany: *http://www.mol-net.de/*

CODESSA (calculation of a series of different descriptors including quantum chemical descriptors); Semichem Inc. – 7204 Mullen, Shawnee, KS 66216, USA: *http://www.semichem.com*

CONCORD; TRIPOS, Inc., 1699 South Hanley Road, St. Louis, MO 63144, USA: *http://www.tripos.com*

CORINA (3D coordinates generator); Molecular Networks GmbH, Nägelsbachstrasse 25, D-91052 Erlangen, Germany: *www.mol-net.de*

DRAGON (calculation of a broad range of descriptors – topological, geometrical, WHIM, 3D-MoRSE descriptors, molecular profiles, etc.); R. Todeschini, distributed by Talete srl, via Pisani 13, 20124 Milano, Italy: *http://www.disat.unimib.it/chm*

GALAXY; AM Technologies, 14785 Omicron Drive, Texas Research Park, San Antonio, TX 78245, USA: *http://www.am-tech.com*

GRID (calculation of the GRID empirical force field at grid points); Peter Goodford, Molecular Discovery Ltd., West Way House, Elms Parade, Oxford OX2 9LL, UK: *http://www.moldiscovery.com/*

HASL: *http://www.eslc.vabiotech.com/hasl/* or *http://www.edusoft-lc.com*

HINT: *http://www.eslc.vabiotech.com/hint*

HYBOT-Plus (hydrogen bonding thermodynamics, calculation of local and molecular physicochemical descriptors): *http://www.timtec.net/software/hybot-plus.htm*

JOELib:*http://joelib.sourceforge.net*

MOE; Chemical Computing Group Inc.: *http://www.chemcomp.com*

Molconn-Z: *http://www.eslc.vabiotech.com/molconn/manuals/310s/preface1.html*

Molinspiration (calculation of log *P* (fragment-based method + correction factors), molecular polar surface area TPSA (Ertl), "Rule of 5" (Lipinski), number of rotatable bonds, and a drug-likeness index): *http://www.molinspiration.com*

PENGUINS; Molecular Discovery Ltd., West Way House, Elms Parade, Oxford OX2 9LL, UK: *http://www.moldiscovery.com/*

PETRA (calculation of physicochemical parameters); Molecular Networks GmbH, Nägelsbachstrasse 25, D-91052 Erlangen, Germany: *http://www.mol-net.de/*

VolSurf; Molecular Discovery Ltd., West Way House, Elms Parade, Oxford OX2 9LL, UK: *http://www.moldiscovery.com/*

Further information about software for computer aided drug design (CADD) is available at the website *http://www.netsci.org/Resources/Software/Modeling/CADD/.*

# References

[1] MDL Information Systems, Inc., 14600 Catalina Street, San Leandro, CA 94577. *http://www.mdli.com*

[2] The definitions of MDL's 166 MACCS search keys can be found in the ISIS/Base help file, section 49.2.4: Specifying searchable keys as a query.

[3] Daylight Chemical Information Systems, Inc., 7401 Los Altos – Suite 360 – Mission Viejo, CA 92691. *http://www.daylight.com*

[4] F. Harary, Graph Theory, Addison-Wesley, Reading, MA, **1971**.

[5] D. M. Cvetković, M. Doob, H. Sachs, *Spectra of Graphs. Theory and Applications*, 3rd edition, Johann Ambrosius Barth Verlag, Heidelberg, **1995**.

[6] H. Wiener, *J. Am. Chem. Soc.* **1947**, *69*, 17–20.

[7] M. Randic, *J. Am. Chem. Soc.* **1975**, *97*, 6609–6615.

[8] G. Moreau, P. Broto, *Nouv. J. Chim.* **1980**, *4*, 359–360.

[9] J. Gasteiger, M. Marsili, *Tetrahedron* **1980**, *36*, 3219–3228.

[10] J. Gasteiger, M. G. Hutchings, *J. Chem. Soc. Perkin 2* **1984**, 559–564.

[11] PETRA Version 3 is available from Molecular Networks GmbH, Nägelsbachstr. 25, 91052 Erlangen, Germany (*http://www.mol-net.de*) and can be tested via the Internet at *http://www2.chemie.uni-erlangen.de/software/petra*

[12] M. Rarey, J. S. Dixon, *J. Comput.-Aided Mol. Des.* **1998**, *12*, 471–490.

[13] L. B. Kier, L. H. Hall, *Eur. J. Med. Chem.* **1977**, *12*, 307–312.

[14] L. B. Kier, L. H. Hall, *Pharm. Res.* **1990**, *7*, 801–807.

[15] R. Todeschini, V. Consonni, *The Handbook of Molecular Descriptors*, in *Methods and Principles in Medicinal Chemistry, Vol. 11*, R. Mannhold, H. Kubinyi, H. Timmerman (eds.), Wiley-VCH, Weinheim, Germany, **2000**, pp. 1–667.

[16] J. Sadowski, *Three-dimensional structure generation: automation*, in *The Encyclopedia of Computational Chemistry Vol. 5*, P. v. R. Schleyer, N. L. Allinger, T. Clark, J. Gasteiger, P. A. Kollman, H. F. Schaefer III, P. R. Schreiner (eds.); John Wiley, Chichester, **1998**; pp. 2976–2988.

[17] R. S. Pearlman, *Chem. Design Auto. News* **1987**, *2*, 1–7.

[18] R. S. Pearlman, *3D Molecular structures: generation and use in 3D-searching*, in *3D QSAR in Drug Design: Theory, Methods and Applications*, H. Kubinyi, (ed.), ESCOM Science Publishers, Leiden, The Netherlands, **1993**, 41–79.

[19] TRIPOS, Inc., 1699 South Hanley Road, St. Louis, MO 63144, USA. *http://www.tripos.com*

[20] J. Sadowski, J. Gasteiger, *Chem. Rev.* **1993**, *93*, 2567–2581.

[21] J. Sadowski, J. Gasteiger, G. Klebe, *J. Chem. Inf. Comput. Sci.* **1994**, *34*, 1000–1008.

[22] J. Sadowski, C. Rudolph, J. Gasteiger, *Tetrahedron Comp. Method.* **1990**, *3*, 537–547.

[23] C. Hiller, J. Gasteiger, *Ein automatisierter Molekülbaukasten*, In *Software-Entwicklung in der Chemie, Vol. 1*, J. Gasteiger (ed.), Springer, Berlin, **1987**, pp. 53–66.

[24] CORINA is available from Molecular Networks GmbH, Nägelsbachstr. 25, 91052 Erlangen, Germany (*http://www.mol-net.de*) and can be tested via the Internet.

[25] M. Wagener, J. Sadowski, J. Gasteiger, *J. Am. Chem. Soc.* **1995**, *117*, 7769–7775.

[26] R. Wierl, Ann. Phys. (Leipzig) **1931**, *8*, 521–564.

[27] L. J. Soltzberg, C. L. Willins, *J. Am. Chem. Soc.* **1977**, *99*, 439–443.

[28] J. H. Schuur, P. Selzer, J. Gasteiger, *J. Chem. Inf. Comput. Sci.* **1996**, *36*, 334–344.

[29] J. Gasteiger, J. Sadowski, J. Schuur, P. Selzer, L. Steinhauer, V. Steinhauer, *J. Chem. Inf. Comput. Sci.* **1996**, *36*, 1030–1037.

[30] V. Steinhauer, J. Gasteiger, *Obtaining the 3D structure from infrared spectra of organic compounds using neural networks*, in *Software-Entwicklung in der Chemie 11*, G. Fels, V. Schubert (eds.), Gesellschaft Deutscher Chemiker, Frankfurt/Main, **1997**.

[31] J. Karle, *J. Chem. Inf. Comput. Sci.* **1994**, *34*, 381–390.

[32] J. Gasteiger, J. Schuur, P. Selzer, L. Steinhauer, V. Steinhauer, *Fresenius J. Anal. Chem.* **1997**, *359*, 50–55.

[33] M. C. Hemmer, V. Steinhauer, J. Gasteiger, *Vib. Spectrosc.* **1999**, *19*, 151–164.

[34] This section was written by J. Aires-de-Sousa, *Ciencias e Technologia*, Portugal.

[35] a) R. S. Cahn, C. Ingold, V. Prelog, *Angew. Chem. Int. Ed. Engl.* **1966**, *5*, 385–415.
b) V. Prelog, G. Helmchen, *Angew. Chem. Int. Ed. Engl.* **1982**, *94*, 614–631.

[36] P. Mata, A. M. Lobo, C. Marshall, A. P. Johnson, *Tetrahedron: Asymmetry* **1993**, *4(4)*, 657–668.

[37] J. Aires-de-Sousa, *Representation of Molecular Chirality*. In *Handbook of Chemoinformatics*, J. Gasteiger (ed.); Wiley-VCH, Weinheim, **2003**.

[38] A. B. Buda, T. Heyde, K. Mislow, *Angew. Chem. Int. Ed. Engl.* **1992**, *31*, 989–1007; *Angew. Chem.* **1992**, *104*, 1012–1031.

[39] D. Avnir, H. Z. Hel-Or, P. G. Mezey, *Symmetry and chirality: continuous measures*, in *The Encyclopedia of Computational Chemistry, Vol. 4*; P. v. R. Schleyer, N. L. Allinger, T. Clark, J. Gasteiger, P. A. Kollman, H. F. Schaefer III, P. R. Schreiner (eds.); John Wiley, Chichester, **1998**; pp. 2890–2901.

[40] H. Zabrodsky, D. Avnir, *J. Am. Chem. Soc.* **1995**, *117*, 462–473.

[41] S. Keinan, D. Avnir, *J. Am. Chem. Soc.* **1998**, *120*, 6152–6159.

[42] J. Aires-de-Sousa, J. Gasteiger, *J. Chem. Inf. Comput. Sci.* **2001**, *41*, 369–375.

[43] J. Aires-de-Sousa, J. Gasteiger, *J. Mol. Graphics Modell.* **2002**, *20(5)*, 373–388.

[44] P. G. Mezey, *Molecular surface*, in K. B. Lipkowitz, D. B. Boyd (eds.), *Reviews in Computational Chemistry, Vol. 2*, VCH, New York, **1990**, pp. 265–294.

[45] M. Karelson, V. S. Lobanov, A. R. Katritzky, *Chem. Rev.* **1996**, *96*, 1027–1043.

[46] A. M. Ferguson, T. W. Heritage, P. Jonathon, S. E. Pack, L. Phillips, J. Rogan, P. J. Snaith *J. Comput.-Aided Mol. Design* **1997**, *11*, 143–152.

[47] D. B. Turner, P. Willett, A. M. Ferguson, T. W. Heritage, *J. Comput.-Aided Mol. Design* **1997**, *11*, 409–422.

[48] R. Todeschini, M. Lasagni, E. Marengo, *J. Chemom.* **1994**, *8*, 263–272.

[49] R. Todeschini, P. Gramatica, *Quant. Struct.–Act. Relat.* **1997**, *16*, 113–119.

[50] P. J. Goodford, *J. Med. Chem.* **1985**, *28*, 849–857.

[51] R. D. Cramer III, D. E. Patterson, J. D. Bunce, *J. Am. Chem. Soc.* **1988**, *110*, 5959–5967.

[52] F. R. Burden, *J. Chem. Inf. Comput. Sci.* **1989**, *29*, 225–227.

[53] A. J. Hopfinger, S. Wang, J. S. Tokarski, B. Jin, M. Albuquerque, P. J. Madhav, C. Duraiswami, *J. Am. Chem. Soc.* **1997**, *119*, 10509–10524.

[54] A. J. Hopfinger, A. Reaka, P. Venkatarangan, J.S. Duca, S. Wang, *J. Chem. Inf. Comput. Sci.* **1999**, *39*, 1151–1160.

[55] A. J. Hopfinger, *J. Am. Chem. Soc.* **1980**, *102*, 7196–7206.

[56] A. J. Hopfinger, *J. Med. Chem.* **1981**, *24*, 818–822.

[57] O. A. Raevsky, V. Y. Grigor'ev, D. B. Kireev, N. S. Zefirov, *Quant. Struct.–Act. Relat.* **1992**, *11*, 49–63.

[58] O. A. Raevsky, V. Y. Grigor'ev, D. B. Kireev, N. S. Zefirov, *J. Chim. Phys.* **1992**, *89*, 1747–1753.

[59] O. A. Raevsky, V. Y. Grigor'ev, E. Mednikova, *QSAR H-bonding descriptors*, in *Trends in QSAR and Molecular Modelling*, C.G. Wermuth (ed.), Escom, Leiden, **1993**, pp. 116–119.

[60] O. A. Raevsky, L. Dolmatova, V. Y. Grigor'ev, S. Bondarev, *Molecular recognition descriptors in QSAR*, in *QSAR and Molecular Modelling: Concepts, Computational Tools and Biological Applications*, F. Sanz, J. Giraldo (eds.), Prous Science, Barcelona, **1995**, pp 241–245.

[61] O. A. Raevsky, *Hydrogen Bond Strength Estimation by means of HYBOT*. In H. van de Waterbeemd, B. Testa, G. Folkers, Eds., *Computer-Assisted Lead Finding and Optimization*, Wiley-VCH, Weinheim, **1997**, pp. 367–378.

[62] H. Briem, I. D. Kuntz, *J. Med. Chem.* **1996**, *39*, 3401–3408.

[63] H. Briem, U. F. Lessel, *Persp. Drug Discov. Des.* **2000**, *20*, 231–244.

[64] U. F. Lessel, H. Briem, *J. Chem. Inf. Comput. Sci.* **2000**, *40*, 246–253.

# 9
# Methods for Data Analysis

*U. Burkard*

## Learning Objectives

- To understand the machine learning process and learning concepts
- To become familiar with the structure and task of decision trees
- To gain insight into chemometric methods such as correlation analysis, Multiple Linear Regression Analysis, Principal Component Analysis, Principal Component Regression, and Partial Least Squares regression/Projection to Latent Structures
- To understand neural networks, especially Kohonen, counter-propagation and back-propagation networks, and their applications
- To know about fuzzy sets and fuzzy logic
- To become familiar with genetic algorithms and their application to descriptor selection
- To understand data mining tasks and methods
- To understand visual data mining and information visualization techniques
- To appreciate the architecture and tasks of expert systems and examples of expert systems in chemistry

## 9.1
## Introduction

In science in general, and in chemistry and in pharmaceutical research in particular, huge amounts of data are produced. The intrinsic information in these data is often difficult to grasp. In many cases, not only are the data themselves interesting, but the intrinsic relationships among the data are of particular interest.

Chemical data contain information about various characteristics of chemical compounds and a wide spectrum of methods are applied to extract the relevant information from the data sets. Data analysis, however, not only deals with the extraction of primary information from data but also with the generation of secondary information, for example the generation of models which can be used for prediction purposes. This task can be accomplished by machine learning methods, among others.

This chapter gives a general introduction into the data analysis methodology.

Section 9.2 presents general information on machine learning and on machine learning strategies. The subsequent sections give a more detailed insight into different data analysis methods such as chemometrics, neural networks, fuzzy sets, and genetic algorithms. Section 9.8 and 9.9 introduce and exemplify data mining, a data analysis approach that encompasses many of these methods. Section 9.10 explains the concept of expert systems.

## 9.2
## Machine Learning Techniques

In recent decades, computer scientists have tried to provide computers with the ability to learn. This area of research was summarized under the umbrella term "machine learning". Today machine learning is defined as "the study of computer algorithms that improve automatically through experience" [1].

Machine Learning is influenced by several disciplines, e.g., computer science, cognitive science, pattern recognition, and statistics: Cognitive science deals with the perception of coherences and the concepts of thinking and learning, and contributes to machine learning through theories about thinking and learning. Computer science and statistics add the technological and methodological support for machine learning. Pattern recognition represents a set of methods from the fields of soft computing that deal with the classification or description of observations and provides these techniques to machine learning.

Figure 9-1 shows the disciplines that contribute to machine learning techniques.

The area of machine learning is thus quite broad, and different people have different notions about the domain of machine learning and what kind of techniques belong to this field. We will meet a similar problem of defining an area and the techniques involved in the field of "data mining", as discussed in Section 9.8. We will use the term "machine learning" in this chapter to collect all the methods that involve learning from data.

One application of machine learning is that a system uses sample data to build a model which can then be used to analyze subsequent data. Learning from exam-

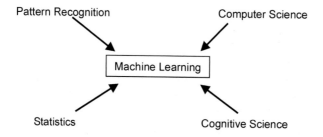

**Figure 9-1.** Machine learning and the disciplines involved in this process.

ples is called inductive learning. A machine, or more generally a system, is said to be learning when it changes its structure in a way that improves its performance.

## 9.2.1
## Machine Learning Process

The machine learning process usually begins with the selection of a data set, which is then divided into two subsets: a training set which is used to train the system and a test set which provides a means to evaluate the results. Machine learning is done by training. The machine learning system uses a training set and tries to learn from these examples. Afterwards the quality of learning is estimated by appraising the ability of the system to predict the outputs from a test set. The capability to predict an output is called generalization. If the system just memorized the input data and did not successfully learn from them, it will not be able to generalize.

The underlying learning process can follow different concepts. The two major learning strategies are unsupervised learning and supervised learning.

## 9.2.2
## Unsupervised Learning

When a machine or a system obtains a set of inputs, the goal of unsupervised learning is to build a representation of the data. This representation can then be used, for example, to detect clusters within the data or to reduce the dimensionality of the data. Common tasks of unsupervised learning strategies are clustering, data compression, or outlier detection. Among the methods used for unsupervised learning are Kohonen networks and conceptual clustering.

Kohonen networks, also known as self-organizing maps (SOMs), belong to the large group of methods called artificial neural networks. Artificial neural networks (ANNs) are techniques which process information in a way that is motivated by the functionality of biological nervous systems. For a more detailed description see Section 9.5.

Conceptual clustering was originally designed as an alternative to statistical clustering techniques [2]. Its goal is to summarize data into groups which represent a concept or an idea of the corresponding data points. Besides clustering, it also gives descriptions of these data subsets.

## 9.2.3
## Supervised Learning

The goal of supervised learning is to make a system learn to associate input data with the corresponding target or output data. In addition to a set of input data, in supervised learning the machine is also given a set of target outputs and its task is to learn to generate the correct output for a newly given input. The output of the

**Table 9-1.** Machine learning methods.

| Unsupervised | Supervised |
|---|---|
| Kohonen network | Decision trees |
| Conceptual clustering | Partial Least Squares (PLS) |
| Principal Component Analysis (PCA) | Multiple Linear Regression (MLR) |
| | Counter-propagation networks |
| | Back-propagation networks |
| | Genetic algorithms (GA) |

system is compared with the correct output and thus an error is obtained. Supervised learing methods try to minimize this error. Besides classification, supervised learning can be used for modeling or prediction. Common methods which use a supervised learning strategy are decision trees (see Section 9.3), genetic algorithms (see Section 9.7), and counter-propagation networks or back-propagation networks (see Section 9.5).

Table 9-1 summarizes common methods for unsupervised and supervised learning.

The following sections present a more detailed description of the methods mentioned above. An overview of machine learning techniques in chemistry is given in Chapter IX, Section 1 in the Handbook.

## 9.3
## Decision Trees

Decision trees constitute a machine learning technique which gives a graphical representation of a procedure for classification. A decision tree is built up from examples, and the tree can then be used for classifying new examples. As plotted in Figure 9-2, a decision tree consists of nodes and branches. An instance is classified by sorting it down the tree from the root to a leaf node. Depending on the value of an attribute on each node, one of the branches is taken. The leaf node finally gives the classification of the instance. Figure 9-2 shows a decision tree which classifies stellar spectra [3]. A star which has a spectrum with few lines, no helium lines, but a weak calcium K line, is thus classified as being of the spectral type A.

## 9.4
## Chemometrics

Chemometrics is the discipline which deals with the application of statistical and, in a more general sense, of mathematical methods to chemical data. Chemometric methods are used for the extraction of chemical information from chemical data,

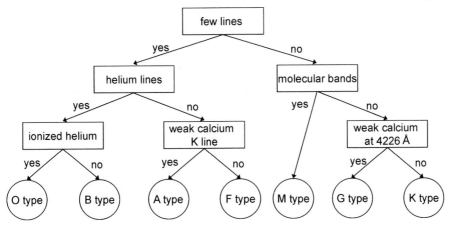

**Figure 9-2.** Decision tree for classifying stellar spectra.

or for the optimization of experimental procedures, e.g., for calibration. A wide area of applications for chemometrics is opened up by analytical chemistry.

To enable the application of electronic data analysis methods, the chemical structures have to be coded as vectors (see Chapter 8). Thus, a chemical data set consists of data vectors, where each vector, i.e., each data object, represents one chemical structure.

Throughout this section $x$ and $y$ are vectors whose components are denoted as $x_1$, ...$x_n$ and $y_1$, ...$y_n$, respectively; $\bar{x}$ is the mean of the components of $x$, $\bar{y}$ the mean for $y$. Data matrices are denoted as $X$ and $Y$, respectively.

### 9.4.1
### Multivariate Statistics

Chemical data are usually multidimensional in nature where one data object is defined through several data components. This data type is called multivariate.

For example, the objects may be chemical compounds. The individual components of a data vector are called features and may, for example, be molecular descriptors (see Chapter 8) specifying the chemical structure of an object. For statistical data analysis, these objects and features are represented by a matrix $X$ which has a row for each object and a column for each feature. In addition, each object will have one or more properties that are to be investigated, e.g., a biological activity of the structure or a class membership. This property or properties are merged into a matrix $Y$. Thus, the data matrix $X$ contains the independent variables whereas the matrix $Y$ contains the dependent ones. Figure 9-3 shows a typical multivariate data matrix.

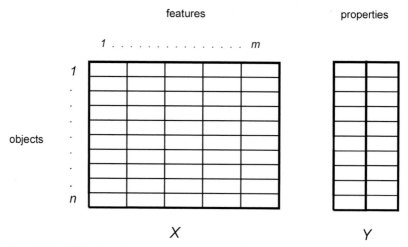

**Figure 9-3.** Multivariate data matrix **X**, containing $n$ objects each represented by $m$ features. The matrix **Y** contains the properties of the objects that are to be investigated.

Multivariate statistics is the discipline to analyze data, to elucidate the intrinsic structure within the data, and to reduce the number of variables needed to describe the data.

Sections 9.4.2–9.4.6 introduce different multivariate data analysis methods, including Multiple Linear Regression (MLR), Principal Component Analysis (PCA), Principal Component Regression (PCR) and Partial Least Squares regression (PLS).

A detailed description of multivariate data analysis in chemistry is given in Chapter IX, Section 1.2 of the Handbook.

### 9.4.2
### Correlation

A first step in a data analysis process is the detection of relationships between variables. This can be achieved through correlation analysis.

The magnitude of dependencies in the variables is determined by the correlation coefficient. The correlation coefficient according to Pearson is given by Eq. (1).

$$r = \frac{\sum\limits_{i=1}^{n} x_i y_i - n\bar{x}\bar{y}}{\sqrt{\left(\sum\limits_{i=1}^{n} x_i^2 - n\bar{x}^2\right)\left(\sum\limits_{i=1}^{n} y_i^2 - n\bar{y}^2\right)}} \tag{1}$$

The value of the correlation coefficient ranges from $r = -1$ to $r = +1$. In those cases where $|r|=1$, the data are completely correlated, either positively or negatively (see Figure 9-4). The smaller the absolute value of $r$, the lower is

A correlation coefficient of $r = +1$ means that there is a completely positive linear relationship. High values on the x-axis are associated with high values on the y-axis.

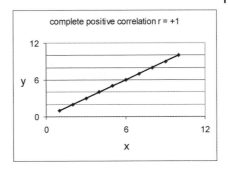

If the correlation coefficient is $r = -1$ high values on the x-axis are associated with low values on the y-axis. The relationship is negatively linear.

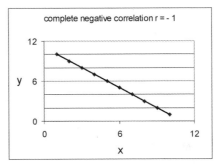

If the x and y values are totally independent of each other the correlation coefficient is $r = 0$.

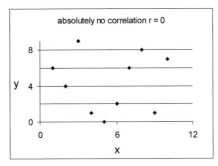

**Figure 9-4.** Correlation analysis: examples of different values of the correlation coefficient.

the correlation among the data. The plots in Figure 9-4 show examples of different correlations.

An application of correlation analysis is the detection of related chemical descriptors: when analyzing chemical data, correlation analysis should be used as a first step to identify those descriptors which are interrelated. If two descriptors are strongly correlated, i.e., the correlation coefficient of two descriptors exceeds a certain value, e.g., $r > 0.90$, one of the descriptors can be excluded from the data set.

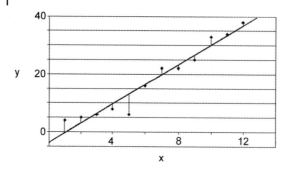

**Figure 9-5.** Linear regression: the sum of the squares of the vertical distances of the points from the line is minimized.

### 9.4.3
**Multiple Linear Regression Analysis (MLRA)**

One task of data analysis is to establish a model which quantitatively describes the relationships between data variables and can then be used for prediction.

Linear regression models a linear relationship between two variables or vectors, $x$ and $y$. Thus, in two dimensions this relationship can be described by a straight line given by the equation $y = ax + b$, where $a$ is the slope of the line and $b$ is the intercept of the line on the $y$-axis.

The goal of linear regression is to adapt the values of the slope and of the intercept so that the line gives the best prediction of $y$ from $x$. This is achieved by minimizing the sum of the squares of the vertical distances of the points from the line. An example of linear regression is given in Figure 9-5.

While simple linear regression uses only one independent variable for modeling, multiple linear regression uses more variables.

Given $n$ input variables, $x_i$, the variable $y$ (Eq. (2)) is modeled analogously to the case with one input variable.

$$y = a_0 + a_1 x_1 + a_2 x_2 + \dots + a_n x_n \tag{2}$$

The $x_i$ should be uncorrelated. If they are correlated, however, one way of finding a solution is stepwise MLR where only those $x_k$ are chosen for the model which are not correlated with already used $x_i$.

### 9.4.4
**Principal Component Analysis (PCA)**

PCA is a frequently used method which is applied to extract the systematic variance in a data matrix. It helps to obtain an overview over dominant patterns and major trends in the data.

The aim of PCA is to create a set of latent variables which is smaller than the set of original variables but still explains all the variance in the matrix $X$ of the original variables.

In mathematical terms, PCA transforms a number of correlated variables into a smaller number of uncorrelated variables, the so-called principal components.

Prior to PCA the data are often pre-processed to convert them into a form most suitable for the application of PCA. Commonly used pre-processing methods for PCA are scaling and mean-centering of the data, which are described in Section 4.3.

From a geometric point of view PCA can be described as follows:

Given a data set or matrix with *n* objects and *m* variables. Each of the variables represents one coordinate axis, so each object can be plotted as a point in *m*-dimensional space. The entire data set is then a swarm of points in this space. In this point-swarm the first principal component (PC1) is the line which gives the best approximation to the data, i.e., it represents the maximum variance within the data. The second principal component (PC2) is orthogonal to the first and again has maximum variance. Further principal components are generated accordingly. Figure 9-6 illustrates the procedure for obtaining the first two principal components.

Thus, the principal components are constructed in order of declining importance: the first principal component comprises as much of the total variation of all variables as possible, the second principal component as much of the remaining variation, and so on.

The coordinate of an object when projected onto an axis given by a principal component is called its score. Scores are usually denoted by *T1, T2,* … . Figure 9-7 is a sketch of a score plot: the points are the objects in the coordinate system

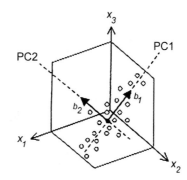

**Figure 9-6.** Procedure to obtain the first two principal components by PCA [4].

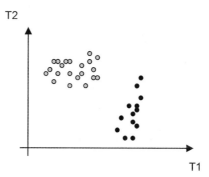

**Figure 9-7.** A score plot. Summary of the relationships among the observations (compounds).

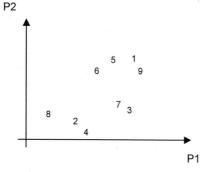

**Figure 9-8.** A loading plot. Summary of the relationships among the variables (descriptors).

given by the first two principal components (*T1* and *T2*). The closer the objects are to one another, the more similar they are.

A second piece of important information obtained by PCA is the loadings, which are denoted by *P1*, *P2*, etc. They indicate which variables influence a model and how the variables are correlated: In algebraic terms the loadings indicate how the variables are combined to build the scores. Figure 9-8 shows a loading plot: each point is a feature of the data set, and features that are close in the plot are correlated.

For most data analysis applications the first three to five principal components give the predominant part of the variance.

In matrix notation PCA approximates the data matrix $X$, which has $n$ objects and $m$ variables, by two smaller matrices: the scores matrix $T$ ($n$ objects and $d$ variables) and the loadings matrix $P$ ($d$ objects and $m$ variables), where $X = TP^T$.

An advantage of PCA is its ability to cope with almost any kind of data matrix, e.g., it can also deal with matrices with many rows and few columns or vice versa.

One widely used algorithm for performing a PCA is the NIPALS (Nonlinear Iterative Partial Least Squares) algorithm, which is described in Ref. [5].

### 9.4.5
### Principal Component Regression (PCR)

The goal of PCR is to extract intrinsic effects in the data matrix $X$ and to use these effects to predict the values of $Y$.

PCR is a combination of PCA and MLR, which are described in Sections 9.4.4 and 9.4.3 respectively. First, a principal component analysis is carried out which yields a loading matrix $P$ and a scores matrix $T$ as described in Section 9.4.4. For the ensuing MLR only PCA scores are used for modeling $Y$. The PCA scores are inherently uncorrelated, so they can be employed directly for MLR. A more detailed description of PCR is given in Ref. [5].

The selection of relevant effects for the MLR in PCR can be quite a complex task. A straightforward approach is to take those PCA scores which have a variance above a certain threshold. By varying the number of PCA components used, the regression model can be optimized. However, if the relevant effects are rather small compared with the irrelevant effects, they will not be included in the first few principal components. A solution to this problem is the application of PLS, which is described in the next section.

### 9.4.6
### Partial Least Squares Regression/Projection to Latent Structures (PLS)

Partial Least Squares Regression, also called Projection to Latent Structures, can be applied to establish a predictive model, even if the features are highly correlated. This makes PLS an attractive method for QSAR (see Section 10.4).

The goal of PLS is to establish a relationship between the two matrices $X$ and $Y$.

The procedure is as follows: first, the principal components for $X$ and $Y$ are calculated separately (cf. Section 9.4.4). The scores of the matrix $X$ are then used for a regression model to predict the scores of $Y$, which can then be used to predict $Y$.

For further information about this method see Chapter IX, Section 1.3 in the Handbook or Ref. [6].

A commonly used algorithm for calculating PLS is SIMPLS [7].

A crucial decision in PLS is the choice of the number of principal components used for the regression. A good approach to solve this problem is the application of cross-validation (see Section 4.4).

The application of PLS in chemoinformatics is described in Chapter IX, Section 1.3 in the Handbook.

### 9.4.7
### Example: Ion Concentrations in Mineral Waters

As described above, PCA can be used for similarity detection: The score plot of two principal components can be used to indicate which objects are similar.

For this example a data set consisting of the concentration of seven ions in 15 samples of mineral water [8] was used; eight of the water samples are shown in Table 9-2.

Initially, the first two principal components were calculated. This yielded the principal components which are given in Figure 9-9 (left) and plotted in Figure 9-9 (right). The score plot shows which mineral water samples have similar mineral concentrations and which are quite different. For example, the mineral waters 6 and 7 are similar while 4 and 7 are rather dissimilar.

### 9.4.8
### Tools: Electronic Data Analysis Service (ELECTRAS)

The electronic data analysis service (ELECTRAS), which was developed at the Computer-Chemie-Centrum of the University of Erlangen-Nürnberg through a project supported by the DFN-Verein and the BMBF, is a web-based application which presents an interface to various kinds of data analysis methods. It offers the methods

**Table 9-2.** Extract from data set containing concentrations of seven minerals in mineral water.

| Water sample | $Ca^{2+}$ | $Na^+$ | $K^+$ | $Mg^{2+}$ | $Cl^-$ | $SO_4^{2-}$ | $HCO_3^-$ |
|---|---|---|---|---|---|---|---|
| 1 | 260 | 332 | 17 | 55 | 55 | 65 | 1720 |
| 2 | 104.1 | 42.7 | 4.2 | 19 | 6.5 | 14 | 494 |
| 3 | 24.05 | 15 | 2.3 | 2.19 | 7.23 | 34 | 66.5 |
| 4. | 116.25 | 750 | 39.2 | 18.95 | 55.6 | 63.8 | 2371 |
| 5 | 114.6 | 292.2 | 15.8 | 24.6 | 126.6 | 5.85 | 1050 |
| 6 | 146.7 | 13.3 | 2.1 | 65.5 | 4.3 | 298.5 | 427.4 |
| 7 | 154 | 1.2 | 0.57 | 38 | 0.6 | 347 | 223 |
| 8 | 39.6 | 8.5 | 1 | 24.5 | 2.3 | 6.8 | 256.2 |

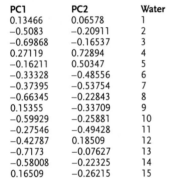

| PC1 | PC2 | Water |
|---|---|---|
| 0.13466 | 0.06578 | 1 |
| −0.5083 | −0.20911 | 2 |
| −0.69868 | −0.16537 | 3 |
| 0.27119 | 0.72894 | 4 |
| −0.16211 | 0.50347 | 5 |
| −0.33328 | −0.48556 | 6 |
| −0.37395 | −0.53754 | 7 |
| −0.66345 | −0.22843 | 8 |
| 0.15355 | −0.33709 | 9 |
| −0.59929 | −0.25881 | 10 |
| −0.27546 | −0.49428 | 11 |
| −0.42787 | 0.18509 | 12 |
| −0.7173 | −0.07627 | 13 |
| −0.58008 | −0.22325 | 14 |
| 0.16509 | −0.26215 | 15 |

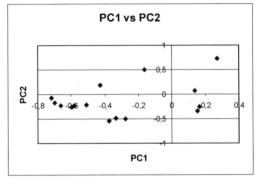

**Figure 9-9.** Left: first two principal components. Right: plot of PC1 against PC2.

described in this chapter (linear regression, correlation, and multivariate methods like PCA, PCR, PLS) via the Internet [9]. It also provides different kinds of neural network methods such as Kohonen networks, counter-propagation networks, and back-propagation networks. Figure 9-10 shows the home page of ELECTRAS.

The data analysis module of ELECTRAS is twofold. One part was designed for general statistical data analysis of numerical data. The second part offers a module for analyzing chemical data. The difference between the two modules is that the module for mere statistics applies the statistical methods or neural networks directly to the input data while the module for chemical data analysis also contains methods for the calculation of descriptors for chemical structures (cf. Chapter 8): Descriptors, and thus structure codes, are calculated for the input structures and then the statistical methods and neural networks can be applied to the codes.

Data input for both modules can be done via file upload, whereby the module for mere statistics reads in plain ASCII files and the module for chemical data analysis takes the chemical structures in the form of SD-files (cf. Chapter 2) as an input. In

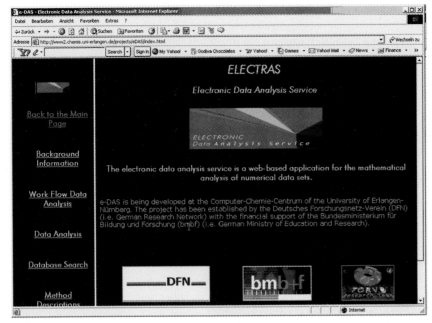

**Figure 9-10.** Electronic Data Analysis Service (ELECTRAS).

addition, the chemical module provides methods for drawing a structure with a molecule editor (see Chapter 2.12).

Figure 9-11 gives an overview of the building blocks of the ELECTRAS system: ELECTRAS was designed for two levels of user experience. The novice part offers a guided data analysis for inexperienced users. Experienced users can analyze their data fast and directly using the expert mode.

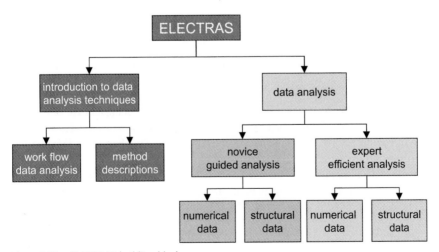

**Figure 9-11.** ELECTRAS building blocks.

An additional feature of ELECTRAS is a module which provides an introduction to various data analysis techniques: One part of this module provides a typical work flow for data analysis. It explains the important steps when conducting a data analysis and describes the output of the data analysis methods. The second part gives a description of the methods offered. This module can be used both as a guideline for novice users and as a reference for experts.

## 9.5
## Neural Networks

This section can only give a short introduction to neural networks. More details can be obtained from Ref. [10] and in Chapter IX, Section 1.4 of the Handbook.

Problems involving routine calculations are solved much faster and more reliably by computers than by humans. Nevertheless, there are tasks in which humans perform better, such as those in which the procedure is not strictly determined and problems which are not strictly algorithmic. One of these tasks is the recognition of patterns such as faces. For several decades people have been trying to develop methods which enable computers to achieve better results in these fields. One approach, artificial neural networks, which model the functionality of the brain, is explained in this section.

### 9.5.1
### Modeling the Brain: Biological Neurons versus Artificial Neurons

The nervous systems and especially the brains of animals and humans work very fast, efficiently, and highly in parallel. They consist of networked neurons which work together and interchange signals with one another. This section describes the functionality of a biological neuron and explains the model of an artificial neuron.

Artificial Neural Networks (ANNs) are information processing units which process information in a way that is motivated by the functionality of the biological nervous system. Just as the brain consists of neurons which are connected with one another, an ANN comprises interrelated artificial neurons. The neurons work together to solve a given problem.

Figure 9-12 shows a sketch of a biological neuron: the dendrites are branching fibers which connect the neuron to different neighboring neurons. They receive the incoming information to convey it to the neuron. This bundle of information is converted into one single information signal in the cell body, the soma of the neuron, and transmitted to the axon if it is larger than a certain threshold value (i.e., the neuron fires). The axon then carries the information to the dendrites of other neurons or to muscle fibers. The information is transmitted from the axon to the dendrites of other neurons through a synapse. The building of synapse strengths is the essence of learning.

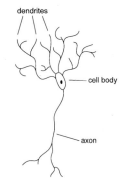

dendrites

cell body

axon

**Figure 9-12.** Biological neuron: the dendrites receive the incoming information and send it to the cell body. The axon carries the information produced in the cell body to neighboring neurons.

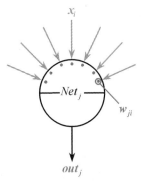

$x_i$

$-Net_j-$

$w_{ji}$

$out_j$

**Figure 9-13.** Artificial neuron: the signals $x_i$ are weighted (with weights $w_i$) and summed to produce a net signal *Net*. This net signal is then modified by a transfer function and sent as an output to other neurons.

An artificial neuron is built in analogy to the biological neuron. Figure 9-13 shows an artificial neuron. The neuron, $j$, receives input signals $x_i$ which are weighted, $w_{ji}$, and summed to give a net signal $Net_j$ (Eq. (3)). The weights model the synapse strength.

$$Net_j = \sum_i w_{ji}x_i \qquad (3)$$

The weights $w_{ji}$ have to be coded as a vector to enable the electronic processing: This is done by gathering all the weights of one neuron in a weight vector (Figure 9-14).

The net signal is then modified by a so-called transfer function and sent as output to other neurons. The most widely used transfer function is sigmoidal: it has two plateau areas having the values zero and one, and between these an area in which it is increasing nonlinearly. Figure 9-15 shows an example of a sigmoidal transfer function.

This function is described by Eq. (4), where $Net_j$ is the net signal, $\vartheta$ gives the threshold value where the neuron starts to react and $1/a$ gives the width of the interval (see Figure 9-15).

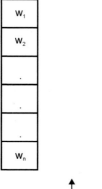

**Figure 9-14.** Weights coded as a vector.

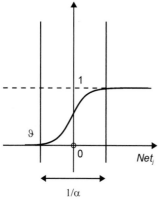

**Figure 9-15.** Sigmoidal transfer function, in which the area between the plateaus does not increase linearly.

$$out_j = \frac{1}{1 + e^{-(\alpha Net_j + \vartheta)}} \qquad (4)$$

## 9.5.2
## Networks

The neurons are linked together to build a neural network. They can be grouped in so-called layers, where the neurons within the same layer all simultaneously produce a set of outputs. An artificial neural network generally consists of several layers. Figure 9-16 gives a typical artificial neural network architecture with a two-layer design. It comprises input units, a so-called hidden layer, and an output layer. The artificial neurons, plotted as circles, are connected just as biological neurons are. The arrows show the direction of flow of signals through the network.

### 9.5.2.1  Training
Just like humans, ANNs learn from examples. The examples are delivered as input data. The learning process of an ANN is called training. In the human brain, the synaptic connections, and thus the connections between the neurons,

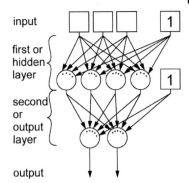

input

first or hidden layer

second or output layer

output

**Figure 9-16.** Artificial neural network architecture with a two-layer design, comprising input units, a so-called hidden layer, and an output layer. The squares inclosing the ones depict the bias which is an extra weight (see Ref. [10] for further details).

are adapted during the learning process. A corresponding procedure is performed when an artificial neural network is trained: the weights of the neurons are adjusted.

### 9.5.2.2 Learning Strategies

The learning process, i.e., the process of adapting the weights of the neurons, can follow different approaches depending on the goals of the application and on the data. Two types of learning strategies are used: unsupervised learning and supervised learning.

#### 9.5.2.2.1 Unsupervised Learning

In unsupervised learning, the network tries to group the input data on the basis of similarities between theses data. Those data points which are similar to each other are allocated to the same neuron or to closely adjacent neurons.

The Kohonen network is a neural network which uses an unsupervised learning strategy. See Section 9.5.3 for a more detailed description.

#### 9.5.2.2.1 Supervised Learning

In supervised learning, the network is trained with the goal of assigning correct target or response signals to the input signals. This procedure is shown in Figure 9-17: The input data vectors $x_S$ are fed into the network system and thus generate an output $Out_S$. This output is compared with the target value or vector, which yields an error $\delta$. The weights of the network are then adapted with the intention of reducing this error.

Supervised learning strategies are applied in counter-propagation and in back-propagation neural networks (see Sections 9.5.5 and 9.5.7).

Besides the artificial neural networks mentioned above, there are various other types of neural networks. This chapter, however, will confine itself to the three most important types used in chemoinformatics: Kohonen networks, counter-propagation networks, and back-propagation networks.

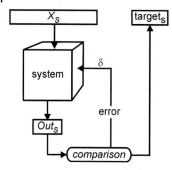

**Figure 9-17.** Outline of the procedure for supervised learning: The output of the network is compared with the target value or vector, which yields the error $\delta$. The weights of the network are then adapted to reduce this error.

## 9.5.3
## Kohonen Network

The Kohonen network or self-organizing map (SOM) was developed by Teuvo Kohonen [11]. It can be used to classify a set of input vectors according to their similarity. The result of such a network is usually a two-dimensional map. Thus, the Kohonen network is a method for projecting objects from a multidimensional space into a two-dimensional space. This projection keeps the topology of the multidimensional space, i.e., points which are close to one another in the multidimensional space are neighbors in the two-dimensional space as well. An advantage of this method is that the results of such a mapping can easily be visualized.

### 9.5.3.1 Architecture
In Figures 9-13 and 9-16 the artificial neurons have been plotted as circles. In order to visualize the weights as well, we will now plot the neurons as columns, consisting of cuboids, where each cuboid represents a weight.

In this illustration, a Kohonen network has a cubic structure where the neurons are columns arranged in a two-dimensional system, e.g., in a square of $n \times l$ neurons. The number of weights of each neuron corresponds to the dimension of the input data. If the input for the network is a set of $m$-dimensional vectors, the architecture of the network is $n \times l \times m$-dimensional. Figure 9-18 plots the architecture of a Kohonen network.

### 9.5.3.2 Training

The training of a Kohonen network is performed in three steps.

1. The network is initialized, and all the weights of the neurons obtain numerical values, in most cases random numbers.
2. An input vector is fed into the network and that neuron is determined whose weights are most similar to the input data vector.

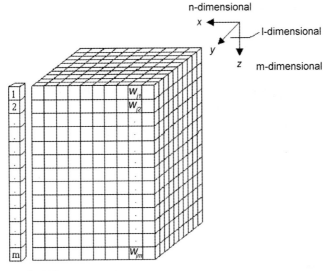

**Figure 9-18.** Architecture of a Kohonen network.

This is done by calculating the Euclidean distance between the input data vector $x_S$ and the weight vectors $w_j$ of all neurons:
The neuron for which this distance gives the minimum is called the winning neuron or central neuron $c$ (see Eq. (5)).

$$out_c, \leftarrow min\left\{\sum_{i=1}^{m}(x_{si} - w_{ji})^2\right\} \tag{5}$$

3. The weights of the winning neuron are further adapted to the input data. The neurons within a certain distance surrounding the winning neuron are also adapted: Their weight adaptation is performed such that the closer a neuron is to the winning neuron the more its weights will be adapted.
Equation (6) gives the weight correction:

$$w_{ji}^{(new)} = w_{ji}^{(old)} + \eta(t)a(d_c - d_j)(x_i - w_{ji}^{(old)}) \tag{6}$$

where $x_i$ is a component of the input vector; the central neuron is named $c$, and the neuron being corrected is $j$; $d_c$–$d_j$ is the topological distance between the central neuron $c$ and the current neuron $j$; $t$ gives the iteration cycle – $\eta(t)$ decreases with each iteration cycle; $a(d_c$–$d_j)$ is a neighborhood-dependent function which decreases with increasing distance from $d_c$ to $d_j$.

Steps 2 and 3 are performed for all input objects. When all data points have been fed into the network one training epoch has been achieved. A network is usually trained in several training epochs, depending on the size of the network and the number of data points.

The Kohonen network adapts its values only with respect to the input values and thus reflects the input data. This approach is unsupervised learning as the adaptation is done with respect merely to the data describing the individual objects.

### 9.5.4
### Tutorial: Application of a Kohonen Network for the Classification of Olive Oils using ELECTRAS [9]

In this example, a Kohonen network is used to classify Italian olive oils on the basis of the concentrations of fatty acids they contain.

The data set, which was provided by Prof. M. Forina, University of Genova (Italy) can be obtained from the webpage *http://www2.chemie.uni-erlangen.de/publications/ ANN-book/datasets/oliveoil/index.html*. It consists of 572 olive oil samples from nine different regions of Italy; each sample is characterized by the concentrations of eight fatty acids. Figure 9-19 shows a part of this data set: each line describes one olive oil sample. The first eight columns contain the concentrations of eight fatty acids, and the ninth column indicates the region of origin, e.g., 4 stands for Sicily. The last column gives the name of the region.

The classification is done in five steps and can be conducted with, for example, ELECTRAS [9]:

1. Input the data set.
2. Select the network type: Kohonen network. Transfer selection by pressing the button "submission".
3. Click the "Select network parameter" button and choose the Kohonen network parameters: topology, width and height of the network, neuron dimension, the index of the class identifier, and the number of training cycles.
4. Start Training.
5. Analyze the Kohonen map. The content of the neurons is given when clicking on the map.

When this data set is input into a Kohonen network, the network projects the oil samples into the various neurons on the basis of their concentrations of the fatty acids. Thus, the oil samples which have similar concentrations of the acids are assigned to neurons which are lying next to one another. After the training has been finished, the entire dataset is again sent through the network and the neurons are then colored according to the region of origin of the majority of the oil samples lying in this neuron. The result of the network, the Kohonen map (see Figure 9-20), shows that the oils which are harvested in the same region cluster in similar regions of the Kohonen map. A comparison with the geographic ar-

| | fatty acid concentrations | | | | | | | | class identifier | regin of origin |
|---|---|---|---|---|---|---|---|---|---|---|
| data vector for first oil sample | 41,3 | 22,6 | 33,2 | 72,2 | 21,9 | 47,2 | 56,3 | 48,2 | 1 | N.Apul. |
| data vector for second oil sample | 62,7 | 46,8 | 35 | 47,3 | 37,6 | 55,6 | 56,3 | 53,6 | 2 | Calabr. |
| data vector for third oil sample | 75 | 63,4 | 19,7 | 35,9 | 55,4 | 34,7 | 42,7 | 30,4 | 3 | S.Apul. |
| . | 54,4 | 44,5 | 33,6 | 53,3 | 36,8 | 47,2 | 65 | 58,9 | 4 | Sicily |
| | 46,1 | 39,6 | 31,4 | 46,1 | 65 | 56,9 | 93,2 | 0 | 5 | I.Sard. |

**Figure 9-19.** Data set for olive oils.

rangement of the different regions in Italy shows that the neural network mirrors the geographic origins of the oils, particularly the arrangement in the north–south direction, rather nicely.

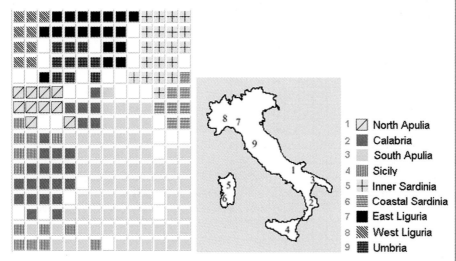

**Figure 9-20.** Left: Kohonen map showing the projection of the olive oil samples. Middle: Map of Italy showing the regions of origin for the olive oils. Right: Key giving the regions and their codes.

## 9.5.5
## Counter-propagation Network

A counter-propagation network is a method for supervised learning which can be used for prediction. It has a two-layer architecture where each neuron in the upper layer, the Kohonen layer, has a corresponding neuron in the lower layer, the output layer (see Figure 9-21). A trained counter-propagation network can be used as a look-up table: a neuron in one layer is used as a pointer to the other layer.

### 9.5.5.1 Architecture
The architecture of a counter-propagation network resembles that of a Kohonen network, but in addition to the cubic Kohonen layer (input layer) it has an additional layer, the output layer. Thus, an input object consists of two parts, the $m$-dimensional input vector (just as for a Kohonen network) plus a second $k$-dimensional vector with the properties for the object.

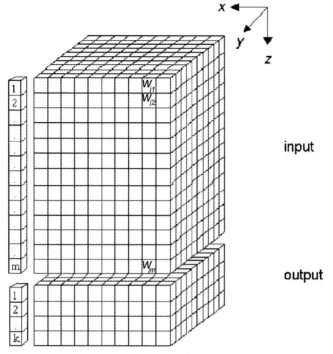

**Figure 9-21.** Counterpropagation network plotted as two boxes. The upper box contains the weights of the input layer, while the lower box contains those of the output layer.

### 9.5.5.2 Training

During training the input layer is adapted as in a regular Kohonen network, i.e., the winning neuron is determined only on the basis of the input values. But in contrast to the training of a Kohonen network, the output layer is also adapted, which gives an opportunity to use the network for prediction.

The output weights are corrected according to Eq. (7), where $\gamma_i$ is a component of the output vector, $c$ is the index of the winning neuron, $j$ is the index of the neighboring neuron which is corrected and $c_j$ contains the output weights for the neuron $j$; $t$, $\eta(t)$, and $a(d_c–d_j)$ are as in Eq. (6).

$$c_{ji}^{(new)} = c_{ji}^{(old)} + \eta(t)a(d_c - d_j)(\gamma_i - c_{ji}^{(old)}) \tag{7}$$

One application of a counterpropagation network is the simulation of infrared (IR) spectra, which is done in the TeleSpec system (cf. Section 10.2, *www2.chemie.uni-erlangen.de/services/*): The counterpropagation network is trained with pairs of molecules and their corresponding IR spectra and thus learns the relationship between molecular structure and IR spectrum. When a new structure is presented to the network after training, the winning neuron for this structure is determined in the upper layer and points to the simulated IR spectrum in the lower layer of the network.

9.5.6
**Tools: SONNIA [12] (Self-Organizing Neural Network for Information Analysis)**

SONNIA is a self-organizing neural network for data analysis and visualization.

It provides unsupervised (Kohonen network) and supervised (counter-propagation network) learning techniques with planar and toroidal topology of the network.

SONNIA can be employed for the classification and clustering of objects, the projection of data from high-dimensional spaces into two-dimensional planes, the perception of similarities, the modeling and prediction of complex relationships, and the subsequent visualization of the underlying data such as chemical structures or reactions which greatly facilitates the investigation of chemical data.

Figure 9-22 shows the user interface of SONNIA presenting the analysis results: The left-hand side gives the Kohonen network, which can be investigated by clicking on the neuron. The contents of the neuron, here the chemical structures, are shown in an additional window plotted on the right-hand side of the figure.

Typical applications of SONNIA include:

- analysis of multidimensional data,
- comparison of compound libraries,
- analysis of the similarity and diversity of combinatorial libraries,
- classification and prediction of biological activity,

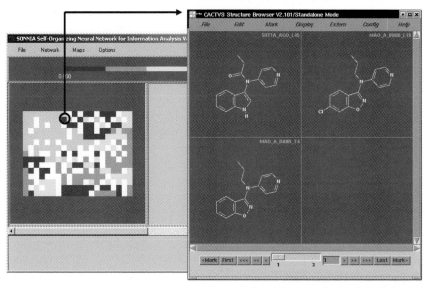

**Figure 9-22.** The user interface of SONNIA presenting the analysis results. The left-hand side gives the Kohonen network, which can be investigated by clicking on each neuron of interest. The contents of the neuron, here the structures, are shown in an individual window illustrated on the right-hand side of the figure.

- analysis of data from high-throughput screening,
- classification of samples from chemical analysis,
- simulation of infrared spectra,
- reaction classification and clustering,
- knowledge extraction from reaction databases.

### 9.5.7
### Back-propagation Network

A back-propagation network, or more precisely, a multilayer neural network trained by the back-propagation algorithm, is also trained by a supervised learning strategy (see Figure 9-17). The general principle of back-propagation learning is to adjust the weights of a neuron depending on the error of its output signal, with the goal of minimizing the output error. A back-propagation network can be used to develop quantitative models which predict a certain output.

#### 9.5.7.1  Architecture
A back-propagation network usually consists of input units, one or more hidden layers and one output layer. Figure 9-16 gives an example of the architecture.

#### 9.5.7.2  Training
The weights of each layer are adapted during training. The final output of the output layer, is then compared with the target vector. This yields the error between the output and the target vector. This error signal is then backpropagated through the layers from the output to the input layer and the weights of each layer are adapted correspondingly. In this way, the network learns the correct classification for a set of inputs. This process of weight correction is plotted in Figure 9-23.

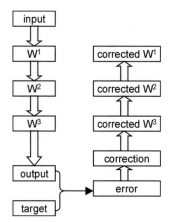

**Figure 9-23.**  Weight correction for a back-propagation network.

A detailed mathematical explanation of the adaptation of the weights is given, e.g., in Ref. [10]. The original publication of back-propagation learning is to be found in Ref. [13].

## 9.5.8
## Tutorial: Neural Networks

The usage of a neural network varies depending on the aim and especially on the network type. This tutorial covers two applications: on the one hand the usage of a Kohonen network for classification, and on the other hand the prediction of object properties with a counter-propagation network.

The following steps give a general outline for the usage of a neural network: The nomenclature is that used in the web-based neural network included in the ELECTRAS system [9]. For other neural networks the parameters may be slightly different.

### Step 1: Choose a training set

For classification purposes (Kohonen network) usually the whole data set is used for training. For prediction (counter-propagation network) a subset which is a characteristic representation of the data set as a whole is selected. The data set usually contains a data matrix where each row gives a structure and each column gives a descriptor.

### Step 2: Select a structure coding scheme

The structure coding depends strongly on the properties which are to be modeled. For an explanation of the different structure codes and their applications see Chapter 8.

### Step 3: Network training

Before the network training is launched various network parameters have to be chosen:

- Network topology – rectangular or toroidal: these two topologies differ in that the neurons on the edges of a rectangular network have only five neighbors (the corners only three), whereas all the neurons in a toroidal network have eight neighbors.
- Network dimension, i.e., the width and height of the network: the numbers of neurons in the network should usually be somewhat smaller than the number of compounds, by a ratio of anything between 1:1 and 1:10.

- Neuron dimension: the neuron dimension is determined by the number of descriptors chosen for an object and should be such that all the important (numerical) columns of the data file are included.
- Class identifier: this gives the column number which contains information about the class membership.
- Counter-propagation network: this network also needs the input dimension. It gives the columns that are used for the upper layer of the network.
- The remainder of the columns contains the properties which are to be predicted and are processed in the lower layer of the network.
- Training cycles: One training cycle is completed when all the input data have once been fed into the network.

**Step 4: Network analysis**

Kohonen network: Classification
   The different classes are identified by different colors and a map shows the classification.

Counterpropagation: Prediction
   A test set is loaded whose object properties are to be predicted. The different property classes are identified by different colors.

9.5.9
**Tasks for Neural Networks and Selection of an Appropriate Neural Network Method**

The possibilities for the application for neural networks in chemistry are huge [10]. They can be used for various tasks: for the classification of structures or reactions, for establishing spectra–structure correlations, for modeling and predicting biological activities, or to map the electrostatic potential on molecular surfaces.

   Depending on the type of problem, different network architectures and different learning strategies can be applied. When addressing a problem decisions on the following questions are crucial:
   What is the goal of the network application (classification, modeling, etc.)?
   What learning strategy should be used?
   Table 9-3 can act as a guideline for the proper selection of a neural network method. It summarizes the different network types and their learning strategy, and lists different types of applications.

**Table 9-3.** Summary of neural network types and different types of applications.

| | Learning strategy | | |
| | unsupervised | supervised | |
| --- | --- | --- | --- |
| | Kohonen network | Counter-propagation | Back-propagation |
| Classification | x | x | x |
| Modeling | | x | x |
| Association | | x | x |
| Mapping | x | x | |

## 9.6
## Fuzzy Sets and Fuzzy Logic

### 9.6.1
### Some Concepts

Conventional computers initially were not conceived to handle vague data. Human reasoning, however, uses vague information and uncertainty to come to a decision. In the mid-1960 this discrepancy led to the conception of fuzzy theory [14]. In fuzzy logic the strict scheme of Boolean logic, which has only two statements (true and false), is extended to handle information about partial truth, i.e., truth values between "absolutely true" and "absolutely false". It thus gives a mathematical representation of uncertainty and vagueness and provides a tool to treat them.

In conventional set theory an element is either a member of a set or not: let $X$ be a set of all the real numbers between 0 and 6, i.e., $X = [0,6]$, and $A$ a subset of the numbers between 2 and 4, thus $A = [2,4]$. In order to give information about which members of $X$ belong to $A$ a membership function can be introduced: It assigns the value $m(x) = 1$ to each element of $A$ and a value $m(x) = 0$ to those elements of $X$ which are not in $A$. Equation (8) gives the membership function, which is plotted in Figure 9-24.

$$m(x) = \begin{cases} 1, & if\ 2 \leq x \leq 4 \\ 0, & else \end{cases} \tag{8}$$

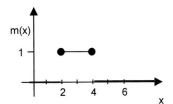

**Figure 9-24.** Membership function for the crisp set $A = [2,4]$.

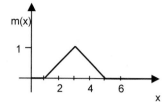

**Figure 9-25.** Membership function for the fuzzy set of numbers close to 3.

A conventional set which contains members that satisfy such precise properties concerning their set membership is called a crisp set.

On the other hand, if we want to characterize objects which are described by the rather fuzzy statement "numbers close to three", we then need a membership function which describes the closeness to three. An adequate membership function could be the one plotted in Figure 9-25: $m(x)$ has its maximum value of $m(x) = 1$ for value $x = 3$. The greater the distance from $x$ to 3 gets, the smaller is the value of $m(x)$. until it reaches its minimum $m(x) = 0$ if the distance from $x$ to 3 is greater than say 2, thus for $x > 5$ or $x < 1$.

An important property of a fuzzy set is its cardinality. While for crisp sets the cardinality is simply the number of elements in a set, the cardinality of a fuzzy set $A$, $CardA$, gives the sum of the values of the membership function of $A$, as in Eq. (9).

$$CardA = \sum_{i=1}^{n} \mu_A(x_i)$$

(9)

### 9.6.2
### Application of Fuzzy Logic in Chemistry

Fuzzy logic and fuzzy set theory are applied to various problems in chemistry. The applications range from component identification and spectral library search to fuzzy pattern recognition or calibrations of analytical methods.

An overview over different applications of fuzzy set theory and fuzzy logic is given in [15]; (see also Chapter IX, Section 1.5 in the Handbook).

Here, the application of fuzzy logic for multicomponent spectral analysis is described.

If a spectrum lacks certain lines or contains extra lines from additional unknown components, or if the true line positions are blurred, fuzzy set theory can improve the matching.

The principle of applying fuzzy logic to matching of spectra is that, given a sample spectrum and a collection of reference spectra, in a first step the reference spectra are unified and fuzzed, i.e., around each characteristic line at a certain wavenumber $k_0$ a certain fuzzy interval $[k_0 - \Delta k, k_0 + \Delta k]$ is laid. The resulting fuzzy set is then intersected with the crisp sample spectrum. A membership function analogous to the one in Figure 9-25 is applied. If a line of the sample spectrum falls into the interval $[k_0 - \Delta k, k_0 + \Delta k]$, it is said to coincide with the line of the reference spectrum, whereby the degree of coincidence is given by the value of the membership function. The fuzzy cardinality is then calculated as a similarity measure [15].

Fuzzy set theory and fuzzy logic and its application to molecular recognition are explained in Chapter IX, Section 1.5 in the Handbook.

## 9.7
## Genetic Algorithms

Genetic algorithms (GA) belong to a set of methods which are subsumed under the umbrella term of evolutionary computation. This field covers optimization methods which simulate evolutionary strategies used in nature [16]: In nature, populations develop over many generations following the principle of "survival of the fittest", which means that those individuals who are able to flourish in their environment are more likely to survive and propagate. Thus, indirectly, the populations adapt to the given living conditions. Other methods in this field are genetic programming (GP) and evolutionary strategies (ES). (see also Chapter IX, Section 1.6 in the Handbook).

The concept of genetic algorithms was developed in the 1970s by John Holland [17].

The evolutionary process of a genetic algorithm is accomplished by genetic operators which translate the evolutionary concepts of selection, recombination or crossover, and mutation into data processing to solve an optimization problem dynamically. Possible solutions to the problem are coded as so-called artificial chromosomes, which are changed and adapted throughout the optimization process until an optimum solution is obtained.

A set of chromosomes is called a population, and creation of a population from a parent population is called generation.

Figure 9-26 shows a typical GA run: in a first step, the original population is created. For each chromosome the fitness is determined and a selection algorithm is applied to choose chromosomes for mating. These chromosomes are then subject to the crossover and the mutation operators, which finally yields a new generation of chromosomes.

GAs or other methods from evolutionary computation are applied in various fields of chemistry. Its tasks include the geometry optimization of conformations of small molecules, the elaboration of models for the prediction of properties or biological activities, the design of molecules de novo, the analysis of the interaction of proteins and their ligands, or the selection of descriptors [18]. The last application is explained briefly in Section 9.7.6.

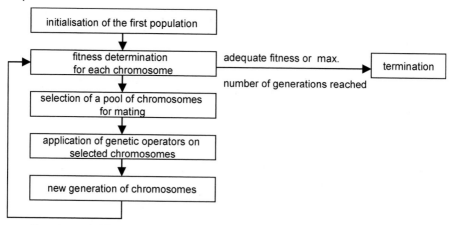

**Figure 9-26.** A GA run.

### 9.7.1
### Representation and Encoding of Chromosomes

When applying a GA to solve a certain problem, the first step is to decide how to represent the possible solutions in the chromosomes for the given problem. Chromosomes are commonly encoded as strings. Depending on the problem, the strings can be binary, and can have integer or even real values. Figure 9-27 shows a binary string representation of an artificial chromosome which consists of blocks of encoded information, called genes.

### 9.7.2
### Initialization of Individuals

Initialization is the first step in a GA run to create the first population of artificial chromosomes. Usually, this is done in a random manner. The number of chromosomes is defined before the run and is usually kept constant throughout it.

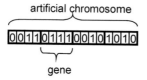

**Figure 9-27.** An artificial chromosome, encoded as a binary string.

## 9.7.3
## Fitness and Objective Function

A key feature of a genetic algorithm is that only the best chromosomes are to pass their features to the next generation during evolution.

In the style of the Darwinian Theory, the quality of a chromosome is called its fitness. The quality or fitness of a chromosome is usually calculated with the help of an objective function, which is a mathematical function indicating how good the solution, and thus the chromosome, is for the optimization problem. This computation of the fitness is done for each chromosome in each population.

## 9.7.4
## Selection Functions

The selection function is used to choose chromosomes which can then mate and thus give their coded information to the next generation. The idea behind the selection mechanism is to obtain more copies of better chromosomes in the next generation.

Frequently used selection functions are roulette wheel selection, tournament selection, and truncation selection:

- *Roulette wheel selection*: In this selection variant the probability for selecting a chromosome is proportional to its fitness. The idea can be illustrated when we imagine a roulette wheel, where a slot is allocated to each chromosome and the size of the slot is chosen with respect to the quality of the chromosome. When the wheel is "spinning" the chromosomes with a better quality are more likely to be chosen than those of a minor quality. Figure 9-28 illustrates this procedure.
- *Tournament selection*: This selection type chooses two or more chromosomes to compete with one another. The best chromosome is determined and selected.
- *Truncation selection*: Prior to the application of this selection variant the chromosomes are ranked according to their fitness. The top $1/x$ of the population are then selected and each of these chromosomes gets $x$ copies for the population, e.g., the top one-third of the chromosomes are each given three copies in the population. All other chromosomes are neglected.

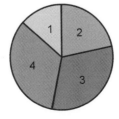

**Figure 9-28.** Roulette wheel selection. The size of each sector is equivalent to the fitness of the corresponding chromosome.

1. Chromosome 1
2. Chromosome 2
3. Chromosome 3
4. Chromosome 4

Selection alone cannot achieve an optimization towards the solution: With mere selection performed over a number of generations, one would get a population which comprises only the best chromosome of the original population. Therefore, an operator has to be applied which causes variance within the population. This is achieved by the application of genetic operators such as the crossover and the mutation operators.

### 9.7.5
### Genetic Operators

Crossover, which is also called recombination, follows the idea that an offspring in nature always holds genes from both its parents. Accordingly, the genetic crossover operator takes parts of two parent chromosomes to create a new offspring.

For the so-called one-point crossover a position within the chromosomes is randomly picked at the same position in both parents and the chromosomes are both cut at this position. Then the first part of chromosome 1 is concatenated with the second part of chromosome 2, and vice versa. This procedure is shown in Figure 9-29.

The mutation operator is the second commonly used operator. Mutation brings new traits into a chromosome. The mutation operator causes a local change in an artificial chromosome.

Thus, if the chromosome is represented as a bit string, the mutation operator will change one bit in the string as illustrated in Figure 9-30. Usually, the mutation operator is applied with only a small probability.

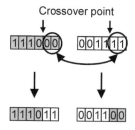

**Figure 9-29.** Crossover: the parent chromosomes are both cut at the crossover position, and the first part of chromosome 1 is concatenated with the second part of chromosome 2, and vice versa.

**Figure 9-30.** Mutation operator, which changes one bit in a chromosome.

original descriptor

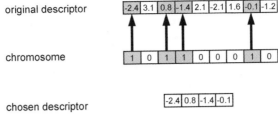

chromosome

chosen descriptor

**Figure 9-31.** Descriptor selection using a GA. The upper part of the figure shows the original set of descriptors, below which is the chromosome. Those genes containing the value 1 are selected for the final set of descriptors, which is given at the bottom.

## 9.7.6
## Tutorial: Selection of Relevant Descriptors in a Structure–Activity Study [19]

As explained in Chapter 8, descriptors are used to represent a chemical structure and, thus, to provide a coding which allows electronic processing of chemical data. The example given here shows how a GA is used to find an optimal set of descriptors for the task of classification using a Kohonen neural network. The chromosomes of the GA are to be used as a means for selecting the descriptors: they indicate which descriptors are used and which are rejected:

The chromosomes are coded as a bit string such that the length of a chromosome equals the length of a set of descriptors. An individual descriptor within a descriptor vector can be identified through its position. Thus, when a gene has the value of 1 within the chromosome the corresponding descriptor is chosen, while a zero indicates that this descriptor is discarded. This coding is shown in Figure 9-31. The original set of descriptors is given in the upper part of the figure. Below it the chromosome is represented. Those genes which contain the value of 1 are selected for the final set of descriptors, which is given in the lower part of the diagram.

### 9.7.6.1 Example: Drug Design
This method of descriptor selection was applied to a problem in drug design (see Section 10.4). The data set was provided by Boehringer Ingelheim Pharma KG. It contained 831 compounds which were synthesized to give antagonists for the platelet activating factor. The compounds were assigned to four classes, on the basis of structural characteristics.

For each structure, topological autocorrelation vectors were calculated, using eight physicochemical properties (atomic identity function, $\sigma$-charges, $\pi$-charges, total charges, $\sigma$-electronegativity, $\pi$-electronegativity, lone-pair electronegativity, and atomic polarizability).

For each property, an autocorrelation vector of length 17 was calculated. This yielded a descriptor vector which contained 136 (8 $\times$ 17) individual descriptors.

The GA was then applied to select those descriptors which give the best classification of the structures when a Kohonen network is used. The objective function was based on the quality of the classification done by a neural network for the reduced descriptors.

Through the application of the GA to the topological autocorrelation vector, the descriptor could be reduced from 136 to nine descriptors and the quality of projection was increased by 9.4%. Thus, besides the improvement in efficiency, an improvement in quality could be obtained.

## 9.8
## Data Mining

Scientific instruments can easily produce and collect huge amounts of data. Nevertheless, it is seldom possible to read the relevant information in these data directly. Therefore more elaborate methods have to be applied for information extraction.

Data mining provides methods for the extraction of implicit or hidden information from large data sets and comprises procedures for the generation of reasonable and dependable secondary information.

It extends the usage of statistical methods and combines it with machine learning methods and the application of expert systems. The visualization of the results of data mining is an important task as it facilitates an interpretation of the results.

Figure 9-32 plots the different disciplines which contribute to data mining.

Data Mining is the core of the more comprehensive process of knowledge discovery in data bases (KDD). However, the term "data mining" is often used synonymously with KDD. KDD describes the process of extracting and storing data and also includes methods for data preparation such as data cleaning, data selection, and data transformation as well as evaluation, presentation, and visualization of the results after the data mining process.

The methods used for data mining also comprise those methods that have already been described in Sections 9.2–9.7.

This section will therefore focus on the aims and tasks of data mining and refer to the methods where applicable. A thorough description of data mining is given in Ref. [20].

Data mining can fulfill various different tasks such as classification, clustering and similarity detection, prediction, estimation, or description retrieval, which are described in Sections 9.8.1–9.8.5.

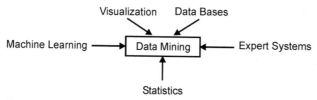

**Figure 9-32.** Disciplines contributing to data mining.

## 9.8.1
## Classification

Classification describes the process of assigning an instance or property to one of several given classes. The classes are defined beforehand and this class assignment is used in the learning process, which is therefore supervised. Statistical methods and decision trees (cf. Section 9.3) are also widely used for classification tasks.

## 9.8.2
## Clustering and Detection of Similarities

In clustering, data vectors are grouped together into clusters on the basis of intrinsic similarities between these vectors. In contrast to classification, no classes are defined beforehand. A commonly used method is the application of Kohonen networks (cf. Section 9.5.3).

One application of clustering could, for example, be the comparison of compound libraries: A training set is chosen which contains members of both libraries. After the structures are coded (cf. Chapter 8), a Kohonen network (cf. Section 9.5.3) is trained and arranges the structures within the Kohonen map in relation to their structural similarity. Thus, the overlap between the two different libraries of compounds can be determined.

## 9.8.3
## Prediction and Regression

Prediction implies the generation of unknown properties. On the basis of example data, a model is established which is able to relate an object to its property. This model can then be used for predicting values for new data vectors.

For this task counter-propagation networks (see Section 9.5.5) can be applied.

The TeleSpec system (cf. Section 10.2) is an application for this data mining task. Its goal is to predict infrared spectra of given structures. It uses pairs of structures and the corresponding infrared spectra to train a counter-propagation network. After training, a new structure can be input into the network and the counter-propagation network will predict the infrared spectrum of this structure.

## 9.8.4
## Association

Association deals with the extraction of relationships among members of a data set. The methods applied for association range from rather simple ones, e.g., correlation analysis, to more sophisticated methods like counter-propagation or back-propagation neural networks (see Sections 9.5.5 and 9.5.7).

### 9.8.5
### Detection of Descriptions

A very important data mining task is the discovery of characteristic descriptions for subsets of data, which characterize its members and distinguish it from other subsets. Descriptions can, for example, be the output of statistical methods like average or variance.

### 9.8.6
### Data Mining in Chemistry

Data mining in chemistry focuses on the extraction and evaluation of information in chemical data sets. In contrast to other fields of data mining applications, chemical data mining does not confine itself to conventional database queries but rather generates new information from the data.

A wide field of applications for chemical data mining is drug design. In short, drug design starts with a compound which has an interesting biological profile and optimizes the compound as well as its activity (see Section 10.4). Thus, the information about the biological activity of a compound is a crucial aspect in drug design. The relationship between a structure and its biological activity is represented by so-called quantitative structure–activity relationships (QSAR) (see Section 10.4). The field of QSAR can be approached via chemical data mining:

Starting from the structure input, e.g., in the form of a connection table (see Section 2.5), a 2D or 3D model of the structure is calculated. Ensuing secondary information, e.g., in the form of physicochemical properties such as charges, is generated for these structures. The enhanced structure model is then the basis for calculating a descriptor, i.e., a structure code in the form of a vector to which computational methods, for example statistical methods or neural networks, can be applied. These methods can then fulfill various data mining tasks such as classification or establishment of QSAR models which can finally be employed for the prediction of properties such as biological activities.

### 9.9
### Visual Data Mining

The results of data analysis applications described within this chapter will usually not be represented in textual or numeric form; rather it will be visualized and interpreted by information visualization techniques. This circumstance has a physiological cause: the visual cortex of human beings allows a much faster and easier processing and interpretation of two-dimensional images and three-dimensional scenes than of text blocks or long tables of numeric data. While fonts and numeric information will be processed sequentially by the human brain, gathering and analysis of graphical information will be handled in a more parallel way. This fact is also recognized in the well-known saying, *"A picture says more than*

*1000 words!"*. Therefore, computer graphical representations allow a very effective and easy analysis of usually high-dimensional, complex, and extremely large chemical data sets. Furthermore, graphical representations facilitate the communication between scientists and decision-makers.

Besides the simple data presentation through computer graphics methods, a new research area has been established within recent years, which is particularly well capable of handling the requirements of data mining – visual data mining [21]. The functions of visual data mining range from visualization and analysis of results from classical data mining approaches, to new methods that allow a complete visual exploration of raw data and thus are an alternative to classical data mining methods.

## 9.9.1
### Advantages of Visual Data Mining Approaches

Although chemical data analysis has become highly automated through the development of computer-aided methods and applications, an effective extraction of new knowledge and information still requires intervention and interaction through the scientist. Especially in the case of complex problems, a successful solution is only warranted if human intuition, flexibility, creativity, and expert knowledge are included in the decision process. However, classical, computer-aided data mining methods are usually so-called "black box" systems that only allow a very limited or no interposition through the scientist. Furthermore, these approaches often require expert knowledge.

The goal of visual data mining methods is the integration of the users and their visual cognitive capabilities into the KDD process. Besides an easier insight into complex data worlds, this approach allows a dynamic rearrangement of the data driven by user interactions and leads to straightforward data analysis and pattern recognition. The most important advantages of visual data mining are:

- *Higher quality of the resulting patterns*: The natural capability of human beings to visually recognize patterns and relations can be used and leads to a more effective data mining process.
- *Usage of expert knowledge*: visual data mining allows the integration of expert knowledge into the data mining process and therefore provides an interactive and direct manipulation of the data analysis and pattern recognition. This can only be achieved in a very difficult way or not at all through classical approaches, because the necessary analysis parameters have to be defined by the scientist before the data analysis process and usually cannot be changed during this process.
- *Increased trust in pattern recognition*: The active user involvement in the data mining process can lead to a deeper understanding of the data and increases the trust in the resulting patterns. In contrast, "black box" systems often lead to a higher uncertainty, because the user usually does not know, in detail, what happened during the data analysis process. This may lead to a more difficult data interpretation and/or model prediction.

- *Easy and intuitive data analysis:* The data analysis process is easy and intuitive, because the pattern recognition only requires the knowledge and intuition of the scientists. Difficult statistical and mathematical methods are not necessary.
- *Handling of complex data sets:* Visual data mining methods especially show huge advantages over classical approaches if only little information about the data is known or if the expected patterns and relationships are not clearly defined. Furthermore, very inhomogeneous data sets or data with a high noise level can still be analyzed by these methods.

### 9.9.2
### Information Visualization Techniques

The explorative analysis of data sets by visual data mining applications takes place in a three-step process: During the first step (*overview*), the user can obtain an overview of the data and maybe can identify some basic relationships between specific data points. In the second step (*filtering*), dynamic and interactive navigation, selection, and query tools will be used to reorganize and filter the data set. Each interaction by the user will lead to an immediate update of the data scene and will reveal the hidden patterns and relationships. Finally, the patterns or data points can be analyzed in detail with specific detail tools.

These tools and techniques can be classified by three different criteria [22]: the data to be visualized, the technique itself, and the interaction and distortion method.

#### 9.9.2.1  Data Types
Large data sets such as screening data or results obtained by combinatorial experiments are made up of a large number of data records. Hence a data record may represent a chemical reaction or substance, for example: its corresponding variables will define the corresponding reaction conditions or biological activities. Depending on the dimensionality or data type of the information, one-, two-, multi-dimensional, or specific data types can be identified.

#### 9.9.2.2  Visualization Techniques
Because of the usually multidimensional character of chemical information, we will only describe information visualization techniques here that have been developed to handle multivariate data. These techniques may be classified into:

- geometrically transformed displays, such as landscapes, parallel coordinates, line graphs, bar-charts/histograms or scatter-plots – these techniques use geometric transformations and projections to generate meaningful visualizations;

- icon- and glyph-based techniques, such as star icons, pies and Chernoff faces – within these displays, each data point is represented by an individual icon and the different data dimensions are taken as features of these icons;
- hierarchical and graph-based techniques, such as dimensional stacking and cone tree techniques – these displays divide and visualize the data in a hierarchical way.

Besides these main categories, a large number of hybrid visualization techniques also exist, which are combinations of the methods described. Well-known hybrid approaches are the 2D or 3D glyph displays. These techniques combine the multi-dimensional representation capabilities of icon-based methods with the easy and intuitive representations of scatter-plot displays. Therefore these techniques can also be frequently found within chemical data analysis applications.

Figure 9-33 shows the three major visualization techniques: geometric display, icon-based display, and hierarchical display, as well as the hybrid technique with 3D glyphs.

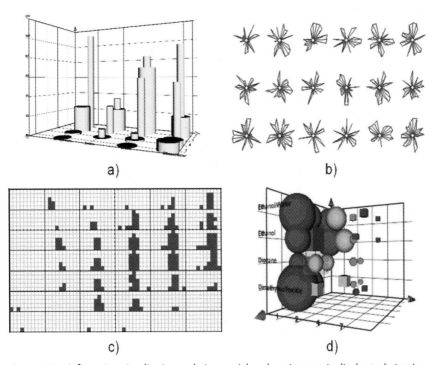

**Figure 9-33.** Information visualization vechniques: a) bar-chart (geometric display technique); b) star display (icon-based display); c) dimensional stacking (hierarchical displays); d) 3D glyphs (hybrid technique).

### 9.9.2.3 **Interaction and Distortion Techniques**

The most important methods within visual data exploration applications are techniques that allow an interactive partitioning and reorganization of the data. These techniques can be categorized into a class of interactive filtering (querying), zooming, linking and detail tools; and a class of interactive distortion techniques. The first class of tools allows the data analyst to make dynamic changes to a visualization according to the exploration objectives. Interactive distortion techniques support the data exploration process by showing interesting portions of data with a high level of detail and the rest of the data with a lower level of detail.

### 9.10
### Expert Systems

Expert systems, which are also called knowledge-based systems, use knowledge which would normally be provided by human experts to solve problems. The idea behind expert systems is to collect and store expert knowledge about a certain field, to draw conclusions from this knowledge, and to offer solutions to problems for a given field. An expert system can thus provide support for human experts in routine tasks. Besides, it can serve as a means to collect and present the knowledge which has been gained by experts through long experience and thus provide a long-term knowledge source in fields of high employee turnover.

Conventional computer programs typically contain only the information they need for solving the specific task for which they were made: the knowledge used is commonly embedded in the programming code and the system has to be rebuilt when the knowledge changes. In expert systems, however, the knowledge is collected in a knowledge base. Whenever a problem is to be solved, a specific part of the knowledge base is used and for different problems the same system can be used without reprogramming.

A detailed description of expert systems is given in Ref. [23].

### 9.10.1
### Architecture of Expert Systems

The classical architecture of an expert system comprises a knowledge base, an inference engine, and some kind of user interface. Most expert systems also include an explanation subsystem and a knowledge acquisition subsystem. This architecture is given in Figure 9-34 and described in more detail below.

- *Knowledge base*: The knowledge base contains the encoded knowledge which is needed to solve a certain problem. The knowledge can be represented in the form of facts or rules.
- *Inference engine*: The inference engine represents the central problem-solving subsystem. It contains strategies for using the information contained in the

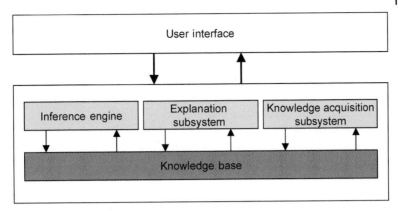

**Figure 9-34.** Expert system architecture.

knowledge base and links the facts which have been extracted from the knowledge base to draw conclusions.

The inference mechanism can be compared with the reasoning of humans: there are different kinds of reasoning, the most popular ones being forward chaining and backward chaining. In forward chaining, data are presented to the expert system which then chains forward to come to a result. Backward chaining starts with a hypothesis that is given to the expert system, which then tracks back to check whether the hypothesis is valid.

- *User Interface*: All expert systems have some kind of user interface which provides the user with a means of interacting with the system.

Besides these three major components, many expert systems also comprise an explanation subsystem and a knowledge acquisition subsystem.

- *Explanation subsystem*: The explanation subsystem supplies information about the reasoning process, for example how the conclusion has been drawn and which facts were used to come to the conclusion.
- *Knowledge acquisition subsystem*: The task of the knowledge acquisition subsystems is to assemble and upgrade the knowledge base. A major task is to verify the data and check for consistency.

Many expert systems also apply fuzzy logic (see Section 9.6) instead of strict Boolean logic and are thus able to cope with rules which are imprecise or incomplete.

## 9.10.2
### Tasks of Expert Systems

Expert systems have been applied in various fields such as medicine, chemistry, engineering, or business. Within all these fields, expert systems are used for similar types of tasks. They can, for example, be used for classification, i.e., to select one decision from a set of given alternatives. Typical applications of an expert system for classification are diagnosis systems which are used to find a errors in technical systems and to give proposals for possible solutions.

Expert systems are widely applied in design and planning where they can give instructions on how to design systems for given criteria or constraints. Expert systems are also used for simulation tasks: Starting from an initial state, these systems deduce subsequent states with the aim of simulating possible outcomes for certain actions.

### 9.10.3
### Expert Systems in Chemistry

In chemistry and chemical engineering, expert systems are used for various tasks ranging from laboratory automation or reaction kinetics to the design of syntheses or the simulations of processes [24]. The application of expert systems in chemistry is described in more detail in Chapter IX, Section 2 of the Handbook.

Two systems will be introduced below: on the one hand the DENDRAL system for automatic structure elucidation [25], which was one of the first expert systems; on the other hand the EROS system [26], which can be used for simulating reactions.

#### 9.10.3.1 **DENDRAL**
DENDRAL [25] was initiated in 1965 at Stanford University by Feigenbaum, Lederberg, and Buchanan. It was designed to help chemists in analyzing the molecular structure of unknown chemical compounds. The system used the molecular formula and the mass spectral data of a compound as an input and deduced the most probable constitution (topology) of the compound. The acronym DENDRAL stands for DENDRitic ALgorithm, which was the core algorithm of the system. This algorithm specified all topologically distinct arrangements of a given set of atoms under the restriction of the rules of chemical valence.

DENDRAL followed a three-stage procedure. In the first phase, the so-called plan, prior knowledge, and heuristics were used to deduce a set of constraints. Constraints could be, for example, the exemption of large sets of candidate solutions or the suggestion for a extensive search over limited classes of solutions.

The second step, the so called generation, created only those structures which complied with the given constraints, and imposed additional restrictions on the compounds such as the number of rings or double bonds. The third and final phase, the tester phase, examined each proposed solution: for each proposed compound a mass spectrum was predicted which was then compared with the actual data of the compound. The possible solutions were then ranked depending on the deviation between the observed and the predicted mass spectra.

A large amount of work was devoted to the development of the DENDRAL system and many highly competent people contributed to this project. In the end, many reasons may have worked together to explain why this project was finally abandoned. Some of them can be guessed from an overview of the entire project that was written when the DENDRAL had already been stopped [27].

### 9.10.3.2 **EROS**

The EROS (Elaboration of Reactions for Organic Synthesis) system [26] is a knowledge-based system which was created for the simulation of organic reactions. Given a certain set of starting materials, EROS investigates the potential reaction pathways. It produces sequences of simultaneous and consecutive reactions and attempts to predict the products that will be obtained in those reactions.

More details can be found in Section 10.3.1.

### Essentials

- *Machine learning* deals with the ability of machines or generally of computer programs to enhance their performance based on previous results. *Machine learning* can follow different learning strategies, e.g., supervised or unsupervised learning.
- *Decision trees* give a graphical representation of a procedure for classification. They consist of nodes and branches; the leaf nodes give the classification of an instance.
- *Correlation analysis* reveals the interdependence between variables. The statistical measure for the interdependence is the *correlation coefficient*.
- *Multiple linear regression (MLR)* models a linear relationship between a dependent variable and one or more independent variables.
- *Principal Component Analysis (PCA)* transforms a number of correlated variables into a smaller number of uncorrelated variables, the so-called principal components.
- *Principal Component Regression (PCR)* is a combination of PCA and MLR: The scores gained by PCA are used for MLR.
- *PLS* is a linear regression extension of PCA which is used to connect the information in two blocks of variables $X$ and $Y$ to each other. It can be applied even if the features are highly correlated.
- *Neural networks* model the functionality of the brain. They learn from examples, whereby the weights of the neurons are adapted on the basis of training data.
- A *Kohonen network* is a neural network which uses an unsupervised learning strategy. It can be used for, e.g., similarity perception, clustering, or classification tasks.
- A *counter-propagation neural network* is a method for supervised learning which can be used for predictions.
- A *back-propagation neural network* is also trained using a supervised learning strategy. The weights of neurons are adjusted so that the error of the output signal is minimized. A *back-propagation neural network* can be used for predictions.
- Fuzzy logic extends the Boolean logic so as to handle information about truth values which are between "absolutely true" and "absolutely false".

- *Data mining* provides methods for the extraction of implicit or hidden information from large data sets, and comprises procedures for the generation of reasonable and dependable secondary information.
- Visual data mining allows the visualization and detection of hidden relationships in sets of data.
- *Expert systems*, also called knowledge-based systems, use a knowledge base of human expertise and a so-called inference engine to solve problems. The inference engine contains strategies for using the information contained in the knowledge base to draw conclusions.

**Selected Reading**

- T. Mitchell, *Machine Learning*, McGraw-Hill, New York, **1996**.
- B.G.M. Vandeginste, L.M.C. Buydens, S. De Jong, P.J. Lewi, J. Smeyers-Verbeke, D.L. Massart, *Handbook of Chemometrics and Qualimetrics*, Elsevier Science, Amsterdam, **1998**.
- M. Otto, *Chemometrics. Statistics and Computer Application in Analytical Chemistry*, Wiley-VCH, Weinheim, **1998**.
- J. Zupan, J. Gasteiger, *Neural Networks in Chemistry and Drug Design*, 2nd Edition, Wiley-VCH, Weinheim, **1999**.
- A.S. Pandya, R.B. Macy, *Pattern Recognition with Neural Networks in C++*, CRC Press, Boca Raton, **1996**.
- D.E. Goldberg, *The Design of Innovation*, Kluwer Academic Publishers, Boston, **2002**.
- D. Goldberg, *Genetic Algorithms in Search, Optimization, and Machine Learning*, Addison Wesley Longman, Reading, **1989**.
- D.H. Rouvray, *Fuzzy Logic in Chemistry*, AP, San Diego, **1997**.
- F. Ehrenteich, Fuzzy Methods in Chemistry, in *Encyclopedia of Computational Chemistry*, P.v.R. Schleyer, N.L. Allinger, T. Clark, J. Gasteiger, P.A. Kollman, H.F. Schaefer III, P.R. Schreiner (Eds.), Wiley, Chichester, UK, **1998**, pp. 1090–1103.
- U.M. Fayyad, G. Piatetsky-Shapiro, P. Smyth, R. Uthurusamy (Eds.), *Advances in Knowledge Discovery and Data Mining*, AAAI/MIT Press, Menlo Park, **1996**.
- B. Thuraisingham, *Data Mining: Technologies, Techniques, Tools, and Trends*, CRC Press, Boca Raton, **1999**.
- P. Jackson, *Introduction to Expert Systems*, Addison Wesley Longman, Harlow, **1999**.

**Interesting Websites**

*http://www2.chemie.uni-erlangen.de/projects/eDAS/index.html* (Electronic Data Analysis Service, Computer-Chemie-Centrum, University of Erlangen-Nürnberg)

**Available Software**

- Weka (Machine Learning Software in Java); Department of Computer Science, University of Waikato; *http://www.cs.waikato.ac.nz/~ml/weka/index.html*
- SNNS (Stuttgart Neural Network Simulator); Wilhelm-Schickard-Institut für Informatik, University of Tübingen; *http://www-ra.informatik.uni-tuebingen.de/SNNS/*
- GAlib (C++ Library of Genetic Algorithm Components), Massachusetts Institute of Technology; *http://lancet.mit.edu/ga/*

# References

[1] T. Mitchell, *Machine Learning*, McGraw-Hill, New York, **1996**.

[2] R. S. Michalski, R. E. Stepp, Learning from Observation: Conceptual Clustering, in *Machine Learning: An Artificial Intelligence Approach*, R.S. Michalski, J. G. Carbonell, T.M. Mitchell (Eds.), Morgan Kauffmann, San Mateo, CA, **1983**, pp. 331–363.

[3] Classifying Stellar Spectra: *http://shiva.uwp.edu/astronomy/stars/spectclass.html*

[4] K. Varmuza, Multivariate data analysis in chemistry, in *Handbook of Chemoinformatics – From Data To Knowledge*, J. Gasteiger (Ed.), Weinheim, Wiley-VCH, **2003**.

[5] M. Otto, *Chemometrie: Statistik und Computereinsatz in der Analytik*, Wiley-VCH, Weinheim, **1997**; M. Otto, *Chemometrics. Statistics and Computer Application in Analytical Chemistry*, Wiley-VCH, Weinheim, **1998**.

[6] S. Wold, M. Sjöström, L. Eriksson, Partial Least Squares Projection to Latent Structures (PLS) in Chemistry, in *Encyclopedia of Computational Chemistry*, P. v. R. Schleyer, N. L. Allinger, T. Clark, J. Gasteiger, P. A. Kollman, H. F. Schaefer III, P. R. Schreiner (Eds.), Wiley, Chichester, UK, **1998**, pp. 2006–2021.

[7] S. de Jong, *Chemometrics and Intelligent Laboratory Systems* **1993**, *18*, 251–263.

[8] H. Lohninger, *Teach/Me Datenanalyse*, Springer-Verlag, Heidelberg, **2001**.

[9] Electronic Data Analysis Service (ELECTRAS), *http://www2.chemie.uni-erlangen.de/projects/eDAS/index.html*

[10] J. Zupan, J. Gasteiger, *Neural Networks in Chemistry and Drug Design*, 2nd Edition, Wiley-VCH, Weinheim, **1999**.

[11] T. Kohonen, *Self-Organizing Maps*, Springer, Berlin, **1997**.

[12] SONNIA (Self Organizing Neural Network for Information Analysis), *http://www2.chemie.uni-erlangen.de/software/kmap/* and *http://www.molnet.de/*

[13] D. Rumelhart, R. Durbin, R. Golden, Y. Chauvin, *Backpropagation: The Basic Theory* in *Mathematical Perspectives on Neural Networks*, P. Smolensky, M. C. Mozer, D. E. Rumelhart (Eds.), Lawrence Earlbaum Assoc, Hillsdale, NJ, **1996**, pp. 533–566.

[14] L. Zadeh, *Fuzzy Sets, Inf. Control,* **1965**, *8,* 338–353.

[15] F. Ehrentreich, Fuzzy Methods in Chemistry, in *Encyclopedia of Computational Chemistry,* P. v. R. Schleyer, N. L. Allinger, T. Clark, J. Gasteiger, P. A. Kollman, H. F. Schaefer III, P. R. Schreiner (Eds.), Wiley, Chichester, UK, **1998**, pp. 1090–1103.

[16] D. Goldberg, *Genetic Algorithms in Search, Optimization, and Machine Learning*, Addison Wesley Longman, Reading, **1989**.

[17] J. Holland, *Adaptation in Natural and Artificial Systems*, University of Michigan Press, Michigan, **1975**.

[18] A. v. Homeyer, Evolutionary Algorithms and Applications in Chemistry, in *Handbook of Chemoinformatics – From Data To Knowledge*, J. Gasteiger (Ed.), Weinheim, Wiley-VCH, **2003**.

[19] *http://www2.chemie.uni-erlangen.de/ services/dissonline/data/dissertation/ Andreas_Teckentrup/html/index.html*

[20] B. Thuraisingham, *Data Mining: Technologies, Techniques, Tools, and Trends*, CRC Press, Boca Raton, **1999**.

[21] U. Fayyad, G. Grinstein, A. Wierse, *Information Visualization in Data Mining and Knowledge Discovery*, Morgan Kaufman Publishers, San Francisco, USA, **2002**.

[22] D. Keim, Information Visualization and Visual Data Mining, *IEEE Trans. Visualization and Computer Graphics*, **2002**, 8,100–107.

[23] P. Jackson, *Introduction to Expert Systems*, Addison Wesley Longman, Harlow, **1999**.

[24] G. Schembecker, Chemical Engineering: Expert Systems, in *Encyclopedia of Computational Chemistry*, P.v.R. Schleyer, N. L. Allinger, T. Clark, J. Gasteiger, P. A. Kollman, H.F. Schaefer III, P. R. Schreiner (Eds.), Wiley, Chichester, UK, **1998**, pp. 323–329.

[25] R. K. Lindsay, B. G. Buchanan, E.A. Feigenbaum, J. Lederberg, *Applications of Artificial Intelligence for Organic Chemistry*, McGraw-Hill, New York, **1980**.

[26] R. Höllering, J. Gasteiger, L. Steinhauer, K-P. Schulz, A. Herwig, *J. Chem. Inf. Comput. Sci.* **2000**, 40, 482–494.

[27] N. A. B. Gray, *Computer-Assisted Structure Elucidation*, Wiley, New York, **1986**.

# 10
# Applications

## 10.1
## Prediction of Properties of Compounds

*Thomas Kleinöder, Aixia Yan, Simon Spycher*

### Learning Objectives

- To understand how to derive a quantitative relationship between property and structure
- To become familiar with the application of the basic principles of the model building process by means of calculating log $P$ and log $S$ values
- To acquire an overview of methods and examples of some pitfalls in modeling log $P$, log $S$, and the toxic effects of compounds

### 10.1.1
### Introduction

The fundamental physical properties of a compound such as its polarity or lipophilicity determine its behavior in chemical, biochemical, or environmental processes and are therefore required for understanding and modeling the action of the compound in fields of high interest such as drug design, reaction prediction, or biodegradation. Although the amount of experimental data is growing rapidly, the number of newly synthesized or designed compounds is increasing even more quickly, especially through high-throughput methods such as parallel synthesis and combinatorial chemistry (CombiChem). With techniques such as virtual screening, compounds are not synthesized at all but their activity against potential drug receptors should nevertheless be modeled. Thus, the need for reliable methods for the prediction of physicochemical properties is evident.

The basic approach to the problem of estimating properties can be written in a very simple form that states that a molecular property $P$ can be expressed as a function of the molecular structure $C$ (Eq. (1)).

$$P = f(C) \tag{1}$$

The function *f(C)* may have a very simple form, as is the case for the calculation of the molecular weight from the relative atomic masses. In most cases, however, *f(C)* will be very complicated when it comes to describe the structure by quantum mechanical means and the property may be derived directly from the wavefunction; for example, the dipole moment may be obtained by applying the dipole operator.

Furthermore, most physicochemical properties are related to interactions between a molecule and its environment. For instance, the partitioning between two phases is a temperature-dependent constant of a substance with respect to the solvent system. Equation (1) therefore has to be rewritten as a function of the molecular structure, *C*, the solvent, *S*, the temperature, *T*, etc. (Eq. (2)).

$$P = f(C, S, T, ...) \tag{2}$$

Obviously, to model these effects simultaneously becomes a very complex task. Hence, most calculation methods treat the effects which are not directly related to the molecular structure as constant. As an important consequence, prediction models are valid only for the system under investigation. A model for the prediction of the acidity constant $pK_a$ in aqueous solutions cannot be applied to the prediction of $pK_a$ values in DMSO solutions. Nevertheless, relationships between different systems might also be quantified. Here, Kamlet's concept of solvatochromism, which allows the prediction of solvent-dependent properties with respect to both solute and solvent [1], comes to mind.

Two approaches to quantify *f(C)*, i.e., to establish a quantitative relationship between the structural features of a compound and its properties, are described in this section: quantitative structure–property relationships (QSPR) and linear free energy relationships (LFER) (cf. Section 3.4.2.2). The LFER approach is important for historical reasons because it contributed the first attempt to predict the property of a compound from an analysis of its structure. LFERs can be established only for congeneric series of compounds, i.e., sets of compounds that share the same skeleton and only have variations in the substituents attached to this skeleton. As examples of a QSPR approach, currently available methods for the prediction of the octanol/water partition coefficient, log *P*, and of aqueous solubility, log *S*, of organic compounds are described in Section 10.1.4 and Section 10.15, respectively.

The partition coefficient and aqueous solubility are properties important for the study of the adsorption, distribution, metabolism, excretion, and toxicity (ADME-Tox) of drugs. The prediction of the ADME-Tox properties of drug candidates has recently attracted much interest because these properties account for the failure of about 60 % of all drug candidates in the clinical phases. The prediction of these properties in an early phase of the drug development process could therefore lead to significant savings in research and development costs.

## 10.1.2
## Linear Free Energy Relationships (LFER)

LFER methods are widely used for the development of quantitative models for energy-based properties such as partition coefficients, binding constants, or reaction rate constants. This is based on the pioneering work of Hammett, who introduced this method for the prediction of chemical reactivity (see Section 3.4). The basic assumption is that the influence of a structural feature on the free energy change of a chemical process is constant for a congeneric series of compounds. A property $\Phi$ that is linearly dependent on a free energy change can then be calculated by the property of the basic element of this series, the so-called parent element, and the constant $\phi_x$ for the structural feature $X$ (see Eqs. (3) – (6)) [2].

$$\Delta G = -2.3RT \log \Phi \tag{3}$$

$$\Delta\Delta G = \Delta G_{R-X} - \Delta G_{R-H} = -2.3RT \log \Phi_{R-X} + 2.3RT \log \Phi_{R-H} \tag{4}$$

$$\log \Phi_{R-X} - \log \Phi_{R-H} = -\frac{1}{2.3RT} \Delta\Delta G = k\Delta\Delta G = \phi_X \tag{5}$$

$$\log \Phi_{R-X} = \log \Phi_{R-H} + \phi_X \tag{6}$$

This basic LFER approach has later been extended to the more general concept of fragmentation. Molecules are dissected into substructures and each substructure is seen to contribute a constant increment to the free-energy based property. The promise of strict linearity does not hold true in most cases, so corrections have to be applied in the majority of methods based on a fragmentation approach. Correction terms are often related to long range interactions such as resonance or steric effects.

Furthermore, QSPR models for the prediction of free-energy based properties that are based on multilinear regression analysis are often referred to as LFER models, especially, in the wide field of quantitative structure–activity relationships (QSAR).

## 10.1.3
## Quantitative Structure–Property Relationships (QSPR)

The general procedure in a QSPR approach consists of three steps: structure representation; descriptor analysis; and model building (see also Chapter X, Section 1.2 of the Handbook).

### 10.1.3.1 Structure Representation
Descriptors have to be found representing the structural features which are related to the target property. This is the most important step in QSPR, and the development of powerful descriptors is of central interest in this field. Descriptors can range from simple atom- or functional group counts to quantum chemical descriptors. They can be derived on the basis of the connectivity (topological or

2D descriptors), the 3D structure, or the molecular surface (3D descriptors) of a structure. Which kind of descriptors should or can be used is primarily dependent on the size of the data set to be studied and the required accuracy; for example, if a QSPR model is intended to be used for hundreds of thousands of compounds, a somehow reduced accuracy will probably be acceptable for the benefit of short processing times. Chapter 8 gives a detailed introduction to the calculation methods for molecular descriptors.

### 10.1.3.2 Descriptor Analysis

In general, the set of calculated descriptors should not be used directly for the model building process, mainly because of three problems: 1) different elements of the descriptor set may intercorrelate, i.e., different descriptors basically encode the same structural aspect; 2) descriptors may encode features that do not contribute to the property at all; or 3) the overall size of a descriptor set may become unmanageably large. Each case requires a pre-processing of the descriptor set such that the essential information is extracted into a reduced descriptor set with higher information density related to the target property. Mainly, two statistical measures are used to judge the quality of the descriptors: the variance and the correlation coefficient among the descriptors. The former is a measure of the variation of a descriptor across a data set. A low variance indicates little information content of a descriptor. The latter is a measure of internal redundancy. Completely independent descriptors have a correlation coefficient of 0.0 and are said to be orthogonal. This ideal case is of course hardly ever found, and the correlation of two descriptors should normally be not greater than 0.6, but reports of acceptable correlation coefficients between descriptors have ranged from less than 0.4 to 0.9 in the literature.

The descriptor set can then be reduced by eliminating candidates that show such bad characteristics. Optimization techniques such as genetic algorithms (see Section 9.7) are powerful means of automating this selection process.

On the other hand, techniques like Principle Component Analysis (PCA) or Partial Least Squares Regression (PLS) (see Section 9.4.6) are used for transforming the descriptor set into smaller sets with higher information density. The disadvantage of such methods is that the transformed descriptors may not be directly related to single physical effects or structural features, and the derived models are thus less interpretable.

### 10.1.3.3 Model Building

The model building step deals with the development of mathematical models to relate the optimized set of descriptors with the target property. Two statistical measures indicate the quality of a model, the regression coefficient, $r$, or its square, $r^2$, and the standard deviation, $\sigma$ (see Chapter 9).

Model building consists of three steps: training, evaluation, and testing. In the ideal case the whole training data set is divided into three portions, the training, the evaluation set, and the test set. A wide variety of statistical or neural network methods can be used to derive QSPR and QSAR models. The most frequently used methods are Multiple Linear Regression Analysis (MLRA) and feedforward neural networks with back-propagation of errors (BPG-NN) (see Sections 9.4 and 9.5). Once a model has been derived with the training set, the evaluation set can be used to test the predictive power of the resulting models, i.e., to predict the target property for compounds yet unknown to the model and to optimize model parameters thereby. In many cases the data set is too small to allow its splitting. Therefore cross-validation techniques are applied for the evaluation step. In $k$-fold cross-validation the training set is split in $k$ subsets. Then $k - 1$ subsets are used as a training set and one subset as test set. This procedure is repeated $k$ times. As we have now a prediction for each compound, we can calculate cross-validated errors of the predictions. Values of $k$ usually range from 5 to 10. If $k$ equals $N$, the number of cases in the training set, the procedure is called leave-one-out cross-validation.

Finally, a model has to be tested using an independent data set with compounds yet completely unknown to the model: the test set. The complete process of building a prediction model is depicted in Figure 10.1-1 as a flow chart.

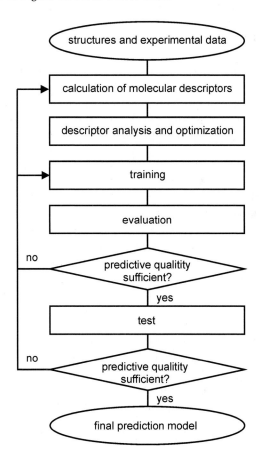

**Figure 10.1-1.** Flow chart for the general model building process in QSPR studies.

10.1.4
**Estimation of Octanol/Water Partition Coefficient (log $P_{OW}$)**

As drug molecules need to have the appropriate lipophilic characteristics to be able to pass through biological membranes but have still to be hydrophilic enough to be water-soluble, lipophilicity is a major determinant in drug design. Since Hansch and Fujita in 1964 derived a LFER equation for the prediction of the biological activity using the octanol/water partition coefficient [3] log $P_{OW}$, this property has been widely used to quantify lipophilicity. Albeit the number of compounds with a measured value for log $P$ was estimated recently to be 30 000 [4], which seems at first glance to be high, this is negligible compared to the rapidly increasing number of compounds for which log $P$ values are desired but missing. Furthermore, the experimental determination is tedious, time-consuming and demands a high purity of the solute [5]; none of these preconditions are compatible with high-throughput techniques. There is, therefore, an ongoing interest in methods for the prediction of log $P$ values.

Fujita et al. were the first to develop a calculation method that was based, analogously to the Hammett approach, on substituent constants $\pi$ [6] (see Eq. (7))

$$\pi_X = \log P_{R-X} - \log P_{R-H} \tag{7}$$

The hydrophobic constant $\pi$ is a measure of the contribution of a substituent $X$ to the lipophilicity of compound R–X compared with R–H. The constant representing the solvent/solvent system, analogously to Hammett's $\rho$ constant for the reaction type, was arbitrarily set to 1 for octanol/water and thus does not appear in Eq. (7). The lipophilicity constant $\pi$ allows the estimation of log $P$ values for congeneric series of compounds with various substituents (see Eq. (8)).

$$\log P_{R-X, Y, Z} = \log_{R-H} + \sum \pi_{X, Y, Z} \tag{8}$$

The obvious drawback of this method is that the parent solute, at least, has to be available or must be synthesized, and its log $P$ value has to be determined experimentally. Nys and Rekker therefore developed a method known as the fragmental constant approach, which is based on the additivity of fragment contributions to the molecular lipophilicity [7] (see Eq. (9), where $a_i$ is the incidence of fragment $i$, $f_i$ the lipophilic fragment constant, $C_M$ a correction factor, and $k_i$ the frequency of $C_M$).

$$\log P = \sum_{i=1}^{n} a_i f_i + \sum_{i=1}^{n} k_i C_M \tag{9}$$

In order to obtain the fragmental constants, sets of molecules for which experimental log $P$ values are available are cut into the predefined fragments and the numerical contribution of each fragment is determined by multiple linear regression analysis.

The correction term in Eq. (9) shows that the basic assumption of additivity of the fragmental constants obviously does not hold true here. Correction has to be applied, e.g., for structural features such as resonance interactions, condensation in aromatics or even hydrogen atoms bound to electronegative groups. Astonishingly, the correction applied for each feature is always a multiple of the constant $C_M$, which is therefore often called the "magic constant". For example, the correction for a resonance interaction is $+2 \cdot C_M$, or per triple bond it is $-1 \cdot C_M$. A detailed treatment of the $\Sigma f$ system approach is given by Mannhold and Rekker [5].

### 10.1.4.1  Other Substructure-Based Methods

The Rekker approach is still used with revised $\Sigma f$ systems, e.g., in the software program $\Sigma f$-SYBYL [8]. Over recent decades various other substructure-based approaches have been developed that are mostly implemented and available as computer programs.

Hansch and Leo developed a fragmental system, which is also based on fragment constants and correction terms [9]. Fragment constants were derived from solutes where the fragment occurs in isolation. Furthermore, the bonding environment was taken into account – alkyl, benzyl, vinyl, styryl, and aromatic neighbors – resulting in five values per fragment. If a fragment in combination with the bonding environment was missing but at least two values for the same fragment with different neighbors could be found, an interpolation was attempted to derive the missing data. The correction factors have been calculated from the corrections required for the specific interactions being modeled. For instance, the interaction of the two hydroxyl groups in diethylene glycol increases the log $P$ value by 0.85 compared with two hydroxyl groups that do not interact. This value is then taken as the correction term for a two-neighbored hydroxyl group. The method has been implemented in the CLOGP program, which is still the most frequently used software for estimating log $P$ values [5].

If a fragment constant is not available from the internal data basis nor can be interpolated from related data, a calculation program has to issue a "missing fragment" error, which is generally a problem with such fragmental approaches. In order to tackle this problem, which could result in 10–20 % calculation failures for databases containing more complex structures, the CLOGP approach was recently extended by an algorithm to estimate fragment constants from scratch [10]; this is based on a test set of about 600 dependably measured fragments having only aliphatic and aromatic bonds.

Several models have been published where the fragments are defined on a purely atomic level. This simplifies both the recognition of fragments and the calculation, as correction substructures are not applied (see Eq. (10)). $N_i$ is the occurrence of the $i$th atom type.

$$\log P = \sum_{i=1}^{n} a_i N_i \tag{10}$$

The most frequently used atomic increment system, ALOGP, was developed by Ghose and Crippen [11]. Atoms are classified by their neighboring environment and carbon atoms additionally by their hybridization. The regression analysis for a training data set of 494 compounds showed a fit with a standard deviation of 0.35. Despite its broad application, the method exhibits larger deviations for more complex compounds and shows a bias toward underestimating log $P$. The method was therefore revised using a training set of 9920 compounds, yielding in a standard deviation of 0.67 [12].

### 10.1.4.2 QSPR Models

Besides these LFER-based models, approaches have been developed using whole-molecule descriptors and learning algorithms other then multiple linear regression (see Section 10.1.2).

Two methods should be mentioned here as examples. Schaper and Samitier reported a model using topological descriptors (see Section 8.3) and back-propagation neural networks (see Section 9.5.7) [13]. Atoms excluding hydrogen were described by an algorithm called "canonical numbering", taking the number of bonds and the mass of the atoms into account. Using this information a set of indicator variables was derived, which stated whether an atom as described by the algorithm was found in a molecule or not. This descriptor set was reduced by simply eliminating the variables that were zero for the entire data set. The remaining 147 descriptors for the training data set were then used for the training of a three-layered back-propagation network, resulting in a standard deviation for the best net of 0.25 (147–3–1 architecture), whereas the test data set of 50 similar structures yielded a standard deviation of 0.66. This indicated that the trained network had not generalized the structural features related to log $P$. The ratio of the training data to the number of adjustable parameters was unfavorably low (~0.6) but should have been about 2.0 (see Section 9.5). On the other hand, the descriptors may not have encoded the important features; after all, the descriptor optimization process was quite rudimentary. This is thus a good example of problems that one may encounter in QSPR studies.

Breindl et. al. published a model based on semi-empirical quantum mechanical descriptors and back-propagation neural networks [14]. The training data set consisted of 1085 compounds, and 36 descriptors were derived from AM1 and PM3 calculations describing electronic and spatial effects. The best results with a standard deviation of 0.41 were obtained with the AM1-based descriptors and a net architecture 16–25–1, corresponding to 451 adjustable parameters and a ratio of 2.17 to the number of input data. For a test data set a standard deviation of 0.53 was reported, which is quite close to the training model.

These few examples are of course a small and arbitrarily chosen set of methods for the calculation of log $P$ values. Nevertheless, it is hoped that they demonstrate some basic principles in the prediction of a physicochemical property.

## 10.1.5
## Estimation of Aqueous Solubility (log S)

Aqueous solubility of organic compounds is a particularly useful property, having many applications in pharmaceutical, environmental, and other chemical disciplines. The solubility of a drug is an important property that determines its bioavailability and biological activity. In the drug design process, it is essential to estimate the solubility of a large number of candidates for a drug before the compound is synthesized. A knowledge of aqueous solubility is also necessary for predicting the general environmental distribution of organic pollutants such as highly toxic, carcinogenic, and other undesirable compounds.

From the standpoint of thermodynamics, the dissolving process is the establishment of an equilibrium between the phase of the solute and its saturated aqueous solution. Aqueous solubility is almost exclusively dependent on the intermolecular forces that exist between the solute molecules and the water molecules. The solute–solute, solute–water, and water–water adhesive interactions determine the amount of compound dissolving in water. Additional solute–solute interactions are associated with the lattice energy in the crystalline state.

The solubility of a compound is thus affected by many factors: the state of the solute, the relative aromatic and aliphatic degree of the molecules, the size and shape of the molecules, the polarity of the molecule, steric effects, and the ability of some groups to participate in hydrogen bonding. In order to predict solubility accurately, all these factors correlated with solubility should be represented numerically by descriptors derived from the structure of the molecule or from experimental observations.

### 10.1.5.1 Solubility Prediction Methods
An extensive series of studies for the prediction of aqueous solubility has been reported in the literature, as summarized by Lipinski et al. [15] and Jorgensen and Duffy [16]. These methods can be categorized into three types: 1) correlation of solubility with experimentally determined physicochemical properties such as melting point and molecular volume; 2) estimation of solubility by group contribution methods; and 3) correlation of solubility with descriptors derived from the molecular structure by computational methods. The third approach has been proven to be particularly successful for the prediction of solubility because it does not need experimental descriptors and can therefore be applied to collections of virtual compounds also.

#### 10.1.5.1.1 Correlation with Descriptors from Experimental Data
A series of studies has been made by Yalkowsky and co-workers. The so-called general solubility equation was used for estimating the solubility of solid non-electrolytes [17, 18]. The solubility log $S$ (logarithm of solubility expressed as mol/L) was formulated with log $P$ (logarithm of octanol/water partition coefficient), and the melting point (MP) as shown in Eq. (11). This equation generally

works well. For 580 compounds, the prediction resulted in an average absolute error of about 0.45 log units.

$$\log S = 0.5 - \log P - 0.01(MP - 25) \tag{11}$$

It is remarkable that only two descriptors were needed in this method. However, this equation is mostly only of historical interest as it is of little use in modern drug and combinatorial library design because it requires a knowledge of the compound's experimental melting point which is not available for virtual compounds. Several methods exist for estimating log $P$ [1–14], but only a few inroads have been made into the estimation of melting points. The melting point is a key index of the cohesive interactions in the solid and is still difficult to estimate.

### 10.1.5.1.2 Group Contribution Method

The group contribution method allows the approximate calculation of solubility by summing up fragmental values associated with substructural units of the compounds (see Section 7.1). In a group contribution model, the aqueous solubility values are computed by Eq. (12), where log $S$ is the logarithm of solubility, $C_i$ is the number of occurrences of a substructural group, $i$, in a molecule, and $G_i$ is the relative contribution of the fragment $i$.

$$\log S = C_0 + \sum_{i=1}^{N} C_i G_i \tag{12}$$

The fragments $i$ are the descriptors and their counts can be obtained from 2D structural diagrams without the need for 3D coordinates. The optimal contribution coefficients $G_i$ and the constant $C_0$ are obtained from regression analyses. With group contribution methods, Kühne et al. developed a solubility model using experimental data on 351 liquids and 343 solids, and compared their models with four other group contribution algorithms [19]. The number of fragments and correction terms was about 50 in the best-performing models, and an additional term for the melting point was added for estimating the solubility of solids. A correlation with $r^2 = 0.95$ and $AAE = 0.38$ log units was obtained ($AAE$: average absolute errors).

The group contribution method is, however, restricted to those functional groups for which it was parameterized. Early studies on group contribution schemes contained a paucity of polyfunctional molecules in the data sets, and the number of fragment types was not large enough to treat drug-like molecules well. In a more recent report by Klopman and Zhu [20], several new models were proposed with an improvement in the accuracy and the scope of previous models by increasing the size and diversity of the training set. The training set consisted of 1168 compounds that covered a large variety of chemical classes, including some complex drugs; 171 fragments were used to yield a correlation via Eq. (12) with $r^2 = 0.95$ and $AAE = 0.49$ log units.

The disadvantages of the group contribution method are that: 1) the groups included must be defined in advance and therefore the solubility of a new compound containing new groups cannot be estimated; and 2) the different effects of a group in different chemical environments are not considered.

### 10.1.5.1.3 Correlation with Calculated Descriptors

Several research groups have built models using theoretical descriptors calculated only from the molecular structure. This approach has been proven to be particularly successful for the prediction of solubility without the need for descriptors of experimental data. Thus, it is also suitable for virtual data screening and library design. The descriptors include 2D (two-dimensional, or topological) descriptors, and 3D (three-dimensional, or geometric) descriptors, as well as electronic descriptors.

Jurs et al. developed QSPR models for the prediction of solubility using multiple linear regression analysis (MLRA) and computational neural networks (CNN) (mainly back-propagation neural networks), relating it to the structures of a diverse set of 332 compounds [21]. A series of topological, geometric, and electronic descriptors were calculated. Genetic algorithm and simulated annealing routines, in conjunction with MLRA and CNN, were used to select subsets of descriptors that relate accurately to aqueous solubility. Nine descriptors, including four topological, one geometric, one electronic, and three polar surface area ones, were selected. The model had the corresponding root mean square (RMS) errors of 0.394, 0.358, and 0.343 for the training set, cross-validation set, and test set, respectively.

Recently, several QSPR solubility prediction models based on a fairly large and diverse data set were generated. Huuskonen developed the models using MLRA and back-propagation neural networks (BPG) on a data set of 1297 diverse compounds [22]. The compounds were described by 24 atom-type $E$-state indices and six other topological indices. For the 413 compounds in the test set, MLRA gave $r^2 = 0.88$ and $s = 0.71$ and neural network provided $r^2 = 0.92$ and $s = 0.60$. Using this data set, some other groups derived new prediction models using different kinds of input descriptors and methods [23–26].

### 10.1.5.1.4 Our Models for the Prediction of Aqueous Solubility

We developed our models for the prediction of aqueous solubility based on the data set of Huuskonen, which (after diminution by two duplicates and the other two compounds that cannot be treated with our method) gave 1293 diverse compounds. The aqueous solubility values were measured at a temperature of 20–25 °C and are expressed as log $S$, where $S$ is the solubility in mol/L, with a minimum value of −11.62 and a maximum value of 1.58. The compounds were described by two different representation methods: 1) with 18 topological descriptors, as discussed in detail in the Tutorial, Section 10.1.5.2 [25]; and 2) with 32 radial distribution function (RDF) codes [27] representing the 3D structure of a molecule and eight additional descriptors mainly to reflect hydrogen bonding (see Section 10.1.5.3) [26]. The data set was divided into a training and a test set based on Kohonen's self-organizing neural network (KNN). Several quantitative models for the prediction of the aqueous solubility were developed by using MLRA and back-propagation (BPG) neural networks.

### 10.1.5.2 Tutorial: Developing Models for Solubility Prediction with 18 Topological Descriptors

#### 10.1.5.2.1 Step 1: Choice of Descriptors

In order to develop a proper QSPR model for solubility prediction, the first task is to select appropriate input descriptors that are highly correlated with solubility. Clearly, many factors influence solubility – to name but a few, the size of a molecule, the polarity of the molecule, and the ability of molecules to participate in hydrogen bonding. For a large diverse data set, some indicators for describing the differences in the molecules are also important.

Thus, a list of 15 descriptors was calculated for these purposes, as described below.

The partition coefficient log $P$ (calculated by a method based on the Ghose/Crippen approach [11]) (see also Chapter X, Section 1.1 in the Handbook) was calculated because it affects the solubility dramatically [17, 18]. All the other descriptors were calculated with the program PETRA (Parameter Estimation for the Treatment of Reactivity Applications) [28].

Mean molecular polarizability expresses how a molecule responds to an external electric field. Molecular weight is correlated with the size of the molecule. The following descriptors can describe the ability of a molecule to participate in hydrogen bonding: the highest hydrogen bond acceptor potential, the highest hydrogen bond donor potential, the number of hydrogen bond donor groups, and the number of atoms of the elements nitrogen, oxygen, and fluorine. The aliphatic and aromatic indicators of a molecule, and the number of atoms of the elements hydrogen, carbon, sulfur, and chlorine, can be used as indicator descriptors.

A set of 2D autocorrelation coefficients vectors (see Section 8.3) was calculated because molecules of different sizes need some mathematical transformation to come up with the same number of descriptors. In these calculations, atomic properties such as electronegativity and charge were used for each atom because they determine the macroscopic properties of a compound. Altogether, seven bond distances (from a distance of $d = 0$ to $d = 6$), and the seven atomic properties polarizability, lone-pair electronegativity, $\pi$-electronegativity, $\sigma$-electronegativity, $\sigma$-charge, $\pi$-charge, and partial atomic charges were considered. Thus, we had 49 autocorrelation descriptors. Altogether, the initial set comprised 64 descriptors.

#### 10.1.5.2.2 Step 2: Selection of Descriptors

In statistical analyses, it was found that the seven 2D autocorrelation coefficients for the atomic polarizability, $\sigma$-electronegativity, and $\pi$-electronegativity properties are highly correlated. The first component of the 2D autocorrelation coefficients had the highest standard deviation for the $\sigma$-charge, $\pi$-charge, partial atomic charge, and lone-pair electronegativity properties. Thus, the first component ($d = 0$) of the autocorrelation coefficients for each of these seven properties was selected for the following analysis. In effect, these values correspond to the sum of the squares of each atomic property for a molecule and they indicate the amount of variability of each of these properties for all the molecules.

These first components of the autocorrelation coefficient of the seven physico-chemical properties were put together with the other 15 descriptors, providing 22 descriptors. Pairwise correlation analysis was then performed: a descriptor was eliminated if the correlation coefficient was equal or higher than 0.90, and four descriptors (molecular weight, the number of carbon atoms, and the first component of the 2D autocorrelation coefficient for the atomic polarizability and $\sigma$-charge) were removed. This left 18 descriptors.

### 10.1.5.2.3 Step 3: Inspection of Separation Capability

The relationship between solubility and the 18 descriptors was investigated by the unsupervised learning algorithm of Kohonen's self-organizing neural network (KNN) (see Section 9.5.3). KNN has the characteristics of projecting the objects with high-dimensional vectors into similar areas of a 2D map. In this case, the distribution of 1293 compounds in the KNN map was in different areas according to their molecular characteristics, as reflected by the 18 input descriptors. From Figure 10.1-2, it was also observed that the compounds located in different areas

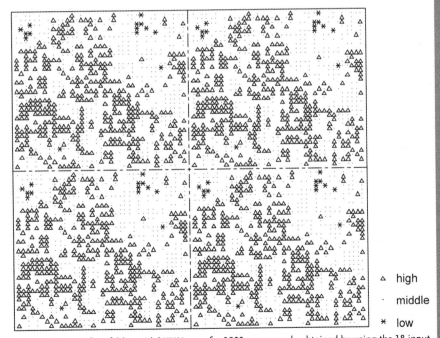

△ high

· middle

∗ low

**Figure 10.1-2.** A four-fold toroidal KNN map for 1293 compounds obtained by using the 18 input descriptors: "high" means compounds with high solubility where log $S$ is in the range −2.82 to 1.58; "middle" means compounds with medium solubility where log $S$ is in the range −7.21 to −2.83; "low" means compounds with low solubility where log $S$ is in the range −11.62 to −7.22.

had different solubilities. This proved that the selected 18 input descriptors were suitable for modeling solubility.

#### 10.1.5.2.4 Step 4: Selection of a Training and of a Test set

A data set can be split into a training set and a test set randomly or according to a specific rule. The 1293 compounds were divided into a training set of 741 compounds and a test set of 552 compounds, based on their distribution in a KNN map. From each occupied neuron, one compound was selected and taken into the training set, and the other compounds were put into the test set. This selection ensured that both the training set and the test set contained as much information as possible, and covered the chemical space as widely as possible.

#### 10.1.5.2.5 Step 5: Building a Multiple Linear Regression Analysis (MLRA) Model

Multiple linear regression analysis is a widely used method, in this case assuming that a linear relationship exists between solubility and the 18 input variables. The multilinear regression analysis was performed by the SPSS program [30]. The training set was used to build a model, and the test set was used for the prediction of solubility. The MLRA model provided, for the training set, a correlation coefficient $r = 0.92$ and a standard deviation of $s = 0.78$, and for the test set, $r = 0.94$ and $s = 0.68$.

#### 10.1.5.5.6 Step 6: Building a Back-Propagation (BPG) Neural Network Model

An artificial neural network is a good tool for simulating complicated nonlinear relationships between input and output variables. A two-layer neural network model trained by the back-propagation algorithm was generated by the SNNS program [31]. A standard back-propagation network was applied to estimate solubility. The architecture consisted of 18 input units representing the descriptors and one output neuron representing log $S$; for the hidden layer different numbers of neurons were investigated. All layers were completely connected. The initial weights were randomly initialized between −0.1 and 0.1. Each input and output value was scaled between 0 and 1. The network was trained following the "standard back-propagation" algorithm as implemented in SNNS, employing a learning rate of 0.2. The number of hidden- layer neurons was varied from 5 to 13. An optimized neural network architecture of 18–10–1 was found. The best number of training epochs was selected by the early stopping method in order to avoid overtraining; it was 6000. For the training set $r = 0.96$ and $s = 0.51$, and for the test set $r = 0.97$ and $s = 0.52$, were obtained. The prediction results for the test set are shown in Figure 10.1-3.

#### 10.1.5.2.7 Step 7: Comparison of the Structural Variety with Another Data Set

The model described above is based on the Huuskonen data set. Later we obtained an additional data set, compiled by the company Merck KGaA. We wanted to know how structurally similar these two data sets are. Thus, the following work was performed.

We know that every QSPR model is limited by the data set that is used for building the model. In order to examine the diversity of this data set (the Huuskonen

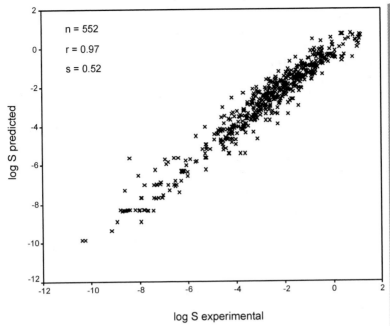

**Figure 10.1-3.** Predicted versus experimental solubility values of 552 compounds in the test set by a back-propagation neural network with 18 topological descriptors.

data set), we put it together with the Merck data set consisting of 2084 diverse compounds. The diversities of the two data sets were compared by investigating their distribution in a 41 × 35 KNN map using the previously reduced 18 descriptors as input vectors. From Figure 10.1-4, one can see that the compounds in the Huus-konen data set are not as diverse as those in the Merck data set.

This then led us to develop a solubility model based on the Merck data set [32].

### 10.1.5.3 Models with 32 Radial Distribution Function Values and Eight Additional Descriptors

The compounds were described by a set of 32 radial distribution function (RDF) code values [27] representing the 3D structure of a molecule and eight additional descriptors. The 3D coordinates were obtained using the 3D structure generator CORINA [33].

This study was done because we wanted to see whether 3D descriptors can improve on the models obtained by 2D descriptors. Futhermore, we wanted to use the descriptor set as initially chosen, without any tedious selection of descriptor as reported in the Tutorial in Section 10.1.5.2.

The radial distribution function (RDF) of an ensemble of $N$ atoms can be interpreted as the probability distribution to find an atom in a spherical volume of

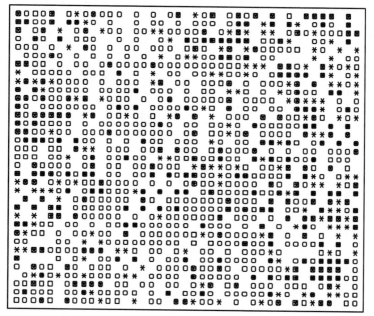

**Figure 10.1-4.** Distribution of compounds from two data sets in the same KNN (Kohonen's self-organizing neural network) map by using 18 topological descriptors as input descriptors, where "1" represents the 1588 compounds in the Merck data set (excluding those compounds that are also in the Huuskonen data set); "2" represents the 799 compounds in the Huuskonen data set (excluding those compounds that are also in the Merck data set), and "3" represents the overlapping part of the Huuskonen data set and the Merck data set.

radius *r*. The RDF function used in this work is as defined in Eqs. (13) and (14), where *f* is a scaling factor and *N* the number of atoms.

$$g(r) = f \sum_{i}^{N-1} \sum_{j>i}^{N} A_i A_j \cdot e^{-B(r-r_{ij})^2} \tag{13}$$

with

$$f = \frac{1}{\sqrt{\sum_{r} [g(r)]^2}} \tag{14}$$

By including characteristic atomic properties, *A*, of atoms *i* and *j*, the RDF code can be used in different tasks to fit the requirements of the information to be represented. The exponential term contains the distance $r_{ij}$ between the atoms *i* and *j* and the smoothing parameter *B*, which defines the probability distribution of the individual distances. The function *g(r)* was calculated at a number of discrete points with defined intervals.

Each molecule was represented by a vector of length 32. The parameter *B* was set to 25 Å$^{-2}$, corresponding to a total resolution of 0.2 Å in the defined distance *r* of [1.0–7.4 Å]. That means, the 32 defined distance intervals are: [1.0–1.2,

1.2–1.4], ... ... [7.2–7.4 Å]. The RDF code for the structure derivations was calculated with the atomic number as the atomic property.

An additional eight descriptors were also calculated by PETRA [28], including the mean molecular polarizability, the aromatic indicator of a molecule, the aliphatic indicator of a molecule, the highest hydrogen bond acceptor potential, the highest hydrogen bond donor potential, the number of hydrogen bond donor groups, and the number of atoms of the elements nitrogen and oxygen.

The 1293 compounds were split into a training set of 797 compounds and a test set of 496 compounds based on a KNN map. Two quantitative models were built by using similar methods to those described above. From the MLRA model, for the training set $r = 0.89$ and $s = 0.93$, and for the test set $r = 0.90$ and $s = 0.79$, were obtained. From a BPG neural network model, for the training set $r = 0.97$ and $s = 0.50$, and for the test set $r = 0.96$ and $s = 0.59$, were obtained. The prediction results for the test set are shown in Figure 10.1-5.

The structure of organic compounds can be described by 2D or 3D descriptors. 2D descriptors are simple and easy to understand, while the 3D descriptors contain more information. The descriptors of both methods were derived from the molecular structures. Thus the models that were developed here are suitable for in-silico data screening and library design.

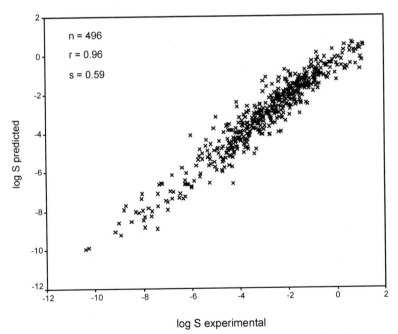

**Figure 10.1-5.** Predicted versus experimental solubility values of 496 compounds in the test set by a back-propagation neural network with 32 radial distribution function codes and eight additional descriptors.

The models are applicable to large data sets; with a rapid calculation speed, a wide range of compounds can be processed. Neural networks provided better models than multilinear regression analysis.

## 10.1.6
## Prediction of the Toxicity of Compounds

### 10.1.6.1 How to Quantify Toxicity

Toxicity may be one of the most difficult properties to model, especially for high-throughput screening in the drug discovery process. The difficulties arise from the following facts: The effects of toxicants are species-specific, organ-specific, and time-dependent (i.e., acute effects differ from chronic effects). This has the consequence that the concentration at which adverse effects occur can vary over several orders of magnitude depending on the species and the type of test. A comprehensive overview of modeling toxicity has been published by Schultz et al. [34], and another one addressing more specific issues of eco-toxicology by Escher and Hermens [35]. The topic of expert systems will not be discussed here; a critical review of all the major systems used today has been published by Benigni and Richard [36].

Toxic effects are measured through a wide variety of tests. Roughly, one can distinguish two types: *in vivo* and *in vitro* tests. *In vivo* tests are carried out with organisms such as rodents, fish, water fleas, earthworms, and algae. In-vitro tests are done mostly with single cells, organelles like mitochondria, or even just enzymes that are affected by a toxicant. The "classical" *in vivo* test for acute toxicity of a chemical is the $LC_{50}$-value. This is the concentration at which 50 % of the test species are killed by the toxic effects of a compound in a given time period. Until now this has also been one of the most common values to be predicted with QSAR equations.

### 10.1.6.2 Modeling Toxicity

A very important issue – disregard of which is a big source of bad modeling studies – is the clear distinction of transport processes (toxicokinetics) and interactions with targets such as membranes, enzymes, or DNA (toxicodynamics). Figure 10.1-6 gives a rather simplified model of a fish to illustrate this distinction.

After a chemical is released and distributed in the environment it might enter our model fish through its gills or to some extent also through food. Then it is distributed in the body (mainly between the aqueous phase, storage lipids, and membrane lipids). At the same time metabolism and excretion processes take place. Metabolism leads in most cases to less toxic compounds, but in some cases the contrary can happen: the product of metabolism is more toxic than the mother compound. Thus, this possibility needs to be kept in mind too, if compounds are tested *in vivo*. The fraction of the compound (and of possible metabolites) reaching the target(s) then causes the adverse effect. Therefore, if toxicity data are modeled,

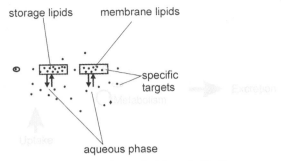

**Figure 10.1-6.** Simplified model of a higher aquatic organism and of the toxicokinetic processes taking place.

one should always have a clear picture of whether one is modeling toxicokinetic or toxicodynamic effects, or both effects together [35].

The fundamental assumption of SAR and QSAR (Structure–Activity Relationships and Quantitative Structure–Activity Relationships) is that the activity of a compound is related to its structural and/or physicochemical properties. In a classic article Corwin Hansch formulated Eq. (15) as a linear free-energy related model for the biological activity (e.g., toxicity) of a group of congeneric chemicals [37], in which the inverse of $C$, the concentration effect of the toxicant, is related to a hydrophobicity term, $\Pi$, an electronic term, $\sigma$ (the Hammett substituent constant). Steric terms can be added to this equation (typically Taft's steric parameter, $E_s$).

$$\log (C^{-1}) = k_1(\Pi) + k_2(\sigma) + k_3 \tag{15}$$

The most widely used descriptor for the hydrophobicity term in toxicology is the distribution coefficient between octanol and water, $\log P_{OW}$ (the environmental scientists would rather call it $\log K_{OW}$). The bulk solvent octanol is of course a rather crude model to approximate uptake and transport in a cell membrane and has shortcomings especially for charged compounds. Some scientists therefore work increasingly with membrane/water distribution coefficients in order to obtain a more realistic picture of how compounds distribute in cell membranes [38].

However, for screening large databases it is still necessary to use the easily calculated $\log P$ value.

Figure 10.1-7 shows that for certain chemicals Eq. (15) can be simplified to the uptake term and indeed can be successful in modeling the acute toxicity for a given species. Inverse logarithms of 96-hour $LC_{50}$ values – the concentration at which 50% of the test organisms die – of the fathead minnow (*Pimephales promelas*) fish species has been taken as a measure of acute toxicity. The $\log P$ was taken as a measure for lipophilicity. The chemicals in this figure represent a quite diverse selection of aliphatic and aromatic hydrocarbons, about half of them with chloro substituents, the rest with either hydroxy or carbonyl groups. However, they have the common feature that they do not have specific effects on vital functions

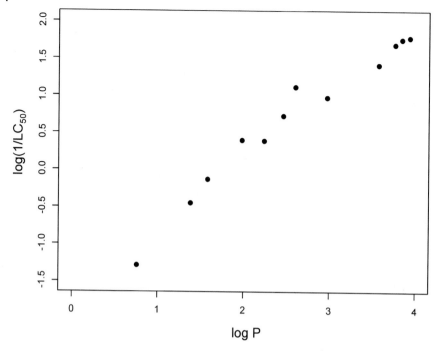

**Figure 10.1-7.** Correlation of LC$_{50}$ values for the fish species *Pimephales promelas* with log $P$ of a series of compounds (a subset of a published data set [39]).

of the organism. The compounds are a subset of a toxicity data set taken from the literature [39].

Figure 10.1-7 shows that there is a strong correlation between lipophilicity and toxicity for these types of compounds. Although the compounds are not congeneric they can all be modeled just by the lipophilicity term of Eq. (15). A multitude of quantitative models have been published over the last 20 years, initiated by Könemann, who proposed a linear relationship between log $P$ and LC$_{50}$ of guppies (*Pocilia reticulata*) for a similar data set with 50 compounds with log $P$ ranging from –1 to 6 [40]. Equations of this type can be applied to model many common industrial chemicals successfully. Figure 10.1-8, however, shows that reality is a little more complicated. Now all compounds of the previously mentioned data set are shown [39]. The data set was actually compiled with the aim of having highly diverse compounds and also a high diversity of toxic effects. All the effects on targets of major importance in acute toxicity are covered – plus receptor-mediated effects through environmental estrogens. These effects can be grouped into modes of action (MOA). A MOA can be defined as a common set of physiological and behavioral signs characterizing a type of adverse biological response, while a toxic mechanism is defined as the biochemical process underlying a given MOA [41].

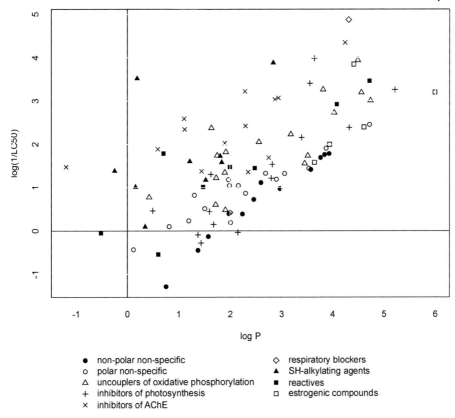

**Figure 10.1-8.** $LC_{50}$ values for the fish species *Pimephalas promelas* versus log $P$ of a highly diverse set of organic chemicals. Chemicals are marked by their mode of action.

Figures 10.1-7 and 10.1-8 differ because the toxicity of the compounds of Figure 10.1-7 is caused solely by their tendency to move into biological membranes and is often referred to as the baseline toxicity – a property that every compound has. However, in addition to that, many compounds can interact with more specific targets. So it is necessary to have QSAR-equations with additional terms that account for electronic and steric effects. However, these are much harder to model and most models are based on the common chemical class of compounds. As a result, we face the problem today that there is a huge number of QSAR equations of a very local nature. The methods applied range from quite simple extension of Eq. (15) to CoMFA models. The descriptors also cover a wide range of types, with global whole-molecule descriptors being used the most often [34]. A major problem is the choice of the appropriate QSAR for the prediction of a new compounds, and there is a great danger of applying inappropriate QSARs. One suggestion of a way to resolve this problem is to establish models that are based on a common MOA and not on chemical classes [42]. The prerequisite of this approach is the correct classification of compounds into their MOA. Different types of descriptors and different classification methods have already been published [39, 43–45]. One possible approach is shown in detail in the following Tutorial, Section 10.1.7.

### 10.1.7
### Tutorial: Classifying Compounds into Different Modes of Action

The previously mentioned data set with a total of 115 compounds has already been studied by other statistical methods such as Principal Component Analysis (PCA), Linear Discriminant Analysis, and the Partial Least Squares (PLS) method [39]. Thus, the choice and selection of descriptors has already been accomplished.

The objective of this study is to show how data sets of compounds for which different biological activities have been determined can be studied. It will be shown how the use of a counter-propagation neural network can lead to new insights [46]. The emphasis in this example is placed on the comparison of different network architectures and not on quantitative results.

The same structure representation as the one taken in the original study [39] is selected in order to show some possibilities evolving from working with a neural network method. Table 10.1-1 gives the ten descriptors chosen for the representation of the 115 molecules of the data set.

The data set was then sent into a counter-propagation (CPG) network consisting of 13 × 9 neurons with 10 layers (one for each descriptor) in the input block and one layer in the output block (Figure 10.1-9), with the output values having nine different values corresponding to the nine different MOA.

The resulting distribution of compounds having different modes of action in the output layer after training the network is shown in Figure 10.1-10.

**Table 10.1-1.** Descriptors of the compounds used for establishing a classification model.

| MW | Molecular weight | $\log P_{OW}$ | Octanol/water distribution coefficient |
|---|---|---|---|
| MR | Molecular Refractivity | $\varepsilon_{HOMO}$ | Energy of Highest Occupied MO |
| $D_{EFF}$ | Effective Diameter | $V^+$ | Potential of positive atomic charges |
| SASA | Solvent-Accessible Surface Area | $Q_{AV}$ | Average of absolute atomic charges |
| SAVOL | Solvent-Accessible Volume | $H^+_{max}$ | Maximum positive charge on hydrogen atom |

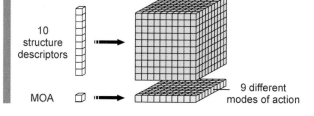

10 structure descriptors

MOA

9 different modes of action

**Figure 10.1-9.** Architecture of the counterpropagation network for the classification of toxicants with one output layer.

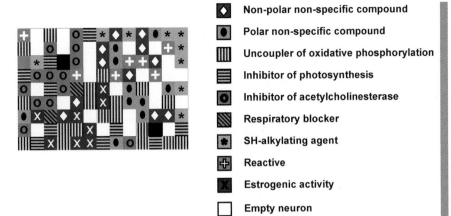

◆ **Non-polar non-specific compound**

▣ **Polar non-specific compound**

▥ **Uncoupler of oxidative phosphorylation**

▤ **Inhibitor of photosynthesis**

◉ **Inhibitor of acetylcholinesterase**

▨ **Respiratory blocker**

✴ **SH-alkylating agent**

⊞ **Reactive**

✕ **Estrogenic activity**

☐ **Empty neuron**

**Figure 10.1-10.** Distribution of the compounds in the output layer according to modes of action (most frequent MOA shown).

Clearly, no pronounced clustering of the compounds according to MOA can be discerned. What is the problem? Is the chosen structure representation not appropriate for this specific problem?

Rather than making this statement, one should consider first whether the representation of the Y-variable is appropriate. What we did here was to take categorical information as a quantitative value. So if we have, for instance, a vector of class 1 and one of class 9 falling into the same neuron, the weights of the output layer will be adapted to a value between 1 and 9, which does not make much sense. Thus, it is necessary to choose another representation with one layer for each biological activity. The architecture of such a counter-propagation network is shown in Figure 10.1-11. Each of the nine layers in the output block corresponds to a different MOA.

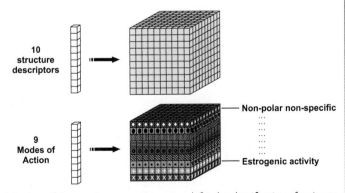

**10 structure descriptors**

**9 Modes of Action**

Non-polar non-specific

Estrogenic activity

**Figure 10.1-11.** Architecture of the counterpropagation network for the classification of toxicants with nine output layers, one for each mode of action.

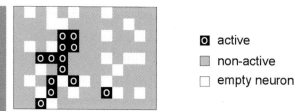

**Figure 10.1-12.** Distribution of compounds in the layer of acetylcholinesterase inhibitors; neurons colored in black and marked with a circle contain inhibitors of acetylcholinesterase, and neurons in light gray contain other compounds.

With this CPG network quite interesting results were obtained. An analysis of the distribution of the compounds in the individual output layers showed that for modes of toxic action that correspond to toxicities associated with receptor binding, a clustering of the compounds could be observed. For example, the layer in the output block corresponding to inhibition of acetylcholinesterase (layer 5 in the output block) showed a clear clustering of the active compounds (Figure 10.1-12).

In a similar manner, compounds that act as estrogenics cluster in the corresponding layer of the output block (Figure 10.1-13).

On the other hand, compounds corresponding to rather general, unspecific modes of toxic action are distributed over a broad area in the respective layer, as shown for polar non-specific toxicants in Figure 10.1-14.

For this kind of toxicity specific structural prerequisites are not required, thus leading to some spread of the compounds in structural space. About five of nine layers showed little tendency to cluster, some of them with quite specifically acting

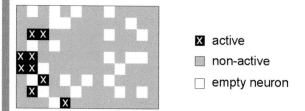

**Figure 10.1-13.** Distribution of compounds in the layer of estrogenic compounds.

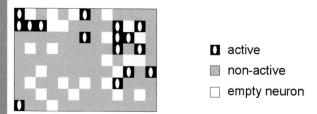

**Figure 10.1-14.** Distribution of compounds in the layer of polar non-specific toxicants.

compounds such as reactives or SH-alkylating agents; this is a a clear indicator that for these MOAs the descriptors used so far are not sufficient to discriminate between these types of compounds and the others in the data set.

However, these results illustrate that the use of a counter-propagation network can lead to new insights when several biological activities are given. Furthermore, a CPG network can also be applied for studying selectivity between different biological activities.

The predictive power of the CPG neural network was tested with leave-one-out cross-validation. The overall percentage of correct classifications was low, with only 33 % correct classifications, so it is clear that there are some major problems regarding the predictive power of this model. First of all one has to remember that the data set is extremely small with only 115 compounds, and has a extremely high number of classes with nine different MOAs into which compounds have to be classified. The second task is to compare the cross-validated classifications of each MOA with the impression we already had from looking at the output layers. This revealed that MOAs that did not form clusters in the corresponding layers were not classified correctly, while compounds like acetylcholinesterase inhibitors could be recognized successfully. If these MOAs with no or just a few correct classifications – for which additional descriptors would apparently be necessary – were removed from the data set, a model with 80 compounds could be established that discriminated between four types of MOA: nonspecific compounds (polar and nonpolar were taken together), oxidative phosphorylation uncouplers, acetylcholinesterase inhibitors, and estrogenic compounds. The cross-validated percentage of the rate of correct classification increased to 70 %. Similar values were obtained with other classification methods such as logistic regression and linear discriminant analysis. So the conclusion of the cross-validation results is that these descriptors – which are commonly applied for the quantitative prediction of activity in toxicology – are not necessarily suited to the qualitative task of classifying highly diverse compounds into different types of activity.

10.1.8
**Conclusion and Future Outlook**

As a conclusion one can say that the distinction of islands of specific activity from within a sea of baseline toxicity, with each island representing a local chemical biological mechanism domain, is still a challenge to be solved by scientists working both experimentally and computationally.

A completely new approach during recent years has been to monitor changes in gene expression at the level of transcription. Typical expression patterns in the presence of known toxicants are saved and can be used as predictions for new compounds. This technology might bring completely new insights, especially for the very subtle art of drug safety evaluation [47].

**Essentials**

- Building a QSPR model consists of three steps: descriptor calculation, descriptor analysis and optimization, and establishment of a mathematical relationship between descriptors and property.
- Successful predictive models in toxicology exist – however, they are of a rather local nature. Effects considered in toxicology can be caused by different mechanisms. Efforts to get away from a class perspective to one that is more consistent regarding modes of toxic action are still a subject of ongoing research.

**Selected Reading (log *P*, log *S*)**

- P. C. Jurs, Quantitative structure–property relationships, in *Encyclopedia of Computational Chemistry, Volume 4*, P. v. R. Schleyer, N. L. Allinger, T. Clark, J. Gasteiger, P. A. Kollman, H. F. Schaefer III and P. R. Schreiner (Eds.), John Wiley & Sons, Chichester, **1998**, pp. 2320–2330.
- R. Mannhold, H. van de Waterbeemd, *J. Comput. Aided Mol. Des.* **2001**, *15*, 337–354.
- Prediction of Physical and Chemical Properties (Chapter X, Section 1) in *Handbook of Chemoinformatics – From Data to Knowledge*, J. Gasteiger (Ed.), Wiley-VCH, Weinheim, **2003**.

**Interesting Websites**

- *http://research.chem.psu.edu/pcjgroup*

**Available Software (log *P*)**

- ALOGP98 : *http://www.accelrys.com*
- MedChem (CLOP): *http://www.daylight.com*
- KOWWIN: *http://esc.syrres.com*
- XLOGP: *ftp://ftp2.ipc.pku.edu.cn/pub/software/xlogp*

# References

[1] M. J. Kamlet, R. W. Taft, *J. Am. Chem. Soc.* **1976**, *98*, 377–383.

[2] N. B. Chapman, J. Shorter, *Advances in Linear Free Energy Relationships*, Plenum Press, London, **1972**.

[3] C. Hansch, T. Fujita, *J. Am. Chem. Soc.* **1964**, *86*, 1616–1626.

[4] R. Mannhold, H. van de Waterbeemd, *J. Comput. Aided Mol. Des.* **2001**, *15*, 337–354.

[5] R. Mannhold, R. F. Rekker, *Perspect. Drug Discovery Des.* **2000**, *18*, 1–18.

[6] T. Fujita, J. Iwasa, C. Hansch, *J. Am. Chem. Soc.* **1964**, *86*, 5175–5180.

[7] G. G. Nys, R. F. Rekker, *Chim. Therap.* **1973**, *8*, 521–535.

[8] R. F. Rekker, K. Dross, G. Bijloo, G. de Vries, *Quant. Struct.–Act. Relat.* **1998**, *17*, 517–537.

[9] A. J. Leo, *Chem. Rev.* **1993**, *93*, 1281–1306.

[10] A. J. Leo, D. Hoekman, *Perspect. Drug Discovery Des.* **2000**, *18*, 19–38.

[11] A. K. Ghose, G. M. Crippen, *J. Comput. Chem.* **1986**, *7*, 565–577.

[12] S. A. Wildman, G. M. Crippen, *J. Chem. Inf. Comput. Sci.* **1999**, *39*, 868–873.

[13] K.-J. Schaper, M. L. R. Samitier, *Quant. Struct.–Act. Relat.* **1997**, *16*, 224–230.

[14] A. Breindl, B. Beck, T. Clark, R. C. Glen, *J. Mol. Model.* **1997**, *3*, 142–155.

[15] C. A. Lipinski, F. Lombardo, B. W. Dominy, P. J. Feeney, *Adv. Drug Delivery Rev.* **1997**, *23*, 3–25.

[16] W. L. Jorgensen, E. M. Duffy, *Adv. Drug Delivery Rev.* **2002**, *54*, 355–366.

[17] S. H. Yalkowski, S. C. Valvani, *J. Pharm. Sci.* **1980**, *69*, 912–922.

[18] N. Jain, S. H. Yalkowsky, *J. Pharm. Sci.* **2001**, *90*, 234–252.

[19] R. Kuhne, R.-U. Ebert, F. Kleint, G. Schmidt, G. Schüürmann, *Chemosphere* **1995**, *30*, 2061–2077.

[20] G. Klopman, H. Zhu, *J. Chem. Inf. Comput. Sci.* **2001**, *41*, 439–445.

[21] B. E. Mitchell, P. C. Jurs, *J. Chem. Inf. Comput. Sci.* **1998**, *38*, 489–496.

[22] J. Huuskonen, *J. Chem. Inf. Comput. Sci.* **2000**, *40*, 773–777.

[23] I. V. Tetko, V. Y. Tanchuk, T. N. Kasheva, A. E. P. Villa, *J. Chem. Inf. Comput. Sci.* **2001**, *41*, 1488–1493.

[24] R. F. Liu, S. S. So, *J. Chem. Inf. Comput. Sci.* **2001**, *41*, 1633–1639.

[25] A. X. Yan, J. Gasteiger, *Quant. Struct.–Act. Relat.* (in press).

[26] A. X. Yan, J. Gasteiger, *J. Chem. Inf. Comput. Sci.*, **2003**, *43*, 429–434.

[27] M. C. Hemmer, V. Steinhauer, J. Gasteiger, *Vibrat. Spectrosc.* **1999**, *19*, 151–164.

[28] J. Gasteiger, Empirical methods for the calculation of physicochemical data of organic compounds. In: *Physical Property Prediction in Organic Compounds*, C. Jochum, M. G. Hicks, J. Sunkel (Eds.); Springer Verlag, Heidelberg, **1988**; pp. 119–138.

[29] PETRA can also be accessed on the web: *http://www2.chemie.uni-erlangen.de/software/petra/index.html*

[30] SPSS for Windows, Rel. 10.0.7, SPSS Inc., Chicago, IL, **2000**.

[31] SNNS: Stuttgart Neural Network Simulator, Version 4.2, developed at the University of Stuttgart, maintained at University of Tübingen, **1995**.

*http://www-ra.informatik.*
*uni-tuebingen.de/SNNS/*

[32] A. X. Yan, J. Gasteiger, *J. Comp.-Aided Mol. Design* (submitted).

[33] J. Sadowski, J. Gasteiger, *Chem. Rev.* **1993**, *93*, 2567–2581. *http://www2. chemie.uni-erlangen.de/software/corina/ index.html*

[34] T. W. Schultz, M. T. D. Cronin, J. D. Walker, A. O. Aptula, *J. Mol. Struct. (Theochem)* **2003**, *622*, 1–22.

[35] B. I. Escher, J. Hermens, *Environ. Sci. Technol.* **2002**, *36*, 4201–4217.

[36] R. Benigni, A. M. Richard, *Methods Enzymol.* **1998**, *14*, 264–276.

[37] C. Hansch, *Acc. Chem. Res.* **1969**, *2*, 232–239.

[38] B. I. Escher, R. P. Schwarzenbach, *Aquat. Sci.* **2002**, *64*, 20–35.

[39] M. Nendza, M. Müller, *Quant. Struct.– Act. Relat.* **2000**, *19*, 581–598.

[40] H. Könemann, *Toxicology* **1981**, *19*, 209–221.

[41] G. Rand, P. Welss, L. S. McCarthy, Introduction to aquatic toxicology, in *Fundamentals of Aquatic Toxicology*, G. Rand (Ed.), Taylor & Francis, Washington, DC, **1995**, pp.3–67.

[42] S. P. Bradbury, *SAR QSAR Environ. Res.* **1994**, *2*, 89–104.

[43] J. Nouwen, F. Lindgren, B. Hansen, W. Karcher, *Environ. Sci. Technol.* **1997**, *31*, 2313–2318.

[44] S. C. Basak, G. D. Grunwald, G. E. Host, G. J. Niemi, S. P. Bradbury, *Environ. Toxicol. Chem.* **1998**, *17*, 1056–1064.

[45] A. O. Aptula, T. I. Netzeva, I. V. Valkova, M. T. D. Cronin, T. W. Schultz, R. Kuhne, G. Schüürmann, *Quant. Struct.–Act. Relat.* **2002**, *21*, 12–22.

[46] S. Spycher, unpublished work.

[47] J. F. Waring, R. G. Ulrich, *Annu. Rev. Pharmacol. Toxicol.* **2000**, *40*, 335–352.

## 10.2
## Structure–Spectra Correlations

*Markus Hemmer, João Aires-de-Sousa*

### Learning Objectives

- To identify the main methods and tools available for the computer prediction of spectra from the molecular structure, and for automatic structure elucidation from spectral data
- To realize that a proper representation of the molecular structure is crucial for the prediction of spectra
- To recognize the main approaches for structure representation in the context of structure–spectra correlations

### 10.2.1
### Introduction

The investigation of molecular structures and of their properties is one of the most fascinating topics in chemistry. Chemistry has a language of its own for molecular structures which has been developed from the first alchemy experiments to modern times. With the improvement of computational methods for chemical information processing, several descriptors for the handling of molecular information have been developed and used in a wide range of applications.

A most important task in the handling of molecular data is the evaluation of "hidden" information in large chemical data sets. One of the differences between data mining techniques and conventional database queries is the generation of new data that are used subsequently to characterize molecular features in a more general way. Generally, it is not possible to hold all the potentially important information in a data set of chemical structures. Thus, the extraction of relevant information and the production of reliable secondary information are important topics.

Finding the adequate descriptor for the representation of chemical structures is one of the basic problems in chemical data analysis. Several methods have been developed in the most recent decades for the description of molecules including their chemical or physicochemical properties [1].

Molecules are usually represented as 2D formulas or 3D molecular models. While the 3D coordinates of atoms in a molecule are sufficient to describe the spatial arrangement of atoms, they exhibit two major disadvantages as molecular descriptors: they depend on the size of a molecule and they do not describe additional properties (e.g., atomic properties). The first feature is most important for computational analysis of data. Even a simple statistical function, e.g., a correlation, requires the information to be represented in equally sized vectors of a fixed dimension. The solution to this problem is a mathematical transformation of the Cartesian coordinates of a molecule into a vector of fixed length. The second point can be overcome by including the desired properties in the transformation algorithm. Examples of such transforms are the descriptors based on Radial Distribution Functions (RDF). RDF descriptors grew out of the research area of structure–spectrum correlation.

## 10.2.2
## Molecular Descriptors

### 10.2.2.1 Fragment-Based Descriptors

A widely used method is fragment-based coding. With this approach, the molecule to be encoded is divided into several substructures that represent the typical information necessary for the task. Many authors have used this method for the automated interpretation of spectra with artificial neural networks [2], in expert systems for structure elucidation [3], with pattern recognition methods [4], and with semi-empirical calculations [5]. For example, a descriptor in the form of a binary vector is used simply to define the presence or absence of functional groups that exhibit important spectral features in the corresponding infrared (IR) spectrum. The main disadvantage of this method is that it imposes a restriction on the number of substructures represented (for the correlation of structure with IR spectra, the number of substructures varies from 40 to 720, depending on the user's more or less subjective view of the problem). Affolter et al. [6] showed that a simple assignment of IR-relevant substructures and corresponding IR signals does not describe spectrum–structure correlation to an adequate accuracy. This is mainly due to the effect of the chemical environment on the shape and position of absorption signals.

### 10.2.2.2 Topological Structure Codes

An enhancement of the simple substructure approach is the "Fragment Reduced to an Environment that is Limited" (FREL) method introduced by Dubois et al. [7] With the FREL method several centers of the molecule are described, including their chemical environment. By taking the elements H, C, N, O, and halogens into account and combining all bond types (single, double, triple, aromatic), the authors found descriptors for 43 different FREL centers that can be used to characterize a molecule.

To characterize the complete arrangement of atoms in a molecule, the entire molecule can be regarded as a connectivity graph where the edges represent the bonds and the nodes represent the atoms. By adding the number of bonds or the sum of bond lengths between all pairs of atoms, it is possible to calculate a descriptor that defines the constitution of a molecule independently of conformational changes. The resulting descriptor is not restricted regarding the number of atoms. Clerc and Terkovics [8] used this method based on the number of bonds for the investigation of quantitative structure–property relationships (QSPR).

One of the most widely used – and successful representations of the constitution, the topology, of a molecule is the HOSE code (Hierarchical Ordered description of the Substructure Environment) [9]. It is an atom-centered code taking into account

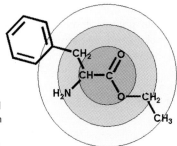

**Figure 10.2-1.** The structural environment is described in hierarchical order by starting from a selected carbon atom and walking through the molecule in spheres. Only the non-hydrogen atoms are considered explicitly.

the spheres of neighbors of each atom. It has proven to be particularly successful for the interpretation and prediction of [13]C NMR spectra (see Section 10.2.3). The rationale for this substructure code is the fact that the structural environment of a certain [13]C atom in a molecule determines the chemical shift of this atom. The length of the code depends on the number of spheres of neighbor atoms used for the structure description around the focus atom. Figure 10.2-1 illustrates the derivation of the HOSE code of the carbon atom in the carboxylic group of the ethyl ester of phenylalanine.

The HOSE code can be determined to various degrees of resolution, depending on how many spheres of neighbor atoms are considered. Figure 10.2-1 shows the first, second, and third spheres of neighbors for the carbon atom being considered. For [13]C NMR spectra a HOSE code for each carbon atom in a molecule has to be determined.

### 10.2.2.3 Three-Dimensional Molecular Descriptors

A descriptor for the 3D arrangement of atoms in a molecule can be derived in a similar manner. The Cartesian coordinates of the atoms in a molecule can be calculated by semi-empirical quantum mechanical or molecular mechanics (force field) methods. For larger data sets, fast 3D structure generators are available that combine data- and rule-driven methods to calculate Cartesian coordinates from the connection table of a molecule (e.g., CORINA [10]).

There are some prerequisites for a 3D structure descriptor. It should be:

- independent of the number of atoms, i.e., the size of a molecule,
- unambiguous regarding the 3D arrangement of the atoms, and
- invariant against translation and rotation of the entire molecule.

Further prerequisites depend on the chemical problem to be solved. Some chemical effects have an undesired influence on the structure descriptor if the experimental data to be processed do not account for them. A typical example is the conformational flexibility of a molecule, which has a profound influence on a 3D descriptor based on Cartesian coordinates. In particular, for the application of structure descriptors with structure–spectrum correlation problems in vibrational spectroscopy two other points are desirable: the descriptor should contain physicochemical information related to vibrational states, and it should be possible to gain structural information or the complete 3D structure from the descriptor.

### 10.2.3
### $^{13}$C NMR Spectra

NMR spectroscopy is probably the singly most powerful technique for the confirmation of structural identity and for structure elucidation of unknown compounds. Additionally, the relatively low measurement times and the facility for automation contribute to its usefulness and industrial interest.

Thus, in the area of combinatorial chemistry, many compounds are produced in short time ranges, and their structures have to be confirmed by analytical methods. A high degree of automation is required, which has fueled the development of software that can predict NMR spectra starting from the chemical structure, and that calculates measures of similarity between simulated and experimental spectra. These tools are obviously also of great importance to chemists working with just a few compounds at a time, using NMR spectroscopy for structure confirmation.

Furthermore, the prediction of $^{1}$H and $^{13}$C NMR spectra is of great importance in systems for automatic structure elucidation. In many such systems, all isomers with a given molecular formula are automatically produced by a structure generator, and are then ranked according to the similarity of the spectrum predicted for each isomer to the experimental spectrum.

$^{1}$H NMR spectra are basically characterized by the chemical shift and coupling constants of signals. The chemical shift for a particular atom is influenced by the 3D arrangement and bond types of the chemical environment of the atom and by its hybridization. The multiplicity of a signal depends on the coupling partners and on the bond types between atom and coupling partner.

However, the situation is much less complex for $^{13}$C NMR spectra. Whereas measurement conditions may have a high impact on $^{1}$H NMR spectra, chemical shift values in $^{13}$C NMR spectra are less sensitive to changes in solvent and experimental conditions. In fact, there is usually a good correlation between the 2D arrangement of atoms and their $^{13}$C chemical shifts. In addition, $^{13}$C NMR spectra are usually measured with the protons decoupled and then show few coupling effects except for fluorine and phosphorus atoms in the vicinity of the carbon atom. These are the reasons why $^{13}$C NMR spectra can be represented quite well as pairs of chemical shift values and intensities.

Several empirical approaches for $^{13}$C NMR spectra prediction are based on the availability of large $^{13}$C NMR spectral databases. By using special methods for encoding substructures that correspond to particular parts of the $^{13}$C NMR spectrum, the correlation of substructures and partial spectra can be modeled. Substructures can be encoded by using the additive model greatly developed by Pretsch [11] and Clerc [12]. The authors represented skeleton structures and substituents by individual codes and calculation rules. A more general additive model was introduced later by Small and Jurs [13], and was enhanced by Schweitzer and Small [14] later. In addition, several methods using neural networks for the processing of substructures have been developed [15–18].

However, one of the most successful approaches to systematically encoding substructures for $^{13}$C NMR spectrum prediction was introduced quite some time ago by Bremser [9]. He used the so-called HOSE (Hierarchical Organization of Spherical Environments) code to describe structures. As mentioned above, the chemical shift value of a carbon atom is basically influenced by the chemical environment of the atom. The HOSE code describes the environment of an atom in several virtual spheres – see Figure 10.2-1. It uses spherical layers (or levels) around the atom to define the chemical environment. The first layer is defined by all the atoms that are one bond away from the central atom, the second layer includes the atoms within the two-bond distance, and so on. This idea can be described as an atom center fragment (ACF) concept, which has been addressed by several other authors in different approaches [19–21].

The spectral signals are assigned to the HOSE codes that represent the corresponding carbon atom. This approach has been used to create algorithms that allow the automatic creation of "substructure–sub-spectrum" databases that are now used in systems for predicting chemical structures directly from $^{13}$C NMR.

The basic HOSE code ignores stereochemical information such as *cis–trans* isomer interaction that can contribute significantly to the chemical shift values. Robien adopted the HOSE code method [22] by extending it with descriptors for three-, four-, and five-bond interactions and with information about axial/equatorial substitution patterns. He used the adopted method in software for identification of organic compounds and automated assignment of $^{13}$C NMR spectra [23]. In the prediction process, the software searches for matches between the HOSE codes of the model and the database of chemical shifts [24, 25]. The software compares all sub-spectra of library fragments with the experimental $^{13}$C NMR spectrum of the analyzed compound. Besides the chemical shifts and multiplicity of the signals, the signal area (i.e., the number of carbon atoms) is needed for automatic assignment. Substructures that have corresponding library sub-spectra coinciding with the query sub-spectra within specified deviation limits are selected and ranked according to their size (i.e., to the number of carbon and skeleton atoms). The chemical structure is generated by superimposing the atoms common to different fragments.

A number of other software packages are available to predict $^{13}$C NMR spectra. The use of large $^{13}$C NMR spectral databases is the most popular approach; it utilizes assigned chemical structures. In an advanced approach, parameters such as solvent information can be used to refine the accuracy of the prediction. A typical application works with tables of experimental chemical shifts from experimental $^{13}$C NMR spectra. Each shift value is assigned to a specific structural fragment. The query structure is dissected into fragments that are compared with the fragments in the database. For each coincidence, the experimental chemical shift from the database is used to compose the final set of chemical shifts for the query structure. If a fragment from the query structure is not found in the internal database, the most similar fragment is used.

For the prediction of $^{13}$C NMR chemical shifts, an approach using neural networks or linear regression modeling has been tried which used topological, physicochemical, or geometric parameters as independent variables [26–28]. Predictions have been reported that are at least in the same range of accuracy as those obtained by the other methods. A trend toward obtaining better results with neural networks than with regression methods could be observed [26, 28].

## 10.2.4
## $^1$H NMR Spectra

### 10.2.4.1 Prediction of Chemical Shifts [29]

#### 10.2.4.1.1 *Ab-Initio and Semi-empirical Calculations*
When a molecule is submitted to a static magnetic field, the nuclear spin energy levels split. An oscillating magnetic field can then induce transitions between these energy levels and produce the NMR spectrum. For a nucleus in a molecule, the magnetic field is due to the applied magnetic field, but also to the magnetic field produced by the electrons and other nuclei. The NMR chemical shift of a nucleus results from the difference in energy between the nuclear spin states, which can be calculated by *ab-initio* quantum mechanical methods, typically by solving the Schrödinger equation with approximations [29–32]. The effects of the external magnetic field on the nucleus of interest are added into the equations as a "perturbation". It is then possible to calculate the chemical shift, which is related to the total molecular energy, the applied magnetic field, and the nuclear magnetic moment.

*Ab-initio* calculations are particularly useful for the prediction of chemical shifts of "unusual species". In this context "unusual species" means chemical entities that are not frequently found in the available large databases of chemical shifts, e.g., charged intermediates of reactions, radicals, and structures containing elements other than H, C, O, N, S, P, halogens, and a few common metals.

The Gaussian program [33] is one of the most popular tools for *ab-initio* calculation of NMR chemical shifts. NMR shielding tensors may be computed with Gaussian 98 using the Gauge-Invariant Atomic Orbital (GIAO) method and the Hartree–Fock, DFT, and MP2 models (see Section 7.4). NMR calculations may include effective core potentials (ECPs). At this high level of theory, the required computation times are considerable, and for most organic compounds the accuracy obtained is only comparable with faster empirical methods, as described in the next sections. Depending on the level of theory, a 100-atom molecule currently requires a CPU time in the order of hours.

$^1$H NMR calculations are based on a given molecular geometry, which means that an experimental 3D structure must be available, or it has to be previously calculated. Because a rigid structure is used for the calculations, different chemical shifts are generally obtained for $^1$H NMR-equivalent nuclei (e.g., the three protons of a methyl group), and so equivalent nuclei have to be specified if their chemical shifts are to be averaged at the end of the calculation.

Using semi-empirical methods, which are also based on approximate solutions of the Schrödinger equation but use parameterized equations, the computation times can be reduced by two orders of magnitude. HyperChem from Hypercube, Inc. [34], is an example of a software package that can calculate 3D geometries, chemical shifts, and coupling constants using semi-empirical approaches (Figure 10.2-2).

A good source of information regarding the scientific background of *ab-initio* and semi-empirical calculations are the manuals that accompany commercial software. Some of the documentation is available for evaluation on the Internet.

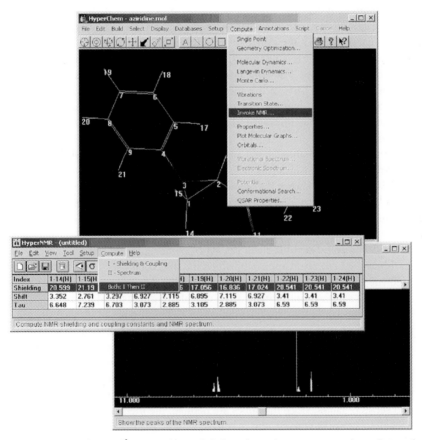

**Figure 10.2-2.** Calculation of $^1$H NMR chemical shifts and coupling constants and simulation of the spectrum with HyperChem 7.

10.2.4.1.2 **Database Approaches [35]**

A useful empirical method for the prediction of chemical shifts and coupling constants relies on the information contained in databases of structures with the corresponding NMR data. Large databases with hundred-thousands of chemical shifts are commercially available and are linked to predictive systems, which basically rely on database searching [35]. Protons are internally represented by their structural environments, usually their HOSE codes [9]. When a query structure is submitted, a search is performed to find the protons belonging to similar (overlapping) substructures. These are the protons with the same HOSE codes as the protons in the query molecule. The prediction of the chemical shift is calculated as the average chemical shift of the retrieved protons.

The similarity of the retrieved protons to those of the query structure, and the distribution of chemical shifts among protons with the same HOSE codes, can be used as measures of prediction reliability. When common substructures cannot be found for a given proton (within a predefined number of bond spheres) interpolations are applied to obtain a prediction; proprietary methods are often used in commercial programs.

The database approaches are heavily dependent on the size and quality of the database, particularly on the availability of entries that are related to the query structure. Such an approach is relatively fast; it is possible to predict the $^1$H NMR spectrum of a molecule with 50–100 atoms in a few seconds. The predicted values can be explained on the basis of the structures that were used for the predictions. Additionally, users can augment the database with their own structures and experimental data, allowing improved predictions for compounds bearing similarities to those added.

Commercial implementations of this general approach are ACD/I-Lab [36], SpecInfo (Chemical Concepts) [37], WINNMR (Bruker), and KnowItAll (Bio-Rad) [38]. Figure 10.2-3 shows the workspace generated by ACD/I-Lab after predicting a $^1$H NMR spectrum. ACD calculations are currently based on over 1 200 000 experimental chemical shifts and 320 000 experimental coupling constants [36].

10.2.4.1.3 **Increment-Based Methods [39, 40]**

In this second empirical approach, which has also been used for $^{13}$C NMR spectra, predictions are based on tabulated chemical shifts for classes of structures, and corrected with additive contributions from neighboring functional groups or substructures. Several tables have been compiled for different types of protons. Increment rules can be found in nearly any textbook on NMR spectroscopy.

In such tables, typical chemical shifts are assigned to standard structure fragments (e.g., protons in a benzene ring). Substituents in these blocks (e.g., substituents in *ortho*, *meta*, or *para* positions) are assumed to make independent additive contributions to the chemical shift. These additive contributions are listed in a second series of tables. Once the tables are defined, the method is easy to implement, does not require databases, and is extremely fast. Predictions for a molecule with 50 atoms can be made in less than a second. On the other hand, it requires that the parent structure and the substituents are tabulated, and it considers no interaction

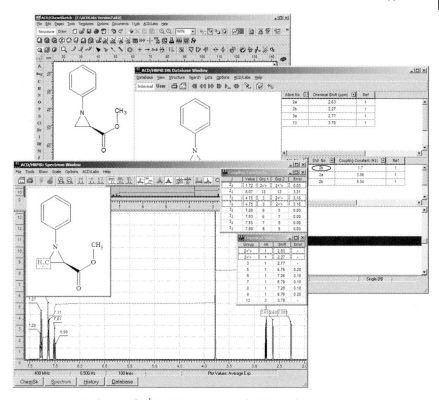

**Figure 10.2-3.** Prediction of a ¹H NMR spectrum with ACD/I-Lab.

between substituents. Usually, slightly less accurate predictions are obtained than with the methods based on large databases. It is estimated that 90 % of $CH_n$ chemical shifts can be predicted with an average error of 0.2–0.3 ppm [41]. Commercial software like ChemDraw [42] and Spectacle [43] integrate such an approach, and Upstream [41] has made a WWW tool freely available. A screen shot of the Upstream interface is shown in Figure 10.2-4.

### 10.2.4.1.4 Predictions by Neural Networks [44]

Neural networks can learn automatically from a data set of examples. In the case of ¹H NMR chemical shifts, neural networks have been trained to predict the chemical shift of protons on submission of a chemical structure. Two main issues play decisive roles: how a proton is represented, and which examples are in the data set.

A proton can be (numerically) represented by a series of topological and physicochemical descriptors, which account for the influence of the neighborhood on its chemical shift. Fast empirical procedures for the calculation of physicochemical descriptors are now easily accessible [45]. Geometric descriptors were added in the case of some rigid substructures, as well as for $\pi$-systems, to account for stereochemistry and 3D effects.

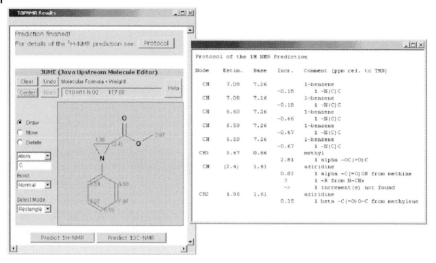

**Figure 10.2-4.** Prediction of $^1$H NMR chemical shifts with the Upstream WWW version.

A relatively small training set of 744 $^1$H NMR chemical shifts for protons from 120 molecular structures was collected from the literature. This set was designed to cover as many situations of protons in organic structures as possible. Only data from spectra obtained in CDCl$_3$ were considered. The collection was restricted to CH$_n$ protons and to compounds containing the elements C, H, N, O, S, F, Cl, Br, or I.

Counterpropagation neural networks (CPG NN) were then used to establish relationships between protons and their $^1$H NMR chemical shifts. A detailed description of this method is given in the "Tools" Section 10.2.4.2.

### 10.2.4.2 Tools: Prediction of $^1$H NMR Chemical Shifts

Fast and accurate predictions of $^1$H NMR chemical shifts of organic compounds are of great interest for automatic structure elucidation, for the analysis of combinatorial libraries, and, of course, for assisting experimental chemists in the structural characterization of small data sets of compounds.

In this section, a fast neural network approach [46] for the predication of $^1$H NMR chemical shifts is explained and exemplified.

#### 10.2.4.2.1 Outline of the Method

A combination of physicochemical, topological, and geometric information is used to encode the environment of a proton. The geometric information is based on (local) proton radial distribution function (RDF) descriptors and characterizes the 3D environment of the proton. Counterpropagation neural networks established the relationship between protons and their $^1$H NMR chemical shifts (for details of neural networks, see Section 9.5). Four different types of protons were treated separately according to their chemical environment: protons belonging to aromatic systems, to non-aromatic $\pi$-systems, to rigid aliphatic substructures, and to non-rigid aliphatic substructures. Each proton was represented by a fixed number of descriptors.

### 10.2.4.2.2 Representation of Protons

An extended set of physicochemical descriptors was used in this study, including, for example, partial atomic charge and effective polarizability of the protons, average of electronegativities of atoms two bonds away, or maximum $\pi$-atomic charge of atoms two bonds away.

Geometric descriptors were based on local RDF descriptors (see Eq. (16)) for the proton $j$.

$$g_H(r) = \sum_{i}^{N_{(4)}} q_i \cdot e^{-B(r-r_{ij})^2} \tag{16}$$

In Eq. (16) $i$ denotes an atom up to four non-rotatable bonds away from the proton and $N_{(4)}$ is the total number of those atoms. A bond is defined as non-rotatable if it belongs to a ring, to a $\pi$-system, or to an amide functional group; $q_i$ is the partial atomic charge of the atom $i$, and $r_{ij}$ is the 3D distance between the proton $j$ and the atom $i$. Figure 10.2-5 shows an example of a proton RDF descriptor.

In the plot of $g_H(r)$ against $r$, each 3D distance contributes to a peak, and the contribution is proportional to $q_i$. Values of $g_H(r)$ at fixed points are used as descriptors of the proton.

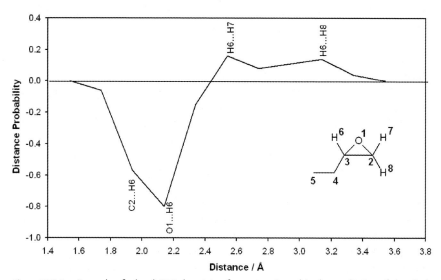

**Figure 10.2.5.** Example of a local RDF descriptor for proton 6 used in the prediction of chemical shifts ($B = 20$ Å$^{-2}$).

In order to represent further geometric features, modifications of Eq. (16) were introduced. The influence of the electronic current in double bonds was incorporated by Eq. (17).

$$g_D(r) = \sum_i^{D_{(7)}} \frac{1}{r_D^2} e^{-B(r-a_D)^2} \tag{17}$$

Here $i$ is now a double bond up to the seventh sphere ($D_{(7)}$) of non-rotatable bonds centered on the proton, $r_D$ is the distance of the proton from the center of the double bond, and $a_D$ is the angle, in radians, between the plane defined by the bond and the distance $r_D$ (Figure 10.2-6a).

Shielding and unshielding by single bonds were encoded using Eq. (18), where $i$ is a single bond up to the seventh sphere ($S_{(7)}$) of non-rotatable bonds centered on the proton, and $r_S$ and $a_S$ are distance and angle, respectively (Figure 10.2-6b).

$$g_S(r) = \sum_i^{S_{(7)}} \frac{1}{r_S^2} e^{-B(r - a_S)^2} \tag{18}$$

In order to account for axial and equatorial positions of protons bonded to cyclohexane-like rings, Eq. (19) was used, where $i$ is an atom three non-rotatable bonds (totally $N_{(3)}$ atoms) away from the proton and belonging to a six-membered ring, and $a_3$ is a dihedral angle in radians (Figure 10.2-6c).

$$g_3(r) = \sum_i^{N_{(3)}} e^{-B(r - a_3)^2} \tag{19}$$

The minimum and maximum bond angles centered on the atom adjacent to the proton were also used as geometric descriptors. For aromatic and non-rigid aliphatic protons, these were the only two geometric descriptors used. In addition to these, for non-aromatic $\pi$-protons the proton RDF descriptor $g_H(r)$ and the number of non-hydrogen atoms at *cis* and *trans* positions were used as geometric descriptors, as we were mostly interested in predicting the influence of *cis–trans* isomerism. For rigid aliphatic protons, all the geometric descriptors were calculated.

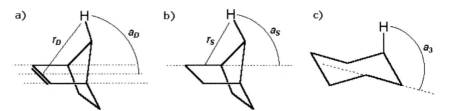

**Figure 10.2.6.** Special distance measures for the characterization of proton environments: a) distance $r_D$ and angle $a_D$ to double bonds; b) distance $r_S$ and angle $a_S$ to single bonds; c) dihedral angle $a_3$ to the third bond from the hydrogen atom.

Topological descriptors were based on the analysis of the connection table and atom properties. Some examples of topological descriptors are the number of carbon atoms in the second sphere centered on the proton, the number of oxygen atoms in the third sphere, and the number of atoms in the second sphere that belong to an aromatic system. A topological $g_i(r)$ function was also used similar to $g_H(r)$, where $i$ is any atom up to the fifth sphere centered on the proton, $a_i$ is the partial atomic charge, and $r_i$ is now the sum of bond lengths on the shortest possible path between the proton and atom $i$.

### 10.2.4.2.3 Selection of Descriptors by Genetic Algorithms

Models developed with selected subsets of descriptors, instead of all possible descriptors, can be more accurate and robust. In order to select adequate descriptors for each of the four classes of protons, genetic algorithms (GA) were used, and the results were compared with those obtained when all the descriptors were used. More information on the GA method is available in Section 9.7.

For the selection of descriptors, GA simulated evolution of a population. Each individual of the population represents a subset of descriptors and is defined by a chromosome of binary values. The chromosome has as many genes as there are possible descriptors (92 for the aromatic group, 119 for non-rigid aliphatic, 174 for rigid aliphatic, and 101 for non-aromatic $\pi$-protons). Each gene takes a value of 1 if the corresponding descriptor is included in the subset, and 0 otherwise.

A population size of 250 individuals was chosen and evolution was allowed over 50 generations. In each generation, half of the individuals (the fittest individuals) mate, and the other half die. Each of the surviving individuals mates with another (randomly chosen) surviving individual, and two new offspring are generated. The evaluation (scoring) of each chromosome is done by a CPG neural network that uses the subset of descriptors encoded in the chromosome for predicting chemical shifts. The score function of one chromosome (fitness function) is the root mean square of errors for the chemical shifts predicted for the cross-validation set. Chromosomes with lower root mean square errors are considered to be fitter than those with higher ones, and are selected for surviving and mating.

### 10.2.4.2.4 Predictions

After selection of descriptors/NN training, the best networks were applied to the prediction of 259 chemical shifts from 31 molecules (prediction set), which were not used for training. The mean absolute error obtained for the whole prediction set was 0.25 ppm, and for 90% of the cases the mean absolute error was 0.19 ppm. Some stereochemical effects could be correctly predicted. In terms of speed, the neural network method is very fast – the whole process to predict the $^1$H NMR shifts of 30 protons in a molecule with 56 atoms, starting from an MDL Molfile, took less than 2 s on a common workstation.

An example of the neural network prediction of $^1$H NMR chemical shifts for a natural product is illustrated in Figure 10.2-7 together with the calculations from other methods. This molecule was chosen as it had been discovered [47]

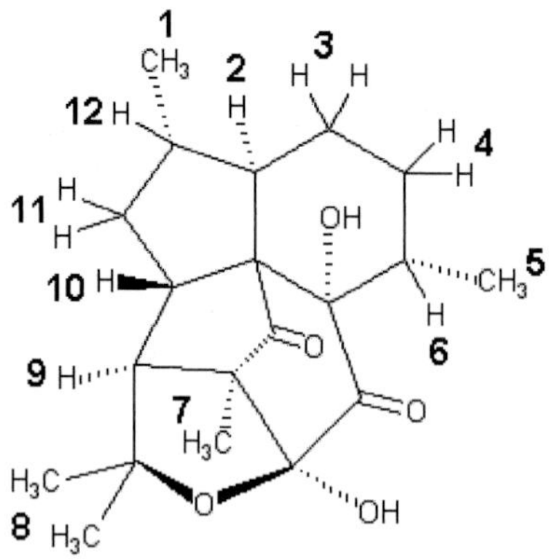

| Atom No. | Observed | NN | ACD I-Lab | Upstream |
|---|---|---|---|---|
| 1 | 0.99 | 0.97 | 0.978 | 1.05 |
| 2 | 2.03 | 2.39 | 1.853 | 1.62 |
| 3α | 1.81 | 2.05 | 2.239* | 1.39 |
| 3β | 1.53 | 1.97 | 1.366* | 1.39 |
| 4α | 1.26 | 1.65 | 1.782* | 1.39 |
| 4β | 2.04 | 1.97 | 1.857* | 1.39 |
| 5 | 0.62 | 0.97 | 1.068 | 1.05 |
| 6 | 2.12 | 2.39 | 2.051 | 1.90 |
| 7 | 1.42 | 1.27 | 1.353 | 1.25 |
| 8 | 1.40 | 1.27 | 1.566 | 1.25 |
| 9 | 1.99 | 2.01 | 2.252 | 2.01 |
| 10 | 3.17 | 2.39 | 2.138 | 1.98 |
| 11α | 0.65 | 1.45 | 1.605 | 1.42 |
| 11β | 1.94 | 1.97 | 2.008* | 1.42 |
| 12 | 1.68 | 2.27 | 2.524* | 1.66 |
| Mean absolute error | | 0.303 | 0.341 | 0.346 |

\* For the calculation of errors, it was assumed that the order of the chemical shifts of a pair of diastereotopic protons (α and β) was correctly predicted by ACD/I-Lab.

**Figure 10.2-7.** Predictions of $^1$H NMR chemical shifts for a complex natural product by neural networks, a database-centered method (ACD), and an increment-based method (Upstream).

shortly before the study and contained an unprecedented scaffold, and was therefore guaranteed to be new to all methods. This case did not pretend to be a comprehensive comparison of all the methods [46]. The strengths and weaknesses of each method can only be accurately established by a large-scale investigation using a large data set of new structures.

The performance of the neural network method is remarkable considering the relatively small data set on which it was based.

SPINUS–WEB

A hands-on experience with the method is possible via the SPINUS web service [48]. This service uses a client–server model. The user can draw a molecular structure within the web browser workspace (the client), and send it to a server where the predictions are computed by neural networks. The results are then sent back to the user in a few seconds and visualized with the same web browser.

Several operations and different types of technology are involved in the system:

• The molecular editor consists of a Java applet that is embedded in the HTML document. It encodes the drawing into a connection table in mol-format, which is sent to the web server.

- At the web server, a 3D model is generated by CORINA [49], and physicochemical parameters are calculated by PETRA [49].
- Using the results from PETRA and CORINA, the required descriptors are calculated for all the protons adjacent to C atoms.
- The descriptors are then submitted to previously trained (and saved) neural networks, which give predictions for the chemical shifts.
- The predictions are sent back to the client web browser and visualized using a combination of HTML, JavaScript [50], MDL Chime [51] scripting, and Java applets.

SPINUS integrates all these procedures into a user-friendly interface that can be accessed by any chemist – even one who is not familiar with chemoinformatics concepts.

## 10.2.5
### Infrared Spectra

#### 10.2.5.1 Overview
Since IR spectroscopy monitors the vibrations of atoms in a molecule in 3D space, information on the 3D arrangement of the atoms should somehow be contained in an IR spectrum. However, the relationships between the 3D structure and the IR spectrum are rather complex, so no general attempt has yet been successful in deriving the 3D structure of a molecule directly from the IR spectrum.

A series of monographs and correlation tables exist for the interpretation of vibrational spectra [52–55]. However, the relationship of frequency characteristics and structural features is rather complicated and the number of known correlations between IR spectra and structures is very large. In many cases, it is almost impossible to analyze a molecular structure without the aid of computational techniques. Existing approaches are mainly based on the interpretation of vibrational spectra by mathematical models, rule sets, and decision trees or fuzzy logic approaches.

Many expert systems designed to assist the chemist in structural problem-solving were based on the approach of characteristic frequencies. Gribov and Orville-Thomas [56] undertook a comprehensive consideration of conditions for the appearance of characteristic bands in an IR spectrum. Gribov and Elyashberg [57] suggested different mathematical techniques in which rules and decisions are expressed in an explicit form and Elyashberg [58] pointed out that in the discrete modeling of the structure–spectrum system symbolic logic [59] is a valuable tool in studying complicated objects of a discrete nature. In agreement with that, Zupan showed [60] that the relationship between the molecular structure and the corresponding IR spectrum could be represented conditionally by a finite discrete model. These relationships can be formulated as if–then rules in the knowledge base of an expert system. Systems based on those logical rules can be found in several reviews [61, 62] and publications [63–66].

Woodruff and co-workers introduced the expert system PAIRS [67], a program that is able to analyze IR spectra in the same manner as a spectroscopist would. Chalmers and co-workers [68] used an approach for automated interpretation of Fourier Transform Raman spectra of complex polymers. Andreev and Argirov developed the expert system EXPIRS [69] for the interpretation of IR spectra. EXPIRS provides a hierarchical organization of the characteristic groups that are recognized by peak detection in discrete frames. Penchev et al. [70] recently introduced a computer system that performs searches in spectral libraries and systematic analysis of mixture spectra. It is able to classify IR spectra with the aid of linear discriminant analysis, artificial neural networks, and the method of $k$-nearest neighbors.

### 10.2.5.2 Infrared Spectra Simulation

Gasteiger and co-authors [71] implemented the following approach for full spectra simulation. The previously mentioned 3D MoRSE descriptor (Section 8.4.3), derived from an equation used in electron diffraction studies, allows the representation of the 3D structure of a molecule by a fixed (constant) number of variables. By using a fast 3D structure generator they were able to study the correlation between any three-dimensional structure and IR spectra using ANN. Steinhauer et al. [72] used RDF codes as structure descriptors. Together with the IR spectrum, a counterpropagation (CPG) neural network was trained to establish the complex relationship between an RDF descriptor and an IR spectrum. After training, the simulation of an IR spectrum is performed using the RDF descriptor of the query compound as the information vector of the Kohonen layer, which determines the central neuron. On input of this query RDF descriptor, the central neuron is selected and the corresponding IR spectrum in the output layer is presented as the simulated spectrum. Selzer et al. [73] described an application of this spectrum simulation method that provides rapid access to arbitrary reference spectra. Kostka et al. described a combined application of spectrum prediction and reaction prediction expert systems [74]. The combination of the reaction prediction system EROS and IR spectrum simulation proved to be a powerful tool for computer-assisted substance identification. More details and the description of a web tool implementing this method can be found in the "Tools" Section 10.2.5.3.

### 10.2.5.3 Tools: TeleSpec – Online Service for the Simulation of Infrared Spectra [75]

The identification of chemical compounds by IR spectroscopy is usually done by comparing an experimental spectrum of the compound with a reference spectrum. However, the number of known chemical structures (ca. 17 000 000) greatly exceeds the number of IR spectra; the largest database of spectra, SpecInfo [76], contains 600 000 spectra.

The TeleSpec system, which was funded by the DFN [77], was developed with the aim to provide a method to relieve this difficulty by simulating an IR spectrum for a given input structure.

### 10.2.5.3.1 Approach

The correlation between a structure and its spectrum is rather complex. Therefore the application of a counterpropagation network, which is a powerful tool for modeling nonlinear correlations, was chosen. The data for training the network are stored in a database comprising 11 000 encoded structures and the corresponding IR spectra.

The structures in the database are encoded using the radial distribution function (RDF) as a descriptor (cf. Section 8.4.4).

This coding is performed in three steps (cf. Chapter 8): First the 3D coordinates of the atoms are calculated using the structure generator CORINA (COoRdINAtes). Subsequently the program PETRA (Parameter Estimation for the Treatment of Reactivity Applications) is applied for calculating physicochemical properties such as charge distribution and polarizability. The 3D information and the physicochemical atomic properties are then used to code the molecule.

### 10.2.5.3.2 Procedure

When a structure is input for spectra simulation this structure is also coded as an RDF descriptor, which allows an easy comparison with the structures in the database. Those 50 structures which are most similar to the input structure are then selected together with their spectra.

These pairs of encoded structures and their IR spectra are used to train a counterpropagation network (see Section 9.5.5). The two-layer network processes the structural information in its upper part and the spectral information in its lower part. Thus the network learns the correlation between the structures and their IR spectra. This procedure is shown in Figure 10.2-8.

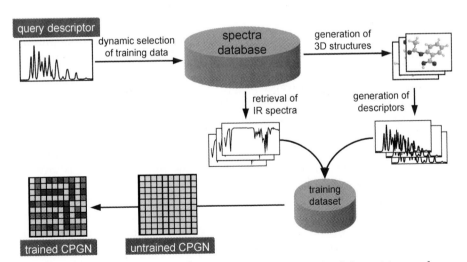

**Figure 10.2-8.** Procedure for spectra simulation: the query structure is coded, a training set of structure–spectra pairs is selected from the database, and the counterpropagation network is trained.

## Query

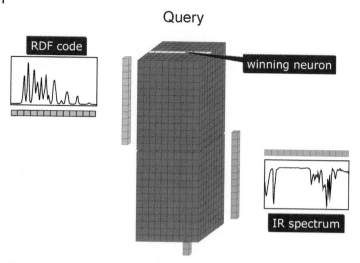

**Figure 10.2-9.** Application of a counterpropagation neural network as a look-up table for IR spectra simulation. The winning neuron which contains the RDF code in the upper layer of the network points to the simulated IR spectrum in the lower layer.

The trained counterpropagation network is then able to predict the spectrum for a new structure when operating as a look-up table (see Figure 10.2-9): the encoded query or the input structure is input into the trained network and the winning neuron is determined by considering just the upper part of the network. The neuron points to the corresponding neuron in the lower part of the network, which then provides the simulated IR spectrum.

### 10.2.5.3.3 Usage

The TeleSpec system offers two ways of reading in a query structure. The structure can be input either directly as a SMILES string (cf. Section 2.3.3) or via a molecule editor which converts the graphical input into the SMILES string. Figure 10.2-10 gives the input form of TeleSpec.

When the structure is submitted its 3D coordinates are calculated and the structure is shown at the left-hand side in the form of a 2D structure as well as a rotatable 3D structure (see Figure 10.2-11). The simulation can then be started: the input structure is coded, the training data are selected, and the network training is launched. After approximately 30 seconds the simulation result is given as shown in Figure 10.2-11.

The right-hand side of the form gives the network map in the upper part and to its right the simulated IR spectrum is plotted, which can be downloaded as a JCAMP file (cf. Section 2.4.5, Section 4.2.4.2). By clicking on the neurons in the map one obtains the RDF code and the spectrum of the corresponding structure in the lower part of the form and compared with those of the winning neuron.

Further information about the TeleSpec system is given in Ref. [73].

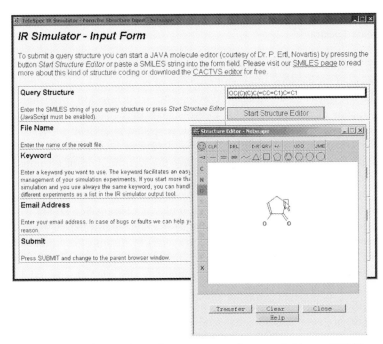

**Figure 10.2-10.** Input form of IR Simulator: the structures can be entered either as SMILES strings or via a molecule editor.

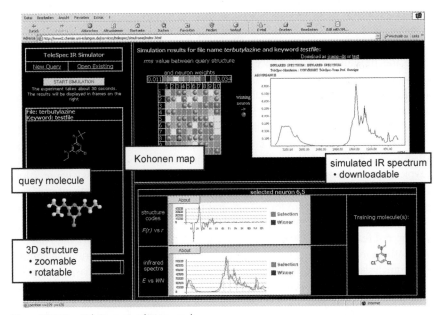

**Figure 10.2-11.** TeleSpec simulation result.

## 10.2.6
### Mass Spectra

In contrast to IR and NMR spectroscopy, the principle of mass spectrometry (MS) is based on decomposition and reactions of organic molecules on their way from the ion source to the detector. Consequently, structure–MS correlation is basically a matter of relating reactions to the signals in a mass spectrum. The chemical structure information contained in mass spectra is difficult to extract because of the complicated relationships between MS data and chemical structures. The aim of spectra evaluation can be either the identification of a compound or the interpretation of spectral data in order to elucidate the chemical structure [78–80].

Identification of mass spectra is typically performed by searching for similarities of the measured spectrum to spectra stored in a library. Several MS databases and corresponding software products are used routinely [81, 82]. A more challenging problem is the interpretation of mass spectra by means of computational chemistry, and different strategies have been applied to the problem of substructure recognition. The use of correlation tables containing characteristic spectral data together with corresponding substructures has been successfully applied to spectroscopic methods. However, because of the complexity of chemical reactions that may occur in mass spectrometers and because of the lack of generally applicable rules, this approach has been less successful with MS.

Other methods consist of algorithms based on multivariate classification techniques or neural networks; they are constructed for automatic recognition of structural properties from spectral data, or for simulation of spectra from structural properties [83]. Multivariate data analysis for spectrum interpretation is based on the characterization of spectra by a set of spectral features. A spectrum can be considered as a point in a multidimensional space with the coordinates defined by spectral features. Exploratory data analysis and cluster analysis are used to investigate the multidimensional space and to evaluate rules to distinguish structure classes.

Multivariate data analysis usually starts with generating a set of spectra and the corresponding chemical structures as a result of a spectrum similarity search in a spectrum database. The peak data are transformed into a set of spectral features and the chemical structures are encoded into molecular descriptors [80]. A spectral feature is a property that can be automatically computed from a mass spectrum. Typical spectral features are the peak intensity at a particular mass/charge value, or logarithmic intensity ratios. The goal of transformation of peak data into spectral features is to obtain descriptors of spectral properties that are more suitable than the original peak list data.

Spectral features and their corresponding molecular descriptors are then applied to mathematical techniques of multivariate data analysis, such as principal component analysis (PCA) for exploratory data analysis or multivariate classification for the development of spectral classifiers [84–87]. Principal component analysis results in a scatter plot that exhibits spectra–structure relationships by clustering similarities in spectral and/or structural features [88, 89].

Knowledge-based methods try to implement spectroscopic knowledge about spectra–structure relationships into computer programs. The most important knowledge-based approach for structure elucidation of organic compounds is based on the DENDRAL (DENDRitic ALgorithm) Project [90–92]. The DENDRAL Project was initiated in 1965 by Edward Feigenbaum, Joshua Lederberg, and Bruce Buchanan. It began as an effort to make the concept of scientific reasoning and the formalization of scientific knowledge available with the computer. DENDRAL used a set of knowledge- or rule-based methods to deduce the molecular structure of organic chemical compounds from chemical analysis and mass spectrometry data. The DENDRAL system for structure elucidation from mass spectral data was the first expert system (ES) developed in chemistry. It uses isomer generator software capable of generating exhaustive sets of isomers from a given molecular formula. Substructures that have to be present in the query structure are collected in the so-called "good list" while forbidden substructures are put into the "bad list". Structural information can be derived from low-resolution mass spectrometry data, whereas high-resolution data can be used to determine the molecular formula.

DENDRAL proved to be fundamentally important in demonstrating how rule-based reasoning could be developed into powerful knowledge-engineering tools [93].

Gasteiger and co-workers followed an approach based on models of the reactions taking place in the spectrometer [94, 95]. Automatic knowledge extraction is performed from a database of spectra and the corresponding structures, and rules are saved. The rules concern possible elementary reactions, and models relating the probability of these reactions with physicochemical parameters calculated for the structures. The knowledge can then be applied to chemical structures in order to predict a) the reactions that occur in the spectrometer, b) the resulting products, and c) the corresponding peaks in the mass spectrum.

10.2.7
**Computer-Assisted Structure Elucidation**

The elucidation of a structure by the use of rule-based systems needs a technique for assembling a complete structure from substructure fragments that have been predicted. Several techniques and computer programs have been proposed under the generic name Computer-Assisted Structure Elucidation (CASE). Lindsay et al. [90] introduced the first program that was able to enumerate all the acyclic structures from a molecular formula. This program was the precursor to the first expert systems for structure elucidation ever published: CONGEN [96] and GENOA [97]. These programs could handle any structure and enumerate the isomers of a molecular formula, and were able to generate structures with more restrictive constraints, e.g., isomers with specified molecular fragments. However, both GENOA and CONGEN use more heuristic than systematic algorithms. Several CASE programs based on a more systematic structure generation technique are the structure generators CHEMICS [98], ASSEMBLE [99], and COMBINE [24].

The problem of overlapping fragments was studied by Dubois and co-workers. They developed the program DARC-EPIOS [100], which could retrieve structural formulas from overlapping $^{13}$C NMR data. Similar techniques have also been applied with the COMBINE program, while GENOA uses a more general technique based on the determination of all possible combinations of non-overlapping molecular fragments.

All the CASE programs described above generate chemical structures by assembling atoms and/or molecular fragments. Another strategy is based on structure generation by removing bonds from a hyperstructure that initially contains all the possible bonds between all the required atoms and molecular fragments (structure reduction). Programs based on the concept of structure generation by reduction are COCOA [101] and GEN [102].

Meiler et al. [103] introduced GENIUS, which implements genetic algorithms to approach the problem of finding structures consistent with an experimental $^{13}$C NMR spectrum and a molecular formula. A population of structures with the given formula is randomly generated. Then a neural network predicts a spectrum for each structure. Each structure is scored according to the similarity between the predicted spectrum and the experimental data. The fittest structures in a population survive, mutate, and mate, while the less fit structures are discarded (die). The procedure is repeated in an iterative way to optimize the constitution of the molecules, until it produces the experimental spectra with a deviation as low as possible. This method avoids the exhaustive generation of all possible isomers (making it much faster) and at the same time it was able to find the correct structures for a set of simple examples.

In recent decades, much attention has been paid to the application of artificial neural networks as a tool for spectral interpretation (see, e.g., Refs. [104, 105]). The ANN approach applied to vibrational spectra allows the determination of adequate functional groups that can exist in the sample, as well as the complete interpretation of spectra. Elyashberg [106] reported an overall prediction accuracy using ANN of about 80 % that was achieved for general-purpose approaches. Klawun and Wilkins managed to increase this value to about 95 % [107].

Neural networks have been applied to IR spectrum interpreting systems in many variations and applications. Anand [108] introduced a neural network approach to analyze the presence of amino acids in protein molecules with a reliability of nearly 90 %. Robb and Munk [109] used a linear neural network model for interpreting IR spectra for routine analysis purposes, with a similar performance. Ehrentreich et al. [110] used a counterpropagation network based on a strategy of Novic and Zupan [111] to model the correlation of structures and IR spectra. Penchev and co-workers [112] compared three types of spectral features derived from IR peak tables for their ability to be used in automatic classification of IR spectra.

In many cases, structure elucidation with artificial neural networks is limited to backpropagation networks [113] and, is therefore performed in a supervised manner. In a recall test with a separate data set, the quality of the training can then be evaluated. Novic and Zupan doubted the benefits of backpropagation networks for IR spectroscopy and introduced the use of Kohonen and counterpropagation networks for the analysis of spectra–structure correlations.

### Essentials

- A proper representation of the molecular structure is crucial for the prediction of spectra. Fragment-based methods, topological descriptors, physicochemical descriptors, and 3D descriptors have been used for this endeavor.
- NMR spectra have been predicted using quantum chemistry calculations, database searches, additive methods, regressions, and neural networks.
- Several methods have been developed for establishing correlations between IR vibrational bands and substructure fragments. Counterpropagation neural networks were used to make predictions of the full spectra from RDF codes of the molecules.
- Correlations between structure and mass spectra were established on the basis of multivariate analysis of the spectra, database searching, or the development of knowledge-based systems, some including explicit management of chemical reactions.
- Robust implementations can currently propose correct structures from spectroscopic data, especially when the molecular formula and $^{13}$C NMR spectrum are available, or from 2D NMR spectra.

### Selected Reading

- R. K. Lindsay, B. G. Buchanan, E. A. Feigenbaum, J. Lederberg, *Applications of Artificial Intelligence for Organic Chemistry – The DENDRAL Project*, McGraw-Hill, New York, **1980**.
- T. L. Clerc, in *Computer-Enhanced Analytical Spectroscopy*, H. L. C. Meuzelaar, T. L. Isenhour (Eds.), Plenum Press, New York, **1987**, pp. 145–162. Automated spectra interpretation and library search systems.
- M. S. Madison, K.-P. Schulz, A. A. Korytko, M. E. Munk, SESAMI: an integrated desktop structure elucidation tool, *Internet Journal of Chemistry*, **1998**, 1, 34.

### Interesting Web Sites

- *http://www.acdlabs.com/ilab*
- *http://www.bio-rad.com*
- *http://www.chemicalconcepts.com/p111.htm*
- *http://www2.chemie.uni-erlangen.de/services/telespec/index.html*
- *http://www2.chemie.uni-erlangen.de/services/spinus*

# References

[1] R. Todeschini, V. Consonni, *Handbook of Molecular Descriptors*, Wiley-VCH, Weinheim, **2000**.

[2] M. E. Munk, M. S. Madison, E. W. Robb, *Mikrochim. Acta [Wien] II* **1991**, 505–514.

[3] H. Huixiao, X. Xinquan, *J. Chem. Inf. Comput. Sci.* **1995**, *35(6)*, 979–1000.

[4] U.-M. Weigel, R. Herges, *J. Chem. Inf. Comput. Sci.* **1992**, *32*, 723–731.

[5] U.-M. Weigel, R. Herges, *Anal. Chim. Acta* **1996**, *331*, 63–74.

[6] C. Affolter, K. Baumann, J. T. Clerc, H. Schriber, E. Pretsch, *Microchim. Acta* **1997**, *14*, 143–147.

[7] J. E. Dubois, G. Mathieu, P. Peguet, A. Panaye, J. P. Doucet, *J. Chem. Inf. Comput. Sci.* **1990**, *30*, 290–302.

[8] J. T. Clerc, A. L. Terkovics, *Anal. Chim. Acta* **1990**, *235*, 93–102.

[9] W. Bremser, *Anal. Chim. Acta* **1978**, *103*, 355–365.

[10] J. Sadowski, J. Gasteiger, *Chem. Rev.* **1993**, *93*, 2567–2573.

[11] E. Pretsch, W. Simon, J. Seibl, *Tables of Spectral Data for Structure Determination of Organic Compounds*, 2nd Edn., Springer-Verlag, Berlin, **1989**.

[12] J. T. Clerc, H. Sommerauer, *Anal. Chim. Acta* **1977**, *95*, 33.

[13] G. W. Small, P. C. Jurs, *Anal. Chem.* **1984**, *56*, 1314.

[14] R. C. Schweitzer, G. W. Small, *J. Chem. Inf. Comput. Sci.* **1996**, *36*, 310.

[15] B. E. Mitchell, P. C. Jurs, *J. Chem. Inf. Comput. Sci.* **1996**, *36*, 58.

[16] G. M. J. West, *J. Chem. Inf. Comput. Sci.* **1993**, *33*, 577.

[17] V. Kvasnicka, S. Skelenak, J. Pospichal, *J. Chem. Inf. Comput. Sci.* **1992**, *32*, 742.

[18] L. S. Anker, P. C. Jurs, *Anal. Chem.* **1992**, *64*, 217.

[19] J. E. Dubois, M. Carabedian, I. Dagane, *Anal. Chim. Acta* **1984**, *158*, 217.

[20] A. Panaye, J.-P. Doucet, B. T. Fan, *J. Chem. Inf. Comput. Sci.* **1993**, *33*, 258.

[21] M. E. Munk, R. J. Lind, M. E. Clay, *Anal. Chim. Acta* **1986**, *184*, 1.

[22] W. Robien, *Mikrochim. Acta* **1986**, *2*, 271.

[23] L. Chen, W. Robien, *Chemom. Intell. Lab. Syst.* **1993**, *19*, 217.

[24] H. Kalchhauser, W. Robien, *J. Chem. Inf. Comput. Sci.* **1985**, *25*,103–8.

[25] V. Schütz, V. Purtuc, S. Felsinger, W. Robien, *Fresenius J. Anal. Chem.* **1997**, *359*, 33–41.

[26] D. L. Clouser, P. C. Jurs, *J. Chem. Inf. Comput. Sci.* **1996**, *36*, 168–172.

[27] O. Ivanciuc, J. P. Rabine, D. Cabrol-Bass, A. Panaye, J. P. Doucet, *J. Chem. Inf. Comput. Sci.* **1997**, *37*, 587–598.

[28] J. Meiler, R. Meusinger, M. Will, *J. Chem. Inf. Comput. Sci.* **2000**, *40*, 1169–1176 and references cited therein.

[29] A. Williams, Current Opinion, in *Drug Discovery & Development* **2000**, June issue (available from ACD web site at *http://www.acdlabs.com/publish/publ_pres.html*).

[30] R. Ditchfield, *Mol. Phys.* **1974**, *27*, 789–811.

[31] K. Wolinski, J. F. Hinton, P. Pulay, *J. Am. Chem. Soc.* **1990**, *112*, 8251–8260.

[32] For examples see: a) Z. Meng, W. R. Carper, *J. Mol. Struct.* **2002**, *588*, 45–53; b) J. Czernek, V. Sklenár, *J. Phys. Chem. A* **1999**, *103*, 4089–4093; c) M. Barfield, P. Fagerness, *J. Am. Chem. Soc.* **1997**, *119*, 8699–8711.

[33] Gaussian, Inc., *http://www.gaussian.com*

[34] Hypercube, Inc., *http://www.hyper.com/*

[35] For a description of the method in the context of $^{13}$C NMR see ref. [24] and W. Robien, NMR data correlation with chemical structure, in *The Encyclopedia of Computational Chemistry*, P. v. R. Schleyer, N. L. Allinger, T. Clark, J. Gasteiger, P. A. Kollman, H. F. Schaefer III, P. R. Schreiner (Eds.), John Wiley & Sons, Chichester, **1998**, pp. 1845–1857.

[36] ACD/ILab – Interactive Laboratory, *http://www.acdlabs.com/ilab*

[37] Chemical Concepts (Wiley-VCH), *http://www.chemicalconcepts.com*

[38] Bio-Rad Laboratories, Inc., *http://www.bio-rad.com*

[39] M. Tusar, L. Tusar, S. Bohanec, J. Zupan, *J. Chem. Inf. Comput. Sci.* **1992**, *32*, 299–303.

[40] a) R. B. Schaller E. Pretsch, *Anal. Chim. Acta* **1994**, *290*, 295–302; b) A. Fürst, E. Pretsch, *Anal. Chim. Acta* **1995**, *312*, 95–105 and references cited therein.

[41] Upstream Solutions GmbH, *http://www.upstream.ch*

[42] CambridgeSoft Corporation, *http://www.cambridgesoft.com*

[43] Creon.Lab.Control AG, *http://www.creonlabcontrol.com*

[44] J. Aires-de-Sousa, M. Hemmer, J. Gasteiger, *Anal. Chem.* **2002**, *74* (1), 80–90.

[45] a) J. Gasteiger, Empirical methods for the calculation of physicochemical data of organic compounds, in *Physical Property Prediction in Organic Chemistry*, C. Jochum, M. G. Hicks, J. Sunkel (Eds.), Springer-Verlag, Heidelberg, **1988**, pp. 119–138; b) PETRA software, Molecular Networks GmbH, *http://www.mol-net.de*

[46] J. Aires-de-Sousa, M. C. Hemmer, J. Gasteiger, *Anal. Chem.* **2002**, *74* (1), 80–90.

[47] A. D. Rodriguez, C. Ramirez, I. I. Rodriguez, C. L. Barnes, *J. Org. Chem.* **2000**, *65*, 1390–1398.

[48] SPINUS Web Service, *http://www2.chemie.uni-erlangen.de/services/spinus/*

[49] Molecular Networks GmbH, *http://www.mol-net.de*

[50] Netscape Communications Corp., *http://www.netscape.com*

[51] MDL Information Systems, Inc., *http://www.mdl.com*

[52] L. J. Bellamy, *The Infrared Spectra of Complex Molecules*, Wiley, Chichester, **1975**.

[53] D. Dolphin, A. Wick, *Tabulation of Infrared Spectral Data*, Wiley, New York, **1977**.

[54] E. Pretsch, T. Clerc, J. Seibl, W. Simon, *Tables of Spectral Data for Structure Determination of Organic Compounds*, Springer-Verlag, Berlin, **1989**.

[55] D. Lin-Vien, N. B. Colthup, W. G. Fately, J. G. Grasselli, *Infrared and Raman Characteristic Frequencies of Organic Molecules*, Academic Press, New York, **1991**.

[56] L. A. Gribov, W. J. Orville-Thomas, *Theory and Methods of Calculation of Molecular Spectra*, Wiley, Chichester, **1988**.

[57] L. A. Gribov, M. E. Elyashberg, *Crit. Rev. Anal. Chem.* **1979**, *8*, 111–220.

[58] M. E. Elyashberg, Infrared spectra interpretation by the characteristic frequency approach, in *The Encyclopedia of Computational Chemistry*, P. v. R. Schleyer, N. L. Allinger, T. Clark, J. Gasteiger, P. A. Kollman, H. F. Schaefer III, P. R. Schreiner (Eds.), John Wiley and Sons, Chichester, **1998**, pp. 1299–1306.

[59] R. Grund, A. Kerber, R. Laue, *MATCH* **1992**, *27*, 87–131.

[60] J. Zupan, *Algorithms for Chemists*, Wiley, New York, **1989**.

[61] H. J. Luinge, *Vib. Spectrosc.* **1990**, *1*, 3–18.

[62] W. Warr, *Anal. Chem.* **1993**, *65*, 1087A–1095A.

[63] M. E. Elyashberg, L. A. Gribov, V. V. Serov, *Molecular Spectral Analysis and Computers*, Nauka, Moskau, **1980** (in Russian).

**[64]** K. Funatsu, Y. Susuta, S. Sasaki, *Anal. Chim. Acta* **1989**, *220*, 155–169.

**[65]** B. Wythoff, H. Q. Xiao, S. P. Levine, S. A. Tomellini, *J. Chem. Inf. Comput. Sci.* **1991**, *31*, 392–399.

**[66]** G. N. Andreev, O. K. Argirov, *J. Mol. Struct.* **1995**, *347*, 439–448.

**[67]** H. B. Woodruff, G. M. Smith, *Anal. Chem.* **1980**, *52*, 2321–2327.

**[68]** M. Claybourn, H. J. Luinge, J. M. Chalmers, *J. Raman Spectrosc.* **1994**, *25*, 115–122.

**[69]** G. N. Andreev, O. K. Argirov, P. N. Penchev, *Anal. Chim. Acta* **1993**, *284*, 131–136.

**[70]** P. N. Penchev, T. Nikolay, N. T. Kotchev, G. N. Andreev, *Traveaux Scientifiques d'Universite de Plovdiv* **2000**, *29(5)*, 21–26.

**[71]** J. Gasteiger, J. Sadowski, J. Schuur, P. Selzer, L. Steinhauer, V. Steinhauer, *J. Chem. Inf. Comput. Sci.* **1996**, *36*, 1030–1037.

**[72]** U. Steinhauer, V. Steinhauer, J. Gasteiger, Obtaining the 3D structure from infrared spectra of organic compounds using neural networks, in *Software-Entwicklung in der Chemie 11*, G. Fels, V. Schubert (Eds.), Gesellschaft Deutscher Chemiker, Frankfurt/Main, **1997**.

**[73]** P. Selzer, R. Salzer, H. Thomas, J. Gasteiger, *Chem. Eur. J.* **2000**, *6*, 920–927.

**[74]** T. Kostka, P. Selzer, J. Gasteiger, Computer-assisted prediction of the degradation products and infrared spectra of *s*-triazine herbicides, in *Software-Entwicklung in der Chemie 11*, G. Fels, V. Schubert (Eds.), Gesellschaft Deutscher Chemiker, Frankfurt/Main, **1997**, pp. 227–233.

**[75]** TeleSpec Website: (*http://www2. chemie.uni-erlangen.de/services/telespec/ index.html*)

**[76]** SpecInfo on the Internet: *http:// www.chemicalconcepts.com/p111.htm*

**[77]** Deutsches Forschungsnetz: *http://www.dfn.de/*

**[78]** T. L. Clerc, *Computer-Enhanced Analytical Spectroscopy*, Plenum Press, New York, **1987**, pp. 145–162. Automated spectra interpretation and library search systems.

**[79]** F. W. McLafferty, S. Y. Loh, D. B. Stauffer, *Computer-Enhanced Analytical Spectroscopy*, H. L. C. Meuzelaar (Ed.), Plenum Press, New York, **1990**, pp. 163–181. Computer identification of mass spectra.

**[80]** K. Varmuza, *Encyclopedia of Spectroscopy and Spectrometry*, J.C. Lindon, G.E. Tranter, J.L. Holmes (Eds.), Academic Press, London, **2000**, pp. 232–243. Chemical structure information from mass spectrometry.

**[81]** F. W. McLafferty, R. H. Hertel, *Org. Mass Spectrom.* **1994**, *8*, 690–702. Probability-based matching of mass spectra.

**[82]** S. E. Stein, D. R. Scott, *J. Am. Soc. Mass Spectrom.* **1994**, 5, 856–866. Optimization and testing of mass spectral library search algorithms for compound identification.

**[83]** M. Jalali-Heravi, M. H. Fatemi, *Anal. Chim. Acta* **2000**, *415*, 95–103.

**[84]** M. J. Adams, *Chemometrics in Analytical Spectroscopy*, The Royal Society of Chemistry, Cambridge, **1995**.

**[85]** K. R. Beebe, R. J. Pell, M. B. Seasholtz, *Chemometrics: A Practical Guide*, John Wiley & Sons, New York, **1998**.

**[86]** D. L. Massart, B. G. M. Vandeginste, L. C. M. Buydens, S. De Jong, J. Smeyers-Verbeke, *Handbook of Chemometrics and Qualimetrics: Part A*, Elsevier, Amsterdam, **1997**.

**[87]** B. G. M. Vandeginste, D. L. Massart, L. C. M. Buydens, S. De Jong, J. Smeyers-Verbeke, *Handbook of Chemometrics and Qualimetrics: Part B*, Elsevier, Amsterdam, **1998**.

**[88]** K. Varmuza, Chemometrics: multivariate view on chemical problems, in *The Encyclopedia of Computational Chemistry*, P. v. R. Schleyer, N. L. Allinger, T. Clark, J. Gasteiger, P. A. Kollman, I. H. F. Schaefer, P. R. Schreiner (Eds.), Wiley, Chichester, **1998**, pp. 346–366.

**[89]** J. Zupan, J. Gasteiger, *Neural Networks in Chemistry and Drug Design*, 2nd Edn., Wiley-VCH, Weinheim, **1999**.

**[90]** R. K. Lindsay, B. G. Buchanan, E. A. Feigenbaum, J. Lederberg, *Applications of Artificial Intelligence for Organic*

*Chemistry – The DENDRAL Project*, McGraw-Hill, New York, **1980**.

[91] B. G. Buchanan, E. A. Feigenbaum, *Artificial Intelligence* **1978**, *11*, 5–24.

[92] J. Lederberg, How DENDRAL was conceived and born, in *ACM Symposium on the History of Medical Informatics*, National Library of Medicine, **1987**. Later published in *A History of Medical Informatics*, B. I. Blum, K. Duncan (Eds.), Association for Computing, Machinery Press, New York **1990**, 14–44.

[93] G. F. Luger, W. A. Stubblefield, *Artificial Intelligence and the Design of Expert Systems*, Benjamin/Cummings Publishing, Redwood City, CA, **1989**.

[94] J. Gasteiger, W. Hanebeck, K.-P. Schultz, *J. Chem. Inf. Comput. Sci.* **1992**, *32*, 264–271.

[95] J. Gasteiger, W. Hanebeck, K.-P. Schultz, S. Bauerschmidt, R. Höllering, Automatic analysis and simulation of mass spectra, in *Computer-Enhanced Analytical Spectroscopy, Vol. 4*, C. L. Wilkins (Ed.), Plenum Press, New York, **1993**.

[96] R. E. Carhart, D. H. Smith, H. Brown, C. Djerassi, *J. Am. Chem. Soc.* **1975**, *97*, 5755–5763.

[97] R. E. Carhart, D. H. Smith, N. A. Gray, J. B. Nourse, C. Djerassi, *J. Org. Chem.* **1981**, *46*, 1708–1718.

[98] K. Funatsu, N. Miyabayaski, S. Sasaki, *J. Chem. Inf. Comp. Sci.* **1988**, *28*, 18–23.

[99] C. A. Shelley, T. R. Hays, M. E. Munk, R. V. Roman, *Anal. Chim. Acta* **1978**, *103*, 121–132.

[100] M. Carabedian, I. Dagane, J. E. Dubois, *Anal. Chem.* **1988**, *60*, 2186–2192.

[101] I. P. Bangov, *J. Chem. Inf. Comput. Sci.* **1994**, *34*, 277–284.

[102] M. L. Contreras, R. Rozas, R. Valdivias, *J. Chem. Inf. Comput. Sci.* **1994**, *34*, 610–619.

[103] J. Meiler, M. Will, *J. Am. Chem. Soc.* **2002**, *124*, 1868–1870.

[104] M. E. Munk, M. S. Madison, E. W. Robb, *Chem. Inf. Comput. Sci.* **1996**, *35*, 231–238.

[105] H. J. Luinge, J. H. van der Maas, T. Visser, *Chemom. Intell. Lab. Syst.* **1995**, *28*, 129–138.

[106] M. E. Elyashberg, *Zh. Anal. Khim.* **1992**, *47*, 698–709 (translated into English).

[107] C. Klawun, C. L. Wilkins, *J. Chem. Inf. Comput. Sci.* **1996**, *36*, 69–81.

[108] R. Anand, K. Mehrotra, C. K. Mohan, S. Ranka, Analyzing images containing multiple sparse patterns with neural networks, in *Proceedings of IJCAI–91* **1991**.

[109] E. W. Robb, M. E. Munk, *Mikrochim. Acta [Wien]* I, **1990**, 131–155.

[110] F. Ehrentreich, M. Novic, S. Bohanec, J. Zupan, Bewertung von IR-Spektrum-Struktur-Korrelationen mit Counterpropagation-Netzen, in *Software-Entwicklung in der Chemie 10*, J. Gasteiger (Ed.), Gesellschaft Deutscher Chemiker, Frankfurt/Main, **1996**.

[111] M. Novic, J. Zupan *J. Chem. Inf. Comput. Sci.* **1995**, *35(3)*, 454–466.

[112] P. N. Penchev, G. N. Andreev, K. Varmuza, *Anal. Chim. Acta* **1999**, *388*, 145–159.

[113] D. Ricard, C. Cachet, D. Cabrol-Bass, T. P. Forrest, *J. Chem. Inf. Comput. Sci.* **1993**, *33*, 202–210.

**10.3**
**Chemical Reactions and Synthesis Design**

**Learning Objectives**

- To be able to define reaction planning, reaction prediction, and synthesis design
- To know how to acquire knowledge from reaction databases
- To understand reaction simulation systems
- To become familiar with a knowledge-based reaction prediction system
- To appreciate the different levels in the evaluation of chemical reactions
- To know how reaction sequences are modeled
- To understand kinetic modeling of chemical reactions
- To become familiar with biochemical pathways
- To recognize the different levels of representation of biochemical reactions
- To understand metabolic reaction networks
- To know the principles of retrosynthetic analysis
- To understand the disconnection approach
- To become familiar with synthesis design systems
- Developing a suitable synthesis strategy for a target compound by searching for synthesis precursors, starting materials and synthesis reactions

10.3.1
**The Prediction of Chemical Reactions**

*Johann Gasteiger*

10.3.1.1  **Introduction**
In dealing with chemical reactions, chemists are faced with a set of different problems (see Figure 10.3-1):

1. I want to transform a given starting material, **A**, into a desired product, **P**; how can I do this (Figure 10.3-1a)?
   This is a question of reaction planning. To answer such a problem reagents and reaction conditions have to be found that allow one to perform the desired transformation. Such a question is best answered by a query into a reaction database (see Section 5.12).
2. I have a set of starting materials **A** and **B**; how will they react, and which product will they give (Figure 10.3-1b)?
   This is a question of reaction prediction. In fact, this is a deterministic system. If we knew the rules of chemistry completely, and understood chemical reactivity fully, we should be able to answer this question and to predict the outcome of a reaction. Thus, we might use quantum mechanical calculations for exploring the structure and energetics of various transition states in order to find out which reaction pathway is followed. This requires calculations of quite a high degree of sophistication. In addition, modeling the influence of solvents on

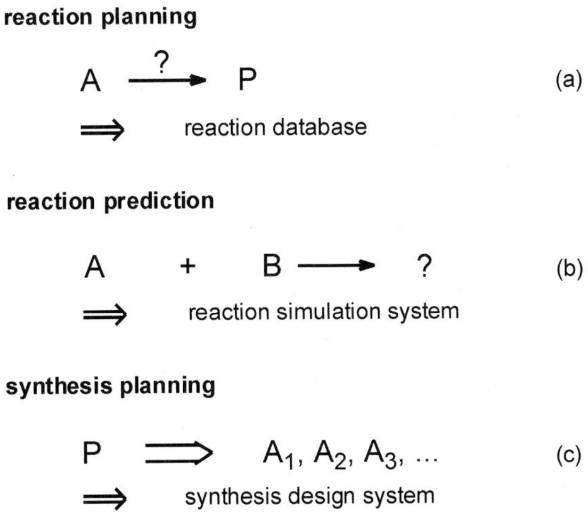

**reaction planning**

$$A \xrightarrow{\text{?}} P \qquad \qquad \text{(a)}$$

$\Longrightarrow$ reaction database

**reaction prediction**

$$A \quad + \quad B \longrightarrow \quad ? \qquad \text{(b)}$$

$\Longrightarrow$ reaction simulation system

**synthesis planning**

**Figure 10.3-1.** Different types of problems encountered when dealing with chemical reactions.

$$P \Longrightarrow A_1, A_2, A_3, \ldots \qquad \text{(c)}$$

$\Longrightarrow$ synthesis design system

reactivity is still in its infancy, to say nothing of the role of reaction conditions such as the catalysts present, temperature, etc.

In this situation, particularly when a broad range of organic reactions has to be predicted, the simulation of reactions based on knowledge gained from experience is the method of choice. This will be the theme of this chapter.

3. I want to obtain a certain chemical compound **P**; how can I make it? Which starting materials, $A_1$, $A_2$, $A_3$, etc. do I need to build this molecule, **P**?

This is the domain of synthesis design (Figure 10.3-1c). The product of the reaction is known and one has to work back from the reaction product to synthesis precursors that provide, on reacting, the desired target compound. This process has to be repeated until one arrives at available starting materials, $A_i$. Synthesis design is the theme of Section 10.3-2.

The prediction of the course and of the products of a chemical reaction is of fundamental interest as it concerns a problem with which chemists are constantly faced in their day-to-day work. They try to solve such questions by making predictions based on analogy, drawing from their experience acquired in their long training or gathered by making a series of experiments.

With the advent of reaction databases chemists now have a treasure trove of information on chemical reactions available at their fingertips (see Sections 5.4, 5.6, 5.10, and 5.12 of this Textbook, and Chapter X, Section 3.1 of the Handbook). Thus, searches in reaction databases might provide an answer to a chemist's question on the product of a specific reaction.

On top of that, reaction databases can also be used to derive knowledge on chemical reactions which can then be used for reaction prediction. The huge amount of information in reaction databases can be processed by inductive learning methods in order to condense these individual pieces of information into essential features and rules on the driving forces of chemical reactions. This potential will be explored in Section 10.3.1.2.

In spite of the importance of reaction prediction, only a few systems have been developed to tackle this problem, largely due to its complexity: it demands a huge amount of work before a system is obtained that can make predictions of sufficient quality to be useful to a chemist. The most difficult task in the development of a system for the simulation of chemical reactions is the prediction of the course of chemical reactions. This can be achieved by using knowledge automatically extracted from reaction databases (see Section 10.3.1.2). Alternatively, explicit models of chemical reactivity will have to be included in a reaction simulation system. The modeling of chemical reactivity is a very complex task because so many factors can influence the course of a reaction (see Section 3.4).

Two systems, CAMEO and EROS, have been developed that approach the task of reaction prediction in a broad and comprehensive manner. Both systems were initiated around 1975 and are presented in more detail in Sections 10.3.1.4 and 10.3.1.5.

### 10.3.1.2 Knowledge Extraction from Reaction Databases

Clearly, the extraction of knowledge on chemical reactions from the information contained in reaction databases (see Section 5.12) is quite a challenging problem. The huge amount of information stored in reaction databases – the larger ones contain several million reactions – renders them quite attractive for knowledge extraction. However, only a few attempts have been made to extract knowledge automatically.

The reasons for this lack of work are manifold: The problem is quite complex and difficult to tackle. The information in reaction databases is inherently biased: only known reactions, no reactions that failed, are stored. However, any learning also needs information on situations where a certain event will not happen or will fail. The quality of information stored in reaction databases often leaves something to be desired: reaction equations are incomplete, certain details on a reaction are often incomplete or missing, the coverage of the reaction space is not homogeneous, etc. Nevertheless, the challenge is there and the merits of success should be great!

One of the first attempts to build a knowledge base for synthetic organic reactions was made by Gelernter's group, through inductive and deductive machine learning [1]. Important work on this topic was also performed by Funatsu and his group [2].

A central – and primary – task in deriving knowledge from information on reaction instances contained in databases is the grouping of such instances into reaction types. The topic of reaction classification has already been treated in Section 3.5, and readers are encouraged to consult this section again. Particularly, the data-driven approaches to reaction classification aim at directly using the information contained in reaction databases for generalizations to allow predictions on the course of chemical reactions by analogy.

The main characteristics of the method developed in our group for reaction classification are 1) the representation of a reaction by physicochemical values calculated for the bonds being broken and made during the reaction, and 2) use of the unsupervised learning method of a self-organizing neural network for the perception of similarity of chemical reactions [3, 4].

The method is incorporated into the CORA (Classification of Organic Reactions for Analysis) system [5]. Here, we want to illustrate the merits of this approach by an example of its application to a specific problem, the prediction of the regioselectivity of a ring closure reaction. This is detailed in the following tutorial.

Before concluding this section it should be emphasized that knowledge extraction from reaction databases is still a challenging problem having many important applications. There is still room for new approaches to this task. Furthermore, great efforts should be made to improve the depth of information stored in reaction databases. With the introduction of electronic lab journals, the primary information on a chemical reaction gained in the laboratory becomes directly available. This could provide a much richer source of information on chemical reactions and thus build a better basis for automatic learning methods.

### 10.3.1.3 Tutorial: Prediction of the Regiochemistry in Pyrazole Synthesis [6]

The application described here shows how a classified data set of reactions producing pyrazole derivatives can be used to predict the correct regioisomer in a pyrazole synthesis before it is carried out practically in the laboratory.

The synthesis of pyrazoles starting from a hydrazine and a 1,3-dicarbonyl compound is a well established reaction in organic synthesis. If a mono-substituted hydrazine is reacted with an unsymmetrically substituted 1,3-dicarbonyl compound, two different pyrazole products which are regioisomers could be formed (see Figure 10.3-2).

In most cases, only one of the two regioisomers is preferentially formed. We will show here how reaction classification by a self-organizing neural network can be used for the prediction of the preferred regioisomer in a pyrazole synthesis.

Any attempt in learning also needs information on situations where the event to be studied will not occur. This principle, if translated to learning on reactions, would require information on reactions that will not proceed. Unfortunately, such information is not contained in reaction databases, and therefore has to be generated. In our case, we also needed information on which regioisomer is *not* formed, or is formed only to a minor extent, in a pyrazole synthesis.

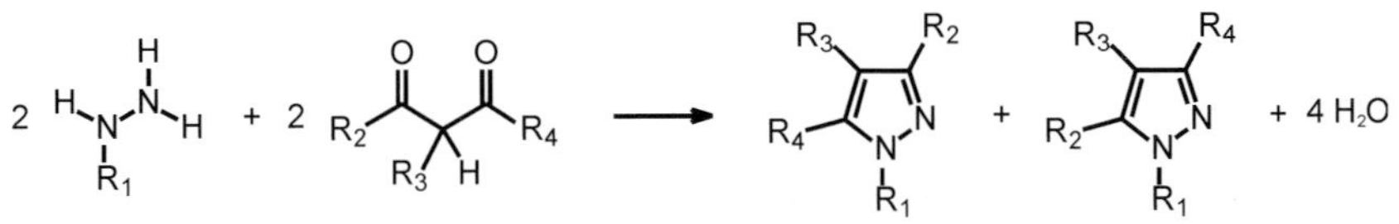

**Figure 10.3-2.** The reaction of a mono-substituted hydrazine with an unsymmetrically substituted 1,3-dicarbonyl compound can lead to two regioisomeric pyrazole products.

A search in the Beilstein database (see Section 5.7 of this Textbook and Chapter V, Section 5 of the Handbook) for reactions where a pyrazole molecule is formed from a mono-substituted hydrazine and an unsymmetrically substituted 1,3-dicarbonyl compound provided a hit list of 313 reactions. For each of these reactions the corresponding reaction resulting in the other regioisomer was generated, and was assigned as not occurring (or as having a low yield).

This reaction data set of 626 reactions was used as a training data set to produce a knowledge base. Before this data set is used as input to a neural Kohonen network, each reaction must be coded in the form of a vector characterizing the reaction event. Six physicochemical effects were calculated for each of five bonds at the reaction center of the starting materials by the PETRA (see Section 7.1.4) program system. As shown in Figure 10.3-3 with an example, the physicochemical effects of the two regioisomeric products are different.

| bond | bo | $\Delta\chi_{AB,\sigma}$ | $\Delta\chi_{AB,\pi}$ | $\Delta q_{AB,tot}$ | $D^-_{AB}$ | $D^+_{AB}$ |
|------|------|------|------|------|------|------|
| O ==> C1 | 0.000 | 0.972 | 0.237 | -0.489 | 0.000 | 0.000 |
| C1 ==> C2 | -1.000 | 0.494 | 0.921 | 0.095 | 0.000 | 0.093 |
| C2 ==> C3 | -1.000 | -0.610 | -0.936 | -0.124 | 0.153 | 0.068 |
| C3 ==> O | 0.000 | -0.882 | -0.103 | 0.595 | 0.480 | 0.463 |
| H2N ==> NH | -1.000 | -0.129 | -0.147 | -0.117 | 0.000 | 0.000 |
| | -5.000 | -5.000 | -5.000 | -5.000 | -5.000 | -5.000 |

| bond | bo | $\Delta\chi_{AB,\sigma}$ | $\Delta\chi_{AB,\pi}$ | $\Delta q_{AB,tot}$ | $D^-_{AB}$ | $D^+_{AB}$ |
|------|------|------|------|------|------|------|
| O ==> C1 | 0.000 | 0.882 | 0.103 | -0.595 | 0.000 | 0.000 |
| C1 ==> C2 | -1.000 | 0.610 | 0.936 | 0.124 | 0.397 | 0.393 |
| C2 ==> C3 | -1.000 | -0.494 | -0.921 | -0.095 | 0.144 | 0.072 |
| C3 ==> O | 0.000 | -0.972 | -0.237 | 0.489 | 0.000 | 0.156 |
| H2N ==> NH | -1.000 | -0.129 | -0.147 | -0.117 | 0.000 | 0.000 |
| | -5.000 | -5.000 | -5.000 | -5.000 | -5.000 | -5.000 |

**Figure 10.3-3.** Two regioisomeric products of the training data set and their corresponding physicochemical effects used as coding vectors; *bo*: bond order; $\Delta\chi_{AB,\sigma}$: difference in $\sigma$-electronegativities; $\Delta\chi_{AB,\pi}$: difference in $\pi$-electronegativities; $\Delta q_{AB,tot}$: difference in total charges; $D^-_{AB}$: delocalization stabilization of a negative charge; $D^+_{AB}$: delocalization stabilization of a positive charge.

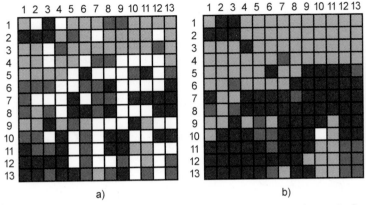

Figure 10.3-4. Trained networks of 626 pyrazole building reactions. a) The trained network after classifying all 626 reactions. Light gray: neurons that have received known reactions; black: neurons that have received reactions considered to have a low yield; dark gray: conflict neurons obtaining both types of reactions; white: empty neurons that have not received a reaction at all; b) Each conflict neuron is assigned to the most populated class within it, and each empty neuron to the most populated class in its neighborhood.

The calculated vectors of physicochemical effects at the reaction center are used as input vectors to a Kohonen neural network of rectangular architecture. The trained network is shown in Figure 10.3-4a. The empty neurons in which no reaction is projected are assigned to that class which is most populated in the neighborhood of the empty neuron. Conflict neurons in which reactions with high and low yields are projected are assigned to that class which is most populated within the conflict neuron. The network with assigned empty neurons is given in Figure 10.3-4b.

The reactions which have been observed are well separated from the reactions which are expected to have no or a low yield. The reactions leading to the preferred regioisomer are projected into the upper part of the Kohonen map, whereas the reactions of the non-preferred regioisomer are projected into the lower part of the Kohonen map.

The trained Kohonen network was validated by using a subset of the 626 reactions for building a training data set and a test data set. The data set was randomly divided into two data sets each of 313 reactions. In a first run, the first subset was used as a training data set, and the reactions of the second subset were predicted by the trained network of 313 reactions. This procedure was repeated using the second subset as the training data set, and the first data set as test data set. With this validation method we found a correctness value of 79.7 % for predicting the correct regioisomer.

The trained and validated network of Figure 10.3-4b can now be used for the prediction of the correct isomer in novel pyrazole syntheses. As a typical example, the reaction shown in Figure 10.3-5 was chosen.

For both reactions leading to the two different regioisomers the corresponding vector was calculated and provided as a test case to the trained network of 626

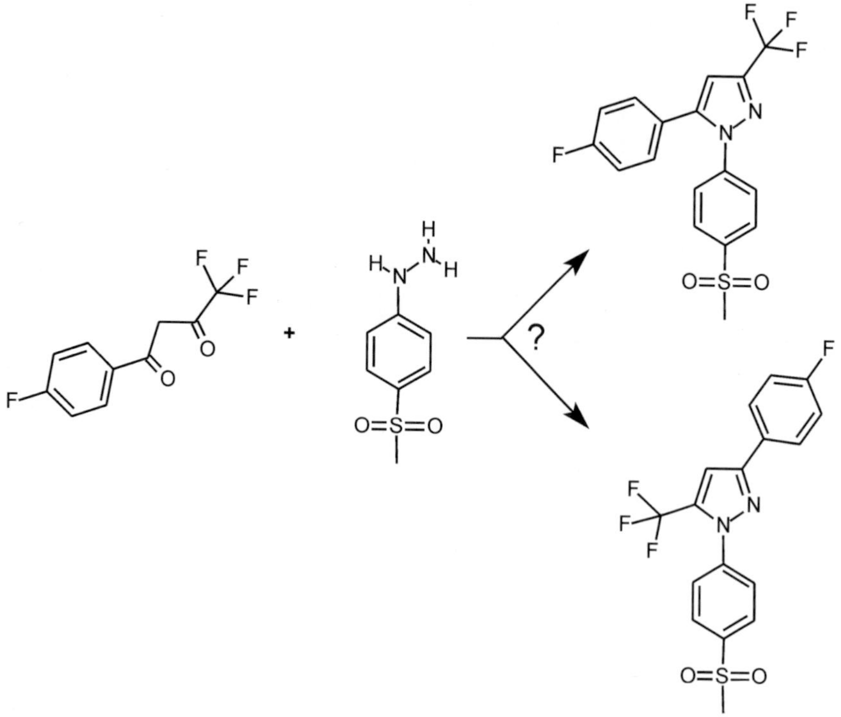

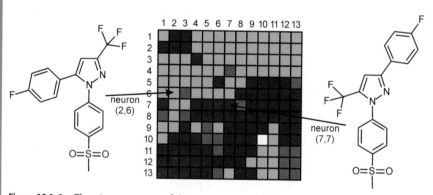

**Figure 10.3-5.** The preferred regioisomer from this reaction should be estimated by using the trained network.

**Figure 10.3-6.** The winner neurons of the two reactions leading to different regioisomers. One of the reactions is projected in that part of the Kohonen network where mostly reactions leading to the preferred regioisomer pyrazole were projected. The other reaction was projected in neuron (7,7), which lies in a region where reactions with low yield are projected.

reactions. One reaction of the two conceivable pyrazoles was projected into neuron (2,6), the other one into neuron (7,7). The two winner neurons for both reactions are shown in Figure 10.3-6.

According to this mapping into the network, the reaction leading to the regio-isomer at the top of Figure 10.3-5 is to be expected.

In reality, this regioisomer was indeed produced in this reaction [7]. It is known as a COX-2 inhibitor with a high selectivity between COX-1 and COX-2 activity (COX-1: $IC_{50} > 100$ µm, COX-2: $IC_{50} = 0.05$ µm).

### 10.3.1.4 **CAMEO**

In the development of the CAMEO system (Computer-Assisted Mechanistic Eva-luation of Organic reactions) [8] it was decided to treat large classes of mechanis-tically related reaction types by separate modules.

Elaborate evaluation procedures were developed for the following mechanistic classes of reactions:

- base-catalyzed and nucleophilic reactions,
- acid-catalyzed and electrophilic reactions,
- pericyclic reactions,
- oxidative and reductive reactions,
- free-radical reactions, and
- carbenoid reactions

Emphasis was put on providing a sound physicochemical basis for the modeling of the effects determining a reaction mechanism. Thus, methods were developed for the estimation of $pK_a$-values, bond dissociation energies, heats of formation, fron-tier molecular orbital energies and coefficients, and steric hindrance.

At the outset of the development of each module for one of the above me-chanistic classes of reactions, a thorough analysis of the literature was performed. On that basis, the developer came up with an evaluation framework that was used to make decisions between various reaction pathways and mechanistic possi-bilities.

For each individual mechanistic reaction type, quite an elaborate heuristic deci-sion scheme was built to arrive at rules that allow one to make predictions on spe-cific reaction queries.

The CAMEO system is a remarkable achievement that is able to model the me-chanism and course of a wide range of organic reactions with a reasonable success rate. In this sense it is also highly valuable as a tool for teaching mechanistic organic chemistry.

A drawback is that the evaluation scheme for modeling the course of chemical reactions, as set up by the initial developer, is difficult to change as any alteration might have unexpected consequences for other types of reactions. Thus, it is a beautiful edifice that has basically not been changed since the early Nineties.

10.3.1.5  **EROS**

The EROS system (Elaboration of Reactions for Organic Synthesis) has gone through a series of versions since its inception in 1973/74. The first versions could be used in a forward search for determining the course of chemical reactions (reaction prediction) as well as in a retrosynthetic search (synthesis design) [9, 10]. With continuing sophistication in the program's evaluation schemes it was decided to separate the two problem tasks, delegating problems of synthesis design to the newly developed (WODCA) system for synthesis design [11] (see Section 10.3.2) and focusing the scope of EROS on reaction prediction [12].

The most recent version of EROS has a clearcut separation of the system proper, which performs all the manipulations on chemical structures and reactions, from the knowledge base, which defines the scope of its application (Figure 10.3-7).

The knowledge base is essentially two-fold; on one hand it consists of a series of procedures for calculating all-important physicochemical effects such as heats of reaction, bond dissociation energies, charge distribution, inductive, resonance, and polarizability effects (see Section 7.1). The other part of the knowledge base defines the reaction types on which the EROS system can work.

This latter part consists of an external file, the reaction rules. First, in the reaction rule header, specifications are made about the applicability of the individual reaction types that are to follow further down in the rule file. This header might specify that the reaction rules apply to degradation reactions in the environment, to biochemical pathways, or to fragmentations and rearrangements in the mass spectrometer. Each reaction rule specifies the bond and electron rearrangement in the course of a reaction type, then gives constraints, i.e., information on atoms and bonds to which the specified reaction scheme is applicable (the reaction center). Figure 10.3-8 shows the two reaction types responsible for the degradation of s-triazine herbicides in soil: hydrolysis and reductive dealkylation. Both reaction types have the same general scheme, breaking two bonds and making two bonds

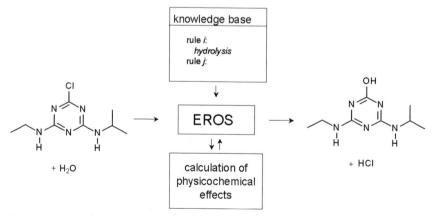

**Figure 10.3-7.**  Outline of the EROS system.

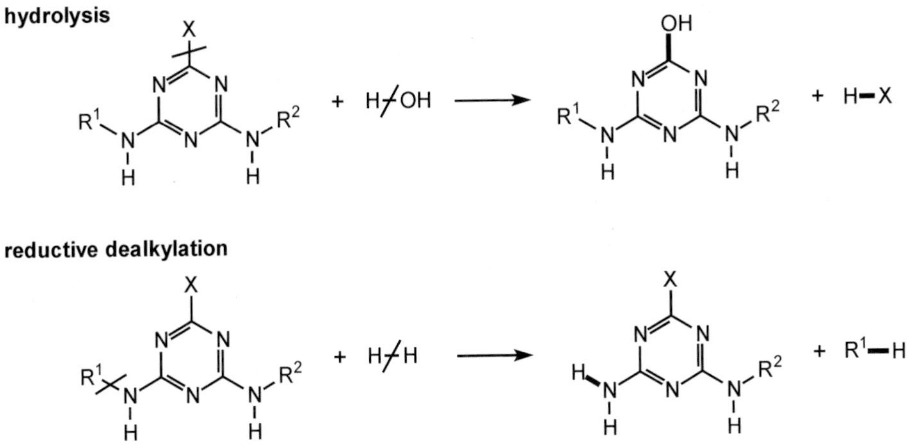

**hydrolysis**

**reductive dealkylation**

**Figure 10.3-8.** The two reaction types necessary for modeling the degradation of s-triazines in soil.

between the four atoms of the reaction center and corresponding to the most important scheme of organic reactions (see Section 3.5). It is the constraints that are imposed on the atoms involved in the reaction center that make the reaction scheme more specific and provide the focus on a certain reaction type.

After the definition of a reaction type, a scheme for the evaluation of the given reaction type can follow in the reaction rule. An entire hierarchy of evaluations can be implemented, from no evaluation at all to a full-fledged estimation of reaction kinetics [12]:

1. *No evaluation*: This allows the exhaustive generation of all conceivable reaction pathways. For example, starting with *n*-octane, all isomeric C$_8$-alkanes can be generated by a reaction type that involves the breaking and making of a C–C and a C–H bond (Figure 10.3-9).
   Clearly, in this case, checks have to be made whether a specific isomer has already been produced by a different pathway. This is achieved by the calculation of hashcodes (see Section 2.7.4).

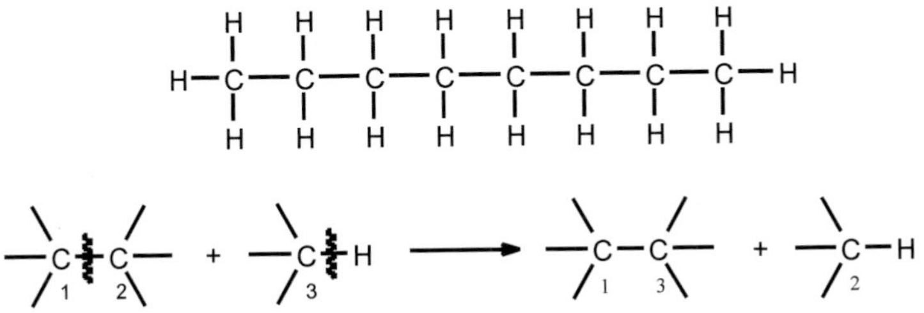

**Figure 10.3-9.** The reaction type for the generation of all isomeric alkanes of a given number of carbon atoms starting with the corresponding *n*-alkane; no evaluation of the various reactions is performed.

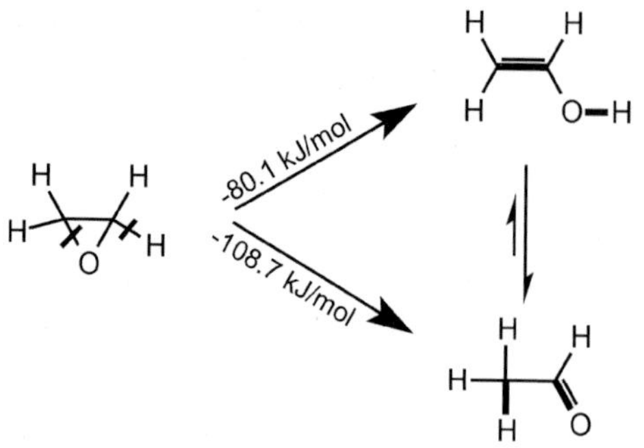

**Figure 10.3-10.** Reaction pathways and the calculated heats of reaction.

2. *Heats of reaction*: Heats of reaction can be obtained as differences between the heats of formation of the products and those of the starting materials of a reaction. In EROS, heats of reaction are calculated on the basis of an additivity scheme as presented in Section 7.1. With such an evaluation, reactions under thermodynamic control can be selected preferentially (Figure 10.3-10).

3. *Physicochemical constraints*: Further constraints can be imposed on the atoms and bonds of the reaction center, such as those physicochemical factors calculated by the PETRA package (see Section 7.1). For example, the partial charges calculated by the PEOE method can be used to extract the chemically feasible reaction from the two conceivable ones as illustrated in Figure 10.3-11.

More elaborate schemes can be envisaged. Thus, a self-organizing neural network as obtained by the classification of a set of chemical reactions as outlined in Section 3.5 can be interfaced with the EROS system to select the reaction that actually occurs from among various reaction alternatives. In this way, knowledge extracted from reaction databases can be interfaced with a reaction prediction system.

4. *Kinetic evaluation*: Clearly, the most in-depth evaluation would be based on the kinetic modeling of a reaction pathway. Unfortunately, in many cases insufficient experimental data are available to develop a full kinetic model of a reaction pathway. Nevertheless, it has been shown with various examples that the development of a kinetic model is possible. This has been performed for the acid-

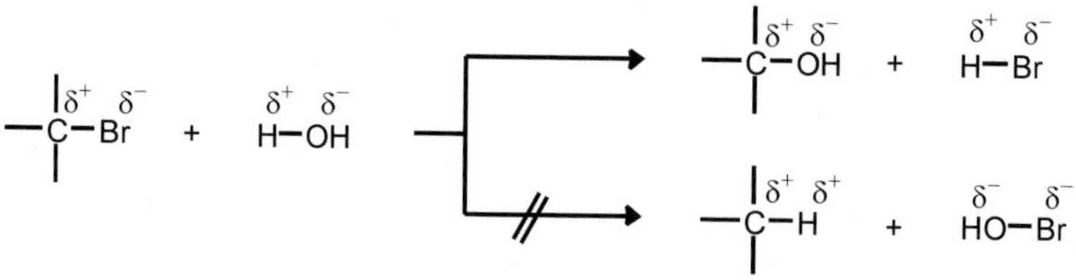

**Figure 10.3-11.** Decision between two conceivable reaction pathways on the basis of partial charges.

and base-catalyzed hydrolysis of amides and the application of the corresponding equations to the prediction of the hydrolysis of benzyl–phenyl–urea herbicides [13]. Similarly, a kinetic modeling of the degradation of *s*-triazine herbicides in soil has been performed (see Tutorial, Section 10.3.1.6).

When reaction rate equations can be given for the individual steps of a reaction sequence, a detailed modeling of product development over time can be made: The reaction rate equations give differential equations that can be solved with methods such as the Runge–Kutta [14] integration or the Gear algorithm [15]. This allows one to predict the change in concentration of all the reaction partners and products over time.

### 10.3.1.5.1 Summary

The advantage of the EROS system is its clearcut separation of the knowledge base on chemical reactions from the system proper. This makes it easy to extend the scope of EROS to new reaction types by specifying new reaction rules. Furthermore, the hierarchy in evaluation procedures allows one to come up with increasingly sophisticated evaluations of reactions, starting from rather cursory constraints to a more detailed kinetic modeling, as deemed necessary or as more data become available. Clearly, the development of reaction rules for the wide range of organic reaction types is a time-consuming process: gathering data, analyzing them for constraints that are to imposed, and developing kinetic models where data are available. Therefore, the automatic extraction of knowledge on chemical reactions from reaction databases is of great interest.

### 10.3.1.6 Tutorial: Modeling the Degradation of *s*-Triazine Herbicides in Soil [16]

It has already been mentioned that the degradation of *s*-triazine herbicides such as atrazine in soil can be described by two reaction types only, hydrolysis and reductive dealkylation (see Figure 10.3-8). Application of these two reaction types to a specific *s*-triazine compound such as atrazine provides the reaction network shown in Figure 10.3-12. This can also be verified by running this example on http://www2.chemie.uni-erlangen.de/services/eros/.

The question is now: Which reaction pathways are followed, and to what extent? This asks for a detailed modeling of the kinetics of the individual reaction steps of this network. This can be achieved on the basis of the half-lives of four *s*-triazine herbicides in soil [17]. Figure 10.3-13 shows the four compounds for which data were found in the literature.

The half-lives for these four compounds taken from the literature allowed the estimation of the four reaction rates necessary to model their degradation [18]. As a first approximation, the rate of hydrolysis of the C–Cl bond of all four *s*-triazine compounds was assumed to be the same and to be $5.0 \times 10^{-9}$ s$^{-1}$ on the basis of literature precedence. This approximation seems reasonable as the four structures differ only in the alkyl groups at a site quite remote from the C–Cl bond. Furthermore, among the four reaction steps hydrolysis is the slowest anyway.

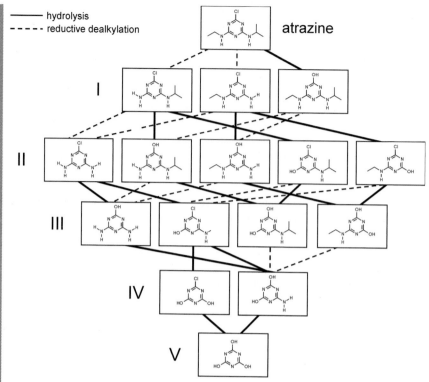

**Figure 10.3-12.** Reaction network obtained for atrazine by application of the two reaction types shown in Figure 10.3-8.

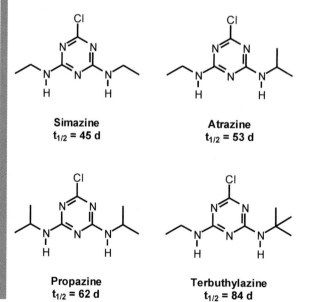

**Figure 10.3-13.** *s*-Triazine herbicides and their half-life in soil [16].

With this assumption, the rate constants for reductive removal of an ethyl, an iso-propyl, and a *tert*-butyl group can be estimated as follows:

$$k_{hydrolysis} = 5.0 \times 10^{-9} \ s^{-1}$$

$$k_{ethyl} = 3.32 \times 10^{-8} \ s^{-1}$$

$$k_{i\text{-prop}} = 2.65 \times 10^{-8} \ s^{-1}$$

$$k_{t\text{-but}} = 2.21 \times 10^{-8} \ s^{-1}$$

With these reaction rate constants, differential reaction rate equations can be constructed for the individual reaction steps of the scheme shown in Figure 10.3-12. Integration of these differential rate equations by the Gear algorithm [15] allows the calculation of the concentration of the various species contained in Figure 10.3-12 over time. This is shown in Figure 10.3-14.

The removal of the two alkyl groups comprises the two primary steps, the abstraction of the ethyl group being the dominant one. These products show up first, as can be seen on the left-hand side of Figure 10.3-14. It can also be seen that it is expected that after one year an appreciable amount of *s*-triazine compounds will still be found in the soil.

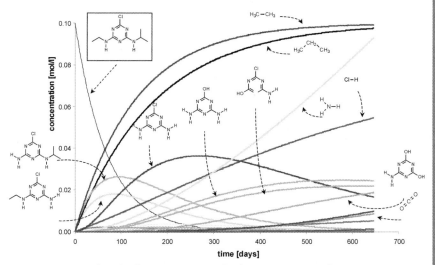

**Figure 10.3-14.** Product development over time for the species contained in Figure 10.3-11.

### 10.3.1.7 **Biochemical Pathways**

In recent years much interest has centered on the analysis and simulation of biochemical pathways. Much of this attention was derived from the realization that one of the primary roles of genes is to regulate biochemical pathways. With the deciphering of the human genome, an understanding of how genes influence individual reactions in our body seemed to have come within reach. However, the deeper we dig into this question the more we have to appreciate its complexity. Nevertheless, it is certainly one of the most challenging problems where chemoinformatics meets bioinformatics. Indeed, we believe that the very fact that scientists from different disciplines meet to unravel the secrets of biochemical pathways offers great promise. In order to emphasize this point, different views of a biochemical reaction will be shown.

For the bioinformatician the emphasis lies on how a gene expresses a protein – an enzyme – that catalyzes a biochemical reaction (Figure 10.3-15).

The starting materials and the products of the reaction are quite often not identified, or identified only by a label, or, at most, by their names.

In a further level of sophistication, the structures of the reaction partners are given, although in many cases only as an image, as a 2D picture such as a gif-file (Figure 10.3-16a). Clearly, this does not allow further processing of chemical structure information. Only when the structures as shown in Figure 10.3-16a are coded as connection tables (Figure 10.3-16b), or any equivalent representation, does the further processing of chemical structure information such as structure or substructure searching, or the exchange of information with other software or databases, become possible.

To emphasize this point, we show in Figure 10.3-17 the reaction given in Figure 10.3-16b after processing the connection tables through the 3D structure generator CORINA. Then, for each one of the molecules involved in this reaction, a 3D molecular model can be generated automatically as shown in Figure 10.3-17.

However, we have seen in Chapter 3 that the representation of the structures of the starting materials and products of a chemical reaction by connection tables is not sufficient information to characterize a reaction. Rather, to really have a

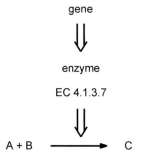

**Figure 10.3-15.** Control of a biochemical reaction through expression of an enzyme by a gene.

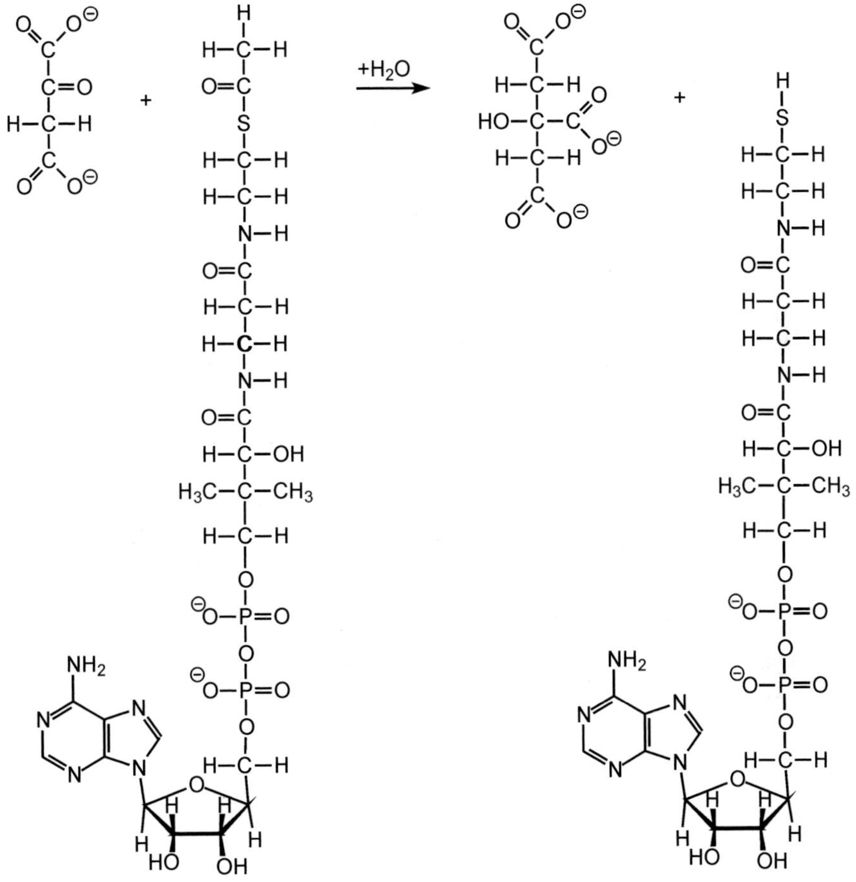

**Figure 10.3-16.** Graphical representation of the chemical structure of the reactants and products of a chemical reaction: a) as a 2D image; b) with structure diagrams showing all atoms and bonds of the reactants and products to indicate how this information is stored in a connection table.

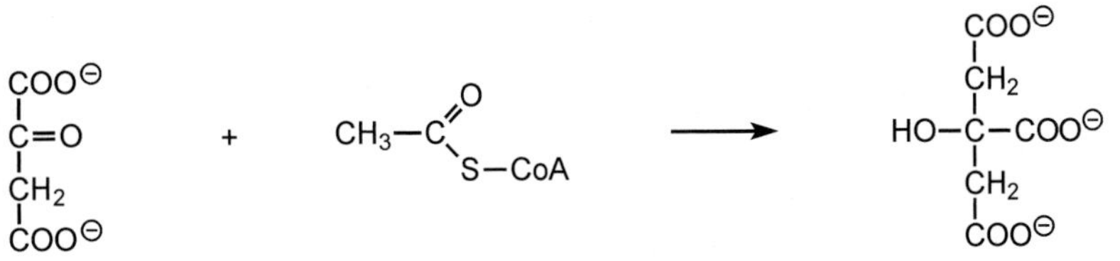

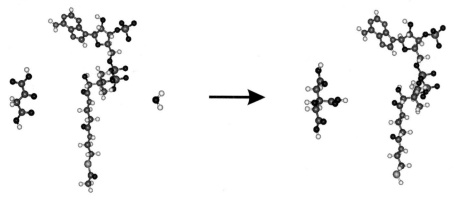

**Figure 10.3-17.** 3D structures of the reactants and products of Figure 10.3-16b as obtained by the 3D structure generator CORINA.

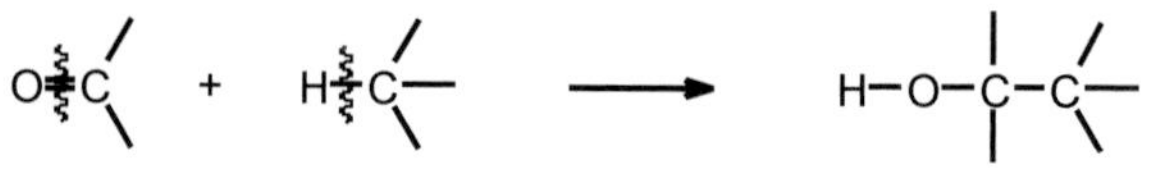

**Figure 10.3-18.** The reaction center, the bonds broken and made, of the reaction shown in Figure 10.3-16.

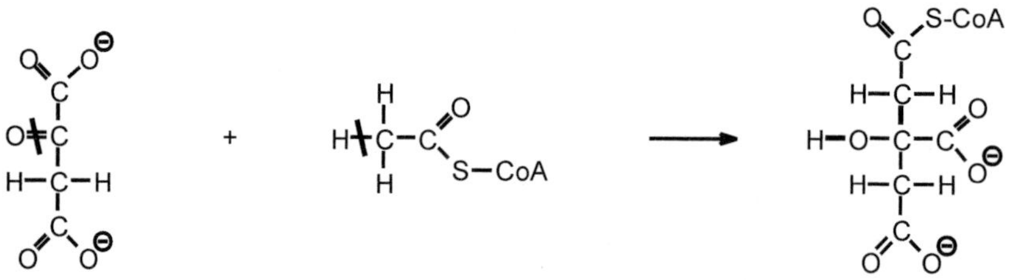

**Figure 10.3-19.** Representation of the reaction shown in Figure 10.3-16, indicating all the atoms and bonds of the chemical structures as well as the reaction center. For the sake of clarity, the coenzyme A has been abbreviated.

genuine reaction representation, the reaction center – the bonds broken and made in a reaction – has to be specified (Figure 10.3-18).

Only then can the full arsenal of processing reaction information, such as reaction center searching, reaction similarity perception, or reaction classification (see Section 3.5) be invoked. Figure 10.3-19 shows such a full-fledged reaction representation.

We aim to show below how an explicit coding of the chemical structures of the starting materials and products of biochemical reactions and their reaction centers might allow us to achieve progress in our understanding of biochemical pathways. Furthermore, it will be shown how a bridge between chemoinformatics and bioinformatics can be built.

For several decades, the poster *Biochemical Pathways*, edited by G. Michal, and initially distributed by Boehringer Mannheim, now Roche Diagnostics, has been a cornerstone for providing information on biochemical reactions. Figure 10.3-20 shows a part of this poster. Building on the poster, an atlas has been issued [19].

This poster indicates the structures of the compounds involved in a reaction, the enzymes catalyzing a reaction, the coenzymes and regulators involved, and whether such a reaction is a general pathway occurring in all species, or a pathway specific to higher plants, animals, or unicellular organisms.

The importance of this information on biochemical reactions is emphasized by the fact that an image of the poster is also accessible on the ExPASy server [20]. However, for this purpose, the poster has only been scanned. The ExPASy server provides a static image of the information on this web site, augmented only by links to additional information.

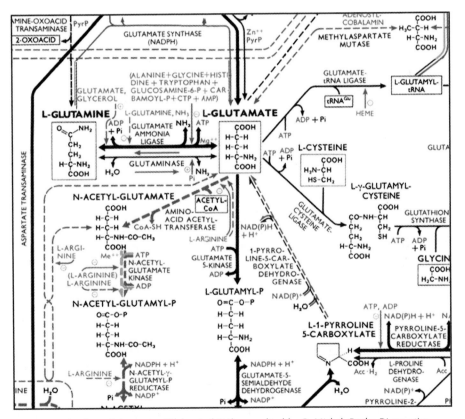

**Figure 10.3-20.** Part of the poster *Biochemical Pathways* edited by G. Michal, Roche Diagnostics (formerly Boehringer Mannheim).

What is needed for progress in achieving a deeper insight into biochemical pathways is to allow all this information to be searchable by electronic means; in other words, it has to be stored in a reaction database.

This is exactly what we have done [21]. For each reaction the constitution and stereochemistry of the reaction partners, the coenzymes, and regulators were stored as connection tables (as far as they were known), and the enzymes by name and EC number.

Furthermore, for each reaction the reaction center was specified, information was given on whether the reaction is reversible or irreversible, and catabolic or anabolic. Finally, it was specified whether a reaction is part of a general pathway or occurs only in unicellular organisms, in higher plants, or in animals (Figure 10.3-21).

Altogether it is not a large reaction database; in January 2003 it comprised only about 1533 unique structures and 2175 reactions. Nevertheless, it is a very important reaction database as it contains the essential processes occurring in living species.

All this information is now directly searchable. All chemical species can be searched not only by name, but also by structure and substructure, and by their role in a reaction. Reactions can be searched by all the species involved but also by the types of atoms and bonds contained in the reaction center.

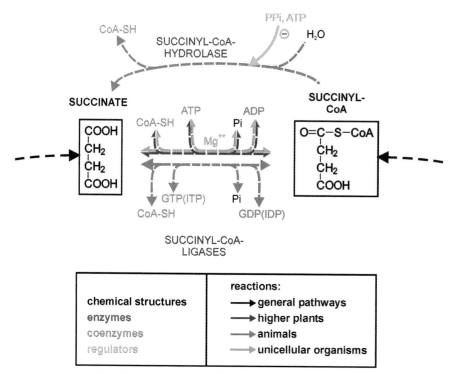

**Figure 10.3-21.** A biochemical reaction and the various pieces of information that have been stored in a reaction database.

The tutorial in Section 10.3.1.8 presents some of the various ways the information in the Biochemical Pathways database can be retrieved. In this tutorial the importance of searching for the reaction center, the atoms and bonds directly involved in the bond rearrangement scheme, is emphasized. It is a prerequisite for getting a deeper understanding of chemical reactions.

Clearly, the next step will be to investigate the physicochemical effects, such as charge distribution and inductive and resonance effects, at the reaction center to obtain a deeper insight into the mechanisms of these biochemical reactions and the finer details of similar reactions. Here, it should be emphasized that biochemical reactions are ruled and driven basically by the same effects as organic reactions. Figure 10.3-22 compares the Claisen condensation of acetic esters to acetoacetic esters with the analogous biochemical reaction in the human body.

The Claisen condensation is initiated by deprotonation of an ester molecule by sodium ethanolate to give a carbanion that is stabilized, mostly by resonance, as an enolate. This carbanion makes a nucleophilic attack at the partially positively charged carbon atom of the ester group, leading to the formation of a C–C bond and the elimination of an ethanolate ion. This Claisen condensation only proceeds in strongly basic conditions with a pH of about 14.

Clearly, such high pH values are not attainable under physiological conditions as the pH has to stay at 7.4 in the blood and in cells, with deviations of 0.1 allowed only for brief moments. Thus, nature must activate the esters in order to run the reaction in milder conditions. First, in the body the reaction is run with the more reactive thioesters instead of esters. Secondly, an additional carboxylic acid group is introduced, making the C–H bond more acid through the possibility of stabilizing the carbanion by two adjacent groups. However, it has to be emphasized that the same kind of physicochemical effects, resonance stabilization of a carbanion and nucleophilic attack at a partially positively charged carbon atom, are operative in both reactions.

This means that the methods developed for the calculation of physicochemical effects can also be used to deepen our understanding of biochemical reactions. Clearly, electronic effects within the substrate molecule are not the only ones determining its reactivity. The binding of the substrate to the enzyme is also influenced

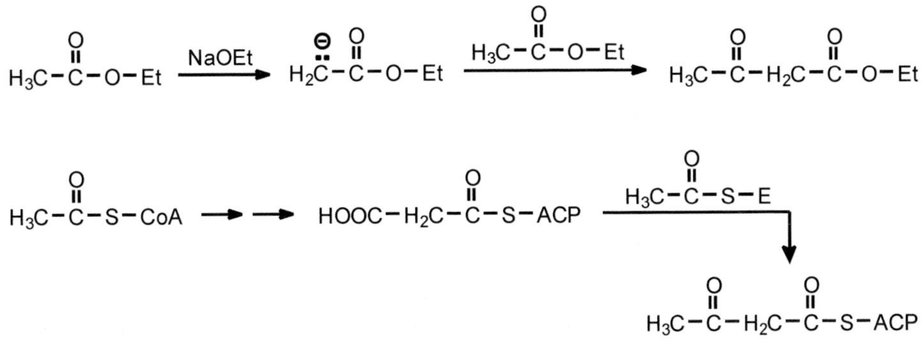

**Figure 10.3-22.** Comparison of a Claisen condensation with its biochemical counterpart.

by the shape of the substrate. The proper positioning of hydrogen bonding, electrostatic and hydrophobic sites in the substrate relative to the enzyme pocket are at least as important and have to be considered in a global attempt to understand biochemical reactions.

### 10.3.1.7.1 Metabolic Reaction Networks

Biochemical pathways consist of networks of individual reactions that have many feedback mechanisms. This makes their study and the elucidation of kinetics of individual reaction steps and their regulation so difficult. Nevertheless, important inroads have already been achieved. Much of this has been done by studying the metabolism of microorganisms in fermentation reactors.

This is not the place to expose in detail the problems and the solutions already obtained in studying biochemical reaction networks. However, because of the importance of this problem and the great recent interest in understanding metabolic networks, we hope to throw a little light on this area. Figure 10.3-23 shows a model for the metabolic pathways involved in the central carbon metabolism of *Escherichia coli* through glycolysis and the pentose phosphate pathway [22].

Again, no further explanation of this reaction scheme is intended here. It should just illustrate the complexity of the problem and present a challenge for further work.

Suffice it to say that a dynamic model of this system was proposed that allowed the estimation of kinetic parameters and gave reasonable agreement with the experimental observations in the bioreactor [22].

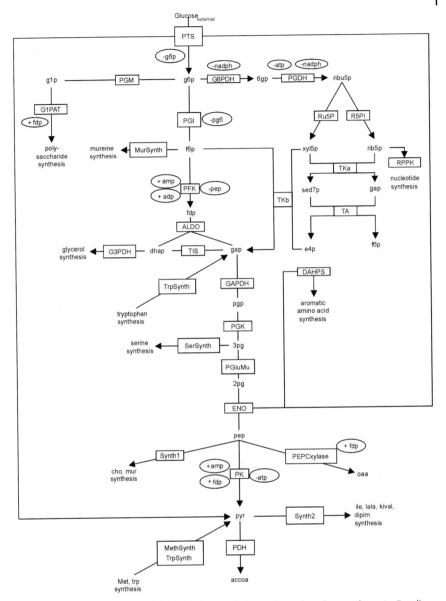

**Figure 10.3-23.** Metabolic model of glycolysis and the pentose phosphate pathway in *E. coli*. Squares indicate enzyme activities; circles indicate regulatory effects.

### 10.3.1.8 Tutorial: Multidimensional Searching in Biochemical Pathways

The very appearance of Figure 10.3-20 – and of the poster Biochemical Pathways as a whole [20] – clearly points out the deficiencies of a two-dimensional medium, a drawing plane, to represent the complexity, the high interconnectivity of biochemical pathways.

Many of the species involved in the endogenous metabolism can undergo a multitude of transformations, have many reaction channels open, and by the same token, can be produced in many reactions. In other words, biochemical pathways represent a multi-dimensional space that has to be explored with novel techniques to appreciate and elucidate the full scope of this dynamic reaction system.

The reaction database compiled on Biochemical Pathways can be accessed on the web and can be investigated with the retrieval system C@ROL (Compound Access and Retrieval On Line) [21] that provides a variety of powerful search techniques. The Biochemical Pathways database is split into a database of chemical structures and a database of chemical reactions that can be searched independently but which have been provided with efficient crosslinks between these two databases.

Figure 10.3-20 shows that *L*-glutamate plays an important role in the metabolism with many reactions leading to this compound and many reactions starting from it. The first example gives a full structure search in order to show how easy it is to find all reactions that a certain compound participates in.

The structure of a compound including its stereochemistry can be specified with the graphical molecule editor JME which converts it into a stereochemically unique SMILES string as search item. Figure 10.3-24a shows how *L*-glutamate can be input as query.

71 reactions were found in which *L*-glutamate participates, 47 reactions leading to *L*-glutamate and 24 reactions starting from *L*-glutamate.

A search for derivatives of *L*-glutamate can be performed by a substructure search (SSS) (see Section 6.3). Figure 10.3-24b shows the query of the SSS; observe that here, too, the stereochemistry can be specified and this is then considered in the SSS. The query provides 25 hits, including *L*-glutamate, N-acetylglutamate, etc. but also compounds such as tetrahydrofolate, folate, etc. which would never be found by a name search.

The structure database has been enriched with the 3D structures of all compounds as generated by the 3D structure generator CORINA (see Section 2.9,

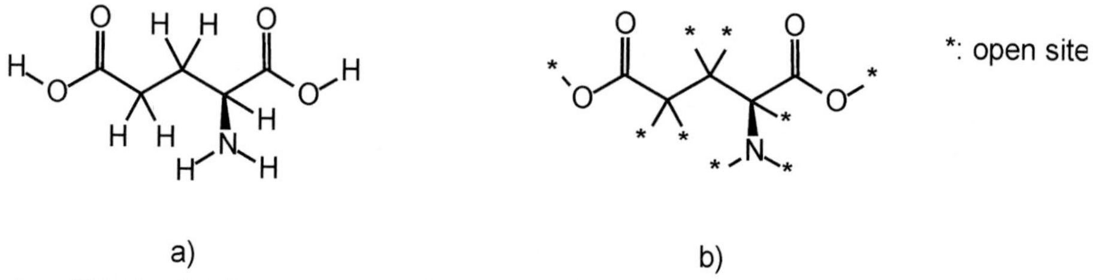

a)                                                                      b)

**Figure 10.3-24.** a) *L*-glutamate as query for a full structure search; b) *L*-glutamate as query for a substructure search; the asterisks mark open sites.

see also Figure 10.3-17). In order to also account for the conformational flexibility of molecular structures, a sample of up to 10 diverse conformations as generated by ROTATE (see Section 2.9) has been included in the database.

As an example for the use of this 3D structure information we present a 3D pharmacophore search. Diethylstilbestrol (DES) (Figure 10.3-25a) is a synthetic estrogen which has come out of use as contraceptive because of its potential carcinogenicity. Clearly, biological activity is strongly associated with the 3D structure of a compound, as the 3D structure determines whether a ligand fits into the binding pocket of a certain receptor. Therefore, in order to find compounds of similar biological activity, 3D structure searches are highly recommendable. In searching for structures with similar biological properties as diethylstilbestrol it was assumed that the phenol substructure and the second oxygen atom are important for biological activity. The 3D structure of DES showed that the two oxygen atoms are at a distance of 11.92 Å. Accordingly, the 3D pharmacophore was specified as consisting of a benzene ring with an OH-group and an additional oxygen atom at a distance of 11.92 ± 1.50 Å from the oxygen atom of the OH-group (Figure 10.3-25b).

The 3D pharmacophore search with C@ROL in the Biochemical Pathways database provided 13 different molecules as hits. To further limit the number of hits, the additional restriction was imposed that the hits should have only two hydrogen

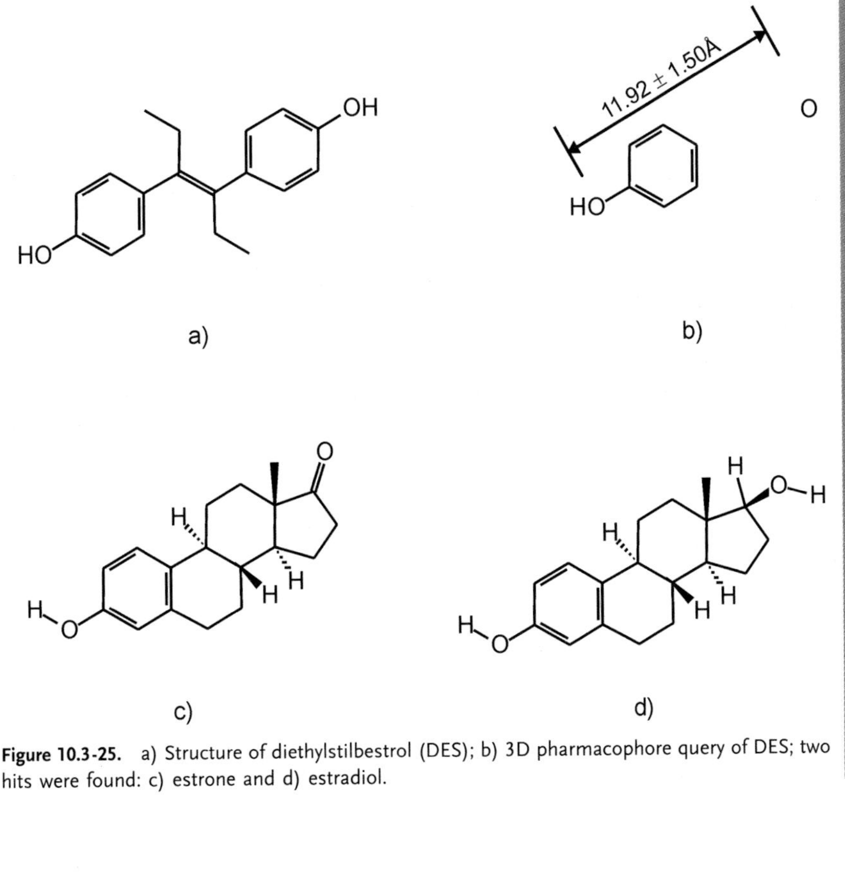

Figure 10.3-25. a) Structure of diethylstilbestrol (DES); b) 3D pharmacophore query of DES; two hits were found: c) estrone and d) estradiol.

| | | |
|---|---|---|
| lycopene | → | ß-carotene |
| acetyl-CoA + acetoacetyl-CoA | → | 2-hydroxy-3-methylglutaryl-CoA |
| 2-ketoglutarate + $CO_2$ | → | oxalosuccinate |
| isopentenyl-PP + dimethylallyl-PP | → | geranyl-PP |
| D-glyceraldehyde-3-P + dihydroxyacetone-P | → | ß-D-fructose-1,6-$P_2$ |

**Figure 10.3-26.** Some of the results of a search for reactions forming a C-C bond.

bonding acceptor atoms just as DES has only two such atoms. This cut the number of hits to two, estrone and estradiol (Figure 10.3-25c and d). Thus, in this pharmacophore search the natural ligands of the estrogen-receptor that also binds the query structure, diethylstilbestrol, were found.

Next, the power and the benefits of reaction center or reaction substructure searching (see Section 3.3) will be illustrated. Figure 10.3-26 shows some of the hits obtained in a search for reactions that form a C-C bond. Intentionally, only the names of the starting materials and products of these reactions are given in order to emphasize that the common feature of these reactions cannot be derived from coding chemical compounds by name. Only a search by reaction center can expose the similarity in these reactions. The next logical steps would then be to explore whether these reactions have more in common than just forming a C-C bond.

The next example shows how different search queries can be combined to shed more light onto a series of related reactions. A reaction substructure search for reactions that break a P-O bond provided 304 reactions as hits. Figure 10.3-27 shows one of the reactions in this hit list.

In order to further classify these reactions, a search for reactions that transform ATP to ADP was made, resulting in 139 reactions; 139 of the above 304 reactions involve the breaking of a P-O bond in ATP, emphasizing the central importance of this bond breaking as a source of energy. An additional three reactions involve the transformation of GTP to GDP. As many reactions transferring a phosphate group

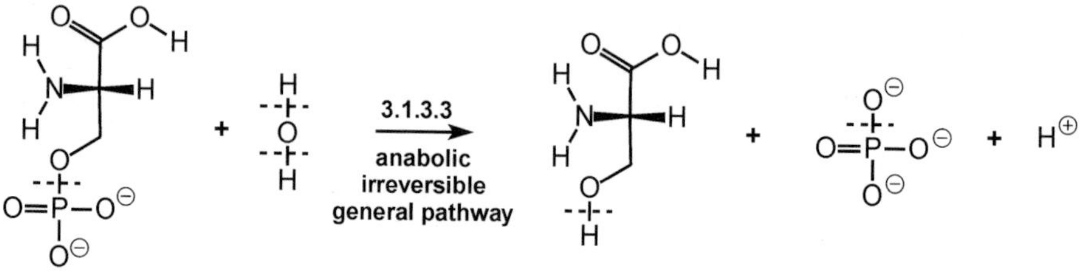

3-phosphoserine + water ⟶ L-serine + phosphate + proton

**Figure 10.3-27:** Example of a reaction which breaks and makes a P-O bond.

involve kinases, a name search on kinase was performed, resulting in 79 reactions.

Reactions can also be searched by enzymes, either by enzyme name or enzyme class (EC notation), both in specific or in generic form. Table 10.3-1 shows the results of searching for EC classes.

These examples served to show that the Biochemical Pathways database provides a rich source of information on these all important reactions that determine the transformation of nutrients into the broad spectrum of compounds contained in living species and the concomitant production of energy to keep these processes going.

**Table 10.3-1.** Classification of enzymes.

| EC class | description | total | % |
|----------|-------------|-------|---|
| 1.x.x.x | oxidoreductases | 206 | 28 |
| 2.x.x.x | transferases | 210 | 29 |
| 3.x.x.x | hydrolases | 113 | 16 |
| 4.x.x.x | lyases | 105 | 15 |
| 5.x.x.x | isomerases | 41 | 6 |
| 6.x.x.x | ligases | 40 | 6 |

## 10.3.2
## Computer-Assisted Synthesis Design

*Markus Sitzmann, Matthias Pförtner*

### 10.3.2.1  Introduction
The chemical synthesis of carbon-containing molecules has been a very important field of scientific work and endeavor for over a century. However, the subject is still far away from being fully developed. One of the major reasons for this is the almost unlimited number of organic structures which can exist as discrete compounds. On the other hand there has been a continuing growth in the ability of chemists to construct increasingly complex molecules.

How do chemists find a pathway to the synthesis of a new organic compound? They try to find suitable starting materials and powerful reactions for the synthesis of the target compound. Thus, synthesis design and chemical reactions are deeply linked, since a chemical reaction is the instrument by which chemists synthesize their compounds; synthesis design is a chemist's major strategy to find the most suitable procedure for a synthesis problem.

The way a synthesis is planned has changed substantially over time. Until the beginning of the 20th century many noteworthy syntheses had been developed, e.g., of alizarin (C. Graebe, C. Liebermann, 1869) and indigo (A. Baeyer, 1878).

In most cases, the syntheses of that time were relatively simple. Only little effort was devoted to their strategic planning. In general, the strategy was directed to finding starting materials, chosen for their resemblance to the target compound, and to choosing synthesis reactions by which the starting materials could easily be connected [23].

The 20th century brought important advances in the field of organic chemistry. In the first decades of the century, the syntheses of increasingly complex molecules were accomplished. Some notable compounds synthesized during that time were *a*-terpinol (W.H. Perkin, 1904), camphor (G. Komppa, 1903), and tropinone (R. Robinson, 1917; Figure 10.3-28).

In the following decades, chemists tried to utilize more and more the knowledge on reactions which had already been gained. A number of landmark syntheses represent the change to modern chemistry, such as the synthesis of the estrogenic steroid equilenin (W. Bachmann, 1939), of pyridoxine (K. Folkers, 1939), and of quinine (R.B. Woodward, W. von E. Doering, 1944) [23].

After the Second World War, our knowledge about chemical reactions and their mechanisms increased tremendously. This provided a major step forward to a more systematic and sophisticated approach to planning organic syntheses, which was made possible by the development of chromatographic methods for product separation and the introduction of spectroscopic methods for structure elucidation. The conformational analysis of organic structures and transition states based on stereochemical principles was introduced. Furthermore, the application of new selective chemical reagents was discovered and improved. Thus, in the 1950s the synthesis of quite complex molecules and natural products was achieved. Some examples are vitamin A (O. Isler, 1949), cortisone (R.B. Woodward, R. Robinson, 1951), morphine (M. Gates, 1956), penicillin (J. Sheehan, 1957), and chlorophyll (R.B. Woodward, 1960).

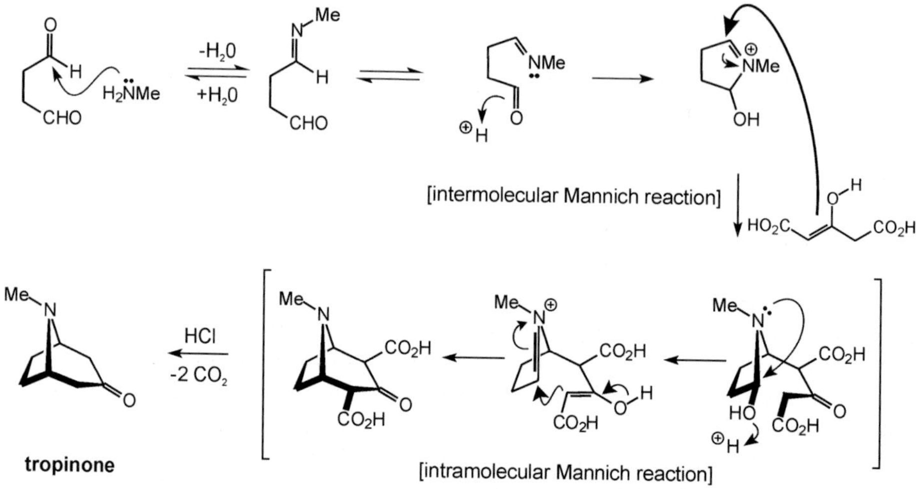

[intermolecular Mannich reaction]

[intramolecular Mannich reaction]

**tropinone**

**Figure 10.3-28.** The total synthesis of (±)-tropinone (Robinson, 1917).

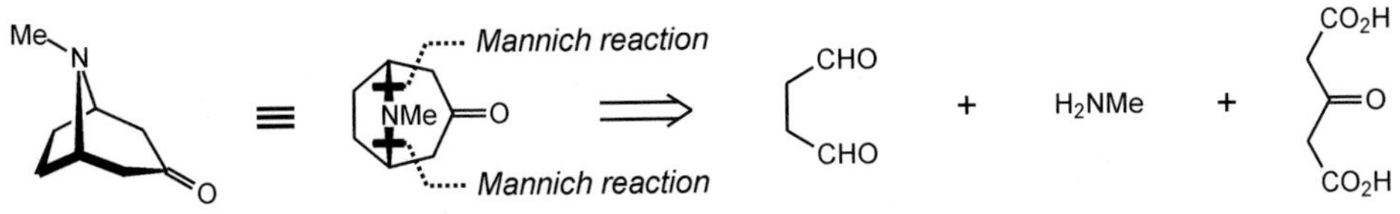

**Figure 10.3-29.** The retrosynthetic analysis of tropinone.

A further advancement in organic synthesis was the accomplishment of multi-step syntheses comprising 20 and more steps and the synthesis of rather unstable organic compounds. It was now feasible to do an experiment on a milligram scale and to separate and identify products from by-products in order to analyze them separately. The application of selective reagents or reaction conditions allowed the synthesis of enantiomerically or diastereomerically pure compounds.

However, the synthesis of each compound was considered as a specific task on its own. A suitable strategy for the synthesis of a target compound was mostly found on the basis of the intuition and experience of the acting chemists, i.e., the planning of a synthesis of a complex organic molecule was considered as an art form. No systematic approach was attempted to handle the strategic design of an organic synthesis. However, the growth of knowledge resulting from the rapid development of analytical methods and other techniques was demanding a more systematic approach for synthesis design.

Then, in 1960, Corey introduced a general methodology for planning organic syntheses. Corey's synthon concept [23–25] was a downright change of the perception of an organic synthesis. The synthesis plan for a target molecule is developed by starting with the target structure (the product of the synthesis!) and working backwards to available starting materials. The retrosynthetic analysis or disconnection of the target molecule in the reverse direction is performed by the systematic use of analytical rules which have been formulated by Corey. For the example of tropinone, this is shown in Figure 10.3-29. Corey's approach is nowadays widely accepted as the disconnection approach and is taught in a number of textbooks (e.g., Ref. [26]).

## 10.3.2.2 Basic Terms

### 10.3.2.2.1 Retrosynthetic Analysis

During the first century of organic chemistry the chemist's view was focused on the structure of organic compounds and their transformations. Reactions were grouped according to the types of compounds which underwent the reaction; thus reaction types were defined such as ester condensation, substitution of aromatic compounds, or glycol cleavage. The view was focused on the direction of the chemical change in a chemical reaction, i.e., from reactants to products. The focus of a retrosynthetic approach, on the other hand, is centered on the perception of the structural features of the reaction products and the modification of structures in the reverse direction from the actual synthesis. This direction has been given the name *retrosynthetic* or *antithetic* [23].

The aim of a retrosynthetic analysis is the transformation of a synthesis target into progressively simpler structures, following a pathway to commercially available starting materials.

### 10.3.2.2.2 Transforms, Retrons, Strategic Bonds, and Synthons

Syntheses are no longer viewed in terms of known name reactions and single steps, but as the global transformation of a skeleton and its pattern of polarities and potential charges obtained through the heterolysis of a bond that can be stabilized by entire classes of substituents.

Since a reaction is considered during a retrosynthetic analysis in a direction opposite to its actual course, it is called a retro-reaction or *transform* (Figure 10.3-30).

If a synthesis is planned according to this reasoning the following steps have to be performed:

1. The target compound is searched for a *retron*. A retron is the structural subunit required to be present in the target in order to apply a transform. In Figure 10.3-30 the retron of a Michael addition is a sequence of five carbon atoms with two carbonyl functions in the 1,5-position. For a Michael addition transform to be applied, it has to be present.
2. A *strategic bond* has to be defined. A strategic bond is a bond within the target molecule which is qualified to be disconnected. Whether a bond is a strategic one or not depends on whether the breaking of the bond leads to a structural simplification in the precursors (topological criteria), and on whether the resultant precursors can be converted easily into the target by a chemical reaction having a broad scope.
3. The specific transformations that disconnect the strategic bonds in the desired fashion are performed.

If a strategic bond is broken, structural fragments called *synthons* are obtained. In most cases, a strategic bond will be broken heterolytically, generating charged syn-

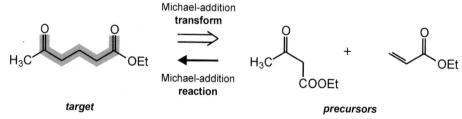

Michael-addition **transform**

Michael-addition **reaction**

*target*                    *precursors*

**Figure 10.3-30.** The retrosynthetic point of view: the transform of a Michael addition. The structure fragment with a gray background is the retron of the Michael addition transform.

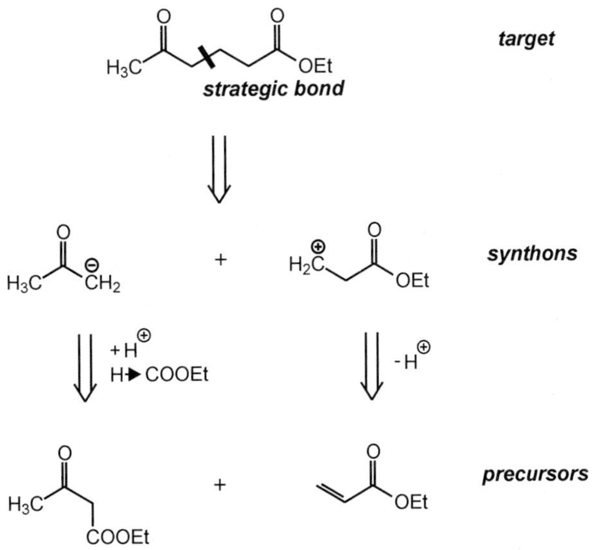

**Figure 10.3-31.** Disconnection of the strategic bond yields (charged) synthons. The synthons are then transformed into neutral, stable reactants.

thons. Synthons are converted into stable molecules by adding or removing atoms, ions, or chemical groups. In this manner, precursor molecules or starting materials are obtained (Figure 10.3-31).

The breaking of a strategic bond and the generation of synthesis precursors defines a synthesis reaction. In the simplest case, the reaction is already known from literature. In most cases, however, the reaction step obtained has to be generalized in order to find any similar and successfully performed reactions with a similar substituent pattern or with a similar rearrangement of bonds. One way of generalizing a reaction is to identify the reaction center and the reaction substructure of the reaction. This defines a reaction type.

### 10.3.2.2.3 Reaction Center and Reaction Substructure
In a reaction, bonds are broken and made. In some cases free electrons are shifted also. The *reaction center* contains all the bonds being broken or made during the reaction as well as all the electron rearrangement processes. The *reaction substructure* is the structural subunit of atoms and bonds around the reaction center that has to be present in a compound in order for the reaction to proceed in the forward (synthesis) direction (Figure 10.3-32). Both characteristics of a reaction can be used to search for reactions with an identical reaction center and reaction substructure but with different structural units beyond the reaction substructure. For example, this can be achieved by searching in a reaction database.

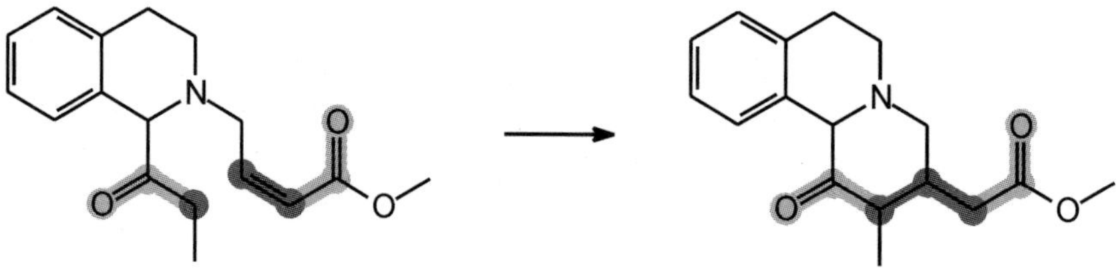

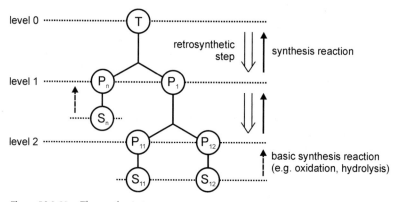

**Figure 10.3-32.** Reaction center and reaction substructure: the parts of the structures with a darker gray background are the reaction center, and those with a lighter gray background are the reaction substructures which must be present to achieve the reaction in the forward direction (in this case, Michael addition).

#### 10.3.2.2.4 Synthesis Tree

A *synthesis tree* is the graphical representation of the result of a retrosynthetic analysis. The target compound (T) is set to the top of a tree that is turned upside down (Figure 10.3-33).

After a strategic bond has been broken and precursor molecules have been obtained, a new level is added to the synthesis tree. Between two levels of the tree the synthesis tree branches out into precursors ($P_n$). In addition, a new retrosynthesis step is defined. If the precursor molecules do not correspond to available starting materials the retrosynthetic analysis has to be continued for these precursors and another level is added to the synthesis tree. For each synthesis precursor, a search is performed for suitable starting materials ($S_n$). If a search is successful the starting material puts an end to the corresponding branch in the synthesis tree.

level 0 ............................( T )...........................................

retrosynthetic step / synthesis reaction

level 1 ..............( $P_n$ )...............( $P_1$ )..........................

( $S_n$ )

level 2 .............................( $P_{11}$ )...............( $P_{12}$ )................

( $S_{11}$ ).........................( $S_{12}$ )...........

basic synthesis reaction (e.g. oxidation, hydrolysis)

**Figure 10.3-33.** The synthesis tree.

### 10.3.2.3 Concepts for Computer-Assisted Organic Synthesis

Program systems for computer-assisted organic synthesis (CAOS) have been under development since the early 1970s [27]. The program systems for computer-assisted synthesis planning can be subdivided into two groups: *information-oriented* and *logic-oriented* systems [28].

The so-called logic-oriented approach generates reactions as bond-breaking and bond-making steps. These steps are often combined with mechanistic or thermodynamic considerations. Logic-oriented systems are mostly based on a clear formal treatment of reactions. In principle, logic-oriented systems should not only be able to predict known reactions but should also generate novel reactions. This is both an advantage and a disadvantage. On one hand, such a system may suggest a reaction nobody has foreseen and thus the system provides a new approach to the synthesis problem being considered. On the other hand, it may generate a huge number of chemically invalid reactions, which an experienced chemist would intuitively avoid. Therefore, the output of a logic-oriented system has to be thoroughly verified by suitable evaluation techniques [29].

Information-oriented systems are based on a library of known retro-reactions which have been collected and evaluated by a group of chemists while coding them in electronic form. In addition, information on the scope and the expected yield under various conditions, as well as a strategic merit, is usually stored. Such a reaction library is called a *knowledge base*. In synthesis design programs the knowledge base normally consists of a database of transforms. Each transform (retro-reaction) has been derived from a number of experimentally performed reactions. Compared with logic-oriented systems, the advantage of information-oriented systems is their ability to also give crude information to the user about regio- and stereoselectivity, strategic merit, expected yield, and the influence on the yield of substituents and other structural features. Such details will not be attained by any logic-oriented system in the near future [29]. A major disadvantage of information-oriented systems is the amount of work that has to go into the compilation and maintenance of the knowledge base. The generation and updating of a knowledge base require the continuous attention of PhD-level chemists. However, there is always the danger that a transform in a knowledge database has been compiled from too few reactions or from the overrated results of an experiment. Furthermore, it is always difficult to balance the merit function of new entries to the knowledge base against those already contained in the database, especially if a team of several specialists with divergent backgrounds and experience is working on the knowledge database.

### 10.3.2.4 Synthesis Design Systems

The use of computers for the design of chemical syntheses was first demonstrated by Corey and Wipke in 1969 with their program OCCS [30]. The successor to OCCS, LHASA [31], is generally considered to be the first synthon-based system. Its development is still going on. Currently, three groups are working on LHASA, one at Harvard University, USA [32], one at the University of Leeds, UK [33], and one at the CMBI (formerly CAOS/CAMM center) of the University of Nijmegen, NL [34].

Another successor to OCCS was SECS [35], which was further developed into the CASP system by a consortium of German and Swiss chemical companies. The development of both the CASP and the SECS systems was discontinued in the early 1990.

Because of the complexity of the problem and the large amount of program development work that has to go into a synthesis design system, only a few groups worldwide have been active in this area. Here, we mention only a selection of the major ideas and achievements in this area. It is not the intention to give a comprehensive overview. For this, interested readers can consult Chapter X, Section 3.2 in the Handbook. This chapter presents one such system, the WODCA program, in greater detail.

Besides LHASA (Section 10.3.2.4.1), other important program systems whose development still continues are SYNCHEM [36, 37], SYNGEN [38, 39], AIPHOS [40], and WODCA [41, 42]. These systems will be presented briefly in the following sections. Zefirov's group has been actively producing a sequence of program systems over the last few decades [43] (see also Chapter X, Section 3.2 in the Handbook). The CHIRON system [44] is a special program as it does not deal explicitly with chemical reactions. Its major focus is on the control of stereochemistry in organic synthesis. It offers a variety of tools to search in a database of compounds from the chiral pool and to match the substructures of the synthesis target stereospecifically on chiral starting materials.

### 10.3.2.4.1 LHASA

LHASA works in a rigorously retrosynthetic manner. It is designed to be used interactively as the user decides which strategies to apply. At present, it contains a knowledge database of about 2200 transforms and 500 so-called tactical combinations [33]. After a target structure is drawn, the user chooses a strategy for the retrosynthetic analysis. LHASA then searches its knowledge base for transforms which can be applied to the target structure, i.e., those for which the retron for the corresponding transform is present in the target structure. The major strategies for the retrosynthetic analysis are [25, 29]:

- *Functional group based or short-range strategies*: The transform selection is guided by the presence and location of functional groups. The various strategies are: unmasking uncommon and/or complex functionality, disconnective (C–C bonds only), reconnective, non-disconnective, and unstrained application of transforms.
- *Topological strategies*: Here, the program applies heuristics to select a strategic bond. The disconnection of a strategic bond should lead to a maximum of simplification. Different strategies can be invoked for different types of bonds, depending on the topology of the target structure (cyclic strategic bonds, polyfused strategic bonds, ring appendages, acyclic strategic bonds, and manually designated bonds).

- *Transform-based or long-range strategies*: The retrosynthetic analysis is directed toward the application of powerful synthesis transforms. Functional groups are introduced into the target compound in order to establish the retron of a certain goal transform (e.g., the transform for the Diels–Alder reaction, Robinson annulation, Birch reduction, halolactonization, etc.).
- *Stereochemical strategies*: The transform selection is guided by stereocenters that have to be removed in retrosynthesis. The user has to select strategic stereocenters.
- *Strategies based on starting materials*: The analysis is directed toward a particular starting material. A database of potential starting materials can be searched for compounds that may be useful starting points for the synthesis of a given target.

Current research in LHASA is focused on developing new methods assisting users in choosing the most appropriate strategy for their target structures. Furthermore, the development of criteria for the evaluation of synthesis routes and/or algorithms for selecting optimal synthesis routes within the tree are discussed [34].

### 10.3.2.4.2 **SYNCHEM**

The SYNCHEM project was initiated by H. Gelernter and his group in the early 1970s. SYNCHEM was designed to offer organic chemists a tool for discovering valid synthesis routes for organic molecules without requiring them to provide on-line guidance. The development of SYNCHEM was also motivated by the conviction that computer-assisted synthesis design provides an environment for the application of artificial intelligence techniques [36]. The user has to draw the target compound and to define some criteria for terminating an analysis.

The major parts of the system are the user interface, a knowledge base, and the so-called inference engine. The knowledge base of SYNCHEM is a library of about 1000 generalized reaction schemata similar to transforms, and a library of 5000 available starting materials selected from the catalogs of several chemical suppliers. The library of starting materials is used to terminate each branch of a generated synthesis tree. The inference engine comprises heuristics which should discover the most suitable and efficient synthesis path within all reasonable synthesis paths that could be generated by a systematic application of the knowledge base to the target compound.

In the SYNCHEM project, an attempt was also made to build the transform library by automated extraction from reaction data. The methodology of inductive and deductive learning for building and redefining of a knowledge database was implemented in the programs BRANGÄNE, ISOLDE, and TRISTAN by Gelernter's group [45].

### 10.3.2.4.3 **SYNGEN**

During the early 1970s J.B. Hendrickson proposed a simple mathematical model for the logical description of structures and reactions [46] which was later on implemented in the program system SYNGEN [38]. SYNGEN generates all conceivable approaches to construct the carbon skeleton of the target structure within defined constraints and without user intervention [47]. A major difference from other synthesis design programs is that no external knowledge base is needed for defining transforms or reaction schemata for the retrosynthetic analysis. Therefore, SYNGEN can be assigned to the group of logic-oriented systems. The central feature of a synthesis strategy generated by SYNGEN is the assembly of the carbon skeleton in the target from the skeletons of available starting materials.

The retrosynthetic analysis is performed in two steps: in a first step, SYNGEN dissects the skeleton to find all fully convergent bondsets which utilize starting material skeletons found in two successive levels of cuts. A bondset is a set of skeletal bonds that is cut during the retrosynthetic analysis or formed in any given synthesis.

In the second step of the analysis, the necessary functionality is generated for each of such bondsets at the respective carbon skeletons in order to make the synthesis feasible. The choice of the bondsets employed for the generation of precursors is controlled by a number of different heuristics, preferably taking care to find precursors of the same size and avoiding fragments that are too small. The construction reactions are generated from mechanistic principles laid down in a generalized description.

### 10.3.2.4.4 **AIPHOS**

The Japanese program system AIPHOS is developed by Funatsu's group at Toyohashi Institute of Technology [40]. AIPHOS is an interactive system which performs the retrosynthetic analysis in a stepwise manner, determining at each step the synthesis precursors from the molecules of the preceding step. AIPHOS tries to combine the merits of a knowledge-based approach with those of a logic-centered approach.

After the user has input the target structure, the program proposes plausible strategic bonds or, using the AIPHOS terminology, strategic sites. For the detection of strategic sites, AIPHOS applies a topological strategy and/or a functional group based strategy. Then, the user has to select one of the proposed strategic sites or to impose a certain synthesis strategy by manually specifying strategic sites. On the basis of the selected strategic sites, possible sets of precursors are generated. After the user has chosen one of these sets, AIPHOS automatically adds appropriate leaving groups, by utilizing its leaving-group knowledge base. Whether the proposed synthesis path can occur or not is evaluated by the reaction knowledge base of AIPHOS. If the evaluation is successful, the proposed retrosynthetic path and the related reaction scheme of the AIPHOS database are displayed to the user.

AIPHOS provides methods for the recognition of synthetic equivalents to incorporate data from a database of available fine chemicals. Forty functional groups are categorized into nine groups of synthetic equivalents. These groups are applied for the abstraction of data, and the resulting library is structured into a hierarchy using abstract graphs. Another interesting derivative of the AIPHOS system is the reaction prediction program SOPHIA, which utilizes the same reaction knowledge base as AIPHOS in the "reverse direction", i.e., in the direction of the synthesis [48].

### 10.3.2.4.5 WODCA

The synthesis design program WODCA has been being developed by J. Gasteiger's group since 1990. Previously, synthesis planning was an integral part of the program system EROS [49–52], which is now focused on reaction prediction. However, since the advancement and refinement of both approaches became more and more sophisticated, it seemed no longer reasonable to treat synthesis planning and reaction prediction in one system. Therefore, around 1990 the decision was made to handle both problems by separate systems. Initially, WODCA had been developed for planning of the syntheses of individual target compounds. Work in recent years now also enables its use in designing entire libraries of compounds [53].

A major concern during the development of WODCA was to break through the conceptual limits of the classical synthesis design programs. Most of the classical systems for synthesis design rely on large transform libraries of generally formulated synthon schemes which are a distilled essence from many sources, from practical experience, or derived from classical name reactions. These systems first search for a certain pattern of substituents or chemical functionality and then systematically apply all corresponding reactions coded in the transform libraries to the target compound.

A drawback of this approach is that it typically generates enormous and unwieldy synthesis trees which contain a large number of dead-end branches which are not worth further consideration. Furthermore, the chemist is forced to follow a rigid scheme during the planning process, alternating between the application of transforms, the derivation of new precursors, and again the application of further transforms to these precursors.

In reality, chemists do not follow such a rigid scheme. At the beginning of a planning process they look for larger structural fragments that might serve as building blocks for the target molecule. Additionally, they evaluate the functionality pattern of the target structure; this does not necessarily mean that their evaluation is done according to the synthon principle. Thus, at quite an early stage of the planning process, they search for long-range relationships with available starting materials. If a set of possible intermediates and starting materials has been collected, the gaps between them are filled with specific reactions or synthon schemes. The entire planning process is neither a classical bottom-up nor a top-down search – it is a complex framework consisting of the target compound, available starting materials, still-imprecise intermediates, and a knowledge of applicable reactions (see Figure 10.3-34) [41].

The most important task of a synthesis planning program is to assist the user during this process. Therefore, WODCA intentionally has been designed as an interactive system to assist the chemist when working on synthesis problems. The system comprises methods for searching for available starting materials, for generating suitable synthesis precursors, and for finding known synthesis

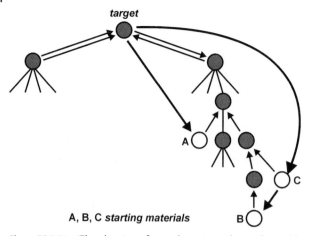

**A, B, C** *starting materials*

**Figure 10.3-34.** The planning of a synthesis is neither a classical bottom-up nor a top-down search, it is an array of different methods for searching and planning, applied in both the retrosynthetic and the synthetic directions.

reactions (Figure 10.3-35). All these methods do not have been used by the chemist in a predefined sequence, but they can be applied at any time to any object within a study. Thus, a study can be expanded in any direction by starting at different points of the synthesis tree.

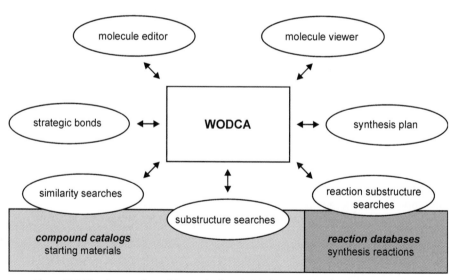

**Figure 10.3-35.** An overview of the methods in WODCA.

Search for Available Starting Materials

To search for available starting materials, similarity searches, substructure searches, and some classical retrieval methods such as full structure searches, name searches, empirical formula searches, etc., have been integrated into the system. All searches can be applied to a number of catalogs of available fine chemicals (e.g., Fluka [54]). In addition, compound libraries such as in-house catalogs can easily be integrated.

Similarity searches are essential methods in WODCA to be used to generate a relationship between the synthesis target or precursors and available starting materials. Similarity searches in WODCA are performed on the basis of similarity criteria that are either based on the presence of substructures or on generalized reactions. Each criterion contains a rule for the transformation of the query structure (target, precursor). The transformed query is then compared with all transformed molecules in the database (the transformed molecules in the database having been computed and stored during the implementation of the catalog). If both the transformed query and the transformed catalog compound are identical, then the database structure is suggested to the user as suitable starting material. In Figure 10.3-36 the principle of similarity searches is illustrated by the criterion "maximum oxidation state".

The comparison for identity of chemical structures is performed on the basis of a hashcode algorithm. Hashcode are unique integer numbers which can be calculated for each chemical structure within the WODCA system [55]. Further exam-

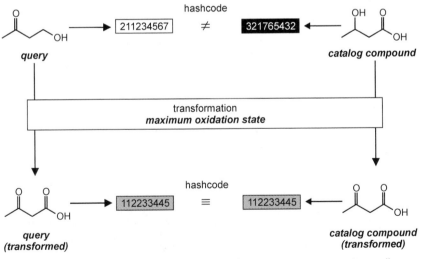

**Figure 10.3-36.** The principle of similarity searches. The query (target, precursor) as well as the catalog compound are transformed by the criterion "maximum oxidation state". Since the transformation for both compounds results in the same transformed structure, the catalog compound is presented to the user as a suitable starting material. The comparison of the structure is performed by a hashcode algorithm.

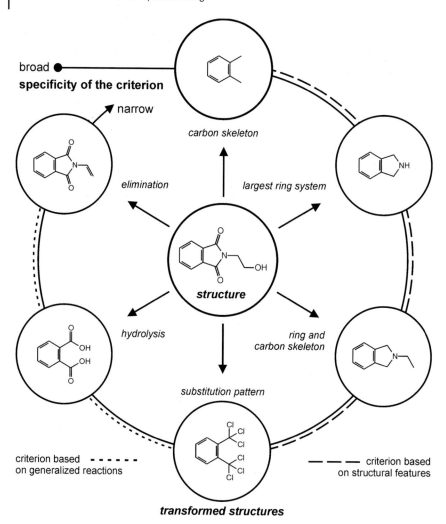

**Figure 10.3-37.** The application of some of the similarity criteria to a query compound (N-(2-hydroxyethyl)phthalimide). The specificity of the similarity criteria applied increases clockwise from criteria with a broad scope to others with quite a narrow scope. The chlorine atoms in the transformed structure of the criterion "substitution pattern" are representing marker atoms which are a generalized form for any kind of substituent. The displayed criterion "substitution pattern" replaces all heteroatoms by chlorine atoms whereby multiple substitution by a heteroatom is counted (i.e., each bond of the double bond to the oxygen atom is counted as substituent and replaced by a chlorine atom). Since the application of this criterion to N-(2-hydroxyethyl)phtha-limide results in two separate carbon skeleton fragments (the structure illustrated and 1,2-dichloroethane), only the larger fragment is considered.

ples of the application of similarity criteria to *N*-(2-hydroxyethyl)phthalimide as the query compound are illustrated in Figure 10.3-37. The criteria are arranged in order of increasing specificity, starting with those that have a broad scope and ending with quite narrowly focused reaction types.

Searching for Building Blocks of a Combinatorial Library

Substructure searches provide another method of searching for available starting materials. They are used primarily for planning the synthesis of combinatorial libraries. After the target compound has been dissected into a set of suitable precursors, substructure searches can provide for each of them a series of representatives of a certain class of compounds. Substructure searches enable the user to specify attributes such as open sites or atom lists at certain positions of the structure. Figure 10.3-38 shows the possible specification elements for the query in a substructure search.

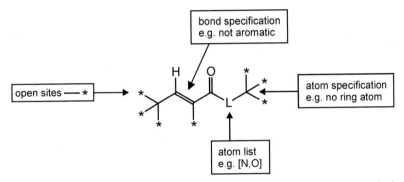

**Figure 10.3-38.** An indication of other specification elements for a substructure query. For both the atom specifications and the bond specifications a vast list of attributes can be set (aromatic/ not aromatic, member of ring with *n* atoms, substituted, etc.).

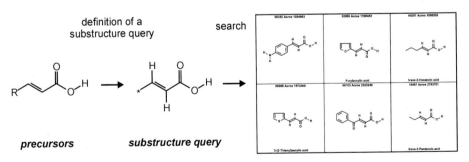

**Figure 10.3-39.** A substructure search in a catalog of fine chemicals. The precursor illustrated may be needed for the synthesis of a library. A substructure is defined by setting an open site at the position of the R group. Some representative examples of the 69 compounds in the Acros catalog which were obtained are shown.

Figure 10.3-39 shows an example of a substructure search in a catalog of fine chemicals. The precursor illustrated maybe needed, for instance, as a building block for a compound library. The substructure search provides the available structural manifoldness for this precursor.

Generating Synthesis Precursors

Only in simple synthesis problems one can expect a search for suitable starting materials to succeed straight away with the target compound. In most cases, the synthesis problem has to be simplified by first generating synthesis precursors. Therefore, methods for the search and evaluation of strategic bonds have been implemented in WODCA whereby the application of a classical synthon- or transform-oriented knowledge base has been avoided. For each bond, the two heterolytic bond-breaking possibilities are explored and evaluated by physico-chemical effects such as bond polarity [56], and the stabilization of charges by inductive [57], resonance [58], and polarizability effects [59]. These methods are calculated by rapid procedures collected in the PETRA system [60, 61]. Furthermore, structural effects such as points of bond branching, positions in ring systems, and stereochemical centers are taken into account. Four different groups of disconnection strategies have been implemented, working on the basis of these parameters:

- carbon–heteroatom bonds,
- aliphatic bonds,
- aromatic substitution,
- polycyclic compounds.

Each bond that matches one of these criteria is rated by a value between 0 (no strategic bond) and 100, which is the maximum rating for a strategic bond. Subsequently, the rating for each strategic bond is scaled relatively to the value of the highest-rated bond in the compound. As an example, the rating scheme for a heterocyclic compound is illustrated in Figure 10.3-40.

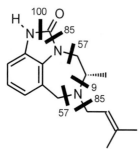

**Figure 10.3-40.** The rating for the disconnection strategy "carbon–heteroatom bonds" is illustrated. Please focus on the nitrogen atom of the tertiary amino group. It is surrounded by three strategic bonds with different values. The low value of 9 for one of these bonds arises because this bond leads to a chiral center. Since its formation requires a stereospecific reaction the strategic weight of this bond has been devalued. In contrast to that, the value of the bond connecting the exocyclic rest has been increased to 85, which may be compared with its basic value as an amine bond.

Searching for Synthesis Reactions

In order to verify a retrosynthetic step suggested by WODCA, a direct connection to reaction databases (e.g., Theilheimer [62]) has been established in the most recent version of WODCA.

Reaction databases contain a wealth of reactions performed in the laboratory and published in the literature, i.e., in contrast to the transform libraries of synthesis design programs they contain raw, uninterpreted reaction information. In Figure 10.3-41 a schematic representation of a reaction in a reaction database is given.

The entry in a reaction database contains the structures of reactants and products as connection tables (in some case also the structures of catalysts, reagents, and/or solvent are included). Each of these structures has to be classified by its proper role (i.e., reactant, product etc.). A crucial feature of a reaction entry in a reaction database is the so-called *atom-to-atom mapping*. The mapping explicitly assigns atoms of the reactant structures to corresponding atoms of the product structures. It enables the user to specify that a particular substructure in the product is derived from a corresponding substructure in the reactants. This implicitly defines the reaction center of a reaction. Additionally, the reaction center is marked by specifying the role of certain bonds during a reaction, i.e., a bond can be annotated as "bond is part of the reaction center or not", "bond is broken or formed", or "bond is changing bond order". Furthermore, general information for a reaction is stored in a reaction database (e.g., literature citation, yield, catalyst, solvent, pressure, etc.).

In WODCA, the connection to reaction databases is implemented on the basis of reaction substructure searches. Thus, the suggested retrosynthetic step has to be generalized by detecting the reaction substructure (see Section 10.3.2.2.3). Through the breaking of one or more strategic bonds, precursors are generated. These precursors, in conjunction with the target compound, define a conceivable synthesis step containig the reaction site and all the participating compounds.

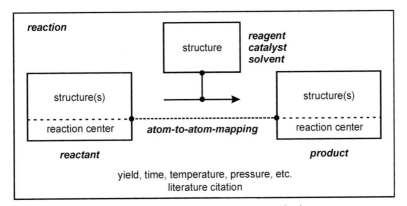

**Figure 10.3-41.** Schematic representation of a reaction in a reaction database.

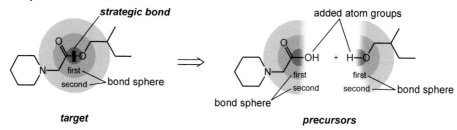

**Figure 10.3-42.** Deriving the reaction substructure. A different number of bond spheres around the strategic bond can be included in the reaction substructure, thereby influencing the specificity of a search in a reaction database.

The position of the chosen strategic bond locates the reaction center. To derive the reaction substructure, the user can select the number of bond spheres around the strategic bond which should be included. The reaction substructure obtained is then used as the query for a reaction substructure search in the database. Figure 10.3-42 illustrates the first and second bond spheres around a selected strategic bond of a retrosynthetic step.

The number of bond spheres chosen influences the specificity of the reaction substructure query and the result of the search. In Figure 10.3-43 the reaction substructure query including the first bond sphere of the retrosynthetic step of Figure 10.3-42 is shown.

The search in the Theilheimer reaction database [62] provides 161 reactions for this query. If the search is performed without any additional bond spheres (covering only atoms of the inner sphere with a dark gray background in Figure 10.3-42 as well as the added atom groups on the precursor side), 705 reactions are obtained in the Theilheimer database. The result of this search is less precise than that of the first search. Additionally, reactions forming any kind of C–O bonds (e.g., making an ether bond instead of an ester bond) are found. However, in both searches too many hits are obtained in order to detect suitable reactions in a reasonable

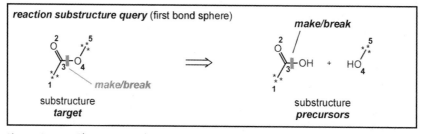

**Figure 10.3-43.** The reaction substructure embracing the strategic bond and first bond sphere of the retrosynthetic step in Figure 10.3-42. For each atom of the reaction substructure the atom-to-atom mapping is set (the number on the atoms). For a suggested retrosynthetic step, the mapping as well as the marking of the reaction center (make/break specification for the corresponding bonds) is set automatically within WODCA. The asterisks represent the numbers of open sites.

manner. A more precise search including the second bond sphere provides 13 reactions in the Theilheimer database. Compared with the previous searches it provides reactions with the reaction center most closely related to the retrosynthetic step of Figure 10.3-42. However, in other cases a search with too specific a query may provide no hits, i.e., the user has to strike a balance between the specificity of the query and the number of hits obtained in a search.

## 10.3.2.5 Tutorial: Synthesis Design with the WODCA Program

### 10.3.2.5.1 Introduction
The following example demonstrates the application of WODCA to a simple target compound, 7-hydroxy-4-methyl-3-phenylcoumarin (**1**) shown in Figure 10.3-44. It belongs to the group of disubstituted 7-hydroxycoumarins which have gained some interest as precursors to anti-AIDS agents [63]. We have intentionally chosen a rather simple synthesis target in order to demonstrate the various features of WODCA's planning tools in a limited amount of space.

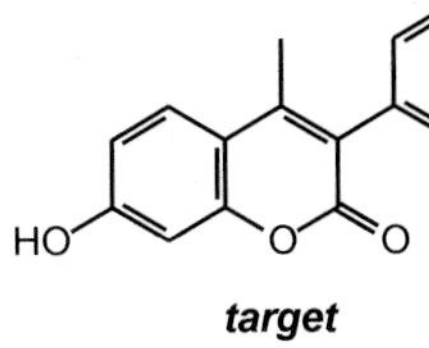

(1)

**Figure 10.3-44.** The target compound **1**.

*target*

### 10.3.2.5.2 Starting WODCA
After the WODCA system has been started the main window appears (Figure 10.3-45). It consists of several parts. The so-called information area, containing icons which display name and size of the currently loaded catalog of chemical, is located on the left-hand side (Figure 10.3-45, **A**). Furthermore, the status of the reaction database (Figure 10.3-45, **B**) is given. WODCA's structure display is on the right-hand side (Figure 10.3-45, **C**). It is used to plot the chemical compound which is actually analyzed. The WODCA console (Figure 10.3-45, **D**) gives additional information about the last action that has been performed.

### 10.3.2.5.3 Define New Target Compound
In order to draw a desired target compound WODCA provides a molecule editor which has to be launched. This can be performed by pressing the "New Compound" button in WODCA's button bar. The molecule editor is then started and afterwards the target compound can be drawn directly with the mouse.

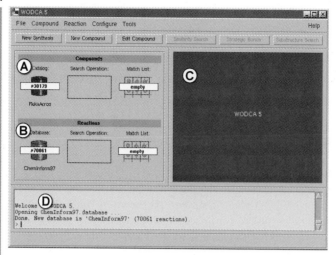

**Figure 10.3-45.** The main window of WODCA.

### 10.3.2.5.4 **Start New Synthesis Tree**

Once a target compound has been loaded into WODCA, a search for this structure in the database of available starting materials is performed automatically in the background. The result of this identity search is given below WODCA's structure display (see the arrow in Figure 10.3-46). The "New Synthesis" button in WODCA's button bar is to be used to generate a new synthesis tree for the target compound, giving a graphical representation of the entire synthesis plan that is to be developed.

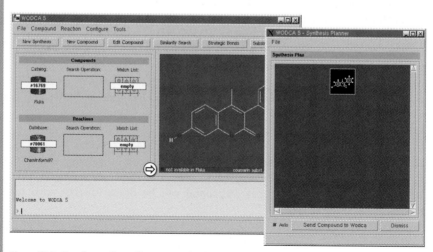

**Figure 10.3-46.** Generation of a new synthesis tree.

##### 10.3.2.5.5 Finding Suitable Starting Materials

Since the target compound is not contained in the catalog of chemicals, the study is continued by performing similarity searches in order to find suitable starting materials in the catalog of chemicals. The similarity search is initiated by pressing the "Similarity Search" button in WODCA's button bar. The window for similarity searches is then displayed. After the desired category of similarity criteria has been selected, all similarity searches of this category are performed automatically. As indicated by the arrow in Figure 10.3-47, the result of each similarity search is then displayed in brackets behind the corresponding similarity criterion.

For the similarity criterion "Largest Ring System" WODCA has found 41 compounds in the Fluka catalog of chemicals which contain exactly the same ring system as the target compound. The hits differ in the number of substituents and their positions in the ring system, since these two characteristics are not considered by this similarity search. A selection of the chemical structures is given in Figure 10.3-48, from which one can see that some potential starting materials, e.g., 7-ethoxycoumarin or 4-methylumbelliferyl sulfate, are found.

A more rigid similarity criterion is "Ring + Substitution Position" in which each substitution position on the largest ring system is marked by a chlorine atom as a place-holder for any type of substituent (see Figure 10.3-49, left-hand side). By this similarity search WODCA has found only one compound in the Fluka catalog of chemicals (see Figure 10.3-49, right-hand side). However, this hit seems not to be a suitable starting material for the desired target compound.

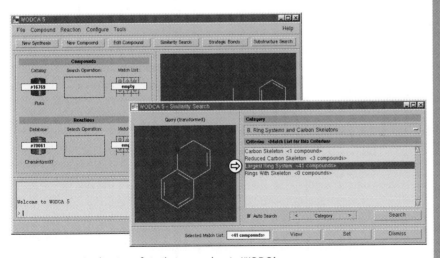

**Figure 10.3-47.** Application of similarity searches in WODCA.

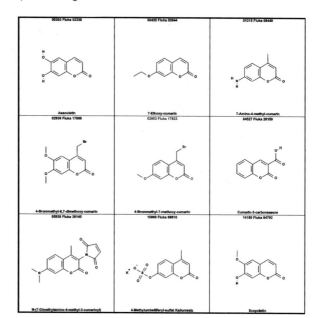

similarity search
**largest ring system**

**Figure 10.3-48.** Nine of 41 hits for the similarity search "Largest Ring System".

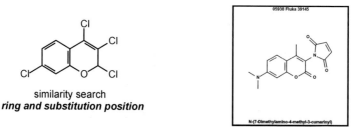

similarity search
**ring and substitution position**

**Figure 10.3-49.** One hit for the similarity search "Ring + Substitution Position".

### 10.3.2.5.6 Disconnection Strategies and Evaluation of Strategic Bonds

After all catalogs of chemicals have been explored for suitable starting materials, the retrosynthetic analysis is continued by studying the existence of strategic bonds within the target compound in order to dissect the target into suitable synthesis precursors. The "Strategic Bonds" button in WODCA's button bar has to be pressed in order to open the window for the retrosynthetic analysis of the target compound. The category "Strategic Bonds" is selected by default. By choosing a suitable disconnection strategy, the retrosynthetic analysis is initiated.

The disconnection strategy "Carbon–Heteroatom Bonds" gives only one strategic bond, which is rated with a value of 100.

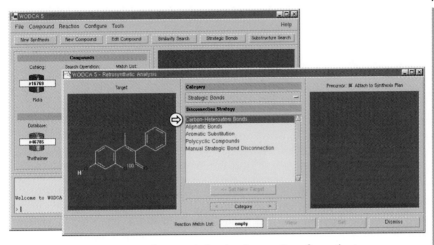

**Figure 10.3-50.** Disconnection of a strategic bond and generation of a synthesis precursor.

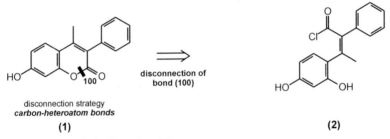

**Figure 10.3-51.** Retrosynthetic dissection of **1**.

The strategic bond is disconnected by just clicking with the mouse on it. A suitable synthesis precursor is then generated automatically by WODCA by adding appropriate atoms or groups to the open valences.

The disconnection of the strategic bond rated 100 corresponds to an esterification in the synthesis direction (Figure 10.3-51). Compound **1** is converted to **2** as a suitable synthesis precursor and, indeed, in the synthesis direction **2** can easily be converted to **1**.

In order to continue the design of a synthesis, **2** is now analyzed further by the disconnection strategy for aliphatic bonds. WODCA detects two strategic bonds which are rated 100 and 67, respectively. The disconnection of the bond with the highest rating forms two precursor compounds, **3** and **4** (Figure 10.3-52).

Both precursors can be used as reactants in an aldol condensation. It has to be emphasized that the chlorine atom in **4** has to be considered as a representative for any electron-withdrawing group; in particular, in the case presented here, it would best be taken as an OEt group. In order to verify this proposal, a reaction substructure search is initiated in the ChemInform reaction database of 1997.

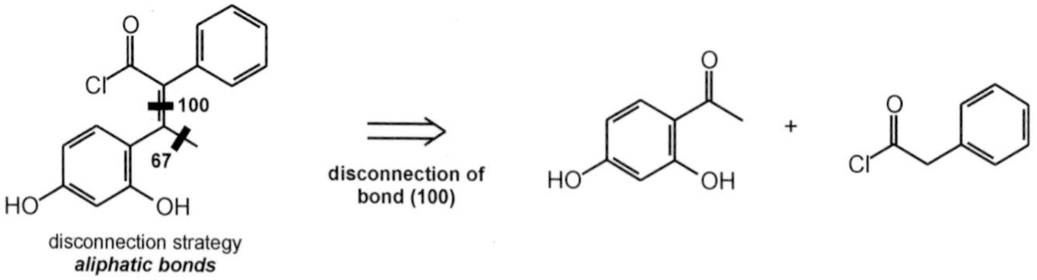

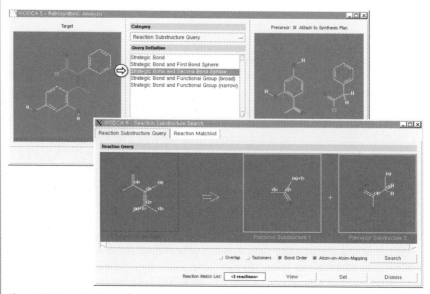

Figure 10.3-52. Disconnection of aliphatic bonds in **2**.

Again, the window for the retrosynthetic analysis is opened, but now the category is changed to "Reaction Substructure Query". After choosing the query definition "Strategic Bond and Second Bond Sphere", WODCA automatically prepares the query for the reaction database opened, and three potential synthesis reactions are found in the ChemInform reaction database (Figure 10.3-53).

Figure 10.3-54 illustrates how the reaction center is derived from the disconnection of the strategic bond and which additional bond spheres are considered in the definition of the reaction substructure search query.

One of the hits found in the ChemInform reaction database is shown in the window for reaction substructure searches in Figure 10.3-55. It fits the synthesis problem perfectly, since in the synthesis direction it forms the coumarin ring system directly, in one step.

Both precursor compounds, **3** and **4**, are available from Fluka, which is checked automatically by WODCA searching in this data file.

Figure 10.3-53. Reaction substructure searches in WODCA.

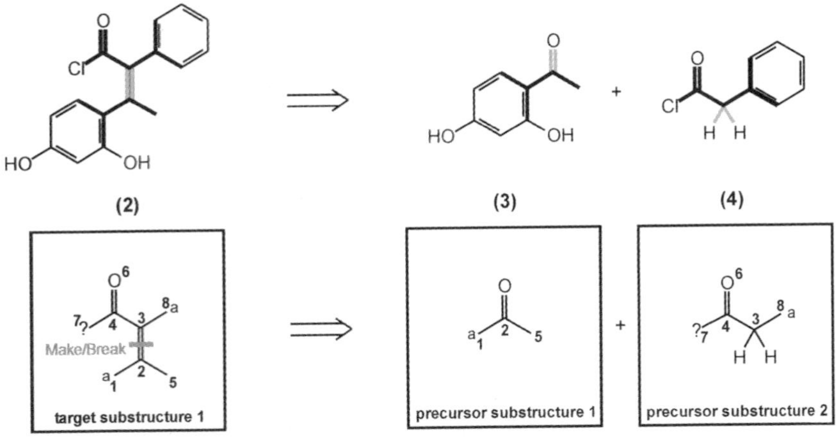

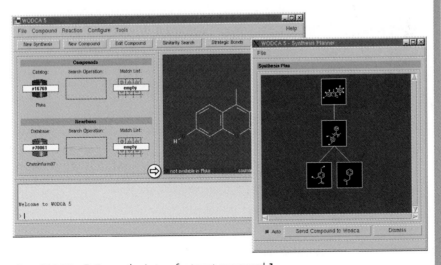

**Figure 10.3-54.** Derivation of a reaction substructure search query. The "a" for atom 1 and 8 indicates that these atoms have to be aromatic in the found reactions.

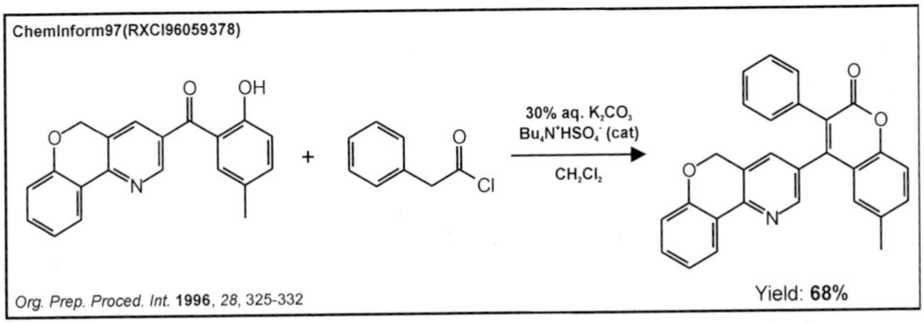

**Figure 10.3-55.** Suitable synthesis reaction found in the ChemInform reaction database.

**Figure 10.3-56.** Entire synthesis tree for target compound 1.

The design of the synthesis for **1** is now finished. The target compound (**1**) can be simplified to commercially available starting materials which can easily be converted to the target compound in the synthesis direction. Figure 10.3-56 shows the entire synthesis tree interactively developed with the WODCA program.

### Essentials

- Reaction prediction treats chemical reactions in their forward direction, and synthesis design in their backward, retrosynthetic direction.
- Reaction databases present a rich source of information for the extraction of knowledge for reaction prediction and synthesis design.
- Reaction prediction, or reaction simulation, has to concentrate on the reaction center, i.e., the bonds broken and made in a reaction.
- A reaction should be modeled as closely as possible to its mechanism.
- The evaluation of chemical reactions can be performed to various levels of sophistication; heats of reaction allow for a consideration of the thermodynamics of a reaction, whereas reaction rates consider its kinetic aspects.
- The more details of a reaction type are available, the better can be the modeling of a reaction of that type.
- The evaluation of the impact of a chemical on the environment has to consider its degradation products also.
- The metabolism of a potential drug has to be considered at an early phase of the drug development process.
- Biochemical pathways and metabolic reaction networks have recently attracted much interest and are an active and rich field for research.
- The disconnection approach has basically changed the view on planning a synthesis.
- The retrosynthetic analysis of a target compound is a systematic approach in developing a synthesis plan starting with the target structure and working backward to available starting materials.
- Several concepts for the implementation of synthesis design system have appeared since the early 1970s.

### Interesting Web Sites

- *http://www2.chemie.uni-erlangen.de/software/eros/index.html*
- *http://www2.chemie.uni-erlangen.de/software/cora/index.html*
- *http://www2.chemie.uni-erlangen.de/software/wodca/index.html*
- *http://www.expasy.ch*
- *http://www.genome.ad.jp/kegg/*
- *http://www2.chemie.uni-erlangen.de/support/biopath/index.html*
- *http://www.mol-net.de/databases/biopath/*
- *http://www.cmbi.kun.nl/cheminf/lhasa*

**Selected Reading**

- R. Höllering, J. Gasteiger, L. Steinhauer, K.-P. Schulz, A. Herwig, *J. Chem. Inf. Comput. Sci.* **2000**, *40*, 482–494.
- *Biochemical Pathways*, G. Michal, Ed., Spektrum Akademischer Verlag, Heidelberg, **1999**.
- W. D. Ihlenfeldt, J. Gasteiger, Angew. Chem. **1995**, *107*, 2807–2829; *Angew. Chem. Int. Ed. Engl.* **1995**, *34*, 2613–2633.

# References

[1] H. Gelernter, J. R. Rose, C. Chen, *J. Chem. Inf. Comput. Sci.* **1990**, *30*, 492–504.

[2] H. Satoh, K. Funatsu, *J. Chem. Inf. Comput. Sci.* **1995**, *35*, 34–44.

[3] L. Chen, J. Gasteiger, *J. Am. Chem. Soc.* **1997**, *119*, 4033–4042.

[4] H. Satoh, O. Sacher, T. Nakata, L. Chen, J. Gasteiger, K. Funatsu, *J. Chem. Inf. Comput. Sci.* **1998**, *38*, 210–219.

[5] http://www2.chemie.uni-erlangen.de/ software/cora/index.html

[6] This section has been provided by O. Sacher. See also: O. Sacher, PhD Thesis, University of Erlangen-Nuremberg, **2001**, pp. 101–130.

[7] J. J. Li et al. *J. Med. Chem.* **1995**, *38*, 4570–4578.

[8] The CAMEO project is outlined in a series of publications; only the first, one somewhat later, and a more recent reference are given: a) T. D. Salatin, W. L. Jorgensen, *J. Org. Chem.* **1980**, *45*, 2043–2051; b) M. G. Bures, B. L. Roos-Kozel, W. L. Jorgensen, *J. Org. Chem.* **1985**, *50*, 4490–4498; c) W. L. Jorgensen, E. R. Laird, A. J. Gushurst, J. M. Fleischer, S. A. Gothe, H. E. Helson, G. D. Paderes, S. Sinclair, *Pure Appl. Chem.* **1990**, *62*, 1921–1932.

[9] J. Gasteiger, C. Jochum, *Topics Curr. Chem.* **1978**, *74*, 93–126.

[10] J. Gasteiger, M. G. Hutchings, B. Christoph, L. Gann, C. Hiller, P. Löw, M. Marsili, H. Saller, K. Yuki, *Topics Curr. Chem.* **1987**, *137*, 19–73.

[11] W.-D. Ihlenfeldt, J. Gasteiger, *Angew. Chem.* **1995**, *107*, 2807–2829; *Angew.*

*Chem. Int. Ed. Engl.* **1995**, *34*, 2613–2633.

[12] R. Höllering, J. Gasteiger, L. Steinhauer, K.-P. Schulz, A. Herwig, *J. Chem. Inf. Comput. Sci.* **2000**, *40*, 482–494.

[13] J. Gasteiger, U. Hondelmann, P. Röse, W. Witzenbichler, *J. Chem. Soc. Perkin 2* **1995**, 193–204.

[14] W. H. Press, B. P. Flannery, S. A. Teukolsky, W. T. Vettering, Numerical Recipes. The Art of Scientific Computing, Cambridge University Press, New York, 1989.

[15] C. W. Gear, Numerical Initial Value Problems in Ordinary Differential Equations, Prentice-Hall, Englewood Cliffs, NJ, 1971; Gear Algorithm, QCPE Program No. QCMP022.

[16] This example is also explained at http://www2.chemie.uni-erlangen.de/ services/eros/

[17] N. Burckhard, J. A. Guth, *Pestic. Sci.* **1981**, *12*, 45.

[18] J. Gasteiger, S. Bauerschmidt, U. Burkard, M. C. Hemmer, A. Herwig, A. von Homeyer, R. Höllering, T. Kleinöder, T. Kostka, C. Schwab, P. Selzer, L. Steinhauer, *SAR & QSAR in Environm. Res.* **2002**, *13*, 89–110.

[19] Biochemical Pathways, G. Michal, Ed., Spektrum Akademischer Verlag, Heidelberg, 1999.

[20] http://www.expasy.ch

[21] a) http://www2.chemie.uni-erlangen. de/support/biopath/index.html; b) http://www.mol-net.de/databases/ biopath/ c) http://www.mol-net.de

[22] C. Chassagnole, N. Noisommit-Rizzi, J. W. Schmidt, K. Mauch, M. Reuss, *Biotechnol. Bioeng.* **2002**, *79*, 53–73.

[23] E. J. Corey, X.-M. Cheng, The Logic of Chemical Synthesis, Wiley, New York, 1989.

[24] E. J. Corey, *Pure Appl. Chem.* **1967**, *14*, 19–37.

[25] E. J. Corey, *Angew. Chem.* **1991**, *103*, 469–479; *Angew. Chem. Int. Ed. Engl.* **1991**, *30*, 455–465.

[26] S. Warren, Organic Synthesis – The Disconnection Approach, Wiley, Chichester, 1982.

[27] R. Barone, M. Chanon, Synthesis design, in The Encyclopedia of Computational Chemistry, Vol. 4, P. v. R. Schleyer, N. L. Allinger, T. Clark, J. Gasteiger, P. A. Kollman, H. F. Schaefer III and P. R. Schreiner (Eds.); Wiley, Chichester, **1998**, 2931–2946.

[28] I. Ugi, J. Bauer, R. Baumgartner, E. Fontain, D. Forstmeyer, S. Lohberger, *Pure Appl. Chem.* **1988**, *60*, 1573.

[29] M. A. Ott, J. H. Noordik, *Recl. Trav. Chim. Pays-Bas* **1992**, *111*, 239–246.

[30] E. J. Corey, W. T. Wipke, *Science* **1969**, *166*, 178–192.

[31] a) E. J. Corey, G. A. Petersson, *J. Am. Chem. Soc.* **1972**, *94*, 460–465; b) E. J. Corey, W. J. Howe, H. W. Orf, D. A. Pensak, G. Petersson, *J. Am. Chem. Soc.* **1975**, *97*, 6116–6124; c) E. J. Corey, A. K. Long, J. Mulzer, H. W. Orf, P. Johnson, A. P. W. Hewett, *J. Chem. Inf. Comput. Sci.* **1980**, *20*, 221–230; d) E. J. Corey, A. P. Johnson, A. K. Long, *J. Org. Chem.* **1980**, *45*, 2051–2057; e) E. J. Corey, A. K. Long, G. I. Lotto, S. D. Rubenstein, *Recl. Trav. Chim. Pays Bas* **1992**, *111*, 304–310; f) A. P. Johnson, C. Marshall, P. N. Hudson, *Recl. Trav. Chim. Pays Bas* **1992**, *111*, 311–317.

[32] LHASA Group homepage, http://lhasa.harvard.edu/ (last update: 2000).

[33] LHASA Limited homepage, http://www.chem.leeds.ac.uk/luk/ (last update: Feb. 26, 1999).

[34] CMBI – Synthesis Planning: LHASA, http://www.cmbi.kun.nl/cheminf/ lhasa/ (last update: Jul 20, 2001).

[35] W. T. Wipke, G. I. Ouchi, S. Krishnan, *Artif. Intell.* **1978**, *11*, 173–193.

[36] H. L. Gelernter, A. F. Sanders, D. L. Larsen, K. K. Agarwal, R. H. Boivie, G. A. Spritzer, J. E. Searleman, *Science* **1977**, *197*, 1041–1049.

[37] The SYNCHEM Group, http:// www.cs.sunysb.edu/~synchem/ (last update: May 15, 2002).

[38] G. J. B. Hendrickson, *Angew. Chem.* **1990**, *102*, 1328–1338; *Angew. Chem. Int. Ed. Engl.* **1990**, *29*, 1286–1296.

[39] http://syngen2.chem.brandeis.edu/ syngen.html (last update: May 21, 2001).

[40] a) K. Funatsu, S. Sasaki, *Tetrahedron: Comput. Methodol.* **1988**, *1*, 27–37; b) K. Satoh, K. Funatsu, *J. Chem. Inf. Comput. Sci.* **1999**, *39*, 316–325.

[41] W. D. Ihlenfeldt, J. Gasteiger, *Angew. Chem.* **1995**, *107*, 2807–2829; *Angew. Chem. Int. Ed. Engl.* **1995**, *34*, 2613–2633.

[42] http://www2.chemie.uni-erlangen.de/ software/wodca

[43] a) N. S. Zefirov, E. Gordeeva, S. S. Tratch, *J. Chem. Inf. Comput. Sci.* **1988**, *28*, 188–193; b) N. S. Zefirov, E. Gordeeva, *J. Org. Chem.* **1988**, *53*, 527–532; c) N. S. Zefirov, S. S. Tratch, M. S. Molchanove, *MATCDY*, **2002**, 253–273; d) S. S. Tratch, M. S. Molchanove, N. S. Zefirov, *MATCDY*, **2002**, 275–301.

[44] a) S. Hanessian, J. Franco, G. Gagnon, D. Laramée, B. Larouche, *J. Chem. Inf. Comput. Sci.* **1990**, *30*, 413–425; b) S. Hanessian, A. M. Faucher, S. Léger, *Tetrahedron* **1990**, *46*, 231–243; c) S. Hanessian, M. Botta, B. Larouche, A. Boyaroglu, *J. Chem. Inf. Comput. Sci.* **1992**, *32*, 718–722.

[45] H. L. Gelernter, J. R. Rose, C. Chen, *J. Chem. Inf. Comput. Sci.* **1990**, *30*, 492–504.

[46] J. B. Hendrickson, *J. Am. Chem. Soc.* **1971**, *93*, 6847–6854.

[47] J. B. Hendrickson, D. L. Grier, A. G. Toczko, *J. Am. Chem. Soc.* **1985**, *107*, 5228–5238.

[48] H. Satoh, K. Funatsu, *J. Chem. Inf. Comput. Sci.* **1995**, *35*, 34–44.

[49] J. Gasteiger, C. Jochum, *Topics Curr. Chem.* **1978**, *173*, 93–126.

[50] J. Gasteiger, M. G. Hutchings, B. Christoph, L. Gann, C. Hiller, P. Löw, M. Marsili, H. Saller, K. Yuki, *Topics Curr. Chem.* **1987**, *137*, 19–73.

[51] P. Röse, J. Gasteiger, *Anal. Chim. Acta* **1990**, *235*, 163–168.

[52] R. Höllering, J. Gasteiger, L. Steinhauer, K.-P. Schulz, A. Herwig, *J. Chem. Inf. Comput. Sci.* **2000**, *40*, 482–494.

[53] J. Gasteiger, M. Pförtner, M. Sitzmann, R. Hollering, O. Sacher, T. Kostka, N. Karg, *Persp. Drug Discov. Design* **2000**, *20*, 245–264.

[54] Sigma-Aldrich (Fluka): http://www.sigmaaldrich.com

[55] W. D. Ihlenfeldt, J. Gasteiger, *J. Comput. Chem.* **1994**, *15*, 793–813.

[56] J. Gasteiger, M. Marsili, *Tetrahedron* **1980**, *36*, 3219–3228.

[57] M. G. Hutchings, J. Gasteiger, *Tetrahedron Lett.* **1983**, *24*, 2541–2544.

[58] J. Gasteiger, H. Saller, *Angew. Chem.* **1985**, *97*, 699–701; *Angew. Chem. Int. Ed. Engl.* **1985**, *24*, 687–689.

[59] J. Gasteiger, M. G. Hutchings, *J. Chem. Soc. Perkin 2* **1984**, 559–564.

[60] J. Gasteiger, Empirical methods for the calculating of physicochemical data of organic compounds, in Physical Property Prediction in Organic Chemistry, C. Jochum, J. Hicks, J. Sunkel (Eds.), Springer Verlag, Heidelberg, **1988**, pp. 119–138.

[61] http://www2.chemie.uni-erlangen.de/software/petra

[62] The Theilheimer reaction database is distributed by MDL Information Systems, Inc., San Leandro, CA, USA. It contains 46 785 selected reactions from the years between 1946 and 1980.

[63] L. Xie, Y. Takeuchi, L. M. Cosentino, A. T. McPhail, K.-H. Lee, *J. Med. Chem.* **2001**, *44*, 664-671.

## 10.4
## Drug Design

*Lothar Terfloth*

### Learning Objectives

- To become familiar with the drug discovery process
- To find out what a lead structure is
- To appreciate the impact of chemoinformatics on the drug discovery process
- To understand the "similar structure-similar property" principle
- To know what virtual screening is
- To become familiar with Lipinski's "Rule of Five"

### 10.4.1
### Introduction

In this chapter we give an overview of the application of chemoinformatics in the drug discovery process. After a brief introduction and some economic considerations some fundamental definitions in the context of drug design are given. Then the drug discovery process is described. Furthermore, we single out some tasks and problems in drug development to be tackled by chemoinformatics. A general workflow for virtual screening is presented. The fundamental methods applied in structure-based and ligand-based drug design are mentioned, although it is beyond the scope of this chapter to explain them in detail. Finally, we report on some applications of virtual screening in drug design.

Readers interested in medicinal chemistry can obtain an excellent overview from the book *The Practice of Medicinal Chemistry*, edited by Wermuth [1]. The first edition is nicknamed "The Green Bible" by medicinal chemists. "The Red Bible" written by Böhm, Klebe, and Kubinyi describes the development of new drugs [2]. An overview of different classes of drugs and their mechanisms of action is given by Mutschler [3].

Many drugs, such as the sulfonamides introduced by Domagk or penicillin by Fleming, were discovered by serendipity and not as a result of rational drug design [4]. Up to the 1970s hypothetical activity models dominated the syntheses of new compounds in drug research. The biological activity of these compounds was verified by experiments with isolated organs or animals. Accordingly, the throughput was limited by the speed of the biological tests. From about 1980 onward, molecular modeling and the development of *in vitro* models for enzyme inhibition and receptor binding studies attained a growing impact on drug research. In these years, the synthesis of compounds became the time-limiting factor. Based on the progress achieved in gene technology, combinatorial chemistry, and high-throughput test models it became feasible to produce the proteins of interest and to conduct a structure-based design of ligands. Nevertheless, it turned out that many compounds obtained by structure-based drug design showed inappropriate ADMET (absorption, distribution, metabolism, excretion, and toxicity) properties; this consequently caused their attrition in the preclinical or clinical phase.

Remarkable progress has been achieved in the different disciplines involved in pharmaceutical research since 1950. It culminated in the elucidation of the sequence of the human genome. Today, genomics, proteomics, bioinformatics, combinatorial chemistry, and ultrahigh-throughput screening (u-HTS) provide an enormous number of targets and data. Hence, the application of data mining methods and virtual screening have a growing impact on the validation of "druggable" targets, lead finding, and the prediction of suitable ADMET profiles.

## 10.4.2
### Some Economic Considerations Affecting Drug Design

The development of a new drug is both a time-consuming and a cost-intensive process. It takes 12 to 15 years and costs up to € 800 million to bring a new drug to the market. As measured by the market capitalization, the pharmaceutical companies play a pivotal role in the global economy. In February 2003 Pfizer was ranked at position five worldwide, with a market capitalization of € 163 billion. Ranking third as far as the market capitalization in Europe is concerned was GlaxoSmithKline, with a current value of € 101 billion. Novartis was number five in Europe with € 82 billion.

Some recent publications from consulting firms 305 million in an ideal case [5, 6]. The still long development periods until a new compound comes to the market are coupled with a high financial effort. Despite of the introduction of combinatorial chemistry and the establishment of high-throughput screening (HTS), the average number of new chemical entities (NCEs) introduced annually to the world market was only about 37 for the period 1991–1999 (1999: 35; 1998: 27; 1997: 39; 1996: 38; 1995: 35; 1994: 44; 1993: 43; 1992: 36; 1991: 36) [7, 8]. In their report, the Boston Consulting Group comes to the conclusion that genomics could reduce the cost of producing a new drug from € 800 million by about € 305 million in an ideal case [5]. The average expected savings across the research and development pipeline under consideration of scientific and market limitations are estimated at about € 73 million.

In particular, *in silico* methods are expected to speed up the drug discovery process, to provide a quicker and cheaper alternative to *in vitro* tests, and to reduce the number of compounds with unfavorable pharmacological properties at an early stage of drug development. Bad ADMET profiles are a reason for attrition of new drug candidates during the development process [9, 10]. The major reasons for attrition of new drugs are:

- lack of clinical efficacy,
- inappropriate pharmacokinetics,
- animal toxicity,

- adverse reactions in humans,
- commercial reasons [11],
- formulation issues.

The estimated attrition rate in development is up to 90 % [9]. The later in the drug discovery process the development of a new compound is discontinued, the higher is the financial loss for a pharmaceutical company. Therefore, the major reasons for the attrition of a new drug have to be addressed in drug design as early as possible. As a consequence, the prediction of pharmacological properties, in addition to lead finding, is a central task of chemoinformatics in drug development.

### 10.4.3
### Definitions of some Terms in the Context of Drug Design

Before we continue with the description of the drug discovery process, we introduce some terms commonly used in drug design and give their definitions:

- *Lead structure*: According to Valler and Green's definition a lead structure is "a representative of a compound series with sufficient potential (as measured by potency, selectivity, pharmacokinetics, physicochemical properties, absence of toxicity and novelty) to progress to a full drug development program" [12].
- *Ligand*: A ligand is a molecule binding to a biological macromolecule.
- *Enzyme*: Enzymes are endogenous catalysts converting one or several substrates into one or several products.
- *Substrate*: A substrate is the starting material of an enzymatic reaction.
- *Inhibitor*: A ligand preventing the binding of a substrate to its enzyme is called an inhibitor.
- *Receptor*: Receptors are membrane-bound or soluble proteins or protein complexes exerting a physiological effect after binding of an agonist.
- *Agonist*: An agonist is a receptor ligand mediating a receptor response (intrinsic effect).
- *Antagonist*: An antagonist is a receptor ligand preventing the action of an agonist, in a direct (competitive) or indirect (allosteric) manner.
- *Ion channel*: A pore formed by proteins allowing the diffusion of certain ions through the cell membrane along a concentration gradient is called an ion channel. The channel opening is either ligand- or voltage-controlled.
- *Transporter*: A transporter is a protein transporting molecules or ions through the cell membrane against a concentration gradient.

10.4.4
**The Drug Discovery Process**

The drug discovery process comprises the following steps:

1. target identification,
2. target validation,
3. lead finding (including design and synthesis of compound libraries as well as screening of compound libraries),
4. lead optimization (acceptable pharmacokinetic profile, toxicity and mutagenicity),
5. preclinical trials,
6. clinical trials,
7. drug approval (e.g., FDA approval).

10.4.4.1  **Target Identification and Validation**

The process of target identification analyzes a complex disease process by dissecting it into its fundamental components. This makes it possible to identify the one that is most integral to the manifestation of the disease. Target identification aims to understand the biological processes related to a disease, and to identify its mechanism and the structure of individual elements of the disease. Commonly these individual elements are receptors, enzymes, etc., which become the target of new drugs.

After target selection, the properties of the target gene are analyzed in disease models, which may be cell models in test tubes and/or animal models. This process is called target validation. A target is validated when a specific action on the target has shown favorable effects in the disease models. A few factors have to be considered during target validation, e.g., for the development of new antibiotics acceptable targets will be either those that are essential for the life of the pathogen, or virulence factors. In addition, the target must be divergent between the pathogen and host, so that modification and/or disruption of the target will attenuate or kill the microbe and/or inhibit its virulence without a detrimental impact on the host.

The target identification and target validation steps are influenced by genomics and functional as well as structural proteomics. The number of drug targets for the existing drugs is currently about 500. An overview of the most important targets and the mechanism of drug action is given in Table 10.4-1. The distribution of different biochemical classes of drug targets is depicted in Figure 10.4-1 [13].

Due to the knowledge of the sequence of the human genome there are over 30 000 potential targets. So far, only little is known about these new targets. The question arises of how many of these targets can be modulated by potent, small, "drug-like" molecules [14], i.e., how many targets are "druggable". Beyond that, what makes a target druggable? Currently, the estimated number of druggable targets is about 3000 [15].

**Table 10.4-1.** Drug targets and mechanisms of drug action.

| Drug target | Mechanism of drug action |
| --- | --- |
| Receptors | agonists and antagonists |
| Enzymes | reversible and irreversible inhibitors |
| Ion channels | blockers and openers |
| DNA | alkylating agents, intercalating agents, wrong substrates (Trojan horses, e.g., 5-fluorouracil) |

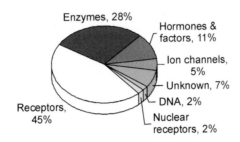

**Figure 10.4-1.** Biochemical classes of the drug targets in current therapies.

### 10.4.4.2  Lead Finding and Optimization

During the lead finding step, compound libraries are designed and synthesized. These compound libraries are screened in assays using the validated functional protein target. Compounds showing activity in a primary assay are labeled as "hits". The hits are validated in another screening experiment. The information generated in the screening assays is used to find compounds which are suitable for further development. Whether a compound is suitable to serve as a new lead structure and for further development depends on several features. Some criteria affecting the search and optimization of a new lead structure are illustrated in Figure 10.4-2.

An "ideal" drug must be safe and effective. The maximum daily dose should not exceed 200–500 mg. Drugs should be well absorbed orally and bioavailable. Metabolic stability ensures a reasonable long half-life. Further on, a drug should be non-

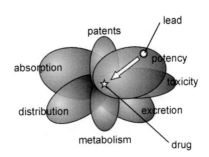

**Figure 10.4-2.** Factors affecting lead identification and optimization.

toxic and cause no adverse effects, or only minimal ones. Finally, an ideal drug will distribute selectively to the target tissues.

The lead optimization step aims to develop compounds with high potency, high selectivity, and an acceptable pharmacokinetic profile, toxicity, and mutagenicity. Lead optimization is an iterative process. The initial lead compound is systematically modified. The analogs thus obtained are then tested in a biological assay. Their spectrum of activity against a broad range of targets has to be evaluated. For successful screening the pharmaceutical companies have libraries containing about two to three million molecularly diverse compounds.

Chemoinformatics is involved in the drug discovery process in both the lead finding and lead optimization steps. Artificial neural networks can play a decisive role of various stages in this process (cf. Section 10.4.7.1).

### 10.4.4.3 Preclinical and Clinical Trials

The approval of a new drug requires preclinical and clinical studies and takes an average of 10 years to complete. The preclinical and clinical trials have to prove the safety and efficacy of every new medicine.

The preclinical trials are performed in *in vitro* and animal studies to assess the biological activity of the new compound. In phase I of the clinical trials the safety of a new drug is examined and the dosage is determined by administering the compound to about 20 to 100 healthy volunteers. The focus in phase II is directed onto the issues of safety, evaluation of efficacy, and investigation of side effects in 100 to 300 patient volunteers. More than 1000 patient volunteers are treated with the new drug in phase III to prove its efficacy and safety over long-term use.

### 10.4.5
### Fields of Application of Chemoinformatics in Drug Design

### 10.4.5.1 Subset Selection and Similarity/Diversity Search

A very fundamental problem encountered in drug design is the search in the huge chemical space. Whereas the estimated number of small, possibly drug-like molecules is in order of magnitude of $10^{200}$, there are only about $10^7$ known compounds. Corporate compound libraries contain $10^6$ compounds. Drug databases comprise $10^4$ compounds. There are $10^3$ compounds on the market as commercial drugs. Among these commercial drugs are about $10^2$ profitable drugs. The probability of selecting a commercial drug from a corporate compound library, assuming that all known drugs are included in this library, is $(10^3/10^6 \times 100) = 0.1\%$. Therefore, powerful techniques for subset selection are necessary for both high-throughput screening and virtual screening. There has been a shift of paradigm in the pharmaceutical industry from screening huge (combinatorial) libraries toward screening focused libraries, allowing hit rates in HTS of up to 10%.

Methods of analyzing the diversity of the selected subset ensure that an appropriate chemical space is covered. Descriptors such as fingerprints, and 2D, and 3D descriptors, as well as molecular surface properties, which can be used for the analysis of the similarity or diversity of compounds have been introduced in Chapter 8. Frequently applied distance measures are the Hamming distance or the Tanimoto coefficient (see Section 8.2). As the distance measures are defined for two compounds, the diversity of a library cannot be calculated directly. One way to quantify the diversity of a library is to determine the centroid of all the compounds and to calculate the mean and standard deviation of the pairwise distance of all compounds within the library from the centroid.

The distance of a compound from a library can be given by the distance of this compound from a) its nearest neighbor in the library, b) its $k$ nearest neighbors in the library, and c) the centroid of the library.

### 10.4.5.2 Analysis of HTS Data

A central problem in chemoinformatics is the establishment of a relationship between a chemical structure and its biological activity. Huge amounts of data are gathered, particularly so through the synthesis of combinatorial libraries and subsequent high-throughput screening (HTS). A chemist synthesizes about 50 compounds per year by traditional, organic synthesis. In combinatorial chemistry a series of homologs are synthesized. A reaction of the type $A^i + B^j \rightarrow A^i\text{–}B^j$, with $i \in \{1; 2; 3;\ldots n\}$ and $j \in \{1; 2; 3;\ldots m\}$ performed in parallel by robots gives access to $n \times m$ products in a single experiment. Thousands of compounds are accessible in a short period of time.

With these massive amounts of data produced in HTS for combinatorial libraries, tools become necessary to make it possible to navigate through these data and to extract the necessary information, to search – as is said quite often – for a needle in a haystack.

A variety of methods have been developed by mathematicians and computer scientists to address this task, which has become known as data mining (see Chapter 9, Section 9.8). Fayyad defined and described the term "data mining" as the "nontrivial extraction of implicit, previously unknown and potentially useful information from data, or the search for relationships and global patterns that exist in databases" [16]. In order to extract information from huge quantities of data and to gain knowledge from this information, the analysis and exploration have to be performed by automatic or semi-automatic methods. Methods applicable for data analysis are presented in Chapter 9.

### 10.4.5.3 Virtual Screening

The term "virtual screening" or "*in silico* screening" is defined as the selection of compounds by evaluating their desirability in a computational model [17]. The desirability comprises high potency, selectivity, appropriate pharmacokinetic properties, and favorable toxicology.

Virtual screening assists the selection of compounds for screening in-house libraries and compounds from external suppliers. Two different strategies can be applied:

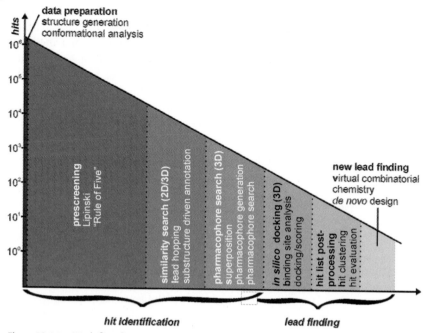

**Figure 10.4-3.** Work flow for virtual screening, from data preparation to finding new leads.

- Diverse libraries can be used for lead finding by screening against several different targets. The selected compounds should cover the biological activity space well.
- Targeted or focused libraries are suited for both lead finding and optimization. If knowledge of a lead compound is available, compounds with a similar structure are selected for the targeted library. Targeted libraries are focused on a single target.

Virtual screening allows the scope of screening to be extended to external databases. When this is done, increasingly diverse hits can be identified. The application of virtual screening techniques before or in parallel with HTS helps to reduce the assay-to-lead attrition rate observed from HTS. In addition, virtual screening is faster and less expensive than experimental synthesis and biological testing.

Both ligand- and structure-based methods can be applied in virtual screening. An example of a workflow chart for virtual screening is depicted in Figure 10.4-3.

### 10.4.5.4 Design of Combinatorial Libraries

HTS data as well as virtual screening can guide and direct the design of combinatorial libraries. A genetic algorithm (GA) can be applied to the generation of combinatorial libraries [18]. The number of compounds accessible by combinatorial synthesis often exceeds the number of compounds which can be synthesized experimentally. To reduce the number of products, a subset of fragments has to be chosen. Sheridan and Kearsley have demonstrated the selection of a subset of amines for the construction of a tripeptoid library with a GA using a similarity measure to a specific tripeptoid target as a scoring function [18a].

#### 10.4.5.5 Further Issues

Except for the aforementioned tasks chemoinformatics can contribute to the following fields in drug design:

- The establishment of QSAR/QSPR models. This process is explained in more detail in Chapter 8. Good QSAR/QSPR models should be interpretable and guide the further development of a new drug. The computer system PASS (prediction of activity spectra for substances) allows to predict simultaneously more than 500 biological activities. Among these activities are pharmacological main and side effects, mechanism of action, mutagenicity, carcinogenicity, teratogenicity, and embryotoxicity [19].
- Flexible 3D alignment of a set of ligands binding to the same target and/or CoMFA analysis allowing the perception of a pharmacophore for this target.
- Prediction of various physicochemical properties (such as solubility, lipophilicity log $P$, $pK_a$, number of H-donor and acceptor atoms, number of rotatable bonds, polar surface area), drug-likeness, lead-likeness, and pharmacokinetic properties (ADMET profile). These properties can be applied as a filter in the prescreening step in virtual screening.
- The synthetic accessibility of ligands designed by *de novo* methods. This topic typically is not considered. Some compounds with unstable or reactive functionalities are eliminated by applying simple rules, but these rules are still insufficient.

The treatment of conformational flexibility of both the ligand and protein, the prediction of the binding affinity of a substrate to an enzyme, the water desolvation of the ligand and of the binding pocket, the modeling of protein–protein interaction, the determination of the geometry as well as the calculation of the strength of hydrogen bonds, and the prediction of the 3D structure of proteins from the sequence are still challenging tasks which have to be addressed and improved in computer-assisted drug design.

#### 10.4.6
#### Ligand- and Structure-based Drug Design

Depending on the information available about the protein structure and the ligands binding to a particular target, four different cases can be distinguished in drug design, as listed in Table 10.4-2.

The lead discovery process is depicted in Figure 10.4-4 and shows how the different methods are interconnected. A lead structure can be discovered by serendipity. In rational drug design all information available about a target serves to direct

**Table 10.4-2.** Cases in the drug discovery process depending on knowledge of the receptor and ligand structure.

|  | Ligand unknown | Ligand known |
| --- | --- | --- |
| Protein structure unknown | combinatorial chemistry and high throughput screening | QSAR, pharmacophore models and hypothesis, similarity search in databases |
| Protein structure known | *de novo* design, receptor-based 3D searching | structure-based design, docking |

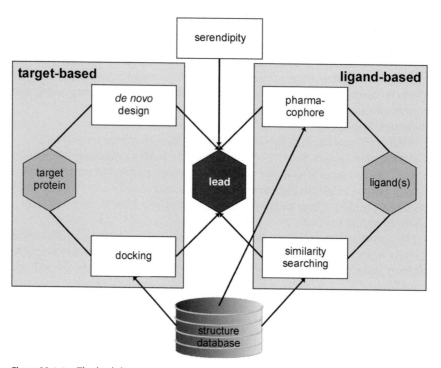

**Figure 10.4-4.** The lead discovery process.

the search for a new lead structure. If the 3D structure (see Section 2.9.7) of the target of interest is known from X-ray crystallography, NMR spectroscopy, or homology modeling, a suite of methods in structure-based drug design can be applied. If an X-ray structure of the protein with a ligand is available, the binding mode of the ligand can be analyzed. Docking of new ligands with the same binding mode, and ranking of these ligands by scoring functions, guide further drug development. Otherwise, one has to apply *de novo* design or perform a 3D search in a ligand database for a compound with a complementary shape and surface properties to the binding site of the receptor.

In an early stage of drug design, without structural information about the target and without any knowledge about ligands binding to the target protein, one is obliged to screen combinatorial and proprietary libraries by HTS. As soon as some of the ligands binding to the target are known, they serve as a starting point for a similarity search in a ligand database and for the perception of a pharmacophore by superposition of the ligands.

Thus there is a distinction between ligand- and structure-based drug design, which are described in more detail in Sections 10.4.6.1 and 10.4.6.2.

### 10.4.6.1 Ligand-Based Drug Design

#### 10.4.6.1.1 Prefiltering in Virtual Screening
In general, the first step in virtual screening is the filtering by the application of Lipinski's "Rule of Five" [20]. Lipinski's work was based on the results of profiling the calculated physical property data in a set of 2245 compounds chosen from the World Drug Index. Polymers, peptides, quaternary ammonium, and phosphates were removed from this data set. Statistical analysis of this data set showed that approximately 90 % of the remaining compounds had:

- a molecular weight of less than 500 g/mol;
- a calculated lipophilicity (log $P$) of less than 5;
- fewer than five H-bond donors;
- fewer than 10 H-bond acceptors (sum of all nitrogen and oxygen atoms).

The cutoff values of this "Rule of Five" (the thresholds are a multiple of five) differ slightly within the pharmaceutical industry. Sometimes the "Rule of Five" is extended by a fifth condition:

- the number of rotatable bonds is less than 10 or one of the four rules can be violated.

In a more recent study the physical properties of drugs in different development phases were compared [21]. It turned out that the molecular weight and lipophilicity are the properties showing the clearest influence on the successful passage of a candidate drug through the development process. The mean molecular weight of orally administered drugs decreases in the course of the development. Further into development, the most lipophilic compounds are discontinued from development.

Other filters used for prefiltering account for lead- [22, 23] or drug-likeness [24–26], an appropriate ADMET profile [27–30], or favorable properties concerning receptor binding [31, 32].

#### 10.4.6.1.2 *In Silico* ADMET
Historically, drug absorption, distribution, metabolism, excretion, and toxicity (ADMET) studies in animal models were performed after the identification of a lead compound. In order to avoid costs, nowadays pharmaceutical companies evaluate the ADMET profiles of potential leads at an earlier stage of the development

process. For the consideration of ADMET properties in virtual screening, computational methods for their prediction are needed. As passive intestinal absorption has been studied by many groups, it is known that lipophilicity is a key property for estimation of the membrane permeability of a molecule. Programs to predict log $P$ are available and give reasonable results [33–37].

Concerning the distribution of a drug, models have been published for log $BB$ (blood/brain partition coefficient) for CNS-active drugs (CNS, central nervous system) crossing the blood–brain barrier (BBB) [38–45] and binding to human serum albumin (HSA) [46].

Metabolism is still a barrier to be overcome. Some QSAR, pharmacophore, protein, and rule-based models are available to predict substrates and inhibitors of a specific cytochrome $P450$ isoenzyme [47–55].

### 10.4.6.1.3 Similarity Searches

Following the "similar structure-similar property" principle, structurally similar molecules are expected to exhibit similar properties or biological activities. In order to select compounds for focused libraries, 2D and 3D similarity searches are performed. The distance of all the ligands in a database from the ligands known to bind to the target of interest is calculated as described in Section 10.4.4.1. Afterwards the ligands are ranked in reverse order of their distance. The ligands with a high rank are selected for the focused library.

### 10.4.6.1.4 Development of a Pharmacophore Model by 3D Structure Alignment

Pharmacophore perception for receptors with an unknown 3D structure can be carried out by comparing the spatial and electronic requirements of a set of ligands that are known to bind to the receptor of interest. The comparison of a set of ligands is performed by the structural alignment of these ligands. A detailed review on structural alignment of molecules is given by Lemmen and Lengauer [56]. A program called GAMMA (Genetic Algorithm for Multiple Molecule Alignment, *http://www2.chemie.uni-erlangen.de/research/drugdesign/ga.phtml*) enables the flexible alignment of multiple molecules combining a GA with a Newton optimizer [57, 58].

### 10.4.6.2 Structure-Based Drug Design

Fitting a ligand from a 3D structure database into the binding site of a target protein is called docking. The iterative building of new molecules in the binding site of a receptor is illustrated in the center and on the right hand side of Figure 10.4-5. These procedures to find new leads are called *de novo* design. The building approach, beginning with a single fragment and proceeding through the stepwise addition of further moieties, is shown in the center. Alternatively, several small molecules are placed in the binding site of the protein and subsequently linked together (linking). To end up with high-affinity ligands from structure-based drug design, a high degree of steric and electronic complementarity of the ligand to the target protein is required. Further on, an appropriate amount of the ligands hydrophobic surface should be buried in the complex. A certain degree of conforma-

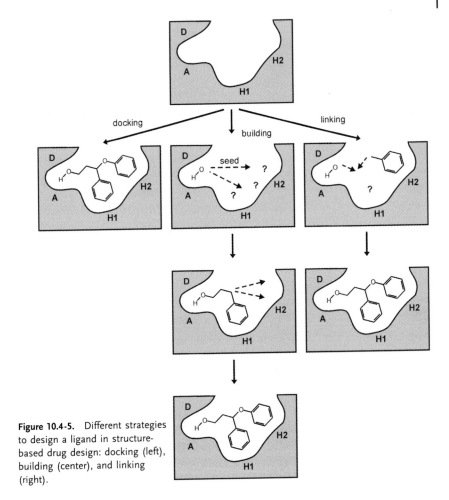

**Figure 10.4-5.** Different strategies to design a ligand in structure-based drug design: docking (left), building (center), and linking (right).

tional rigidity is essential to ensure that the loss of entropy upon ligand binding is acceptable.

An overview of docking programs is given in Table 10.4-3. Depending on the way the conformational flexibility of the ligand is treated, docking can be either rigid or flexible.

The suitability of these methods for large-scale virtual screening depends on how ligand flexibility is addressed. The methods using a genetic algorithm, an exhaustive search, a Monte Carlo simulation, or a pseudo-Brownian sampling are less suitable for large-scale virtual screening because they require considerably more computer power.

Recently Schneider and Böhm reviewed programs for *de novo* design [66]. A list of these is compiled in Table 10.4-4.

**Table 10.4-3.** Common docking tools for virtual screening.

| Method | Ligand flexibility sampling | Scoring function |
|---|---|---|
| Dock [59] | incremental build | force field or contact score |
| FlexX [60] | incremental build | empirical score |
| Slide [61] | conformational ensembles | empirical score |
| Fred (Openeye Software) | conformational ensembles | Gaussian or empirical score |
| Gold [62] | genetic algorithm | empirical score |
| Glide (Schrödinger) | exhaustive search | empirical score |
| AutoDock [63] | genetic algorithm | force field |
| LigandFit (Accelrys) | Monte Carlo | empirical score |
| ICM [64] | pseudo-Brownian sampling and local minimization | mixed force field and empirical score |
| QXP [65] | Monte Carlo | force field |

**Table 10.4-4.** Overview of available programs for *de novo* design.

| Program | Description |
|---|---|
| GENSTAR [67] | atom-based, grows molecules in situ based on an enzyme contact model |
| GROUPBUILD [68] | fragment-based, sequential growth, combinatorial search |
| GROW [69] | peptide design, sequential growth |
| GROWMOL [70] | fragment-based, sequential growth, stochastic search |
| HOOK [71] | linker search for fragments placed by MCSS |
| LEGEND [72] | atom-based, stochastic search |
| LUDI [73] | fragment-based, combinatorial search |
| MCSS [74] | fragment based, stochastic sampling |
| MOLMAKER [75] | graph-theoretical 3D design |
| PRO-LIGAND [76] | fragment-based search |
| PRO-SELECT [77] | fragment-based, scaffold-linker approach |
| SKELGEN [78] | small-fragment based, Monte-Carlo search |
| SPROUT [79] | fragment-based, sequential growth, combinatorial search |

The prediction of the binding affinity of a ligand to a receptor is a challenging task. This becomes evident when the thermodynamics of ligand binding is analyzed in more detail (Figure 10.4-6). The following enthalpic and entropic terms ($\Delta H$, $\Delta S$) and other factors influence the free energy ($\Delta G$) of ligand binding:

- the binding enthalpy of the ligand–protein complex $\Delta H_{LR}$;
- the desolvation enthalpy $\Delta H_{LW}$, $\Delta H_{RW}$ and entropy (of both ligand and protein);

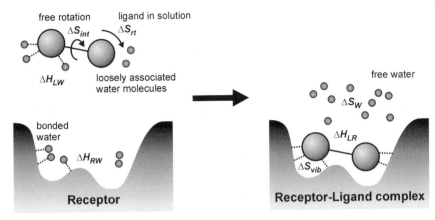

free rotation   ligand in solution

$\Delta S_{int}$   $\Delta S_{rt}$

$\Delta H_{LW}$   loosely associated water molecules

free water

$\Delta S_W$

bonded water

$\Delta H_{RW}$

$\Delta H_{LR}$

$\Delta S_{vib}$

**Receptor**     **Receptor-Ligand complex**

**Figure 10.4-6.** Thermodynamics of ligand binding.

- the entropic loss of translational and rotational degrees of freedom of the ligand $\Delta S_{rt}$ and $\Delta S_{int}$ respectively;
- the entropic contribution $\Delta S_w$ from the freed water molecules;
- the distortion energy of the ligand and its binding site $\Delta H_{dist}$;
- the protonation state of the ligand and the binding site;
- the molecular electrostatic potential and dielectric constant at the binding site;
- the dipole moment of the ligand and the local dipole moment at the binding site;
- repulsive effects;
- inserted water molecules;
- the solvation enthalpy and entropy of the complex.

The calculation of the binding affinity with consideration of all these effects for virtual screening is not possible. In order to circumvent thus difficulty, scoring functions are used instead, e.g., the Ludi scoring function [80], or consensus scoring functions derived from FlexX score, DOCK score, GOLD score, ChemScore, or PMF score [81].

Rognan published the scoring function FRESNO (fast free energy scoring function), which considers a hydrogen-bond term, a lipophilic term, a repulsive term for the buried polar surface, a rotational term, and a desolvation term [82].

In a recent review the pharmacophore identification programs Catalyst, DISCO, and GASP have been compared [83].

Interested readers are referred to Chapter X, Section 4 in the Handbook, dealing with drug design.

## 10.4.7
## Applications

### 10.4.7.1 Distinguishing Molecules of Different Biological Activities and Finding a New Lead Structure – An Example of Ligand-Based Drug Design

The task of the study presented here was to separate a data set of 172 molecules into benzodiazepine agonists (60 compounds) and dopamine agonists (112 compounds) [84].

Figure 10.4-7 shows some of the structures contained in the data set, emphasizing the problem to represent molecules with different numbers of atoms by the same numbers of descriptors.

The structures were represented by topological autocorrelation. Thus, $d$ in Eq. (19) in Chapter 8 was the number of bonds between the two atoms $i$ and $j$. The distance $d$ was kept running from 2 to 8 (seven distances altogether). As a first shot, as no specific requirement of the receptor was available, we decided to use a rather broad structure representation, including a variety of physicochemical effects in the autocorrelation vector. Separate seven-dimensional vectors (seven distance intervals) were constructed with $\sigma$-atomic charges, $(\sigma+\pi)$ atomic charges, $\sigma$-electronegativity, $\pi$-electronegativity, lone-pair electronegativity, atomic polariz-

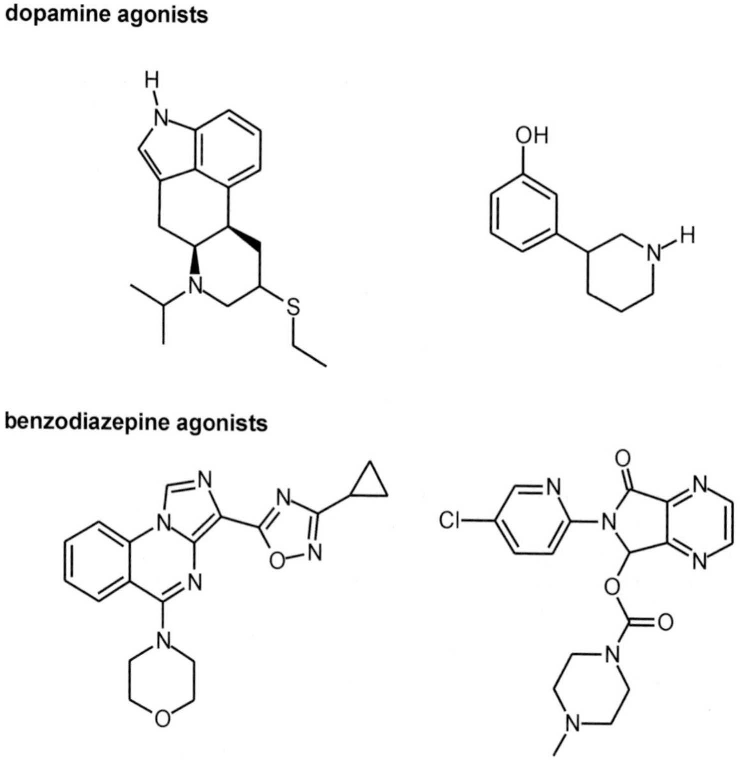

**dopamine agonists**

**benzodiazepine agonists**

**Figure 10.4-7.** Representative 2D structures of dopamine and benzodiazepine agonists.

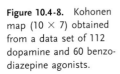 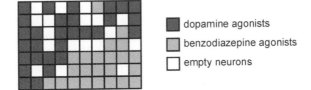

**Figure 10.4-8.** Kohonen map (10 × 7) obtained from a data set of 112 dopamine and 60 benzodiazepine agonists.

ability, and an atomic property of 1 (just to represent the molecular graph). These seven autocorrelation vectors were then concatenated to give a 49-dimensional representation of the 172 molecules in the data set. Training a 10 × 7 Kohonen network with the entire data set gave a map that was then marked by assigning colors to the network depending on whether a neuron contained a dopamine or a benzodiazepine agonist (Figure 10.4-8) [84].

As can be seen, the two sets of molecules separate quite well. This is even more remarkable as the class membership was not used in training the network but only in visualizing the results of training (unsupervised learning). This attests to the relevance of the chosen structure representation for reproducing effects that are responsible for the different binding of dopamine and benzodiazepine agonists.

Next we turned our attention to the question of whether we could still see the separation of the two sets of molecules when they were buried in a large data set of diverse structures. For this purpose we added this data set of 172 molecules to the entire catalog of 8223 compounds available from a chemical supplier (Janssen Chimica). Now, having a larger data set one also has to increase the size of the network: a network of 40 × 30 neurons was chosen. Training this network with the same 49-dimensional structure representation as previously described, but now for all 8395 structures, provided the map shown in Figure 10.4-9.

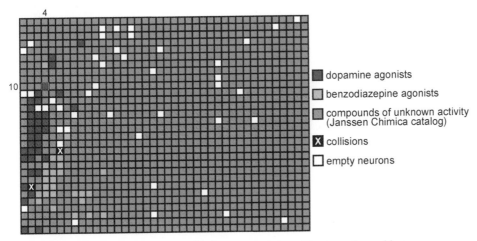

**Figure 10.4-9.** Kohonen map (40 × 30) of a data set consisting of the dopamine and benzodiazepine agonists of Figure 10.4-8 and 8323 compounds in a chemical supplier's catalog.

Even in this fairly diverse data set of structures, the dopamine and benzodiazepine agonists could be separated quite well; only two neurons had collisions
between these two types of compounds. Even more importantly, however, we now
know in which chemical space one would have to search for new lead structures
for dopamine or for benzodiazepine agonists.

To illustrate this point, Figure 10.4-10 shows the contents of the neuron at position (4,10) (see Figure 10.4-9). This neuron obtained two dopamine agonists and,

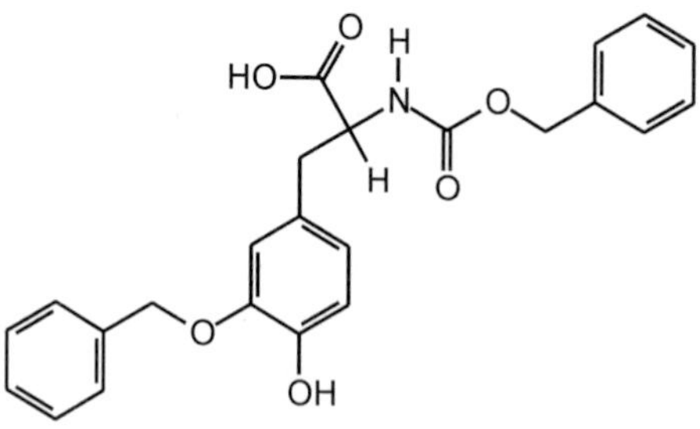

dopamine 62

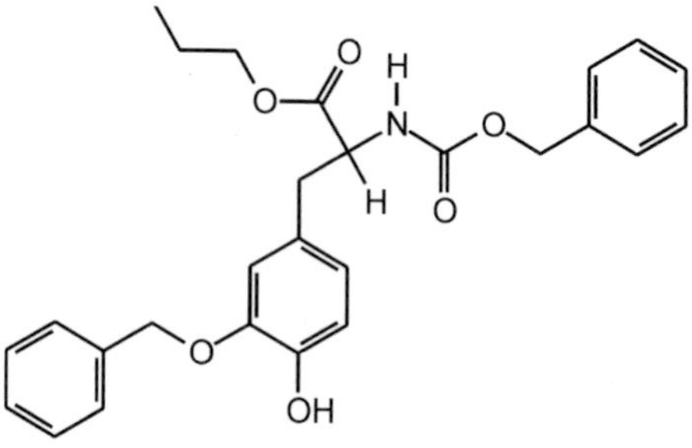

dopamine 63

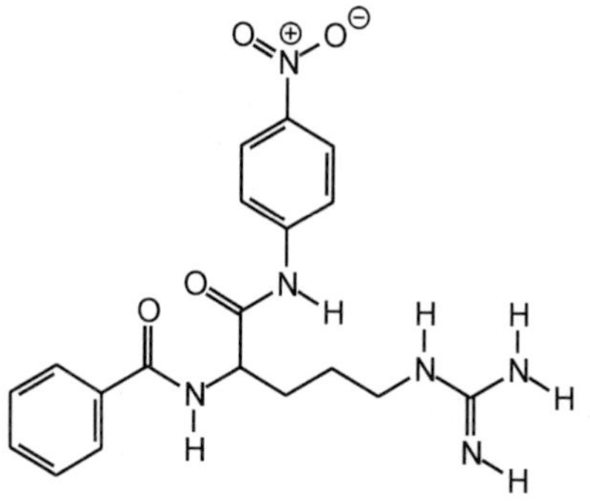

Janssen 935

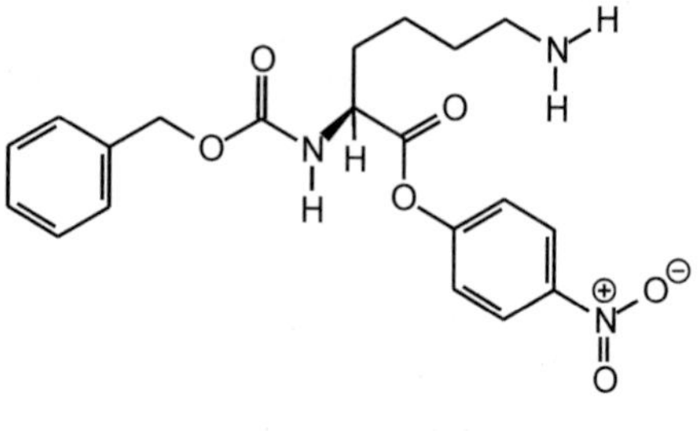

Janssen 1685

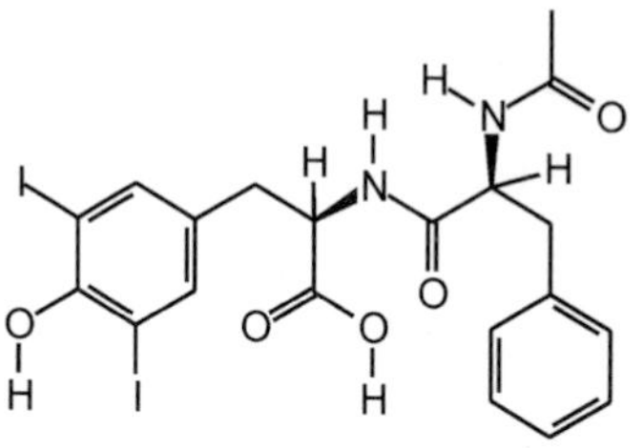

Janssen 139

**Figure 10.4-10.** Structures that were mapped into the neuron at position (4,10) of the Kohonen
map in Figure 10.4-9.

from the Janssen Chimica catalog, three compounds of unknown biological activity which might be taken as lead structures for developing dopamine agonists. Lead hopping can be done by considering not only a single neuron but also its neighborhood, in order to retrieve potential new leads having a different scaffold.

The results presented here imply that a similar approach can be used for comparing two different libraries, for determining the degree of overlap between the compounds in these two libraries. Examples of the application of artificial neural networks or GA in drug design are given in [57, 58, 84, 85].

### 10.4.7.2 Examples of Structure-Based Drug Design

Virtual screening of chemical databases has been applied successfully to finding new leads in structure-based drug design [81, 86]. Some examples are briefly described below. A compilation of recent successes in structure-based virtual screening for various targets is given in Table 10.4-5. Wang and co-workers found a novel potent dopamine transporter inhibitor by 3D database pharmacophore searching and subsequent chemical modification [87]. The analog has a high affinity and is a promising lead for the treatment of cocaine abuse.

Another approach followed by Wang et al. led to an inhibitor of *Tritrichomonas foetus* hypoxanthine-guanine-xanthine phosphoribosyltransferase (HGXPRT) [88]. The X-ray structure of this enzyme is available and was used in a search for novel scaffolds. The Available Chemicals Directory (ACD) was screened with the program DOCK. Isatin and phthalic anhydride were capable of mimicking the substrate purine base and acting as competitive inhibitors. A virtual library of substituted 4-phthalimidocarboxanilides was constructed and synthesized on solid supports. The most active compound was used as lead for the further opti-

**Table 10.4-5.** Recent successes of structure-based virtual screening approaches.

| Target | Method | Ref. |
| --- | --- | --- |
| Carbonic anhydrase II | FlexX | 93 |
| Protein tyrosine phosphatase 1B | Dock | 94 |
| Estrogen receptor | PRO_LEADS | 95 |
| Thrombin | PRO_LEADS | 95 |
| Factor Xa | PRO_LEADS | 95 |
| Thymidylate synthase | Dock | 96 |
| Retinoic acid receptor | ICM | 97 |
| Farnesyl transferase | EUDOC | 91 |
| Kinesin | Dock | 98 |
| Hypoxanthine-guanine-xanthine phosphoribosyl transferase | Dock | 88 |
| DNA gyrase | LUDI | 90 |
| HIV-1 RNA transactivation response element | ICM | 99 |

mization providing [(4-phthalimido)carboxamido-3-(4-bromobenzyloxy)benzene] as a selective submicromolar inhibitor for *T. foetus* HGXPRT.

Zhang and co-workers worked on the structure-based, computer-assisted search for low molecular weight, non-peptidic protein tyrosine phosphate 1B (PTP1B) inhibitors, also using the DOCK methodology [89]. They identified several potent and selective PTP1B inhibitors by screening the ACD.

Researchers from Hoffmann–La Roche discovered new inhibitors of DNA gyrase by combining several key techniques [90]. On the basis of the 3D structural information of the binding site they started with an *in silico* screening for potential low molecular weight inhibitors followed by a biased high-throughput DNA gyrase screen. The hits of the screening were validated by biophysical methods and finally optimized in a 3D guided optimization process. The lead optimization of an indazole derivative resulted in a DNA gyrase inhibitor that was 10 times more potent than novobiocin.

Pang et al. also performed a virtual screening of the ACD with a newly developed docking program (EUDOC) [91]. Their interest was the search for new farnesyl-transferase inhibitor leads. Among the compounds identified by the virtual screening, four showed $IC_{50}$ values in the range from 25 to 100 μm.

Docking also provided insight into the 3D hydrophobic requirements for binding of nonpeptide molecules in the SH2 domain of pp60src at ARIAD Pharmaceuticals [92].

## 10.4.8
## Outlook – Future Perspectives

Nowadays a broad range of methods is available in the field of chemoinformatics. These methods will have a growing impact on drug design. In particular, the discovery of new lead structures and their optimization will profit by virtual screening [17, 66, 100–103]. The huge amounts of data produced by HTS and combinatorial chemistry enforce the use of database and data mining techniques.

Integrated systems suitable for processing the sometimes quite complex workflows more easily or automatically and for optimizing new compounds in parallel for their potency, selectivity, and ADMET profile have to be developed. Structure-based drug design will profit from faster algorithms making it possible to perform high-throughput docking of virtual libraries with much accuracy. Improved information management techniques may facilitate the connection and analysis of different sources of data.

The reliability of the *in silico* models will be improved and their scope for predictions will be broader as soon as more reliable experimental data are available. However, there is the paradox of predictivity versus diversity. The greater the chemical diversity in a data set, the more difficult is the establishment of a predictive structure–activity relationship. Otherwise, a model developed based on compounds representing only a small subspace of the chemical space has no predictivity for compounds beyond its boundaries.

**Essentials**

- The drug discovery process comprises the following steps: a) target identification; b) target validation; c) lead finding (including design and synthesis of compound libraries as well as screening of compound libraries); d) lead optimization (searching for an acceptable pharmacokinetic profile, toxicity and mutagenicity); e) preclinical studies; f) clinical studies; g) drug approval (FDA approval).
- A lead structure is "a representative of a compound series with sufficient potential (as measured by potency, selectivity, pharmacokinetics, physicochemical properties, absence of toxicity, and novelty) to progress to a full drug development program".
- Chemoinformatics is primarily used for the steps of lead finding and lead optimization within the drug discovery process. In particular the following tasks are involved:
  - analysis of HTS data,
  - similarity search,
  - design of combinatorial libraries,
  - design of focused libraries,
  - comparison of the similarity/diversity of libraries,
  - virtual screening,
  - docking,
  - *de novo* design,
  - pharmacophore perception,
  - prediction of binding affinities, physicochemical properties (such as solubility, $\log P$, $pK_a$), and pharmacokinetic properties (ADMET profile),
  - establishment of QSAR models which can be interpreted and guide the further development of a new drug.
- Following the "similar structure-similar property" principle, structurally similar molecules are expected to exhibit similar properties or biological activities.
- The term "virtual screening" or "*in silico* screening" is defined as the selection of compounds by evaluating their desirability in a computational model. The desirability comprises high potency, selectivity, appropriate pharmacokinetic properties, and favorable toxicology.
- Lipinski's "Rule of Five":
  - molecular weight $< 500$,
  - lipophilicity ($\log P$) $< 5$,
  - number of H-bond donors $< 5$,
  - number of H-bond acceptors (number of $N + O$) $< 10$,
  - (number of rotatable bonds $< 10$ or one of the four rules can be violated).

## Selected Reading

- C. G. Wermuth (Ed.), *The Practice of Medicinal Chemistry*, Academic Press, London, **1996** (1$^{st}$ Edition), **2003** (2$^{nd}$ Edition).
- H.-J. Böhm, G. Klebe, H. Kubinyi, *Wirkstoffdesign*, Spektrum Akademischer Verlag, Heidelberg, **1996**.
- E. Mutschler, *Arzneimittelwirkungen*, Wissenschaftliche Verlagsgesellschaft, Stuttgart, **1996**.
- H.-J. Böhm, G. Schneider (Eds.), Virtual screening for bioactive molecules, in *Methods and Principles in Medicinal Chemistry*, *Vol. 10*, R. Mannhold, H. Kubinyi, H. Timmerman (Eds.), Wiley-VCH, Weinheim, **2000**, pp. 1–307
- S. Anzali, J. Gasteiger, U. Holzgrabe, J. Polanski, J. Sadowski, A. Teckentrup, M. Wagener, *The Use of Self-Organizing Neural Networks in Drug Design*, in *3D QSAR in Drug Design*, Vol. 2, H. Kubinyi, G. Folkers, Y. C. Martin (Eds.), Kluwer/ESCOM, Dordrecht, NL, **1998**, 273–299.
- Chapter X, Section 4 in *Handbook of Chemoinformatics – From Data to Knowledge*, J. Gasteiger (Ed.), Wiley-VCH, Weinheim, **2003**.

## Interesting Websites

- *http://home.t-online.de/home/kubinyi/lectures.html*
- *http://www.netsci.org/Science/Compchem/index.html*

# References

[1] C. G. Wermuth (Ed.), *The Practice of Medicinal Chemistry*, Academic Press, London, **1996** (1st Edition), **2003** (2nd Edition).

[2] H.-J. Böhm, G. Klebe, H. Kubinyi, *Wirkstoffdesign*, Spektrum Akademischer Verlag, Heidelberg, **1996**.

[3] E. Mutschler, *Arzneimittelwirkungen*, Wissenschaftliche Verlagsgesellschaft, Stuttgart, **1996**.

[4] H. Kubinyi, *J. Receptor & Signal Transduction Res.* **1999**, *19*, 15–39.

[5] P. Tollman, P. Guy, J. Althuler, A. Flanagan, M. Steiner, *Boston Consulting Group Report* **2001**, 1–60. www.bcg.de

[6] M. J. Veverka, *Accenture Report* **2000**, *23*, 1–53.

[7] T. Olsson, T. I. Oprea, *Curr. Opin. Drug Discov. Develop.* **2001**, *4*, 308–313.

[8] B. Gaudillière, P. Berna, *Annu. Rep. Med. Chem.* **2000**, *35*, 331–356.

[9] R. A. Prentis, Y. Lis, S. R. Walker, *Br. J. Clin. Pharm.* **1988**, *25*, 387–396.

[10] T. Kennedy, *Drug Discov. Today* **1997**, *2*, 436–444.

[11] S. Venkatesh, R. A. Lipper, *J. Pharm. Sci.* **2000**, *89*, 145–154.

[12] M. J. Valler, D. Green, *Drug Discov. Today* **2000**, *5*, 286–293.

[13] J. G. Hardman, L. E. Limbird, A. Goodman Gilman (Eds.), *Goodman and Gilman's The Pharmacological Basis of Therapeutics*, McGraw-Hill, New York, *9*, **1996**.

[14] a) J. Sadowski, H. Kubinyi, *J. Med. Chem.* **1998**, *41(8)*, 3325–3329;
b) W. P. Walters, Ajay, M. A. Murcko, *Curr. Opin. Chem. Bid.* **1999**, *3(4)*, 384–387.

[15] a) A. L. Hopkins, C. R. Groom, *Nature Rev. Drug Disc.* **2002**, *1*, 727–730;
b) J. Drews, *Science* **2000**, *287*, 1960–1963.

[16] U. M. Fayyad, G. Piatetsky-Shapiro, P. Smyth, From data mining to knowledge discovery: An overview, in *Advances in Knowledge Discovery and Datamining*, U. M. Fayyad, G. Piatetsky-Shapiro, P. Smyth, R. Uthurusamy (Eds.), AAAI Press, Menlo Park, CA, USA, **1996**, pp. 1–37.

[17] W. P. Walters, M. T. Stahl, M. A. Murcko, *Drug Discov. Today* **1998**, *3*, 160–178.

[18] a) R. P. Sheridan, S. K. Kearsley, *J. Chem. Inf. Comput. Sci.* **1995**, *35*, 310–320;
b) R. D. Brown, D. E. Clark, *Expert Opin. Ther. Pat.* **1998**, *8*, 1447–1460.

[19] a) D. A. Folimonov, V. V. Poroikov, E. I. Karaicheva, R. K. Kazaryan, A. P. Boudunova, E. M. Mikhailovsky, A. V. Rudnitskih, L. V. Goncharenko, Y. V. Burov, *Exp. Clin. Pharmacol. (Rus)* **1995**, *58* 56–62;
b) D. A. Filimonov, V. V. Poroikov, *PASS: Computerized prediction of biological activity spectra for chemical substances. In Bioactive Compound Design: Possibilities for Industrial Use*; BIOS Scientific Publishers: Oxford, **1996**, 47–56;
c) V. V. Poroikov, D. A. Filimonov, A. V. Stepanchikova, A. P. Boudunova, E. V. Shilova, A. V. Rudnitskih, T. M. Selezneva, L. V. Goncharenko, *Chim.-Pharm. J. (Rus)* **1996**, *30*, 20–23;
d) S. Anzali, G. Barnickel, B. Cezanne,

M. Krug, D. Filimonov, V. Poroikov, *J. Med. Chem.* **2001**, *44*, 2432–2437.
e) Web site: http://www.ibmh.msk.su/ PASS.

[20] C. A. Lipinski, F. Lombardo, B. W. Dominy, P. J. Feeny, *Adv. Drug Deliv. Rev.* **1997**, *23*, 3–25.

[21] M. C. Wenlock, R. P. Austin, P. Barton, A. M. Davis, P. D. Leeson, *J. Med. Chem.* **2003**, *46*, 1250–1256.

[22] S. J. Teague, A. M. Davis, P. D. Leeson, T. Oprea, *Angew. Chem. Int. Ed. Engl.* **1999**, *38*, 3743–3747.

[23] T. I. Oprea, A. M. Davis, S. J. Teague, P. D. Leeson, *J. Chem. Inf. Comput. Sci.* **2001**, *41*, 1308–1315.

[24] D. E. Clark, S. D. Pickett, *Drug Discov. Today* **2000**, *5*, 49–58.

[25] A. Ajay, W. P. Walters, M. A. Murcko, *J. Med. Chem.* **1998**, *41*, 3314–3324.

[26] J. F. Blake, *Curr. Opin. Biotechnol.* **2000**, *11*, 104–107.

[27] A. P. Li, M. Segall, *Drug Discov. Today* **2002**, *7*, 25–27.

[28] A. P. Li, *Drug Discov. Today* **2001**, *6*, 357–366.

[29] T. N. Thompson, *Curr. Drug Metabol.* **2000**, *1*, 215–241.

[30] G. M. Keserü, L. Molnár, *J. Chem. Inf. Comput. Sci.* **2002**, *42*, 437–444.

[31] P. R. Andrews, D. J. Craik, J. L. Martin, *J. Med. Chem.* **1984**, *27*, 1648–1657.

[32] R. S. Bohacek, C. McMartin. *J. Med. Chem.* **1992**, *35*, 1671–1684.

[33] V. N. Viswanadhan, M. R. Reddy, R. J. Bacquet, D.M. Erion, *J. Comput. Chem.* **1993**, *9*, 1019–1026.

[34] G. Klopman, J.-Y. Li, S. Wang, M. Dimayuga, *J. Chem. Inf. Comput. Sci.* **1994**, *34*, 752–781.

[35] R. Wang, Y. Fu, L. Lai, *J. Chem. Inf. Comput. Sci.* **1997**, *37*, 615–621.

[36] B. Beck, A. Breindl, T. Clark, *J. Chem. Inf. Comput. Sci.* **2000**, *40*, 1046–1051.

[37] W. J. Egan, K. M. Merz Jr., J. J. Baldwin, *J. Med. Chem.* **2000**, *43*, 3867–3877.

[38] F. Lombardo, J. F. Blake, W. J. Curatolo, *J. Med. Chem.* **1996**, *39*, 4750–4755.

[39] E. G. Chikhale, K. Y. Ng, P. S. Burton, R. T. Borchardt, *Pharm. Res.* **1994**, *11*, 412–419.

[40] R. C. Young, R. C. Mitchell, T. H. Brown, C. R. Ganellin, R. Griffith,

M. Jones, K. K. Rana, D. Saunders, I. R. Smith, N. E. Sore, T. J. Wilks, *J. Med. Chem.* **1988**, *31*, 656–671.

[41] P. Seiler, *Eur. J. Med. Chem.* **1974**, *9*, 473–479.

[42] H. van de Waterbeemd, M. Kansy, *Chimia* **1992**, *46*, 299–303.

[43] M. H. Abraham, H. S. Chadha, R. C. Mitchell, *J. Pharm. Sci.* **1994**, *83*, 1257–1268.

[44] H. S. Chadha, M. H. Abraham, R. C. Mitchell, *Bioorg. Med. Chem. Lett.* **1994**, *4*, 2511–2516.

[45] M. H. Abraham, *Chem. Soc. Rev.* **1993**, *22*, 73–83.

[46] G. Colmenarejo, A. Alvarez-Pedraglio, J.-L. Lavandera, *J. Med. Chem.* **2001**, *44*, 4370–4378.

[47] S. Ekins, G. Bravi, S. Blinkley, J. S. Gillespie, B.J. Ring, J. H. Wikel, S. A. Wrighton, *J. Pharm. Exp. Ther.* **1999**, *290*, 429–438.

[48] S. Ekins, G. Bravi, S. Blinkley, J. S. Gillespie, B.J. Ring, J. H. Wikel, S. A. Wrighton, *Pharmacogenetics* **1999**, *9*, 477–489.

[49] M. J. De Groot, N. P. Vermeulen, *Drug Metab. Rev.* **1997**, *29*, 747–799.

[50] B. Testa, G. Cruciani, Structure–metabolism relations and the challenge of predicting biotransformation, in *Pharmacokinetic Optimization in Drug Research*, B. Testa, H. van de Waterbeemd, G. Folkers, R. Guy (Eds.), Wiley-VCH, Weinheim, New York, **2001**, pp. 65–84.

[51] G. M. A. Keseru, *J. Comput.-Aided Mol. Des.* **2001**, *15*, 649–657.

[52] A. Boobis, U. Gundert-Remy, P. Kremers, P. Macheras, O. Pelkonen, *Eur. J. Pharm. Sci.* **2002**, *17*, 183–193.

[53] J. Langowski, A. Long, *Adv. Drug Deliv. Rev.* **2002**, *54*, 407–415.

[54] S. Ekins, S. A. Wrighton, *J. Pharmacol. Toxicol.* **2001**, *45*, 65–69.

[55] A. M. ter Laak, N. P. E. Vermeulen, Molecular-modeling approaches to predict metabolism and toxicity, in *Pharmacokinetic Optimization in Drug Research*, B. Testa, H. van de Waterbeemd, G. Folkers, R. Guy (Eds.), Wiley-VCH, Weinheim, **2001**, pp. 551–588.

[56] C. Lemmen, T. Lengauer, *J. Comput.-Aided Mol. Design* **2000**, *14*, 215–232.

[57] S. Handschuh, J. Gasteiger, *J. Mol. Model.* **2000**, *6*, 358–378.

[58] S. Handschuh, M. Wagener, J. Gasteiger, *J. Chem. Inf. Comput. Sci.* **1998**, *38*, 220–232.

[59] T. J. A. Ewing, I. D. Kuntz, *Comput. Chem.* **1997**, *18*, 1175–1189.

[60] M. Rarey, B. Kramer, T. Lengauer, G. Klebe, *J. Mol. Biol.* **1996**, *261*, 470–489.

[61] V. Schnecke, L. A. Kuhn, *Perpect. Drug Des. Discov.* **2000**, *20*, 171–190.

[62] G. Jones, P. Willett, R. C. Glen, A. R. Leach, R. Taylor, *J. Mol. Biol.* **1997**, *267*, 727–748.

[63] G. M. Morris, D. S. Goodsell, R. S. Halliday, R. Huey, W. E. Hart, R. K. Belew, A. J. Olson, *J. Comput. Chem.* **1998**, *19*, 1639–1662.

[64] R. A. Abagyan, M. Totrov, D. Kuznetsov, *J. Comput. Chem.* **1994**, *15*, 488–506.

[65] C. McMartin, R. S. Bohacek, *J. Comput. Aided Mol. Des.* **1997**, *11*, 333–344.

[66] G. Schneider, H.-J. Böhm, *Drug Discov. Today* **2002**, *7*, 64–70.

[67] S. H. Rotstein, M. A. Murcko, *J. Comput.-Aided Mol. Des.* **1993**, *7*, 23–43.

[68] S. H. Rotstein, M. A. Murcko, *J. Med. Chem.* **1993**, 1700–1710.

[69] J. B. Moon, W. J. Howe, *Proteins* **1991**, *11*, 314–328.

[70] R. S. Bohacek, C. McMartin, *J. Am. Chem. Soc.* **1994**, *116*, 5560–5571.

[71] M. B. Eisen, D. C. Wiley, M. Karplus, R. E. Hubbard, *Proteins* **1994**, *19*, 199–221.

[72] Y. Nishibata, A. Itai, *J. Med. Chem.* **1993**, *36*, 2921–2928.

[73] H.-J. Böhm, *J. Comput.-Aided Mol. Des.* **1992**, *6*, 61–78.

[74] A. Miranker, M. Karplus, *Proteins* **1991**, *11*, 29–34.

[75] D. E. Clark, M. A. Firth, C. W, Murry, *J. Chem. Inf. Comput. Sci.* **1996**, *36*, 137–145.

[76] D. E. Clark, D. Frenkel, S. A. Levy, J. Li, C.W. Murray, B. Robson, B. Waszkowycz, D. R. Westhead, *J. Comput.-Aided Mol. Des.* **1995**, *9*, 13–32.

[77] C. W. Murray, D. E. Clark, T. R. Auton, M. A. Firth, J. Li, R. A. Sykes, B. Waszkowycz, D. R. Westhead, S. C. Young, *J. Comput.-Aided Mol. Des.* **1997**, *11*, 193–207.

[78] N. P. Todorov, P. M. Dean, *J. Comput.-Aided Mol. Des.* **1997**, *11*, 175–192.

[79] V. Gillet, A. P. Johnson, P. Mata, S. Sike, P. Williams, *J. Comput.-Aided Mol. Des.* **1993**, *7*, 127–153.

[80] H.-J. Böhm, *J. Comput.-Aided Mol. Des.* **1994**, *8*, 243–256.

[81] C. Bissantz, G. Folkers, D. Rognan, *J. Med. Chem.* **2000**, *43*, 4759–4767.

[82] D. Rognan, S. L. Lauemoller, A. Holm, S. Buus, V. Tschinke, *J. Med. Chem.* **1999**, *42*, 4650–4658.

[83] Y. Patel, V. J. Gillet, G. Bravi, A. R. Leach, *J. Comput.-Aided Mol. Des.* **2002**, *16*, 653–681.

[84] H. Bauknecht, A. Zell, H. Bayer, P. Levi, M. Wagener, J. Sadowski, J. Gasteiger, *J. Chem. Inf. Comput. Sci.* **1996**, *36*, 1205–1213.

[85] a) S. Anzali, J. Gasteiger, U. Holzgrabe, J. Polanski, J. Sadowski, A. Teckentrup, M. Wagener, *The Use of Self-Organizing Neural Networks in Drug Design* in: *3D QSAR in Drug Design – Volume 2*, H. Kubinyi, G. Folkers, Y. C. Martin (Eds.), Kluwer/ESCOM, Dordrecht, NL, **1998**, 273–299;
b) J. Gasteiger, A. Teckentrup, L. Terfloth, S. Spycher, *J. Phys. Org. Chem.* **2003**, *16*, 232–245;
c) L. Terfloth, J. Gasteiger, *Drug Discovery Today* (Supplement: Genomics) **2001**, *15*, 102–108;
d) J. Gasteiger, *Data Mining in Drug Design* in: *Rational Approaches to Drug Design* H.-D. Höltje, W. Sippl (Eds.), Prous Science, Barcelona, E, **2001**, 459–474;
e) J. Polanski, F. Zouhiri, L. Jeanson, D. Desmaële, J. d'Angelo, J.-F. Mouscadet, R. Gieleciak, J. Gasteiger, M. LeBret, *J. Med. Chem.*, **2002**, *45*, 4647–4654;
f) S. Anzali, G. Barnickel, M. Krug, J. Sadowski, M. Wagener, J. Gasteiger, J. Polynaski, *J. Comput.-Aided Mol. Design*, **1996**, *10*, 521–534;
g) M. Wagener, J. Sadowski, J. Gastei-

ger, *J. Am. Chem. Soc.*, **1995**, *117*, 7769–7775;
h) S. Handschuh, M. Wagener, J. Gasteiger, *J. Chem. Inf. Comput. Sci.*, **1998**, *38*, 220–232.

[86] a) G. Klebe (Ed.), *Virtual screening: an alternative or complement to high throughput screening*, in *Persp. Drug Des. Discov.* **2000**, *20*, 1–287;
b) H.-J. Böhm, G. Schneider, *Virtual screening for bioactive molecules*. In *Methods and Principles in Medicinal Chemistry*, Vol. 10, R. Mannhold, H. Kubinyi, H. Timmerman (Eds.), Wiley-VCH, Weinheim, **2000**, 1–307.

[87] S. Wang, S. Sakamuri, I. J. Enyedy, A. P. Kozikowski, O. Deschaux, B. C. Bandyopadhyay, S. R. Tella, W. A. Zaman, K. M. Johnson, *J. Med. Chem.* **2000**, *43*, 351–360.

[88] A. M. Aronov, N. R. Munagala, P. R. Ortiz de Montellano, I. D. Kuntz, C. C. Wang, *Biochemistry* **2000**, *39*, 4684–4691.

[89] M. Sarmiento, L. Wu, Y.-F. Keng, L. Song, Z. Luo, Z. Huang, G.-Z. Wu, A.-K. Yuan, Z.-Y. Zhang, *J. Med. Chem.* **2000**, *43*, 146–155.

[90] H. Böhm, M. Böhringer, D. Bur, H. Gmünder, W. Huber, W. Klaus, D. Kostrewa, H. Kühne, T. Lübbers, N. Meunier-Keller, F. Müller, *J. Med. Chem.* **2000**, *43*, 2664–2674.

[91] E. Perola, K. Xu, T. M. Kollmeyer, S. H. Kaufmann, F. G. Prendergast, Y.-P. Pang, *J. Med. Chem.* **2000**, *43*, 401–408.

[92] C. A. Metcalf III, C. J. Eyermann, R. S. Bohacek, C. A. Haraldson, V. M. Varkhedkar, B. A. Lynch, C. Bart

[93] S. Grüneberg, B. Wendt, G. Klebe, *Angew. Chem. Int. Ed.* **2001**, *40*, 389–393.

[94] T. N. Doman, S. L. McGovern, B. J. Witherbee, T. P. Kasten, R. Kurumail, W. C. Stallings, D. T. Connolly, B. K. Shoichet, *J. Med. Chem.* **2002**, *45*, 2213–2221.

[95] C. A. Baxter, C. W. Murray, B. Waszkowycz, J. Li, R. A. Sykes, R. G. A. Bone, T. D. J. Perkins, W. Wylie, *J. Chem. Inf. Comput. Sci.* **2000**, *40*, 254–262.

[96] D. Tondi, U. Slomczynska, M. P. Costi, D. M. Watterson, S. Ghelli, B. K. Shoichet, *Chem. Biol.* **1999**, *6*, 319–331.

[97] M. Schapira, B. M. Raaka, H. H. Samuels, R. Abagyan, *Proc. Natl. Acad. Sci. U.S.A.* **2000**, *97*, 1008–1013.

[98] S. C. Hopkins, R. D. Vale, I. D. Kuntz, *Biochemistry* **2000**, *39*, 2805–2814.

[99] A. V. Filikov, V. Mohan, T. A. Vickers, R. H. Griffey, P. D. Cook, R. A. Abagyan, T.L. James, *J. Comput.-Aided Mol. Des.* **2000**, *14*, 593–610.

[100] P. W. Walters, M. T. Stahl, M. A. Murcko, High-throughput, virtual chemistry, in *Encyclopedia of Computational Chemistry, Vol. 2,* P. v. R. Schleyer, N. L. Allinger, T. Clark, J. Gasteiger, P. A. Kollman, H. F. Schaefer III, P. R. Schreiner (Eds) John Wiley & Sons, New York, **1998**, pp. 1225–1237.

[101] A. C. Good, S. R. Krystek, J. S. Mason, *Drug Discov. Today* **2000**, *5* (Suppl.), S61–S69.

[102] B. Waszkowycz, T. D. J. Perkins, R. A. Sykes, J. Li, *IBM Syst. J.* **2001**, *40*, 360–376.

[103] T. I. Oprea, *Molecules* **2002**, *7*, 51–62.

# 11
# Future Directions

*J. Gasteiger*

Foremost we hope – and believe – that chemoinformatics will become of increasing importance in the teaching of chemistry. The instruments and methods that are used in chemistry will continue to swamp us with data and we have to manage these data to increase our chemical knowledge. We have to understand more deeply, and exploit, the results of our experiments. Concomitantly, demands on the properties of the compounds that are produced by the chemical and pharmaceutical industries will continue to rise. We will need materials that are better; we need them to be more selective, have fewer undesirable properties, able to be broken down easily in the environment without producing toxic by-products, and so on. This asks for more insight into the relationships between chemical structures and their properties. Furthermore, we have to plan and perform fewer and more efficient experiments.

Thus, chemoinformatics specialists will continue to be in high demand. This asks both for the training of chemoinformatics specialists and for incorporation of the essential features of chemoinformatics into regular chemistry – and informatics – curricula.

In order to achieve the goals of making more efficient use of the information that is produced and of planning and performing better experiments, chemoinformatics will have to be more integrated into the daily work processes of the chemist, and into the work of the bench chemist. Certainly, many chemists still have to overcome high barriers to using the computer for assistance in the solution of their daily scientific problems.

However, better use of spectral information for more rapid elucidation of the structure of a reaction product, or of a natural product that has just been isolated, requires the use of computer-assisted structure elucidation (CASE) systems. The CASE systems that exist now are far away from being routinely used by the bench chemist. More work has to go into their development.

The planning of organic syntheses is a great intellectual challenge that an organic chemist will hesitate to delegate to a computer. However, computer-assisted synthesis design (CASD) can facilitate the planning of more efficient syntheses. Chemists will eventually accept that CASD systems provide methods that can assist them in the design of syntheses, and still let them enjoy the intellectual fun. It has to be realized that the chemist and the computer can form a perfect team for planning organic syntheses, the chemist's mind being good at lateral thinking and having spontaneous ideas, the computer working rapidly and tirelessly through many data to sift out the relevant and interesting ones.

It is clear that the power of computers will continue to increase rapidly, as we have observed since the mid-20th century. This will allow us to tackle more complicated problems and larger systems, and do so with higher accuracy.

Concomitantly with the increase in hardware capabilities, better software techniques will have to be developed. It will pay us to continue to learn how nature tackles problems. Artificial neural networks are a far cry away from the capabilities of the human brain. There is a lot of room left from the information processing of the human brain in order to develop more powerful artificial neural networks. Nature has developed over millions of years efficient optimization methods for adapting to changes in the environment. The development of evolutionary and genetic algorithms will continue.

With better hardware and software, more exact methods can be used for the representation of chemical structures and reactions. More and more quantum mechanical calculations can be utilized for chemoinformatics tasks. The representation of chemical structures will have to correspond more and more to our insight into theoretical chemistry, chemical bonding, and energetics. On the other hand, chemoinformatics methods should be used in theoretical chemistry. Why do we not yet have databases storing the results of quantum mechanical calculations? We are certain that the analysis of the results of quantum mechanical calculations by chemoinformatics methods could vastly increase our chemical insight and knowledge.

The understanding and simulation of chemical reactions is one of the great challenges of chemoinformatics. Each day millions of reactions are performed, sometimes with rather poor results because of our limited understanding of chemical reactivity and the influence of solvents, catalysts, temperature, etc. This problem has to be tackled by both deductive and inductive learning methods.

A particularly challenging problem is the understanding and modeling of biochemical and metabolic reactions, and even more so of metabolic reaction networks. Much work will go into this field in the next few years.

With the increase in hardware and software, larger systems can be handled with higher accuracy. Much work will continue to be devoted to the study of proteins and polynucleotides (DNA and RNA), and particularly their interactions with more sophisticated methods. Remember: proteins and genes are chemical compounds and sophisticated theoretical and chemoinformatics methods should be applied to their study – in addition to the methods developed by bioinformaticians.

This highlights another field of great activity to be expected: the merging of bioinformatics and chemoinformatics. These fields will be bridged and will overlap more and more, allowing the study of the problems in genomics, proteomics, metabolomics, and drug design, with a battery of methods coming from both bioinformatics and chemoinformatics.

Drug design is at present by far the major playground of chemoinformatics. This situation will certainly change in the future.

Chemistry produces many materials, other than drugs, that have to be optimized in their properties and preparation. Chemoinformatics methods will be used more and more for the elucidation and modeling of the relationships between chemical structure, or chemical composition, and many physical and chemical properties, be they nonlinear optical properties, adhesive power, conversion of light into electrical energy, detergent properties, hair-coloring suitablity, or whatever.

Moreover, multivariate optimization, the simultaneous optimization of several properties, will increasingly come into focus. A drug should have high selectivity in binding to different receptors and minimal toxicity, good solubility and penetration, and so on. A hair color should have a brilliant shine, be absorbed well, not be washed out, not damage the hair, not be toxic, and be stable under sunlight, etc.

We will also see large changes in the dissemination of information. Bench chemists write their observations into laboratory journals, then they write manuscripts which are evaluated by referees, then printed, then read by colleagues; some of the information may go into a monograph or a handbook, and eventually some of it may also be stored in a database. Many steps! And at nearly each step we might have a break in the medium; there is no continuous, or automatic, flow of information, which is picked up from one medium (e.g., a journal) by a scientist (usually not the one who produced the information initially!) and input into a new medium (e.g., a book or a database). At each one of these breaking points the information might become distorted, and errors might creep in.

This does not have to be so! Why not build an uninterrupted stream of information from the producer (the bench chemist) to the consumer (the reader of a journal or book, or the scientist that puts a query into a database)? It is quite clear that the producers of information knows best what experiments were done, what observations were made, what results have been obtained. They should put this information into electronic laboratory books, augmented with spectral data (that they can obtain directly from the analytical laboratory). From this electronic repository all other information sources –manuscripts, journals, books, databases – could be filled, clearly sometimes by manual selection, but not by changing data!

And last not least, we will have to see further improvements in the graphical user interfaces of software systems and the retrieval systems of databases in order to make software and databases more acceptable to the chemical community at large. Software and databases should speak the language a chemist is used to, with hand-drawn chemical structures and reaction equations, or even understand the spoken word – and only provide the desired information selectively, not buried in a pile of unnecessary output.

# Appendix

## A.1
## Software Development

*Michael Wünstel*

The requirements for scientific software development have continually increased. Besides the algorithmic core functionality, nowadays there is often a demand for a graphical user interface. In addition to the increasing importance of this visible component, which may still be seen as just an add-on, the software development itself has to fulfill stronger demands on software engineering requirements, such as maintainability and recoverability.

The scientific area is often characterized by a highly fluctuating and international workforce. This makes it necessary to document the source code and to represent e. g. the class structure in order to introduce new staff to the system easily. Conversely, non-existent documentation may lead to the necessity of reimplementation.

This section cannot give a comprehensive overview of software programming. Instead, it will present a brief introduction to programming languages, also covering object-oriented programming, the Universal Modeling Language (UML) and design patterns. Furthermore, graphical user interfaces are briefly addressed and the fields of source code documentation and version control are introduced.

### A.1.1
### Programming Languages

The object-oriented programming language C++ is the predominant one [1] within the scientific area. A C++ compiler that translates the source code into machine code is available for every common operating system. An alternative to C++ that also is often used is Java™ [2] (*http://java.sun.com*). It is also an object-oriented programming language which has the advantage that its resulting byte code can be run on various operating systems. Another advantage is that Java™ provides much functionality, such as the import of images, that facilitates the generation of a running program. The advantage of platform independence has to be traded for a slower processing time compared with a corresponding machine code resulting from C++ Code. The reason is that the byte code has to be interpreted by the

Java™ Virtual Machine running on the system. Mechanisms like JIT (Just in Time Compiler – parts of the byte code were compiled into machine code) or JNI (Java™ Native Interface – enables the use of fast libraries resulting from compiled C++ Code within a Java program) can be used to reduce this characteristic disadvantage.

## A.1.2
### Object-Oriented Programming

Object-oriented programming combines abstract data types, which means types consisting of variables and operations on these variables, with the concept of heredity. The prototype is a class, which is the basis of an arbitrary number of objects. Heredity means that a new class can be derived from a class that already-exists. The derived class owns the variables and operations from the class from which it originates, but further variables or operations can be added to them. This enables the adaptation of a basic class to that what is actually needed. When the code is being compiled it is often clear which class within the class hierarchy has to be used. But this is not necessarily the case in a dynamic program flow. Imagine a molecule viewer that draws a molecule alternatively as van der Waals radii or a ball-and-stick diagram depending on the user's choice. The corresponding basic class molecule has two child classes with drawing functions: the van der Waals radii class, and the ball-and-stick class. The goal is that, depending on the user's choice, the correct drawing function is used for the molecule, either the one for the charge distribution or the one for the ball-and-stick diagram. The technique that can be used to realize this flexible behavior is called dynamic polymorphy (late binding). It can be used to decide, at run time, which object has to be shown and which corresponding object-function has to be used. Another advantage is that this principle eases the expandability. In our example a new molecule class can be realized by adding a new child class of the basic class molecule, whereas most parts of the code do not have to be touched.

## A.1.3
### Universal Modeling Language (UML)

The Universal Modeling Language is used to describe a software system [4, 5]. Several kinds of diagrams exist to model the diverse properties of the system. Thus a description of the system can be developed that enables the systematic and uniform documentation of the system. The class diagram, for example, represents the classes and their relationships. But also interacting diagrams exist, to describe the dynamic behavior of the system and its objects.

There are several tools to support UML usage:

Rational Rose® (*http://www.rational.com/products/rose*),

Together® ControlCenter™ (*http://www.togethersoft.com/products/controlcenter*),

Microsoft® Visio® (*http://www.microsoft.com/office/visio/*) or

ArgoUML (*http://argouml.tigris.org*).

### A.1.4
### Design Patterns

Object-Oriented Design Patterns are software elements which fulfill certain tasks [3]. As there are recurring requirements for software, a design pattern offers a proven solution for a certain problem. Thus the developer does not need to redevelop code for standard problems. An example is the design pattern "Singleton", which ensures that only one object from a certain class can be generated. Besides the direct technical advantage, it also supports communication with other developers as the design patterns can be seen as a common foundation that facilitates the conceptual discussion.

### A.1.5
### Graphical User Interface

The need or wish for platform-independent software development, which means developing programs that can be ported directly to different operating systems, particularly affects the graphical user interface as this is an operating system oriented area. Besides the capabilities of Java™ that have already been described, the use of graphic libraries for C++ is an alternative approach to create platform-independent programs. Two common examples are Q't® (*http://www.trolltech.com*) and WxWindows (*http://www.wxwindows.org/*). Although Q't® is more powerful, it is available free of charge only under certain conditions, whereas WxWindows is an open source project (see the license conditions for further details).

### A.1.6
### Source Code Documentation

As reuse and maintenance are very important issues in software development, the documentation of the source code is a really crucial point. Java™ offers, with *Javadoc*™ (*http://java.sun.com/javadoc*) a possibility of extracting a separate HTML-documentation directly from the source code. For other languages, such as C++, tools exist which provide the possibility. to generate the documentation directly derived from the source code. An example is *Doxygen*, which is also a tool available free of charge (*http://www. doxygen.org*). The source code has to be expanded by the software developer with certain tags such as class description. The documentation is then generated in a further processing step in several possible formats such as HTML or PostScript. Additionally, a graphical representation of the class hierarchy can be extracted, similarly to UML.

A.1.7
**Version Control**

The development of scientific software is often characterized by multiple steps of changing and improving algorithms done by several developers. The different code versions can be handled with special version control programs that save these versions in a database, which enables comparison, for example, of the actual version with an older one. To prevent simultaneous changes a file can be blocked. Different versions can also be merged together. Examples of commercial programs include Rational® Clear Case© (*http://www. rational.com/products/clearcase/*) and Microsoft® Visual Source Safe® (*http://msdn.microsoft. com/SSAFE/*). There is also open source software available such as CVS (*http://www.gnu.org/software/cvs*).

# References

[1] S. B. Lippman, J. LaJoie, *C++ Primer*, Addison Wesley Professional, Reading, Mass **1998**.

[2] K. Arnold, J. Gosling, D. Holmes, *The Java™ Programming Language*, Addison Wesley Professional, Reading, Mass **2000**.

[3] E. Gamma, R. Helm, R. Johnson, *Design Patterns*, Addison Wesley, Reading, Mass **1997**.

[4] G. Booch, J. Rumbaugh, I. Jacobson, *Unified Modelling Language User Guide*, Addison Wesley Longman, Reading, Mass **1999**.

[5] B. Oestereich, *Developing Software with UML*, Addison Wesley Longman, Reading, Mass **2002**.

**A.2**

**Mathematical Excursion into Matrices and Determinants**

*Thomas Engel*

A matrix can be defined as a two-dimensional arrangement of elements (numbers, variables, vectors, etc.) set up in rows and columns. The elements $a_{nm}$ are indexed as follows:

$$M = \begin{pmatrix} a_{11} & a_{12} & a_{13} & \cdots & a_{1m} \\ a_{21} & a_{22} & a_{23} & \cdots & a_{2m} \\ a_{31} & a_{32} & a_{33} & \cdots & a_{3m} \\ \vdots & \vdots & \vdots & & \vdots \\ a_{n1} & a_{n2} & a_{n3} & \cdots & a_{nm} \end{pmatrix}$$

The dimension of a matrix is defined by the number of rows ($m$) and columns ($n$) and is called an $n \times m$ matrix.

A square matrix has the eigenvalue $\lambda$ if there is a vector $x$ fulfilling the equation: $Ax = \lambda x$. The result of this equation is that indefinite numbers of vectors could be multiplied with any constants. Anyway, to calculate the eigenvalues and the eigenvectors of a matrix, the characteristic polynomial can be used. Therefore $(A - \lambda E)x = 0$ characterizes the determinant $(A - \lambda E)$ with the identity matrix $E$ (i.e., the $n \times n$ matrix). Solutions can be obtained when this determinant is set to zero.

$$\det(A - \lambda E) = \begin{vmatrix} a_{11} - \lambda & a_{12} & a_{13} & \cdots & a_{1n} \\ a_{21} & a_{22} - \lambda & a_{23} & \cdots & a_{2n} \\ a_{31} & a_{32} & a_{33} - \lambda & \cdots & a_{3n} \\ \vdots & \vdots & \vdots & & \vdots \\ a_{n1} & a_{n2} & a_{n3} & \cdots & a_{nn} - \lambda \end{vmatrix} = 0$$

The inside terms of the determinant represent the characteristic polynomial $P(\lambda)$ of $A$:

$$P(\lambda) = A_n\lambda^n + A_{n-1}\lambda^{n-1} + \dots + A_0\lambda^0$$

whose zero values correspond to the eigenvalues of $A$.

The characteristic polynomial can be determined according to Sarrus: the columns of the matrix, except the last one, are written after the matrix. Then, the sum of all the products of all the diagonal elements from the last row is subtracted from the sum of all the products of all the diagonal elements from the first row.

$$
\begin{vmatrix}
a_{11} & a_{12} & a_{13} & a_{11} & a_{12} \\
a_{21} & a_{22} & a_{23} & a_{21} & a_{22} \\
a_{31} & a_{32} & a_{33} & a_{31} & a_{32}
\end{vmatrix} =
$$

$$
(a_{11}a_{22}a_{33} + a_{12}a_{23}a_{31} + a_{13}a_{21}a_{32}) - (a_{31}a_{22}a_{13} + a_{32}a_{23}a_{11} + a_{33}a_{21}a_{12}) = 0
$$

$$
\begin{vmatrix}
a_{11} & a_{12} & a_{13} & a_{11} & a_{12} \\
a_{21} & a_{22} & a_{23} & a_{21} & a_{22} \\
a_{31} & a_{32} & a_{33} & a_{31} & a_{32}
\end{vmatrix} =
$$

$$
(a_{11}a_{22}a_{33} + a_{12}a_{23}a_{31} + a_{13}a_{21}a_{32}) - (a_{31}a_{22}a_{13} + a_{32}a_{23}a_{11} + a_{33}a_{21}a_{12}) = 0
$$

## A.2.1
## Mathematical Example

Given is a matrix $A = \begin{pmatrix} 2 & 2 & 1 \\ 1 & 3 & 1 \\ 1 & 2 & 2 \end{pmatrix}$

with its characteristic polynomial:

$$
P(\lambda) = \begin{vmatrix}
2-\lambda & 2 & 1 \\
1 & 3-\lambda & 1 \\
1 & 2 & 2-\lambda
\end{vmatrix} = -\lambda^3 + 7\lambda^2 - 11\lambda + 5 = 0
$$

This third-degree polynomial has the value zero with two eigenvalues:
$\lambda_{1/2} = 1$ and $\lambda_3 = 5$.

## A.2.2
## Chemical Example of an Atom Connectivity Matrix:

The characteristic polynomial of HCN is:

$$
\begin{vmatrix}
H & 1 & 0 \\
1 & C & 3 \\
1 & 3 & N
\end{vmatrix} = H \begin{vmatrix} C & 3 \\ 3 & N \end{vmatrix} - 1 \begin{vmatrix} 1 & 0 \\ 3 & N \end{vmatrix} + 0 \begin{vmatrix} 1 & 0 \\ C & 3 \end{vmatrix} = HCN - 9H - N
$$

The characteristic polynomial of HCN (derived according to Sarrus) is:

$$
\begin{vmatrix}
H & 1 & 0 & H & 1 \\
1 & C & 3 & 1 & C \\
1 & 3 & N & 1 & 3
\end{vmatrix} = (HCN + 0 + 0) - (0 + 9H + N) = HCN - 9H - N
$$

# Index